D0207473

The Institute of Mathematics
and its Applications
Conference Series

The Institute of Mathematics
and its Applications
Conference Series

Previous volumes in this series were published by
Academic Press to whom all enquiries should be addressed.
Forthcoming volumes will be published by
Oxford University Press throughout the world.

NEW SERIES

1. *Supercomputers and parallel computation* Edited by D. J. Paddon
2. *The mathematical basis of finite element methods* Edited by David F. Griffiths
3. *Multigrid methods for integral and differential equations* Edited by D. J. Paddon and H. Holstein
4. *Turbulence and diffusion in stable environments* Edited by J. C. R. Hunt
5. *Wave propagation and scattering* Edited by B. J. Uscinski
6. *The mathematics of surfaces* Edited by J. A. Gregory
7. *Numerical methods for fluid dynamics II* Edited by K. W. Morton and M. J. Baines
8. *Analysing conflict and its resolution* Edited by P. G. Bennett
9. *The state of the art in numerical analysis* Edited by A. Iserles and M. J. D. Powell

The State of the Art in Numerical Analysis

Proceedings of the joint IMA/SIAM conference on The State of the Art in Numerical Analysis held at the University of Birmingham, 14–18 April 1986

Edited by

A. ISERLES
Lecturer in Applied Mathematics
University of Cambridge

and

M. J. D. POWELL
Professor of Applied Numerical Analysis
University of Cambridge

CLARENDON PRESS · OXFORD · 1987

Oxford University Press, Walton Street, Oxford OX2 6DP
Oxford New York Toronto
Delhi Bombay Calcutta Madras Karachi
Petaling Jaya Singapore Hong Kong Tokyo
Nairobi Dar es Salaam Cape Town
Melbourne Auckland
and associated companies in
Beirut Berlin Ibadan Nicosia

Oxford is a trademark of Oxford University Press

Published in the United States
by Oxford University Press, New York

British Library Cataloguing in Publication Data
The State of the art in numerical analysis: proceedings of the joint IMA/SIAM conference on the state of the art in numerical analysis held in Birmingham in April 1986.—
(The Institute of Mathematics and its Applications conference series. New series; 9)
1. Numerical analysis
I. Iserles, A. II. Powell, M.J.D. III. Series
519.4 QA297
ISBN 0–19–853614–3

Library of Congress Cataloging-in-Publication Data
The State of the art in numerical analysis.
(The Institute of Mathematics and Its Applications conference series; new ser., 9)
1. Numerical analysis—Congresses. I. Iserles, A.
II. Powell, M. J. D.
III. Institute of Mathematics and Its Applications.
IV. Society for Industrial and Applied Mathematics.
V. Series.
QA297.S78 1987 511 87–1537
ISBN 0–19–853614–3

Printed by St Edmundsbury Press,
Bury St Edmunds, Suffolk

J.H. Wilkinson 1919-1986

Jim Wilkinson, through his profound and timely publications, through his stimulating lectures, and through his willingness to share his knowledge, has for many years led research in linear algebra, zeros of functions, error analysis and mathematical software. His learned contribution to this volume is witness to the fact that at his death on October 5th, 1986 his concern for the future development of our subject was undiminished by his retirement from the National Physical Laboratory a few years ago; indeed we have lost a friend whose scholarly work was at its height. As a token of our respect and gratitude for Jim and his achievements we dedicate to him our contributions to this book.

C.T.H. Baker
C. de Boor
H. Brunner
M.G. Cox
C.W. Cryer
A.R. Curtis
I.S. Duff
R. Fletcher
P.D. Frank
A. Griewank
N.J. Higham
A. Iserles
A.D. Jepson
W. Kahan
J.D. Lambert
K.W. Morton
S. Osher
M.J.D. Powell
J.K. Reid
R.B. Schnabel
A. Spence
G.W. Stewart
P.K. Sweby
J. Walsh
G. Wanner
G.A. Watson
W.L. Wendland

Preface

For the last few decades numerical analysis has experienced an era of truly prodigious growth. Both our skill in devising efficient and robust algorithms and our ability to analyse their behaviour on a firm mathematical footing are expanding at a fast pace. Further, the sheer number of both active numerical analysts and of users of numerical techniques is increasing too. Thus the *raison d'être* of the State of the Art conferences is to provide an opportunity (once a decade) to pause and take stock of current developments in the discipline.

The 1986 Birmingham conference, third to be held (and the first under the joint auspices of the Institute of Mathematics and its Applications and the Society for Industrial and Applied Mathematics) was a unique occasion for its many participants to update their knowledge across the whole spectrum of numerical mathematics and methods. As the volume of scientific production increases, it becomes progressively more difficult for even a knowledgeable individual to keep track of new developments and to maintain the right sense of proportion between the necessarily narrow domain of specialisation and the overall intellectual output of the whole discipline. Thus, the 23 talks at the conference, attempting to cover the main developments in all of numerical analysis throughout the decade and presented by acknowledged authorities in their subjects, were of real benefit to all participants. This volume is an attempt to extend this benefit to the whole numerical community.

A comparison of the contents of this volume with the proceedings of the two previous State of the Art conferences (*Numerical Analysis: An Introduction*, Walsh, J., ed., Academic Press (London), 1966; *The State of the Art in Numerical Analysis*, Jacobs, D.A.H., ed., Academic Press (London), 1977) reveals both continuity and progress. Continuity follows from the fact that scientific computation remains focussed on a relatively static set of problems — linear algebra, optimization, ordinary differential equations, integral equations, partial differential equations, and approximation theory. Moreover, many of the techniques that are currently in use have been around, at least in embryo, for a long time. However, even if the objects of study remain unchanged, our understanding advances greatly. Many of the topics of the papers in this volume are artifacts of the last decade and virtually all the papers report substantial — frequently crucial — new developments.

So what are the motive forces of numerical analysis *circa* 1986? First, there is contemporary pure mathematics. The relationship between pure mathematics and numerical analysis (and, by implication, pure mathematicians and numerical analysts) is not always easy and free of misunderstanding. It is sometimes heard from the first domain that numerical analysts strip mathematics of its rigour and take shortcuts that are unjustified from the purely formal point of view. On the other hand, some numerical analysts reserve towards their "pure" brethren the sort of attitude of front-line troops toward arm-chair strategists. Although an unending discussion of these matters will (probably for ever) help pass many a dull hour in a common-room, it obscures what should be almost self-evident: contemporary numerical analysis is one of the largest, most diversified and most demanding customers of pure mathematics. Almost no modern mathematical discipline is exempt from the attention of some numerical analysis specialists, and important work in this book is dependent on linear algebra, graph theory, differential equations, Banach algebras, dynamics, topology, functional analysis, group theory, harmonic analysis, special functions, analytic functions, convexity theory etc.

The role of mathematics is a source of yet another (hopefully creative) tension, this time inside the numerical community. There will always be those who strive to prove and analyse everything in a rigorous and formal framework, whereas others claim that the value of an algorithm can be assessed by experimentation and that, as a rule, a method having unaccountably good performance is to be preferred to a method that is well understood and analysed but performs less well. The lesson of modern numerical analysis is that both approaches are fruitful and, indeed, they complement each other. There is no unique path of advance in numerical analysis. Frequently a breakthrough is achieved by sheer algorithmic ingenuity and the theorists lag behind, trying to explain *a posteriori* why a given method is so successful. However, equally frequently we are guided forwards by a thorough theoretical understanding, and algorithms are constructed from a firm mathematical basis.

The second motive force of contemporary numerical analysis, unlike pure mathematics, is very recent — we refer here to advances in computers and computer architectures. The introduction of digital computers laid the basis for the first revolution in numerical analysis. There is every reason to believe that presently we are witnessing the second revolution in our discipline, as a consequence of ongoing developments in computer design. Until a few years ago new developments on the hardware side were in size and speed, essentially on a linear scale, giving faster mainframe serial machines and much more powerful storage facilities. Because these improvements were taken for granted by numerical analysts, the impact of computers on the suitability of algorithms was small. All this has changed radically during the last decade, due to the twin developments of microcomputers and supercomputers. The impact of microcomputers creates widespread demand for algorithms that are suited to that particular environment: relatively low on memory and number-crunching speed but fully interactive and with excellent graphic capabilities. The growing importance of computer aided design is just one instance of that phenomenon.

Even more important is the impact of supercomputers with new, nonserial architectures. The revolution is not just in scale and in speed but also in the very essence of "good" algorithms, for what was efficient on a serial computer may well be suboptimal on a parallel machine. Thus, the whole way we "think numerical" undergoes changes that can hardly be over-stated. This calls for collaboration with computer scientists and intelligent awareness by numerical analysts of the scope of new architectures.

Powerful computers of novel design and thorough mathematical analysis together bring progressively more intricate and more "real" problems within the realm of feasible scientific computation, so our third motive force is interaction and collaboration with users of numerical algorithms. In working on the difficult problems of engineering and science, with all their nonlinearities, singularities and — frequently — vagueness, and without too many oversimplifying assumptions, numerical analysts are bound to be out on a limb, outside the safe and cosy shelter of mathematical certainty. Numerical results may then (at least for the time being) be beyond the range of careful analytic scrutiny, being verified instead in a wind tunnel or on a construction site. This poses a different kind of challenge, that of close cooperation — not just one-way communication — with scientists and engineers.

Contacts with users, however, are not always blissful. Virtually every experienced numerical analyst can recall gory details of misuse and abuse of computational techniques by individuals who are competent professionals in their own disciplines, but who seem to be attracted by numerical methods that are unstable, ill-posed, inefficient, or plainly wrong. Such algorithms may give exciting and welcome numerical results, but also they can waste human time, they can contribute to the collapse of a suspension bridge or two, and they can bring the science of numerical computation into disrepute. This state of affairs calls, in our opinion, for some self-criticism by the numerical analysis community, because all too often we are content with inventing algorithms, underpinning them with mathematical analysis that is constrained by our competence, and even producing software, but we seldom give enough attention to the way the fruits of our labours fare in the "outside world". Instead it is more usual to be concerned with the reactions of fellow numerical analysts to our wares. This is an old problem that shows clearly the importance of understanding and communicating the actual limitations of numerical algorithms without stifling their usefulness. Here two-way communication is much more successful, but requires more total effort, than a subroutine library interface.

These three motive forces ensure that numerical analysis is

an exciting discipline that blends relevance with intellectual depth and practicality with beauty. Thus, little wonder that even this rather weighty volume can present only some of the wide-ranging developments of the last decade, but the nature of recent research and important highlights and landmarks are clear. Of course, hundreds of references throughout the book enable readers to follow developments in detail, and professional numerical analysts will be able to find their own research interests as part and parcel of the entire subject.

Any conference is at least as good as the quality of its talks. It was the good fortune of the participants of the State of the Art in Numerical Analysis conference to listen to a string of speakers who were both knowledgeable and eloquent and who frequently imbued the audience with their excitement in their work. We are all grateful to the IMA for providing such a successful meeting, to SIAM for its support, and to the University of Birmingham for its hospitality.

Likewise, a volume of proceedings is as good as its collective authorship. It is a pleasure on our part, as editors, to thank all the authors for their goodwill, their efforts, their expertise, and their resilient enthusiasm, that have provided the contents of this showcase for contemporary numerical analysis. On the production side, the cooperation of Oxford University Press and assistance from the Department of Applied Mathematics and Theoretical Physics of Cambridge University are acknowledged warmly, and especially we are grateful to Ina Godwin for her expert typing. We have learned much from the contributions to this book, and we recommend strongly to readers that they seize this opportunity to share our new-found knowledge.

Cambridge
October, 1986

A. Iserles
M.J.D. Powell

Contents

Contributors

C.T.H. Baker
Department of Mathematics, University of Manchester, Manchester, M13 9PL, U.K.

Carl de Boor
Mathematics Research Center, 610 Walnut Street, Madison, Wisconsin, 53705, U.S.A.

H. Brunner
Department of Mathematics and Statistics, Memorial University of Newfoundland, St. John's, Newfoundland A1C 5S7, Canada.

M.G. Cox
Division of Information Technology and Computing, National Physical Laboratory, Teddington, Middlesex, TW11 0LW, U.K.

C.W. Cryer
Institut für Numerische und Instrumentelle Mathematik, Westfaelischen Wilhelms-Universität, Einsteinstrasse 62, D-4400 Münster, Germany FDR.

A.R. Curtis
Computer Science and Systems Division, UKAEA, Harwell Laboratory, Didcot, Oxon, OX11 0RA, U.K.

I.S. Duff
Computer Science and Systems Division, UKAEA, Harwell Laboratory, Didcot, Oxon, OX11 0RA, U.K.

R. Fletcher
Department of Mathematical Sciences, University of Dundee, Dundee, DD1 4HN, Scotland, U.K.

P.D. Frank
Boeing Computer Services, Rukwila, Washington, 98042, U.S.A.

A. Griewank
Mathematics Department, Southern Methodist University, Dallas, Texas, 75275, U.S.A.

N.J. Higham
Department of Mathematics, University of Manchester, Manchester, M13 9PL, U.K.

A.D. Jepson
Department of Computer Science, University of Toronto, Toronto, M5S 1A7, Canada.

W. Kahan
Elect. Eng. and Computer Science and Mathematics Departments
University of California, Berkeley, California, 94720, U.S.A.

J.D. Lambert
Department of Mathematical Sciences, University of Dundee, Dundee, DD1 4HN, Scotland, U.K.

K.W. Morton
Oxford University Computing Laboratory, 8-11 Keble Road, Oxford, OX1 3QD, U.K.

S. Osher
Department of Mathematics, University of California, Los Angeles, California, 90024, U.S.A.

M.J.D. Powell
Department of Applied Mathematics and Theoretical Physics, University of Cambridge, Silver Street, Cambridge, CB3 9EW, U.K.

J.K. Reid
Computer Science and Systems Division, UKAEA, Harwell Laboratory, Didcot, Oxon, OX11 0RA, U.K.

R.B. Schnabel
Department of Computer Science, University of Colorado, Boulder, Colorado, 80309, U.S.A.

A. Spence
School of Mathematics, University of Bath, Claverton Down, Bath, BA2 7AY, U.K.

G.W. Stewart
Department of Computer Science, University of Maryland, College Park, Maryland, 20742, U.S.A.

P.K. Sweby
Department of Mathematics, University of Reading, Whiteknights, Reading RG6 2AS, U.K.

J. Walsh
Department of Mathematics, University of Manchester, Manchester, M13 9PL, U.K.

G. Wanner
Section de Mathématiques, Université de Genève, C.P. 240, CH-1211 Genève 24, Switzerland.

G.A. Watson
Department of Mathematical Sciences, University of Dundee, Dundee, DD1 4HN, Scotland, U.K.

J.H. Wilkinson (deceased)
40, Atbara Road, Teddington, Middlesex, TW11 9PD, U.K.

W.L. Wendland
Mathematical Institute A, University of Stuttgart, Pfaffenwaldring 57, D-7000 Stuttgart 80, Germany FDR.

1 Eigenvalue Problems

J. H. WILKINSON

1. INTRODUCTION

Apart from work in the important areas of parallel and vector processing the focus of attention in the last decade has shifted decisively away from the development of algorithms for individual eigenvalues and eigenvectors of the standard problem. The main research areas have been invariant subspaces for the standard problem, the generalized problems $A\underline{x} = \lambda B\underline{x}$ and $(A_r\lambda^r + \cdots + A_1\lambda + A_0)\underline{x} = \underline{0}$, inverse eigenvalue problems and error bounds for the computed solutions to all of these.

Although the background theory is covered to a considerable extent in the classical literature it is not generally expressed in terms which are appropriate for numerical analysts. Much of the work of the last decade has been devoted to adapting this material to our needs so that inevitably it has had a more theoretical flavour. It has a good deal in common with the early work on digital computers which was much concerned with the sensitivity of eigenvalues and eigenvectors and its relevance to the numerical stability of algorithms.

Some of the more influential papers were published earlier than 1976 and inevitably the material covered in them would be treated slightly differently now. In this article we give a review of the main areas of work and we attempt to update earlier work in the light of later developments.

2. THE KRONECKER PRODUCT

Much of the analysis connected with invariant subspaces may be conveniently expressed in terms of the Kronecker product.

DEFINITION. If $A \in \mathbb{C}^{p\times q}$ and $B \in \mathbb{C}^{r\times s}$ then the Kronecker product, $A \otimes B \in \mathbb{C}^{pr\times qs}$ and may be regarded as a $p \times q$ block matrix in which the i,j block is $a_{ij}B$.

The following results follow quite simply from this definition.

(a) $(A\otimes B)(C\otimes D) = AC\otimes BD$, whenever the dimensions are such that AC and BD are defined.

(b) If $A\underline{x}_i = \alpha_i \underline{x}_i$ and $B\underline{y}_j = \beta_j \underline{x}_j$ then $(A\otimes B)(\underline{x}_i\otimes \underline{y}_j) = \alpha_i\beta_j(\underline{x}_i\otimes\underline{y}_j)$.

(c) When A and B are square then

$$\det(A\otimes B-\lambda I) = \prod_{i,j}(\alpha_i\beta_j-\lambda)$$

where the α_i and β_j are the eigenvalues of A and B. This says a little more than (b) about the eigenvalues of $A\otimes B$ since it deals with the case when A and/or B are defective.

(d) $(A\otimes I-I\otimes B)(\underline{x}_i\otimes\underline{y}_j) = (\alpha_i-\beta_j)(\underline{x}_i\otimes\underline{y}_j)$.

(e) $$\det(A\otimes I-I\otimes B-\lambda I) = \prod_{i,j}(\alpha_i-\beta_j-\lambda).$$

In particular $A\otimes I-I\otimes B$ is singular iff $\alpha_i=\beta_j$ for some i,j. Again this deals with the case when A and/or B are defective.

(f) $A\otimes B = P(B\otimes A)Q$ where P and Q are permutation matrices. When A and B are square $Q=P^T$, so that $A\otimes B$ and $B\otimes A$ are similar.

(g) When A and B are square $A\otimes I-I\otimes B = P(I\otimes A-B\otimes I)P^T$ for some permutation matrix. Hence the relevant matrices have the same eigenvalues and the same singular values.

Matrix equations of the type

$$AX-XB = C, \tag{2.1}$$

where $A \in \mathbb{C}^{r\times r}$ and $B \in \mathbb{C}^{s\times s}$, commonly occur in work associated with invariant subspaces. Equation (2.1) can be written as a set of ordinary linear algebraic equations in the elements $x_{i,j}$ of X. In fact, equating corresponding columns in (2.1), we have

$$(I \otimes A - B^T \otimes I)\,\mathrm{Vec}(X) = \mathrm{Vec}(C)\,, \tag{2.2}$$

where $\mathrm{Vec}(X)$ denotes the vector of dimension rs in which the elements of X occur in sequence column by column. The system will be singular iff

$$K = I \otimes A - B^T \otimes I \tag{2.3}$$

is a singular matrix, i.e. from (e), iff $\alpha_i = \beta_j$ for some i,j since B^T has the same eigenvalues as B. From (2.2)

$$\|X\|_F = \|\mathrm{Vec}(X)\|_2 \leq \|K^{-1}\|_2\, \|\mathrm{Vec}(C)\|_2 = \|K^{-1}\|_2\, \|C\|_F\,, \tag{2.4}$$

equality being attained when $\mathrm{Vec}(C)$ is a singular vector of K corresponding to the smallest singular value. This smallest singular value is denoted by $\mathrm{sep}(A,B)$, an appropriate name since $\mathrm{sep}(A,B) = 0$ iff A and B have an eigenvalue in common and hence, in some sense, must be related to the 'separation' of the eigenvalues. However, since a small singular value does not imply a correspondingly small eigenvalue (Golub and Wilkinson, 1976), sep can be small even when the minimum separation of eigenvalues of A and B is not unduly so.

3. INVARIANT SUBSPACES

Since all the material will be presented in matrix terms it will be convenient to use abbreviations of the type 'the subspace X_1 (columns)' and 'the subspace Y_1^H (rows)' to refer to the subspace spanned by the columns of X_1 and the subspace spanned by the rows of Y_1^H respectively.

The subspace X_1 (columns) will be said to be a right-hand invariant subspace of A if the columns of AX_1 lie in the subspace X_1. This implies that

$$AX_1 = X_1 A_{11} \tag{3.1}$$

for some $A_{11} \in \mathbb{C}^{r\times r}$, where $X_1 \in \mathbb{C}^{n\times r}$. If the columns of X_1 are linearly independent and if X_2 completes a basis for the n-dimensional space then we may write $X=[X_1, X_2]$, where X is nonsingular and

$$X_1 = XI_{nr}. \tag{3.2}$$

Here I_{nr} consists of the first r columns of I. Equations (3.1) and (3.2) then give

$$AXI_{nr} = XI_{nr}A_{11} \Rightarrow (X^{-1}AX)I_{nr} = I_{nr}A_{11}, \tag{3.3}$$

which states that the first r columns of $X^{-1}AX$ are of the form $I_{nr}A_{11}$. Hence we may write

$$AX = X \begin{bmatrix} A_{11} & A_{12} \\ 0 & A_{22} \end{bmatrix} \tag{3.4}$$

reflecting the fact that as yet we know nothing about the subspace AX_2 (columns) except that it lies of course in the subspace $[X_1, X_2]$ (columns). From (3.4) we have

$$\det(A-\lambda I) = \det(A_{11}-\lambda I)\det(A_{22}-\lambda I) \tag{3.5}$$

so that the spectrum of A is the union of the spectra of A_{11} and A_{22}. If we replace the basis X_1 with <u>any</u> equivalent basis $\tilde{X}_1$ we have

$$\tilde{X}_1 = X_1 M \tag{3.6}$$

for some nonsingular M, and (3.1) and (3.6) give

$$A\tilde{X}_1 = \tilde{X}_1(M^{-1}A_{11}M) = \tilde{X}_1\tilde{A}_{11}. \tag{3.7}$$

Thus $\tilde{A}_{11}$ is similar to A_{11} and hence has the same elementary divisors.

It is convenient to write

$$X^{-1} = Y^H, \quad Y = [Y_1, Y_2], \quad Y^H X = XY^H = I \tag{3.8}$$

and expressed in these terms (3.4) gives

$$Y_1^H A = (A_{11}Y_1^H + A_{12}Y_2^H), \quad Y_2^H A = A_{22}Y_2^H. \tag{3.9}$$

The subspace Y_2 (columns) is the orthogonal complement of X_1 (columns) and it automatically provides a left-hand invariant subspace of A corresponding to A_{22}. Although there is a wide choice of the basis X_2 all choices give the same invariant subspace Y_2^H (rows).

If the original basis X_1 has orthonormal columns then it is natural to complete the basis by taking the columns of X_2 to be a complementary orthonormal set. In this case X is unitary and $Y^H = X^H$; further X_2 automatically provides an orthonormal basis for the left-hand invariant subspace associated with A_{22}.

From now on we restrict ourselves to the case when A_{11} and A_{22} have disjoint spectra. The significance of this is apparent if we consider a Hermitian matrix of order three with eigenvalues λ_1, $\lambda_2 = \lambda_1$, λ_3 with orthogonal eigenvectors $\underline{x}_1$, $\underline{x}_2$, $\underline{x}_3$. A right-hand invariant subspace associated with λ_1 is defined by the single vector $\underline{x}_1$ or $\underline{x}_2$ (or any linear combination of $\underline{x}_1$ and $\underline{x}_2$). These are obviously different subspaces.

Continuing now with the reduction of A we have

$$\begin{bmatrix} I & R \\ 0 & I \end{bmatrix} \begin{bmatrix} A_{11} & A_{12} \\ 0 & A_{22} \end{bmatrix} \begin{bmatrix} I & -R \\ 0 & I \end{bmatrix} = \begin{bmatrix} A_{11} & A_{12} - A_{11}R + RA_{22} \\ 0 & A_{22} \end{bmatrix} \tag{3.10}$$

and hence a further similiarity transformation reduces A to $\operatorname{diag}(A_{11}, A_{22})$ provided there exists a solution to the equation

$$A_{12} = A_{11}R - RA_{22}. \tag{3.11}$$

From Section 2 it has a unique solution when A_{11} and A_{22} have disjoint spectra which we are currently assuming. (There may of course be a solution even when this assumption is not true.) We now have the left-hand invariant subspace $\tilde{Y}_1^H = Y_1^H + RY_2^H$ (rows) associated with the spectrum of A_{11} and the right-hand invariant subspace $\tilde{X}_2 = X_2 - X_1R$ (columns) associated with the spectrum of A_{22}. Writing

$$\tilde{Y} = [Y_1 + Y_2R^H, Y_2], \quad \tilde{X} = [X_1, X_2 - X_1R] \tag{3.12}$$

we have

$$\tilde{X}^{-1} = \tilde{Y}^H \quad \text{and} \quad \tilde{Y}^H A \tilde{X} = \text{diag}(A_{11}, A_{22}) \tag{3.13}$$

and the decomposition is complete.

When X_1 and X_2 are orthonormal, the bases X_1 and X_2^H provide orthonormal bases for right-hand and left-hand invariant subspaces but the other two bases $\tilde{X}_2$ and $\tilde{Y}_1^H$ are not orthonormal. In fact we have

$$\|\tilde{X}_2\|_2^2 = \|\tilde{Y}_1^H\|_2^2 = 1 + \|R\|_2^2 . \tag{3.14}$$

Orthonormal bases are given explicitly by

$$\tilde{X}_2 [I + R^H R]^{-1/2} \quad \text{and} \quad [I + RR^H]^{-1/2} \tilde{Y}_1^H \tag{3.15}$$

but in practice may be more conveniently determined by QR factorizations.

The larger the value of $\|R\|$, the less satisfactory the determination of these two orthonormal bases is likely to be; from (2.4) and (3.11) we see that $\|R\|$ can be large only if sep is small though of course, if Vec (A_{12}) is deficient in all the very small singular values of $I \otimes A_{11}^T - A_{22} \otimes I$, the corresponding R will not be large even then.

4. KRONECKER'S CANONICAL FORM

Motivation for canonical forms associated with the generalized eigenvalue problem $A\underline{x} = \lambda B\underline{x}$ is provided by considering the related system of first order differential equations

$$B\dot{\underline{u}} = A\underline{u} + \underline{f}(t) \tag{4.1}$$

where $\underline{f}(t)$ is a vector function of t. For the standard problem B is I and the corresponding system of differential equations is

$$\dot{\underline{u}} = A\underline{u} + \underline{f}(t) . \tag{4.2}$$

We refer to this as an explicit system. Complete information on the structure of the general solution of (4.2) is provided by the Jordan canonical form of A or the Smith canonical form of the matrix pencil $A-\lambda I$ (Wilkinson, 1965, Chapter I). Corresponding information for the generalized problem is provided by the Kronecker form of the pencil $A-\lambda B$ (Gantmacher, 1959, Chapter XII). The Smith canonical form of $A-\lambda B$ is still relevant. It is interesting to note that up to a decade ago the Kronecker canonical form had scarcely been mentioned in the numerical analysis literature.

Basic features of the explicit system (4.2) are:

(i) $\underline{u}=\underline{x}e^{\lambda t}$ is a solution of the related homogeneous system if

$$\lambda\underline{x} = A\underline{x}\,. \tag{4.3}$$

This leads to the characteristic equation $\det(A-\lambda I)=0$ which is *always* of degree n. Hence there are always n eigenvalues if full account is taken of multiplicities.

(ii) There exists a unique solution corresponding to any set of initial values $\underline{u}^{(0)}$, provided $\underline{f}(t)$ satisfies quite mild integrability conditions. This unique solution is

$$\underline{u} = e^{At}\underline{u}^{(0)} + e^{At}\int_0^t e^{-A\tau}\underline{f}(\tau)\,d\tau\,. \tag{4.4}$$

For the implicit system (4.1) $\underline{u}=\underline{x}e^{\lambda t}$ is a solution of the related homogeneous system if

$$\lambda B\underline{x} = A\underline{x} \tag{4.5}$$

leading again to the characteristic equation $\det(A-\lambda B)=0$ but this can now be of any degree $\leqslant n$. In particular it can be the null polynomial. Ignoring this last case for the moment we may write

$$\det(A-\lambda B) = c_p\lambda^p + c_{p+1}\lambda^{p+1} + \cdots + c_q\lambda^q \tag{4.6}$$

where $0\leqslant p\leqslant q\leqslant n$. The characteristic equation therefore has p zero roots and $q-p$ finite nonzero roots. The deficiency is covered if we regard the equation as having $n-q$ infinite roots. The justification for this is that $A\underline{x}=\lambda B\underline{x} \Rightarrow \mu A\underline{x}=B\underline{x}$ where $\mu=\lambda^{-1}$, and the equation

$$0 = \det(\mu A - B) = \mu^n \det(A - \mu^{-1}B) = c_q \mu^{n-q} + \cdots + c_p \mu^{n-p} \tag{4.7}$$

has $n-q$ zero roots.

Since $c_0 = \det(A)$ and $c_n = \det(-B)$, zero (infinite) roots can occur iff $\det(A) = 0$ $(\det(B) = 0)$. If $\operatorname{rank}(A) = r$ the determinantal expression shows that $p \geqslant n-r$. Similarly if $\operatorname{rank}(B) = s$ we have $n-q \geqslant n-s$. Either of these inequalities may be strict. From the point of view of the differential equations, deficiency of rank (A) is not of fundamental importance since under the simple transformation $\underline{u} = \underline{v}e^{\alpha t}$ the system becomes

$$B\dot{\underline{v}} = (A - \alpha B)\,\underline{v} + e^{-\alpha t}\underline{f}(t) = (A - \alpha B)\,\underline{v} + \underline{g}(t). \tag{4.8}$$

Unless $\det(A - \lambda B) \equiv 0$, a case we are explicitly excluding for the moment, $\det(A - \alpha B)$ will be nonzero for all α except the set of finite eigenvalues. Hence we can always think in terms of a nonsingular A.

Rank deficiency in B on the other hand is quite fundamental. When $\operatorname{rank}(B) = s < n$ there are $n-s$ linearly independent vectors $\underline{y}$ such that $\underline{y}^H B = 0$. For each such vector $\underline{y}$ we have

$$\underline{y}^H B\dot{\underline{u}} = \underline{y}^H A\underline{u} + \underline{y}^H \underline{f}(t) \Rightarrow 0 = (\underline{y}^H A)\,\underline{u} + \underline{y}^H \underline{f}(t). \tag{4.9}$$

This gives $n-s$ independent relations between the elements of $\underline{u}$ and the forcing function $\underline{f}$, not involving derivatives. This feature is well illustrated by the simple example

$$A = \begin{bmatrix} 1 & 0 & 0 \\ 1 & 0 & 2 \\ 0 & 1 & 0 \end{bmatrix}, \qquad B = \begin{bmatrix} 0 & 0 & 1 \\ 0 & 0 & 2 \\ 1 & 0 & 0 \end{bmatrix} \tag{4.10}$$

for which $\operatorname{rank}(B) = 2$ but $\det(A - \lambda B) = \lambda - 2$, giving $p = 0$, $q = 1$. The differential system is

$$\dot{u}_3 = u_1 + f_1(t), \quad 2\dot{u}_3 = u_1 + 2u_3 + f_2(t), \quad \dot{u}_1 = u_2 + f_3(t). \tag{4.11}$$

There is only the one purely algebraic relation

$$u_1 - 2u_3 = f_2(t) - 2f_1(t) \tag{4.12}$$

although there are two infinite eigenvalues, from which it is obvious that we cannot assign the $u_i^{(0)}$ arbitrarily. The general solution of the homogeneous system is

$$u_1 = 2ae^{2t}, \quad u_2 = 4ae^{2t}, \quad u_3 = ae^{2t}. \tag{4.13}$$

Provided $\det(A - \lambda B) \not\equiv 0$ the relevant canonical form can be deduced quite simply from Jordan's canonical form. When B is nonsingular we have

$$\dot{\underline{u}} = B^{-1}A\underline{u} + B^{-1}\underline{f}(t) \tag{4.14}$$

which is an explicit system. If J is the Jordan canonical form of $B^{-1}A$ then

$$P^{-1}(B^{-1}A)P = J \tag{4.15}$$

for some nonsingular P and writing $P^{-1}B^{-1} = Q$ this gives

$$QAP = J, \qquad QBP = I. \tag{4.16}$$

Hence under the linear transformation $\underline{u} = P\underline{v}$ the system becomes

$$\dot{\underline{v}} = J\underline{v} + Q\underline{f}(t) = J\underline{v} + \underline{g}(t) \tag{4.17}$$

(say), which is the standard form for an explicit system.

When B is singular but A is not, we may write the system in the form

$$A^{-1}B\dot{\underline{u}} = \underline{u} + A^{-1}\underline{f}(t), \tag{4.18}$$

and now if $\tilde{J}$ is the canonical form of $A^{-1}B$ we have

$$P^{-1}(A^{-1}B)P = \tilde{J} \tag{4.19}$$

for a nonsingular P. Writing $P^{-1}A^{-1} = Q$, $\underline{u} = P\underline{v}$ we have

$$QBP = \tilde{J}, \quad QAP = I, \quad \tilde{J}\dot{\underline{v}} = \underline{v} + Q\underline{f}(t). \tag{4.20}$$

This does not give an explicit system for $\underline{v}$ but it may readily be verified that (4.20) is just as convenient for solving the differential equations. The eigenvalues associated with $\tilde{J}$ are the reciprocals of the eigenvalues λ_i of $A\underline{x} = \lambda B\underline{x}$; some of the λ_i are infinite and these appear as blocks in $\tilde{J}$ associated with zero μ_i.

There cannot be nonsingular transformations which reduce the system to pure explicit form since infinite eigenvalues are associated with features which are not present in any explicit system. Those eigenvalues associated with singular blocks in $\tilde{J}$ must therefore stay. However, corresponding to a nonsingular Jordan block $\tilde{J}(\mu_i)$ in $\tilde{J}$, the equations are

$$\tilde{J}(\mu_i)\,\dot{\underline{v}}^{(i)} = \underline{v}^{(i)} + (Q\underline{f})^{(i)} \tag{4.21}$$

where $\underline{v}^{(i)}$ denotes the corresponding subset of components of $\underline{v}$. By the earlier case this can be reduced to an equivalent explicit system

$$\dot{\underline{w}}^{(i)} = J(\lambda_i)\,\underline{w}^{(i)} + \underline{h}^{(i)} . \tag{4.22}$$

Hence the original system leads to explicit subsystems in standard form associated with the noninfinite eigenvalues and subsystems of the form

$$J^{(k)}(0)\,\dot{\underline{w}}^{(k)} = \underline{w}^{(k)} + \underline{h}^{(k)} \tag{4.23}$$

associated with the infinite eigenvalues. The corresponding canonical form of $A-\lambda B$ under equivalence transformations follows immediately.

The case when A and B are both singular but $\det(A-\lambda B)\not\equiv 0$ is not essentially different since a preliminary transformation $\underline{u} = \underline{v}e^{\alpha t}$ reduces the system to the form (4.8) in which $A-\alpha B$ is nonsingular. From the point of view of solving the differential equations there is no virtue in returning to the original matrix A. If we are thinking in terms of canonical forms under equivalence transformations we have derived a canonical form for $A-\alpha B$ and B. Thus we can immediately deduce a canonical form for A and B. The final result is that we have nonsingular matrices P and Q such that PAQ and PBQ are conformally block diagonal. Corresponding to noninfinite eigenvalues λ_i there will be a number of blocks $J^{r_j}(\lambda_i)$ (of appropriate dimensions r_j) in A, each associated with a conformal block I^{r_j} in B. Corresponding to the infinite eigenvalues there will be a number of blocks

$J^{s_j}(0)$ (of appropriate dimensions s_j) in B each associated with a conformal block I^{s_j} in A. This is the Weierstrass canonical form of the pencil $A-\lambda B$.

Turning now to the case when $\det(A-\lambda B)\equiv 0$, simple examples show that entirely new phenomena arise in the corresponding differential systems. Thus for

$$\dot{u}_1+2\dot{u}_2 = u_1+u_2+f_1(t) \tag{4.24}$$

$$2\dot{u}_1+4\dot{u}_2 = 2u_1+2u_2+f_2(t), \tag{4.25}$$

equation 2 — (2 × equation 1) gives

$$0 = f_2(t)-2f_1(t) . \tag{4.26}$$

Hence the equations will be incompatible unless $f_2(t)=2f_1(t)$. When this condition is satisfied the second equation is merely twice the first and we are left with a single equation in the two variables u_1 and u_2. We may choose u_2 to be an arbitrary function of t taking a prescribed initial value and u_1 is uniquely determined via the relation

$$\dot{u}_1 = u_1+(u_2-2\dot{u}_2+f_1(t)) \tag{4.27}$$

Since the solution of this equation involves e^t it gives the impression that $\lambda=1$ is in some sense 'an eigenvalue', indeed the only one. This is an illusion. We could just as well have chosen u_1 arbitrarily and then determined u_2 via the relation

$$2\dot{u}_2 = u_2+(u_1-\dot{u}_1+f_1(t)) \tag{4.28}$$

and now $\lambda=1/2$ appears to be the unique eigenvalue.

As a second example consider the system

$$\dot{u}_1+2\dot{u}_2 = 2u_1+4u_2+f_1(t) \tag{4.29}$$

$$2\dot{u}_1+4\dot{u}_2 = 3u_1+6u_2+f_2(t) . \tag{4.30}$$

Writing $v_1=u_1+2u_2$ the equations become

$$\dot{v}_1 = 2v_1+f_1(t) , \quad 2\dot{v}_1 = 3v_1+f_2(t) , \tag{4.31}$$

i.e. two equations in the single variable v_1. From (4.31) we deduce that

$$v_1 = f_2(t) - 2f_1(t) \quad \text{and} \quad \dot{f}_2(t) - 2\dot{f}_1(t) = 2f_2(t) - 3f_1(t). \tag{4.32}$$

Again we have a compatability condition but it now involves the first derivatives of the forcing function; v_1 is given without integration and we observe that its initial value cannot be prescribed arbitrarily. The system determines v_1 (i.e. $u_1 + 2u_2$) only; either u_1 or u_2 may be taken to be an arbitrary function of t and the other is then determined.

In these two examples there are no 'true eigenvalues' associated with the solution. Is this always true when $\det(A - \lambda B) \equiv 0$? The answer is 'no'. If we add a third equation

$$\dot{u}_3 = ku_3 + f_3(t) \tag{4.33}$$

to (4.24) and (4.25) then the exponential e^{kt} is 'intimately connected' with the solution and $\lambda = k$ must be, in some sense, an eigenvalue. Thus although $A\underline{x} = \lambda B\underline{x}$ has a solution $\underline{x} \neq \underline{0}$ for all λ, some values of λ can still be 'special'.

A canonical form for $A - \lambda B$ should expose all of these features of the corresponding differential system. It should reveal how many compatability conditions there are, whether they involve derivatives of the $f_i(t)$ and, if so, of what orders, how many arbitrary functions of t are involved in the general solution and whether there are any 'genuine eigenvalues' including infinite eigenvalues. The Kronecker canonical form does precisely that. Before discussing it we comment that although our simple examples are both 2×2 systems the first reduces to a 1×2 system and the second to a 2×1 system. Hence it is vital that we should have a canonical form covering rectangular A and B. To be specific we require a convenient canonical form under equivalence transformations $P(A - \lambda B)Q$ where $P \in \mathbb{C}^{m \times m}$ and $Q \in \mathbb{C}^{n \times n}$ are nonsingular. The differential systems

$$B\underline{\dot{u}} = A\underline{u} + \underline{f}(t) \quad \text{and} \quad (PBQ)\underline{\dot{v}} = (PAQ)\underline{v} + P\underline{f}(t) \tag{4.34}$$

are, of course, fully equivalent. A matrix pencil $A-\lambda B$ is said to be regular if A and B are square and $\det(A-\lambda B)\not\equiv 0$. It is singular if either

(a) A and B are square and $\det(A-\lambda B)\equiv 0$, or

(b) A and B are rectangular.

For square nonsingular pencils elementary Jordan blocks provide the only diagonal elements required in a satisfactory canonical form. For rectangular A and B this can no longer be true since the Jordan blocks are square. Some form of elementary rectangular block is essential. It was no coincidence that our two simple systems reduced to 1×2 and 2×1 systems. Additional elementary blocks of dimensions $k\times(k+1)$ and $(k+1)\times k$ (where k is arbitrary and can be zero) prove to be adequate to deal with all cases.

The elementary $k\times(k+1)$ blocks and the associated differential system are illustrated when $k=2$ by

$$\begin{bmatrix} -\lambda & 1 & 0 \\ 0 & -\lambda & 1 \end{bmatrix} \quad \text{and} \quad \begin{aligned} \dot{u}_1 &= u_2+f_1(t) \\ \dot{u}_2 &= u_3+f_2(t) \end{aligned} \tag{4.35}$$

respectively; u_3 can be taken to be an arbitrary function of t having a prescribed initial value $u_3^{(0)}$, and u_2 and u_1 are then uniquely determined for prescribed initial values. There is just one arbitrary function associated with a $k\times(k+1)$ block of any order.

The corresponding $(k+1)\times k$ blocks are illustrated by

$$\begin{bmatrix} -\lambda & 0 \\ 1 & -\lambda \\ 0 & 1 \end{bmatrix} \quad \text{and} \quad \begin{aligned} \dot{u}_1 &= f_1(t) \\ \dot{u}_2 &= u_1+f_2(t) \\ 0 &= u_2+f_3(t)\,. \end{aligned} \tag{4.36}$$

u_2 is determined uniquely by the third equation and u_1 is then derived from the second equation. Notice that one is not free to prescribe $u_2^{(0)}$ and $u_1^{(0)}$ and no integration is required. Substituting u_1 in the first equation gives

$$f_1(t) + \dot{f}_2(t) + \ddot{f}_3(t) = 0 \tag{4.37}$$

a compatability condition involving derivatives up to the second order in the forcing function. In general there is one compatability condition associated with each $(k+1)\times k$ block and it involves derivatives up to the kth order.

The two cases when $k=0$ may at first sight seem puzzling since they involve 0×1 and 1×0 matrices respectively, They should not be confused with null matrices. A $p\times q$ matrix has pq elements. In consistency with this a 1×0 matrix or a 0×1 matrix has 1×0 or 0×1 elements respectively, i.e. it has no elements at all. (These matrices are illustrated in (4.40) and (4.41).) They have in fact already been encountered in the two simple examples. Premultiplication of (4.24) and (4.25) by the matrix

$$\begin{bmatrix} 1 & 0 \\ -2 & 1 \end{bmatrix} \quad \text{gives} \quad \begin{aligned} \dot{u}_1 + 2\dot{u}_2 &= u_1 + u_2 + f_1(t) \\ 0\dot{u}_1 + 0\dot{u}_2 &= 0u_1 + 0u_2 + f_2(t) - 2f_1(t). \end{aligned} \tag{4.38}$$

The homogeneous part of the second equation is null and the current transformed $A-\lambda B$ is

$$\begin{bmatrix} 1-\lambda & 1-2\lambda \\ 0 & 0 \end{bmatrix}. \tag{4.39}$$

Postmultiplication with

$$\begin{bmatrix} -1 & 2 \\ 1 & -1 \end{bmatrix} \quad \text{gives} \quad \begin{bmatrix} -\lambda & 1 \\ 0 & 0 \end{bmatrix} \quad \text{i.e.} \quad \begin{array}{|cc|c} \hline -\lambda & 1 & \\ \hline & & \vert \end{array}. \tag{4.40}$$

This is in block diagonal form with a 1×2 block followed by a 1×0 block, the latter being represented by the vertical line in the bottom right-hand corner of the final entry of (4.40). In much the same way the second system reduces to

$$\begin{bmatrix} -\lambda & 0 \\ 1 & 0 \end{bmatrix} \quad \text{i.e.} \quad \begin{array}{|c|c} \hline -\lambda & \\ 1 & \\ \hline & \end{array} \tag{4.41}$$

which is in block diagonal form with a 2×1 block followed by a

0×1 block, the latter being represented by the horizontal line in (4.41).

To sum up, the K.c.f. of a matrix pencil $A - \lambda B$ consists of two main parts, the singular part and the regular part, either of which may be missing; it is of block diagonal form and in general some of the blocks are rectangular. The regular part consists of the Weierstrass canonical form of a regular pencil $A_0 - \lambda B_0$ (say). It identifies the number and character of both infinite and finite 'eigenvalues' (or more properly 'elementary divisors').

The singular part consists of a number of blocks of dimensions $k \times (k+1)$ and $(k+1) \times k$ illustrated when $k = 2$ by (4.35) and (4.36) respectively. A $k \times (k+1)$ block of this type is denoted by L_k and accordingly the corresponding $(k+1) \times k$ block is denoted by L_k^T. The dimensions k associated with these blocks are denoted by $\varepsilon_1, \varepsilon_2, \cdots$ and $\eta_1, \eta_2, \cdots$ respectively, a somewhat uncomfortable notation since ε and η are traditionally associated with small quantities. (See, for example, Gantmacher, 1959, Chapter XII.) Any or all of the ε_i and η_i may be zero. These integers are called the 'minimal indices'. Two pencils $A_1 - \lambda B_1$ and $A_2 - \lambda B_2$ having the same K.c.f. are strictly equivalent, i.e. there are nonsingular P and Q such that

$$P(A_1 - \lambda B_1)\, Q = A_2 - \lambda B_2 .$$

For many purposes it is advantageous to think in terms of homogeneous pencils $\mu A - \lambda B$.

5. KRONECKER'S FORM IN NUMERICAL ANALYSIS

In the 1950's and 1960's the attitude of numerical analysts to the J.c.f. was somewhat equivocal. In some ways it was regarded as irrelevant since it depends in a discontinuous way on the elements of the matrix. Moreover the eigenvalues associated with a nonlinear elementary divisor are often extremely sensitive to small perturbations in the matrix. Thus a perturbation of

$\varepsilon^3 \underline{e}_3 \underline{e}_1^T$ in $J_3(\lambda_i)$ converts the cubic divisor $(\lambda_i - \lambda)^3$ into three linear elementary divisors $(\lambda_i + \varepsilon - \lambda)$, $(\lambda_i + \omega\varepsilon - \lambda)$ and $(\lambda_i + \omega^2\varepsilon - \lambda)$, where $\omega = e^{2i\pi/3}$. If $\varepsilon = 10^{-4}$ this means that a perturbation of order 10^{-12} produces separations of order 10^{-4} in the eigenvalues. The explicit system of differential equations corresponding to the original matrix has solutions that include components of the form $e^{\lambda_i t}$, $te^{\lambda_i t}$ and $t^2 e^{\lambda_i t}$ while the perturbed system has pure exponential solutions. The analytical character of the explicit solution is completely altered by what might well be regarded in a practical situation as a negligible perturbation. Many felt that I should have ignored the J.c.f. in the Algebraic Eigenvalue Problem (Wilkinson, 1965) and I had misgivings about including a discussion of it. Gradually the attitude has changed and it is now recognized that the range of possible structures of the J.c.f.'s of all matrices in the ball $(A + F : \|F\| \leqslant \eta)$ is highly relevant to the understanding of the underlying physical problem. Here η is the uncertainty in our knowledge of A.

Now for the simple pencil $A - \lambda I$ the K.c.f. is precisely $J - \lambda I$ where J is the J.c.f. of A. Hence the K.c.f. has all the instabilities associated with the J.c.f. in addition to others associated with the infinite eigenvalues and the singular part. Nevertheless the range of possible structures of all pencils $(A + F) - \lambda(B + G)$ within the domain of uncertainty is just as relevant. Work in this area has been particularly extensive in the last decade.

In an early paper (Golub and Wilkinson, 1976), the elementary Jordan blocks associated with a given eigenvalue λ_i of a matrix A were determined in a backward stable method by identifying the structure of the nullities associated with $A - \lambda_i I$. Backward stability was assured by using singular value decompositions (S.V.D.'s) to identify these nullities. Analogous methods may be used to determine the minimal indices and the structures of the elementary Jordan blocks associated with the infinite eigenvalues of $A - \lambda B$.

The nullities are exposed by applying a sequence of elementary equivalence transformations defined by

$$B^{(i+1)} = P^{(i)}B^{(i)}Q^{(i)}, \quad A^{(i+1)} = P^{(i)}A^{(i)}Q^{(i)}. \qquad (5.1)$$

When the S.V.D. is used for this purpose the $P^{(i)}$ and $Q^{(i)}$ are unitary. There are two main stages, the first being concerned with row nullities and the second with column nullities. The two stages differ only in that the roles of rows and columns are interchanged. Each stage consists of a number of major steps, each major step being divided into two minor steps. At the beginning of the ith major step the current state of the B and A matrices may be expressed in the forms

$$\begin{bmatrix} B_{11}^{(i)} & B_{12}^{(i)} \\ 0 & B_{22}^{(i)} \end{bmatrix} \text{ and } \begin{bmatrix} A_{11}^{(i)} & A_{12}^{(i)} \\ 0 & A_{22}^{(i)} \end{bmatrix}, \qquad (5.2)$$

where $B_{22}^{(i)}$ and $A_{22}^{(i)}$ have structures which will be exposed by a description of the ith major step. This step is based on the rows and columns of $B_{11}^{(i)}$ and $A_{11}^{(i)}$ though the transformations are applied to the full B and A matrices.

In the first minor step the row nullity n_i of $B_{11}^{(i)}$ is exposed by premultiplications and postmultiplications; these are appplied to both B and A. The matrices $B_{11}^{(i)}$ and $A_{11}^{(i)}$ are thereby reduced to

$$\begin{bmatrix} C^{(i)} \\ 0 \end{bmatrix} \}n_i \text{ and } \begin{bmatrix} G^{(i)} \\ F^{(i)} \end{bmatrix} \}n_i, \qquad (5.3)$$

where the rows of $C^{(i)}$ are linearly independent. In the second minor step the rank r_i of $F^{(i)}$ is exposed by premultiplications and postmultiplications. Since these involve only the rows of $F^{(i)}$ itself, the null matrix in the current B is unchanged. Clearly $r_i \leqslant n_i$. The resulting forms of $B_{11}^{(i)}$ and $A_{11}^{(i)}$ are

$$\begin{bmatrix} B_{11}^{(i+1)} & K_{12}^{(i+1)} \\ 0 & 0 \\ 0 & \underbrace{0}_{r_i} \end{bmatrix} \}\, r_i \quad \text{and} \quad \begin{bmatrix} A_{11}^{(i+1)} & L_{12}^{(i+1)} \\ 0 & M_{22}^{(i+1)} \\ 0 & \underbrace{0}_{r_i} \end{bmatrix} \}\, r_i \tag{5.4}$$

where $M_{22}^{(i+1)}$ is nonsingular when $r_i > 0$. Some of the partitions will be absent when $r_i = n_i$ or $r_i = 0$. The matrices $B_{11}^{(i+1)}$ and $A_{11}^{(i+1)}$ determine the next major step while the remaining matrices contribute to the parts denoted by $B_{12}^{(i+1)}$, $B_{22}^{(i+1)}$, $A_{12}^{(i+1)}$ and $A_{22}^{(i+1)}$ in the obvious way.

The first stage terminates with $B_{11}^{(s+1)}$ when $n_{s+1} = 0$. Notice that this termination may occur because $B_{11}^{(s+1)}$ is of zero dimensions, i.e. if all rows and columns have been exhausted by the first s major steps. The nullity of a 0×0 matrix is taken to be 0. The second stage continues with $B_{11}^{(s+1)}$ and $A_{11}^{(s+1)}$, the roles of rows and columns being reversed. We denote the nullities and ranks determined in the second stage by $\tilde{n}_i, \tilde{s}_i$ $(i = 1, \cdots, t)$. It terminates when $\tilde{n}_{t+1} = 0$, and this could occur because all rows and columns have been exhausted. The remaining pencil $A_{11}^{(s+t+1)} - \lambda B_{11}^{(s+t+1)}$ will be regular and indeed will have no infinite eigenvalues. It will, of course, be absent if there is no regular part with finite eigenvalues.

Thinking in terms of differential equations this algorithm reduces the original system to a form in which the explicit solution may be determined by a process analogous to back-substitution. Thus, for example, the first major step of the first stage shows that the last $n_1 - r_1$ transformed equations have no entries in either A or B. Hence the associated transformed forcing functions must be zero, giving $n_1 - r_1$ compatibility conditions which do not involve derivatives. The last r_1 variables are given directly in terms of the forcing functions without integrations. Similarly the first major step of the second stage shows that $\tilde{n}_1 - \tilde{r}_1$ of the transformed variables are not involved at all in the transformed equations and hence can be taken to be arbitrary functions of t.

It is therefore not surprising that the integers n_i, r_i, $\tilde{n}_i$ and $\tilde{r}_i$ determine the minimal indices and the nature of the infinite elementary divisors. It can be shown (Wilkinson, 1978) that there are $n_i - r_i$ blocks of the form L_{i-1}^T and $\tilde{n}_i - \tilde{r}_i$ blocks of the form L_{i-1}; there are $r_i - n_{i+1}$ infinite elementary divisors of degree i.

Several comments are in order.

(i) We could equally well have determined column nullities in the first stage and row nullities in the second. In either case the infinite eigenvalues are determined in the first stage because row and column nullities are involved in a symmetrical way in $I - \lambda J_k(0)$.

(ii) The asymmetry in the information provided by the first and second stages could be removed if in the second stage we reverse the roles of B and A as well as those of rows and columns. The second stage then determines the zero eigenvalues as well as the L_k but, as we have seen, zero eigenvalues have far less significance than infinite eigenvalues for differential systems.

(iii) For backward stability the S.V.D. should be used in every step. However, when the elements of A and B are modest sized integers the elimination could be done in exact integer arithmetic using 'cross-multiplication' without exceeding computer capacity. Such an algorithm is useful for dealing with small illustrative examples.

(iv) If we are merely interested in the minimal indices and the infinite eigenvalues we can dispense with everything except $B_{11}^{(i)}$ and $A_{11}^{(i)}$ in the ith major step.

(v) If we are interested in solving a 'differential' system explicitly we must retain full information on the current B and A matrices (and the transformed forcing functions) but there is no need to proceed to the K.c.f. itself. The equations will be in an appropriate form for determining the explicit solution.

(vi) Although the determination of the n_i, r_i, $\tilde{n}_i$, $\tilde{r}_i$ can be done using unitary transformations only, if we wish to produce the K.c.f. itself then the further transformations and the final P and Q cannot, in general, be unitary (Wilkinson, 1978).

Some properties of the K.c.f. which are of fundamental importance to numerical analysts were not immediately appreciated. They may be introduced via the J.c.f.. The eigenvalues of a matrix A are generally determined by transforming it to a simple form; commonly this is the Schur triangular form T. Although this can be done in a backward stable manner rounding errors will usually destroy multiple eigenvalues. Hence, even if A has nonlinear elementary divisors, T will almost certainly have linear divisors and the relevant eigenvalues may not be particularly close. In much the same way, if the pencil $A-\lambda B$ is square and has non-linear elementary divisors associated with infinite eigenvalues, these will become finite and distinct. (In fact it is more convenient to think in terms of zero eigenvalues of $B-\mu A$ since this avoids the very large variations in perturbed infinite eigenvalues.) Moreover any singular part will in general be destroyed by the rounding errors made in the equivalence transformations. This is true of any property that depends on rank. On the other hand if the pencil is rectangular it will inevitably have true row or column nullity.

One of the most effective algorithms for finding the eigenvalues of a square pencil is the QZ algorithm which reduces the pencil to the form $T_1-\lambda T_2$ (where T_1 and T_2 are upper triangular) by unitary equivalence transformations. If the pencil were singular exact transformations would give a T_1 and T_2 in which $t_{ii}^{(1)}=t_{ii}^{(2)}=0$ for at least one value of i since $\det(T_1-\lambda T_2) = k\det(A-\lambda B) \equiv 0$, where k is independent of λ. Rounding errors will mean that the two relevant diagonal elements will not be zero. It is tempting to imagine that if, for example, a computed triangular pencil on a 10-digit computer is

$$\begin{bmatrix} 1 & 2 & 3 \\ 0 & 3\times10^{-10} & 4 \\ 0 & 0 & 2 \end{bmatrix} - \lambda \begin{bmatrix} 2 & 1 & 1 \\ 0 & 10^{-10} & 1 \\ 0 & 0 & 1 \end{bmatrix} \tag{5.5}$$

then, although the small diagonal elements are effectively zero and their ratio does not therefore determine an eigenvalue, the other two ratios give well-determined eigenvalues 1/2 and 2. This is an illusion. Assuming that the small diagonal elements really are zero then there are no true eigenvalues. In fact replacing them by zero and then adding column 2 to column 1 and rwo 2 to row 3 we have

$$\begin{bmatrix} 3 & 2 & 3 \\ 0 & 0 & 4 \\ 0 & 0 & 6 \end{bmatrix} - \lambda \begin{bmatrix} 3 & 1 & 1 \\ 0 & 0 & 1 \\ 0 & 0 & 2 \end{bmatrix} \tag{5.6}$$

and the apparently well-determined ratios have changed to 1 and 3. If we take two upper triangular matrices having $t_{ii}^{(1)} = t_{ii}^{(2)} = 0$ for some i and choose the remaining elements at random the probability is unity that the K.c.f. will consist of an $L_{i-1}(\lambda)$ and an $L_{n-i}^{T}(\lambda)$; there will in general be no regular part. For a regular part to exist the remaining elements must satisfy certain relations. Similarly for an $m \times n$ rectangular pencil the probability is unity that the K.c.f. will be determined uniquely by m and n and be independent of the elements of A and B.

6. NEAREST DEFECTIVE MATRIX

The structures of pencils in the neighbourhood of a given pencil are often of great practical significance. Perhaps the simplest problem of this kind is the determination of the nearest defective matrix to a given matrix A and considerable effort has been devoted to it. Viewed in retrospect much of this work has suffered from 'oversophistication'. In many respects the most important results have remained two simple theorems which have been known to numerical analysts for many years.

THEOREM A. If λ_i is a simple eigenvalue of A with left and right eigenvectors $\underline{y}_i^H$ and $\underline{x}_i$ then, for all sufficiently small η, the matrix $A+\eta E$, where E is any matrix with $\|E\|=1$, has a simple eigenvalue $\lambda_i(\eta)$ such that

$$d\lambda/d\eta = \underline{y}_i^H E\underline{x}_i/\underline{y}_i^H\underline{x}_i \quad \text{at} \quad \eta = 0 \tag{6.1}$$

(Wilkinson, 1965, p.69).

Although it gives only a local rate of change, this theorem is invaluable in providing an indication of the direction in which the closest defective matrix lies and rough estimates of the distance to it. In any given norm we have

$$|d\lambda/d\eta| \leqslant \|\underline{y}_i\|_D\|\underline{x}_i\|/|\underline{y}_i^H\underline{x}_i| = \rho \tag{6.2}$$

where $\|\cdot\|_D$ denotes the dual norm. This maximum is attained for a rank one E equal to $\underline{p}\underline{q}^H$ where $\|\underline{p}\|=1$, $\|\underline{q}\|_D=1$, $\underline{p}$ is the vector dual to $\underline{y}_i$ and $\underline{q}$ the vector dual to $\underline{x}_i$. From the definition of the dual norm we have $\rho \geqslant 1$. Since the l_2 norm is self dual we have in this case

$$\rho = \|\underline{y}_i\|_2\|\underline{x}_i\|_2/|\underline{y}_i^H\underline{x}_i| = 1/|s_i| \tag{6.3}$$

where, in the usual notation, s_i denotes the cosine of the angle between $\underline{y}_i$ and $\underline{x}_i$. Though the use of the l_2 norm has dominated applications of this result, the l_∞ and l_1 norms are in some ways more convenient. When $\underline{y}_i$ and $\underline{x}_i$ are real we have for the l_∞ norm

$$\underline{p}^T = (\pm 1, \pm 1, \cdots, \pm 1), \quad \underline{q}^T = \pm \underline{e}_k^T, \tag{6.4}$$

where the sign of p_j is that of the jth component of $\underline{y}_i$ and where $\underline{q}$ is such that $\underline{q}^T\underline{x}_i = \|\underline{x}_i\|_\infty$. (For complex $\underline{y}_i$ the jth element of $\underline{p}_i$ is $\exp(i\theta_j)$ where θ_j is the phase of the jth element of $\underline{y}_i$.) This has the advantage that in the real case the optimal E is usually constant for a wide range of variations in A.

The second theorem is an immediate consequence of the fact that for singularity of $B+F$ (i.e. of $B(I+B^{-1}F)$) we must have $\|B^{-1}\|\,\|F\| \geq 1$. Since λ is an eigenvalue of $A+F$ iff $A+F-\lambda I$ is singular this simple result expressed in terms of eigenvalues gives the following:

THEOREM B. If λ is an eigenvalue of $A+\eta E$ ($\eta > 0$, $\|E\| = 1$) then

$$\eta \geq 1/\|(A-\lambda I)^{-1}\| . \tag{6.5}$$

Notice that we can take η to be positive since $(\eta e^{i\theta})E = \eta(e^{i\theta}E)$. For any $E = -\underline{p}\underline{q}^H$ the required perturbation which induces λ as an eigenvalue is

$$\eta E = -\underline{p}\underline{q}^H / \underline{q}^H (A-\lambda I)^{-1} \underline{p} \tag{6.6}$$

and we have

$$\eta = 1/\underline{q}^H (A-\lambda I)^{-1} \underline{p} \cdot \tag{6.7}$$

Taking $\underline{p}$ to be a unit vector for which

$$(A-\lambda I)^{-1}\underline{p} = \|(A-\lambda I)^{-1}\|\underline{w} \quad (\|\underline{w}\| = 1) \tag{6.8}$$

(i.e. $\underline{p}$ is a maximal vector for $(A-\lambda I)^{-1}$) and $\underline{q}$ to be the dual vector to $\underline{w}$ we have

$$\underline{q}^H (A-\lambda I)^{-1}\underline{p} = \|(A-\lambda I)^{-1}\| , \tag{6.9}$$

and hence equality is achieved in (6.5) with a rank-one perturbation.

The inclusion domain defined by (6.5) gives the set of λ's which may be induced as eigenvalues by perturbations of norm $\leq \eta$. As we have shown, any λ in this domain can indeed be induced by a rank-one perturbation and the theorem shows how it is determined via $(A-\lambda I)^{-1}$.

For the l_2 norm the relevant perturbation is $-\sigma_n \underline{u}_n \underline{v}_n^H$ where σ_n is the smallest singular value of $(A-\lambda I)$ and $\underline{u}_n$ and

$\underline{v}_n$ are the corresponding singular vectors; we have, of course, $\sigma_n = 1/\|(A-\lambda I)^{-1}\|_2$. The formal simplicity of this result is deceptive. The computations of σ_n, $\underline{u}_n$ and $\underline{v}_n$ are formidable undertakings. The results for the l_∞ and l_1 norms are considerably simpler. In the real case $\underline{p} = (\pm 1, \pm 1, \cdots, \pm 1)^T$ and $\underline{q} = \underline{e}_k$ where the signs in $\underline{p}$ are determined by the elements in the maximal row of $|(A-\lambda I)^{-1}|$ and k is the index of an element of maximum modulus in $(A-\lambda I)^{-1}\underline{p}$. It is clear from this that the same $\underline{p}$ and $\underline{q}$ will be optimal for inducing a whole range of values of λ; only the value of η will be changed.

This last feature often makes Theorem B suitable for determining a defective matrix which really is at the minimum distance from A. (We say 'a' matrix since there may be others equidistant from A.) Thus, if

$$A = \begin{bmatrix} 0 & 1 & 0 \\ 0 & \varepsilon & 1 \\ 0 & 0 & 3\varepsilon \end{bmatrix}, \tag{6.10}$$

the nearest defective matrix has a double eigenvalue lying between 0 and ε. The maximum row of $(A-\lambda I)^{-1}$ for all relevant λ is the first. This is

$$[-1/\lambda,\ 1/\lambda(\varepsilon-\lambda),\ -1/\{\lambda(\varepsilon-\lambda)(3\varepsilon-\lambda)\}] \tag{6.11}$$

and the inclusion domain is therefore

$$\eta \geqslant 1/\|(A-\lambda I)^{-1}\|_\infty = |\lambda(\varepsilon-\lambda)(3\varepsilon-\lambda)|/\{1+|3\varepsilon-\lambda|(1+|\varepsilon-\lambda|)\}. \tag{6.12}$$

It can be deduced from the continuity property of eigenvalues that the maximum value of the R.H.S. for $0 \leqslant \lambda \leqslant \varepsilon$ gives the minimal η inducing a double eigenvalue and the location of that eigenvalue. When $\varepsilon \ll 1$ the denominator is approximately 1 for all λ in the interval and we merely have to locate the maximum value of $|\lambda(\varepsilon-\lambda)(3\varepsilon-\lambda)|$. When $\varepsilon < 1$ but not negligible the maximum may be determined numerically in an obvious way; the maximum of the numerator gives a good starting value for the required λ. Notice that the direction

$$E = [-1, 1, -1]^T \underline{e}_1^T \tag{6.13}$$

is the optimal direction for inducing an eigenvalue anywhere between 0 and ε.

The matrix in this example had simple eigenvalues but if we change λ_3 from 3ε to ε it then has a double eigenvalue at ε. We might seek the smallest perturbation which makes the eigenvalue $\lambda = 0$ coalesce with one of the double eigenvalues (which are separated by the perturbation). Clearly exactly the same method may be used; in fact the presence of the double root makes the algebra formally simpler.

Theorem B extends immediately to the more general problems $A\underline{x} = \lambda B\underline{x}$ and $(A_r\lambda^r + \cdots + A_1\lambda + A_0)\,\underline{x} = 0$. An eigenvalue of the first of these may be induced by a rank-one perturbation $\eta\underline{p}\underline{q}^H$ in A, where (assuming $\|\underline{p}\| = \|\underline{q}\|_D = 1$)

$$\eta = 1/\|(A - \lambda B)^{-1}\|. \tag{6.14}$$

If we regard perturbations in both A and B as equally acceptable then we have

$$(1 + |\lambda|)\eta = 1/\|(A - \lambda B)^{-1}\|. \tag{6.15}$$

When A and B and the eigenvalues which move to coalescence are all real one has merely to find the point at which

$$1/\|(A - \lambda B)^{-1}\| \quad \text{or} \quad 1/\{(1 + |\lambda|)\|(A - \lambda B)^{-1}\|\} \tag{6.16}$$

achieves its maximum in the relevant range of λ.

If $f(\lambda)$ is any real function of λ such that

$$f(\lambda) \geq \|(A - \lambda I)^{-1}\| \tag{6.17}$$

then the domain

$$1/f(\lambda) \leq \eta \tag{6.18}$$

certainly includes the fundamental domain $D(\eta)$ of values of λ that satisfy (6.5). A number of theorems giving 'simplified' inclusion domains for eigenvalues are really based on the above principle though this fact seems not to have been appreciated. Proofs based explicitly on the principle are not only simpler but give greater insight into the shortcomings of the theorems. Thus if A has simple eigenvalues we have

$$(A-\lambda I)^{-1} = \sum \underline{x}_i \{(\lambda_i - \lambda)\, \underline{y}_i^H \underline{x}_i\}^{-1} \underline{y}_i^H \tag{6.19}$$

and hence

$$\|(A-\lambda I)^{-1}\| \leqslant \sum \|\underline{x}_i\|\, \|\underline{y}_i\|_D / (|\lambda_i - \lambda|\, |\underline{y}_i^H \underline{x}_i|) . \tag{6.20}$$

For the l_2 norm this gives

$$\|(A-\lambda I)^{-1}\|_2 \leqslant \sum 1/|(\lambda_i - \lambda)s_i| \tag{6.21}$$

where $1/|s_i|$ is the usual l_2 condition number of λ_i. In general $\|\underline{x}_i\|\, \|\underline{y}_i\|_D / |\underline{y}_i^H \underline{x}_i|$ gives the condition number corresponding to any given vector norm. Hence (6.21) is true with this generalized definition of the s_i and all eigenvalues in $D(\eta)$ are certainly contained in the domain

$$\{\lambda :\, 1/\{\sum (1/|(\lambda_i - \lambda)s_i|)\} \leqslant \eta\} . \tag{6.22}$$

If one thinks in terms of numerical methods there is little difficulty in determining the domain defined by (6.22) for various values of η. However, there has been a tendency to simplify (6.22), weakening it further. (It is already weaker than (6.5).) Thus in (6.22) we may replace every $|\lambda_i - \lambda|$ by $\min|\lambda_i - \lambda|$ and we then have the following theorem.

All eigenvalues which can be induced by perturbations bounded by η lie in at least one of the circular discs

$$|\lambda_j - \lambda| \leqslant \eta\{\sum 1/|s_i|\} , \quad j = 1,2,\cdots,n. \tag{6.23}$$

On the other hand, if we replace every term $1/|(\lambda_i - \lambda)s_i|$ by $1/\min|(\lambda_i - \lambda)s_i|$, then instead of (6.23) we have the circular discs

$$\{\lambda_j - \lambda| \leqslant \eta(n/|s_j|), \quad j = 1,2,\cdots,n. \tag{6.24}$$

The weakness of (6.23) is that all discs are of the same radius and this is determined primarily by the most sensitive eigenvalues;

(6.24) has the advantage that the radius of the disc based on λ_j is dependent only on its own sensitivity but the factor n is in general much too large. The discs based on very sensitive eigenvalues overlap those based on the insensitive eigenvalues for much too small a value of η.

For a similarity reduction to block diagonal form we have

$$X^{-1}AX = Y^H AX = \operatorname{diag}(A_{ii}) \quad (i = 1, \cdots, s) \tag{6.25}$$

where

$$X^{-1} = Y^H, \quad X = [X_1, X_2, \cdots, X_s], \quad Y = [Y_1, Y_2, \cdots, Y_s]. \tag{6.26}$$

X_i (columns) and Y_i^H (rows) give the invariant subspaces corresponding to the eigenvalues of A_{ii}. Accordingly we have

$$(A-\lambda I)^{-1} = \sum X_i (A_{ii}-\lambda I)^{-1} Y_i^H \tag{6.27}$$

giving

$$\|(A-\lambda I)^{-1}\| \leqslant \sum \|X_i\| \, \|Y_i\| \, \|(A_{ii}-\lambda I)^{-1}\| . \tag{6.28}$$

In this relation one cannot have orthonormal X_i <u>and</u> Y_i since $Y^H X = I$ but we can always take either X_i or Y_i to be orthonormal; the other member of the pair is then of norm $\|P_i\|_2$ where P_i is the associated projector. Hence the domain

$$1/\{\sum \|P_i\|_2 \|(A_{ii}-\lambda I)^{-1}\|\} \leqslant \eta \tag{6.29}$$

certainly contains all eigenvalues in $D(\eta)$. Again this may be simplified (but weakened) to give the following two theorems.

All λ in $D(\eta)$ lie in at least one of the domains

$$\text{(I)} \quad 1/\|(A_{jj}-\lambda I)^{-1}\|_2 \leqslant \eta \{\sum \|P_i\|_2\}, \quad j = 1, 2, \cdots, s \tag{6.30}$$

$$\text{(II)} \quad 1/\|(A_{jj}-\lambda I)^{-1}\|_2 \leqslant s\eta \|P_j\|_2, \quad j = 1, 2, \cdots, s. \tag{6.31}$$

These theorems reduce to (6.23) and (6.24) when $s = n$. In the particular case when

$$A = \begin{bmatrix} A_{11} & A_{12} \\ 0 & A_{22} \end{bmatrix} \tag{6.32}$$

and the block diagonal form is $\operatorname{diag}(A_{ii})$ we have from (3.10)

$$X_1^T = [I, 0], \quad Y_1^H = [I, R], \quad X_2^T = [-R^T, I], \quad Y_2^H = [0, I]. \tag{6.33}$$

Hence X_1 and Y_2 are automatically orthonormal while, as in (3.14),

$$\|P_1\|_2 = \|P_2\|_2 = \|P\|_2 = (1 + \|R\|_2^2)^{1/2}. \tag{6.34}$$

(I) and (II) therefore both give

$$1/\|(A_{jj} - \lambda I)^{-1}\|_2 \leq 2\|P\|_2 \eta, \quad j = 1,2, \tag{6.35}$$

and a minor refinement gives $\|P\|_2 + \|R\|_2$ in place of $2\|P\|_2$. The domain (6.29) reduces to

$$1/\{\|(A_{11} - \lambda I)^{-1}\|_2 + \|(A_{22} - \lambda I)^{-1}\|_2\} \leq \|P\|_2 \eta \tag{6.36}$$

which is better than (6.35) particularly for λ in a neighbourhood of the spectra of A_{11} or A_{22}. Unfortunately all of these bounds give inclusion domains which are much too large except when $\|P\| = O(1)$ and in this case one scarcely needs them.

In recent work (unpublished) I have used Theorem B to provide a basis for what might be called 'backward perturbation theory'. For many purposes it is simpler and more illuminating than classical perturbation theory.

7 DEFLATING SUBSPACES

For the standard eigenvalue problem an r dimensional invariant subspace X_1 (columns) provides a generalization of the one dimensional subspace determined by an eigenvector $\underline{x}_1$. For the problem $A\underline{x} = \lambda B\underline{x}$ we may proceed analogously to a generalization as follows. The relation $A\underline{x}_1 = \lambda_1 B\underline{x}_1$ states that $A\underline{x}_1$ and $B\underline{x}_1$ belong to the same one dimensional subspace. Accordingly we consider a subspace X_1 (columns) such that AX_1 (columns) and BX_1 (columns) belong to the same subspace Z_1 (columns), where the latter is also r dimensional. Hence we have

$$AX_1 = Z_1 A_{11}, \quad BX_1 = Z_1 B_{11} \tag{7.1}$$

where $A_{11}, B_{11} \in \mathbb{C}^{r \times r}$. If we extend X_1 and Z_1 to full bases

X and Z this gives

$$AXI_{nr} = ZI_{nr}A_{11}, \quad BXI_{nr} = ZI_{nr}B_{11} \tag{7.2}$$

showing that $Z^{-1}AX$ and $Z^{-1}BX$ are of the forms

$$\begin{bmatrix} A_{11} & A_{12} \\ 0 & A_{22} \end{bmatrix} \text{ and } \begin{bmatrix} B_{11} & B_{12} \\ 0 & B_{22} \end{bmatrix}. \tag{7.3}$$

For this reason Stewart (1973) refers to such an X_1 (columns) and Z_1 (columns) as deflating subspaces. Clearly when B is I we may take Z_1 to be X_1 itself and B_{11} becomes I_r. Since

$$\det(A-\lambda B) = k \det(A_{11}-\lambda B_{11}) \det(A_{22}-\lambda B_{22}), \tag{7.4}$$

where $k = \det(Z)/\det(X)$, we see that the eigenvalues of $A-\lambda B$ are the union of those of $A_{ii}-\lambda B_{ii}$ $(i=1,2)$. Writing

$$Y^H = Z^{-1}, \quad W^H = X^{-1} \tag{7.5}$$

we have (in an obvious notation)

$$Y_2^H A = A_{22}W_2^H, \quad Y_2^H B = B_{22}W_2^H \tag{7.6}$$

so that Y_2^H and W_2^H are automatically left-hand deflating subspaces corresponding to the eigenvalues of $A_{22}-\lambda B_{22}$. From (7.5) Y_2 and W_2 are orthogonal complements of Z_1 and X_1 respectively. When X_1 and Z_1 provide orthonormal bases then it is natural to take X_2 and Z_2 to be their orthogonal complements. In this case $W^H = X^{-1} = X^H$ and $Y^H = Z^{-1} = Z^H$.

When the eigenvalues of $A_{ii}-\lambda B_{ii}$ $(i=1,2)$ are disjoint we can continue the deflation of A and B to block diagonal form. We observe that

$$\begin{bmatrix} A_{11} & A_{12} \\ 0 & A_{22} \end{bmatrix} \begin{bmatrix} I & P \\ 0 & I \end{bmatrix} = \begin{bmatrix} I & Q \\ 0 & I \end{bmatrix} \begin{bmatrix} A_{11} & 0 \\ 0 & A_{22} \end{bmatrix} \tag{7.7}$$

and

$$\begin{bmatrix} B_{11} & B_{12} \\ 0 & B_{22} \end{bmatrix} \begin{bmatrix} I & P \\ 0 & I \end{bmatrix} = \begin{bmatrix} I & Q \\ 0 & I \end{bmatrix} \begin{bmatrix} B_{11} & 0 \\ 0 & B_{22} \end{bmatrix}, \tag{7.8}$$

if P and Q satisfy the relations

$$A_{12} = QA_{22} - A_{11}P \quad \text{and} \quad B_{12} = QB_{22} - B_{11}P \, . \tag{7.9}$$

The equivalent system of linear equations is

$$\left[\begin{array}{c|c} A_{22}^T \otimes I & -I \otimes A_{11} \\ \hline B_{22}^T \otimes I & -I \otimes B_{11} \end{array}\right] \begin{bmatrix} \mathrm{Vec}(Q) \\ \mathrm{Vec}(P) \end{bmatrix} = \begin{bmatrix} \mathrm{Vec}(A_{12}) \\ \mathrm{Vec}(B_{12}) \end{bmatrix}, \tag{7.10}$$

and since the determinant of the matrix on the left is equal to $\det(B_{22}^T \otimes A_{11} - A_{22}^T \otimes B_{11})$ it is easy to show that it is nonsingular provided $A_{ii} - \lambda B_{ii}$ $(i = 1, 2)$ do not have a common finite or infinite eigenvalue. Hence A and B can be simultaneously deflated to $\mathrm{diag}(A_{11}, A_{22})$ and $\mathrm{diag}(B_{11}, B_{22})$.

8. ERROR BOUNDS

Work in the eigenvalue field on error bounds may be described in terms of a typical problem. Suppose we have an orthonormal approximation X_1 (columns) to an invariant subspace and $X = [X_1, X_2]$ reduces A to

$$X^{-1}AX = \begin{bmatrix} A_{11} & A_{12} \\ E_{21} & A_{22} \end{bmatrix} \tag{8.1}$$

where E_{21} would be null if X_1 were exact. When the spectra of A_{11} and A_{22} are disjoint and $\|E_{21}\|$ is small enough the equation

$$\begin{bmatrix} A_{11} & A_{12} \\ E_{21} & A_{22} \end{bmatrix} \begin{bmatrix} I \\ Z \end{bmatrix} = \begin{bmatrix} I \\ Z \end{bmatrix} \tilde{A}_{11} \tag{8.2}$$

will have a solution. Indeed $\|Z\|$ and $\tilde{A}_{11}$ tend to zero and A_{11} respectively as $\|E_{21}\|$ tends to zero. From (8.2) we have

$$\tilde{A}_{11} = A_{11} + A_{12}Z, \quad E_{21} + A_{22}Z = Z\tilde{A}_{11} \tag{8.3}$$

giving

$$E_{21} = ZA_{11} - A_{22}Z + ZA_{12}Z \,. \tag{8.4}$$

When (8.4) has a solution, $X_1 + X_2Z$ (columns) defines an exact invariant subspace of A corresponding to the spectrum of $\tilde{A}_{11}$. Expressed as a standard set of linear equations (8.4) becomes

$$\{A_{11}^T \otimes I - I \otimes A_{22}\}\,\mathrm{Vec}(Z) = \mathrm{Vec}(E_{21}) - \mathrm{Vec}(ZA_{12}Z) \tag{8.5}$$

which for convenience may be written in the simpler form

$$B\underline{z} = \underline{g} + \underline{h}(\underline{z}) \tag{8.6}$$

where

$$\left.\begin{aligned} B &= A_{11}^T \otimes I - I \otimes A_{22}\,, \quad \underline{z} = \mathrm{Vec}(Z)\,, \\ \underline{g} &= \mathrm{Vec}(E_{21})\,, \quad \underline{h}(\underline{z}) = -\mathrm{Vec}(ZA_{12}Z)\,. \end{aligned}\right\} \tag{8.7}$$

The vector $\underline{h}(\underline{z})$ has components which are purely quadratic in the z_i.

Equation (8.6) may be solved by the simple iterative procedure

$$\underline{z}^{(0)} = \underline{0}, \quad B\underline{z}^{(k+1)} = \underline{g} + \underline{h}(\underline{z}^{(k)}) \tag{8.8}$$

or by Newton's method. If we write

$$\underline{F}(\underline{z}) = B\underline{z} - \underline{g} - \underline{h}(\underline{z}) \tag{8.9}$$

the latter is defined by

$$\tilde{\underline{z}}^{(0)} = \underline{0}, \quad \tilde{\underline{z}}^{(k+1)} = \tilde{\underline{z}}^{(k)} - \{\underline{F}'(\tilde{\underline{z}}^{(k)})\}^{-1}\,\underline{F}(\tilde{\underline{z}}^{(k)}). \tag{8.10}$$

We have

$$\underline{z}^{(1)} = B^{-1}\underline{g}\,, \quad \tilde{\underline{z}}^{(1)} = B^{-1}\underline{g} \tag{8.11}$$

so that both algorithms give the same first iterate.

Either of these iterative procedures may be used in two different ways.

(i) We may regard it as a method of getting an accurate invariant subspace of A by continuing to iterate until some convergence criterion is met.

(ii) We may have little interest in the iterative procedure

per se, merely using the theory of the convergence to show that there really is a solution $\underline{z}$ and to obtain a bound for $\|\underline{z}-\underline{z}^{(1)}\|$. In this second case we are expecting that X_1 (columns) is a good approximation to an invariant subspace.

The behaviour of either set of iterates may be described in terms of the corresponding iterates for the simple, real scalar equation

$$\beta\xi = \beta\|B^{-1}\underline{g}\| + (\alpha/2)\,\xi^2\,, \quad \xi^{(0)} = 0\,, \tag{8.12}$$

where

$$\alpha = \|F''(\underline{z})\|\,, \quad \beta = 1/\|B^{-1}\| \tag{8.13}$$

with appropriate definitions of the norm in each case. Since F is quadratic α is constant. Clearly

$$\tilde{\xi}_1 = \xi_1 = \|B^{-1}\underline{g}\|\,. \tag{8.14}$$

Equation (8.12) has real solutions iff

$$\rho = 2\alpha\|B^{-1}\underline{g}\|/\beta \leqslant 1\,, \tag{8.15}$$

and it is easy to show that either of the two sets of iterates then converges monotonically to the smaller of the two roots, σ_1 and σ_2 say. Since

$$\begin{aligned}\sigma_1 &= 2\|B^{-1}\underline{g}\|/\{1+(1-\rho)^{1/2}\}\\ &= 2\xi_1/\{1+(1-\rho)^{1/2}\}\end{aligned} \tag{8.16}$$

we have

$$\xi_1 \leqslant \sigma_1 \leqslant 2\xi_1\,. \tag{8.17}$$

The behaviour of the iterates for the scalar equation is intuitively obvious geometrically and, knowing what one has to prove, it is almost trivial to provide rigorous algebraic demonstrations.

For the equation (8.6) simple inductive proofs show that all key quantities in the $\underline{z}^{(k)}$ and $\tilde{\underline{z}}^{(k)}$ sequences such as $\|\underline{z}^{(k+1)}-\underline{z}^{(k)}\|$ are majorized by the corresponding quantities in the $\xi^{(k)}$ and $\tilde{\xi}^{(k)}$ sequences (see, for example, Ortega and Rheinboldt, 1970, Chapter XII). Hence if $\rho\leqslant 1$ both $\underline{z}_k$ and $\tilde{\underline{z}}_k$

tend to the unique solution $\underline{z}$ of (8.6) in the domain

$$\|\underline{z}-\underline{z}_1\| = \|\underline{z}-\tilde{\underline{z}}_1\| \leq \sigma_1 - \xi_1 = \xi_1\rho/\{1+(1-\rho)^{1/2}\}^2 . \quad (8.18)$$

When ρ is small we have

$$\|\underline{z}-\underline{z}_1\| < (\rho/4)\,\xi_1 + 0(\rho^2) \quad (8.19)$$

so that the first iteration takes us very close to the solution.

The key quantities governing the rate of convergence and error bounds for the first iterate are α, β and $\|B^{-1}\underline{g}\|$. Of these α is given directly since $\underline{h}(\underline{z}) = -\text{Vec}\,(ZA_{12}Z)$ and $B^{-1}\underline{g}$ will normally be computed explicitly by solving the system $B\underline{z} = \underline{g}$. One step of iterative refinement will give very reliable information on the computed $B^{-1}\underline{g}$. On the other hand $\beta = 1/\|B^{-1}\|$, and a rigorous bound involves considerable computation. Observe that

$$\beta = \text{sep}\,(A_{11}, A_{22}) \quad (8.20)$$

since $A_{11}^T \otimes I - I \otimes A_{22}$ may be derived from $I \otimes A_{11}^T - A_{22} \otimes I$ by permutations. In practice one is perhaps more likely to determine sep from the iterates than to compute sep in order to forecast their behaviour!

The algorithms used for achieving a reduction to the form (8.1) might well be designed to give with exact computation triangular A_{11} and A_{22}. In practice one will then have almost triangular A_{11} and A_{22} and we may write

$$A_{11} = T_{11} + E_{11}, \quad A_{22} = T_{22} + E_{22}, \quad (8.21)$$

where E_{11} and E_{22} are small. In order to take advantage of this it will be convenient to modify (8.4) to give the iterative procedure

$$Z^{(k+1)}T_{11} - T_{22}Z^{(k+1)} = -Z^{(k)}E_{11} + E_{22}Z^{(k)} + E_{21} - Z^{(k)}A_{12}Z^{(k)}, \quad (8.22)$$

and we could modify this relation still further by using 'latest values' as in Gauss–Seidel iteration. Little purpose is served in doing a detailed error analysis of minor variations of this kind.

Exactly analogous techniques may be used either for refining approximate deflating subspaces or for obtaining error bounds for them in the generalized eigenvalue problems. They are based on solving the equations

$$\begin{bmatrix} A_{11} & A_{12} \\ E_{21} & A_{22} \end{bmatrix} \begin{bmatrix} I \\ Z \end{bmatrix} = \begin{bmatrix} I \\ W \end{bmatrix} \tilde{A}_{11}; \quad \begin{bmatrix} B_{11} & B_{12} \\ F_{21} & B_{22} \end{bmatrix} \begin{bmatrix} I \\ Z \end{bmatrix} = \begin{bmatrix} I \\ W \end{bmatrix} \tilde{B}_{11}, \tag{8.23}$$

where $\|E_{21}\|$ and $\|F_{21}\|$ are small when the approximations are good.

The above material is often discussed in very sophisticated terms with appeals to Kantorovic's theorem in Banach space. Although of course the resulting algorithms are identical, proofs of inequalities satisfied by the iterates are usually unnecessarily complicated. Further the insight gained by thinking directly in terms of the underlying scalar equations is sacrificed.

9. INVERSE EIGENVALUE PROBLEMS

The general inverse eigenvalue problem (i.e.p.) may be defined in the following terms.

Let $A(c_1, \cdots, c_n)$ be an $n \times n$ matrix having elements which are functions of n parameters $c_1, \cdots, c_n$. Determine $(c_i, \; i = 1, 2, \cdots, n)$ such that $A(\underline{c})$ has n given eigenvalues $\lambda_1^*, \cdots, \lambda_n^*$.

The literature is now so extensive that we content ourselves with illustrating the nature of the problem for the restricted class

$$A(\underline{c}) = A_0 + \sum c_k A_k \tag{9.1}$$

where the A_i are real and symmetric. $A(\underline{c})$ then has real eigenvalues for all real $\underline{c}$. We refer to this as the symmetric i.e.p.. An important problem in this class, known as the additive i.e.p., is obtained when

$$A_k = \underline{e}_k \underline{e}_k^T . \qquad (9.2)$$

This problem is usually described in the following terms. Given a symmetric matrix A, determine $\mathrm{diag}(c_k)$ so that $A + \mathrm{diag}(c_k)$ has eigenvalues $(\lambda_i^*, i = 1, 2, \cdots, n)$.

A closely related problem, known as the multiplicative i.e.p., is to determine a positive diagonal matrix C such that $\mathrm{diag}(c_k) A$ has eigenvalues (λ_i^*). Written in a form analogous to (9.1) this becomes

$$A(\underline{c}) = \sum c_k \underline{e}_k \underline{a}_k^T \quad (\text{i.e. } A_k = \underline{e}_k \underline{a}_k^T) \qquad (9.3)$$

where $\underline{a}_k^T$ is the kth row of A. The A_k are no longer symmetric but, since $C^{-1/2}(CA)C^{1/2}$ is real and symmetric, $A(\underline{c})$ certainly has real eigenvalues for all positive C.

Algorithms for solving the symmetric i.e.p. have mainly been based on Newton's method for solving the nonlinear system of n algebraic equations which may be written in the form

$$\underline{f}(\underline{c}) = \underline{\lambda}(\underline{c}) - \underline{\lambda}^* = \underline{0} . \qquad (9.4)$$

The general step of the Newton iteration may be described in the following terms.

We denote the current $\underline{c}$ by $\underline{c}^{(r)}$ and the corresponding eigenvalues and normalized eigenvectors by $\lambda_i^{(r)}$ and $\underline{x}_i^{(r)}$ respectively. Then elementary perturbation theory shows that the first order perturbations in the eigenvalues corresponding to a perturbation $\underline{\delta}^{(r)}$ in $\underline{c}^{(r)}$ are given by $J^{(r)}\underline{\delta}^{(r)}$ where

$$J_{ik}^{(r)} = \underline{x}_i^{(r)T} A_k \underline{x}_i^{(r)} . \qquad (9.5)$$

Hence we have the Newton iteration

$$J^{(r)}\underline{\delta}^{(r)} = \underline{\lambda}^* - \underline{\lambda}^{(r)}, \quad \underline{c}^{(r+1)} = \underline{c}^{(r)} + \underline{\delta}^{(r)} \qquad (9.6)$$

and since $A(\underline{c}^{(r)})\,\underline{x}_i^{(r)} = \lambda_i^{(r)}\,\underline{x}_i^{(r)}$ this can be written in the form

$$J^{(r)}\underline{c}^{(r+1)} = \underline{\lambda}^* + J^{(r)}\underline{c}^{(r)} - \underline{\lambda}^{(r)} = \underline{\lambda}^* - \underline{b}^{(r)}\ , \quad \text{where}$$

$$b_i^{(r)} = \underline{x}_i^{(r)T} A_0 \underline{x}_i^{(r)}\ . \tag{9.7}$$

The $\lambda_i^{(r+1)}$ and $\underline{x}_i^{(r+1)}$ must then be determined from $A(\underline{c}^{(r+1)})$.

Various minor modifications of this algorithm have been proposed such as updating the current set of eigenvectors by using one step of inverse iteration for each of them rather than solving $A(\underline{c}^{(r+1)})$ completely at each stage.

These algorithms are on a reasonably firm theoretical basis when the (λ_i^*) are distinct, at least as far as local convergence is concerned, since when a solution $\underline{c}^*$ to the symmetric i.e.p. exists the eigenvalues are differentiable in the neighbourhood of $\underline{c}^*$. Convergence of the algorithms is therefore locally quadratic. When there are multiple eigenvalues among the (λ_i^*) they are, in general, no longer differentiable. Nevertheless, convergence of algorithms based on the above is generally quadratic. Investigation of this case is currently an active field of research.

10. RECOMMENDED READING

Since we have been able to give only a brief sketch of current research we conclude this article with recommendations to the reader wishing to pursue in depth the topics discussed here.

It has not proved possible to find a discussion of the Kronecker product which deals with those aspects which are of particular relevance to numerical analysts. It is recommended that the reader extends results of the type given in Section 2 and familiarizes himself with their use.

The seminal papers on invariant subspaces, deflating subspaces and the determination of error bounds by Stewart (1971, 1972, 1973) had already appeared before the previous State of the Art meeting but they have continued to influence work in all of these areas. A stimulus to research in this area has been the very fruitful collaboration between control engineers and numerical analysts. It is primarily this which has been responsible for the current interest in Kronecker's canonical form.

Early work in this area is due to Wilkinson (1978). It was undertaken in connection with linear systems of differential equations and its primary object was to show the structure of the general solution of such systems without appealing to Kronecker's canonical form. Inevitably, it led implicitly to a determination of the K.c.f.. Quite independently Van Dooren (1979) provided a backward stable method for determining the K.c.f. which was based on the paper by Golub and Wilkinson (1976). Current research in this area is concerned with the 'pole assignment' problem and is typified by the papers by Demmel (1985), and Kautsky, Nichols and Van Dooren (1985).

The determination of the nearest defective matrix has intrigued numerical analysts for many years and widely held misconceptions about this problem have persisted. The best antidote to these misconceptions is consideration of the inclusion domains resulting from perturbations bounded by η. Recent work on these lines is given in Wilkinson (1984) and in papers submitted for publication by Demmel (1986) and by Wilkinson (1986).

Two books in this area are recommended reading. The first by Parlett (1980) is an elegant exposition of the state of the art on the symmetric eigenvalue problem. The second book, by Golub and Van Loan (1983), covers a wide field in numerical linear algebra much of which is of great interest to those working on eigenvalue problems.

REFERENCES

Boley, D.L. (1981), *Computing the Controllability/Observability Decomposition of a Linear Time Invariant Dynamic System*, Ph.D. Thesis, Dept. of Computer Science, Stanford University.

Campbell, S.L. (1980, 1982), *Singular Systems of Differential Equations, Vols 1 and 2*, Pitman (Marshfield, Mass.).

Campbell, S.L. (1985), "The numerical solution of higher index linear time varying singular systems of differential equations", *SIAM J. Sci. Statist. Comput.*, *6*, pp.334-348.

Demmel, J. (1983), "The condition number of equivalence transformations that block diagonalize matrix pencils", *SIAM J. Numer. Anal.*, *20*, pp.599-610.

Demmel, J. (1985),"On the conditioning of pole assignment", Computer Science Dept. Report No. 150, Courant Institute of Mathematical Sciences, New York.

Demmel, J. (1986), "Computing stable eigendecompositions of matrices", *Linear Algebra Appl.*, *79*, pp.163-193.

Demmel, J. and Kågström, B. (1985), "Computing stable eigendecompositions of matrix pencils", Technical Report No. 164, Courant Institute of Mathematical Sciences, New York.

Gantmacher, F.R. (1959), *The Theory of Matrices*, *Vols. I and II*, Chelsea (New York).

Golub, G.H. and Van Loan, C.F. (1983), *Matrix Computations*, Johns Hopkins University Press (Baltimore).

Golub, G.H. and Wilkinson, J.H. (1976), "Ill-conditioned eigensystems and the computation of the Jordan canonical form", *SIAM Rev.*, *18*, pp.578-619.

Kautsky, J., Nichols, N.K. and Van Dooren, P. (1985), "Robust pole assignment in linear state feedback", *Internat. J. Control*, *41*, pp.1129-1155.

Kublanovskaya, V.N. (1984), "AB-algorithm and its modifications for the spectral problems of linear pencils of matrices", *Numer. Math.*, *43*, pp.329-342.

Kågström, B. (1986), "RGSVD — An algorithm for computing the Kronecker structure and reducing subspaces of singular $A-\lambda B$ pencils", *SIAM J. Sci. Statist. Comput.*, *7*, pp.185-211.

Kågström, B. and Ruhe, A. (1980), "An algorithm for numerical computation of the Jordan normal form of a complex matrix", *ACM Trans. Math. Software*, *6*, pp.398-419.

Kågström, B. and Ruhe, A. (Eds.) (1983), *Matrix Pencils*, *Lecture Notes in Mathematics*, *973*, Springer-Verlag (Berlin).

Ortega, J.M. and Rheinboldt, W.C. (1970), *Iterative Solution of Nonlinear Equations in Several Variables*, Academic Press (New York).

Parlett, B.N. (1980), *The Symmetric Eigenvalue Problem*, Prentice-Hall (Englewood Cliffs, New Jersey).

Stewart, G.W. (1971), "Error bounds for approximate invariant subspaces of closed linear operators", *SIAM J. Numer. Anal.*, *8*, pp.796-808.

Stewart, G.W. (1972), "On the sensitivity of the eigenvalue problem $A\underline{x} = \lambda B\underline{x}$, *SIAM J. Numer. Anal.*, *9*, pp.669-686.

Stewart, G.W. (1973), "Error and perturbation bounds for subspaces associated with certain eigenvalue problems", *SIAM Rev.*, *15*, pp.727-764.

Stewart, G.W. (1978), "Perturbation theory for the generalized eigenvalue problem", in *Recent Advances in Numerical Analysis*, eds. de Boor, C. and Golub, G.H., Academic Press (New York), pp.193-206.

Sun, J.G. (1983), "Perturbation analysis for the generalized eigenvalue and the generalized singular value problem", in *Matrix Pencils, Lecture Notes in Mathematics*, *973*, eds. Kågström, B. and Ruhe, A., Springer-Verlag (Berlin), pp.221-244.

Van Dooren, P. (1979), "The computation of Kronecker's canonical form of a singular pencil", *Linear Algebra Appl.*, *27*, pp.103-140.

Van Dooren, P. (1981), "The generalized eigenstructure problem in linear system theory", *IEEE Trans. Automat. Control*, *26*, pp.111-129.

Wilkinson, J.H. (1965), *The Algebraic Eigenvalue Problem*, Oxford University Press (Oxford).

Wilkinson, J.H. (1978), "Linear differential equations and Kronecker's canonical form", in *Recent Advances in Numerical Analysis*, eds. de Boor, C. and Golub, G.H., Academic Press (New York), pp.231-265.

Wilkinson, J.H. (1979), "Kronecker's canonical form and the QZ algorithm", *Linear Algebra Appl.*, *28*, pp.285-303.

Wilkinson, J.H. (1984), "Sensitivity of eigenvalues", *Utilitas Math.*, *25*, pp.5-76.

Wilkinson, J.H. (1986), "Sensitivity of eigenvalues II", *Utilitas Math.* (to be published).

J.H. Wilkinson
40 Atbara Road
Teddington
Middlesex TW11 9PD
England

(dec. October 5th, 1986)

2 Numerical Linear Algebra in Statistical Computing

N. J. HIGHAM and G. W. STEWART

ABSTRACT

Some of the factors to be considered when applying the techniques of numerical linear algebra to statistical problems are discussed with reference to three particular examples: the use of the normal equations in regression problems; the use of perturbation theory to assess the effects of errors in regression matrices; and the phenomenon of benign degeneracy, in which the numerical problem becomes more difficult even as the associated statistical problem becomes easier.

1. INTRODUCTION

Although statistics contains a wealth of problems for the practitioner of numerical linear algebra, their solution is not as straightforward as it might at first seem. Some of the verities of our field appear in a curious light when we attempt to adapt them to the realities of the statistical world. In this paper we shall give three examples, each pertaining to the classical linear regression model

$$\underline{y} = X\underline{b} + \underline{e}, \tag{1.1}$$

where

$$\underline{y} \in \mathbb{R}^n, \quad X \in \mathbb{R}^{n\times p}, \quad \underline{b} \in \mathbb{R}^p, \quad n \geqslant p,$$

and the random vector $\underline{e} \in \mathbb{R}^n$ is normally distributed according to

$$\underline{e} \sim N(\underline{0}, \sigma^2 I).$$

X is referred to as the regression matrix and $\underline{b}$ the vector of regression coefficients.

In Section 2 we appraise the role of the normal equations in regression problems and offer some explanations as to why the normal equations method has been used very satisfactorily by statisticians for a long time, despite its shortcomings when compared to the orthogonalisation methods preferred by numerical analysts.

In Section 3 we consider the use of perturbation theory to assess the effects of errors in the regression matrix on the regression coefficients. Standard perturbation results tend to be too crude in the context of statistical problems and their sensitivity to the scaling of the problem is unsatisfactory. We indicate how finer bounds can be obtained and show that certain "collinearity coefficients" can provide useful information about the sensitivity of a regression problem.

It is not uncommon in statistics to find the phenomenon of benign degeneracy, in which the numerical problem becomes more difficult even as the associated statistical problem becomes easier. This phenomenon is examined in Section 4, using the Fisher discriminant for illustration.

Throughout this paper we will assume that the regression matrix X in (1.1) has full rank. Pertinent discussions concerning rank deficient regression problems may be found in Stewart (1984) and Hammarling (1985).

We shall use $\|\cdot\|$ to denote the vector 2-norm,

$$\|\underline{x}\| = (\underline{x}^T\underline{x})^{\frac{1}{2}} ,$$

or the induced matrix norm,

$$\|X\| = \max_{\|\underline{x}\|=1} \|X\underline{x}\| = \rho(X^TX)^{\frac{1}{2}} ,$$

where ρ denotes the spectral radius (the largest eigenvalue in modulus). Some feel for the size of $\|X\|$ may be obtained from the relation

$$\|X\| \leq \left(\sum_i \sum_j x_{ij}^2\right)^{\frac{1}{2}} \leq \sqrt{p}\|X\| .$$

We shall also make use of the matrix condition number, defined by

$$\kappa(X) = \|X\| \ \|X^+\|$$

where X^+ is the pseudo-inverse of X.

2. THE NORMAL EQUATIONS

For the regression problem (1.1) the least squares estimate of the regression coefficients is the unique vector $\hat{\underline{b}}$ satisfying

$$\|\underline{y} - X\hat{\underline{b}}\| = \min_{\underline{b}} \|\underline{y} - X\underline{b}\| . \qquad (2.1)$$

There are two methods commonly used for solving the least squares problem (2.1). The first is based on the readily derived normal equations, which were known to Gauss,

$$A\hat{\underline{b}} = \underline{c} , \qquad (2.2a)$$

where

$$A = X^T X , \ \underline{c} = X^T \underline{y} . \qquad (2.2b)$$

Since X has full rank, the cross-product matrix $A = X^T X$ is symmetric positive definite. Hence the normal equations may be solved by computing a Choleski decomposition

$$A = T^T T ,$$

where T is upper triangular with positive diagonal elements (Golub and Van Loan, 1983, p.88), and performing a forward substitution followed by a backward substitution.

The second popular approach is to make use of a QR factorisation of the matrix X,

$$X = Q \begin{bmatrix} R \\ 0 \end{bmatrix},$$

where $Q \in \mathbb{R}^{n \times n}$ is orthogonal and $R \in \mathbb{R}^{p \times p}$ is upper triangular. Since rank $(R) = $ rank (X), and

$$\|\underline{y}-X\underline{b}\| = \left\|Q^T\left(\underline{y}-Q\begin{bmatrix}R\\0\end{bmatrix}\underline{b}\right)\right\| = \left\|\begin{bmatrix}Q_1^T\\Q_2^T\end{bmatrix}\underline{y} - \begin{bmatrix}R\\0\end{bmatrix}\underline{b}\right\|,$$

where $Q=[Q_1, Q_2]$, the least squares estimate $\hat{\underline{b}}$ is obtained by solving the nonsingular triangular system

$$R\hat{\underline{b}} = Q_1^T\underline{y} \,. \tag{2.3}$$

The QR factorisation may be computed in several ways which we now summarise; for further details see Golub and Van Loan (1983, Chapters 3 and 6). The preferred approach for general problems is orthogonal reduction of X to triangular form using Householder transformations, as first described in detail by Golub (1965). The reduction takes the form

$$H_s H_{s-1} \cdots H_1 X = \begin{bmatrix}R\\0\end{bmatrix}, \qquad s = \min\{n-1, p\},$$

where the Householder matrix H_k satisfies

$$H_k = I - 2\underline{u}_k\underline{u}_k^T, \qquad \|\underline{u}_k\| = 1,$$

and where the first $k-1$ components of $\underline{u}_k$ are zero. The last $n-k+1$ components of $\underline{u}_k$ are chosen so that pre-multiplication by H_k creates zeros below the diagonal in the kth column of the partially triangularised matrix $H_{k-1} \cdots H_1 X$.

An alternative elimination technique is one based on Givens rotations. A Givens rotation is an orthogonal matrix which differs from the identity matrix only in one submatrix of order 2, which takes the form

$$\begin{matrix} & i & k \\ i & c & s \\ k & -s & c \end{matrix}, \qquad c^2 + s^2 = 1 \,.$$

The reduction of X to upper triangular form may be accomplished by pre-multiplying X by a sequence of Givens rotations, each of which is chosen so as to introduce one new zero into the lower triangular part of the matrix product. Although up to twice as expensive as the Householder approach for full matrices, the

Givens QR reduction has two redeeming features. First, zeros are introduced in a selective fashion, so that a suitably tailored implementation can be more efficient on sparse or structured problems. Second, the rows of X can be processed one at a time, which is desirable if X is too large to fit into main storage, or if X is generated row-wise, as is often the case in statistical computations.

Other techniques of interest for historical reasons, though of less practical importance nowadays, are the Gram-Schmidt and modified Gram-Schmidt algorithms.

A relationship between the normal equations and the QR factorisation methods can be seen by noting that

$$X^T X = R^T Q^T Q R = R^T R = R^T \operatorname{diag}(\operatorname{sign}(r_{ii}))^2 R ,$$

so that R is the Choleski factor of $X^T X$ up to scaling of each row by ± 1. It is interesting to note that equation (2.3) can be derived by substituting $X = Q_1 R$ into the normal equations (2.2) and pre-multiplying by R^{-T}.

The normal equations method is almost universally used by statisticians while the QR factorisation method is almost universally recommended by numerical analysts. On the surface the numerical analysts would seem to have the better of it. In the first place the QR equations have a favourable backward error analysis which the normal equations do not. For the QR factorisation method using Householder transformations it can be shown (Stewart, 1973, p.240) that the solution $\bar{\underline{b}}$ computed in floating point arithmetic with rounding unit ε_M is the true least squares estimate for the perturbed regression equation

$$\underline{y} + \underline{f} = (X + E)\underline{b} + \underline{e} , \tag{2.4}$$

where the perturbations $\underline{f}$ and E are bounded by

$$\|E\| \leqslant \phi_1(n,p)\varepsilon_M \|X\| ,$$

$$\|\underline{f}\| \leqslant \phi_2(n,p)\varepsilon_M \|\underline{y}\| ,$$

where ϕ_1 and ϕ_2 are low degree polynomials in n and p. Thus the QR factorisation method solves a "nearby" regression problem, and if the computed solution is unsatisfactory, the blame can be placed on the provider of the problem rather than the numerical method.

From the backward error analysis for the Choleski decomposition (Golub and Van Loan, 1983, p.89) it follows that the computed solution $\tilde{\underline{b}}$ for the normal equations method satisfies

$$(A+G)\tilde{\underline{b}} = \underline{c}\,, \tag{2.5}$$

where

$$\|G\| \leqslant c_p \, \varepsilon_M \, \|A\|\,,$$

c_p being a small constant depending on p. Here, for simplicity, we have assumed that the normal equations are formed exactly.

It is possible to translate this backward error result into one of the form of (2.4), but inevitably the perturbation G is magnified in the process. To see this, assume $A+G$ is symmetric positive definite and consider the equation

$$(X+H)^T(X+H) = A+G\,,$$

where the smallest of the many solutions H is to be determined (the one that minimises $\sum_i \sum_j h_{ij}^2$, say). By passing to the singular value decomposition we may assume that X has the form

$$X = \begin{bmatrix} \Psi \\ 0 \end{bmatrix},$$

where

$$\Psi = \operatorname{diag}(\psi_1, \cdots, \psi_p)\,, \qquad \psi_1 \geqslant \cdots \geqslant \psi_p > 0\,.$$

Partitioning H conformally as

$$H = \begin{bmatrix} H_1 \\ H_2 \end{bmatrix},$$

the equation to solve becomes

$$\Psi^2 + \Psi^T H_1 + H_1^T \Psi + H^T H = \Psi^2 + G\,.$$

Ignoring second order terms and writing the equation in scalar form, we get

$$\psi_i h_{ij} + h_{ji}\psi_j = g_{ij}\ , \tag{2.6}$$

where $H_1 = (h_{ij})$, $G = (g_{ij})$. Since we require the smallest solution we minimise $h_{ij}^2 + h_{ji}^2$ subject to (2.6). Remembering that G is symmetric, the solution is easily seen to be

$$h_{ij} = \frac{\psi_i g_{ij}}{\psi_i^2 + \psi_j^2},$$

which for $i = j = p$ reduces to

$$h_{pp} = \frac{g_{pp}}{2\psi_p}\ .$$

Thus, to first order, part of the error is magnified by a factor proportional to ψ_p^{-1}. Since $\|G\|$ is proportional to $\|X\|^2$, this means that $\|H\|/\|X\|$ will be proportional to $\kappa(X) = \psi_1/\psi_p$. Clearly, then, the QR method is superior from the point of view of backwards stability.

Turning to the forward error, $\hat{\underline{b}} - \tilde{\underline{b}}$, a bound for the normal equations may be obtained directly from (2.5) on using standard perturbation theory for square linear systems (Golub and Van Loan, 1983, p.27). We have

$$\frac{\|\hat{\underline{b}} - \tilde{\underline{b}}\|}{\|\hat{\underline{b}}\|} \leqslant \kappa(A)\ \frac{\|G\|}{\|A\|} + 0\left(\frac{\|G\|}{\|A\|}\right)^2, \tag{2.7}$$

where

$$\kappa(A) = \kappa(X^T X) = \kappa(X)^2\ . \tag{2.8}$$

A forward error bound for the QR method can be obtained by making use of standard least squares perturbation theory. From Golub and Wilkinson (1966), provided $X + E$ has full rank,

$$\frac{\|\hat{\underline{b}} - \bar{\underline{b}}\|}{\|\hat{\underline{b}}\|} \leqslant \varepsilon\kappa(X)\left(1 + \frac{\|\underline{y}\|}{\|X\|\ \|\hat{\underline{b}}\|}\right) + \varepsilon\kappa(X)^2\ \frac{\|\hat{\underline{r}}\|}{\|X\|\ \|\hat{\underline{b}}\|} + 0(\varepsilon^2), \tag{2.9}$$

where

$$\varepsilon = \max\left\{\frac{\|E\|}{\|X\|},\ \frac{\|\underline{f}\|}{\|\underline{y}\|}\right\},$$

$$\hat{\underline{r}} = \underline{y} - X\hat{\underline{b}}\ .$$

Comparing (2.7) and (2.9) we see that while both bounds contain the term $\kappa(X)^2$, in (2.9) a small residual vector mitigates the effect of this term. Hence the bounds suggest that the QR method will produce more accurate solutions than the normal equations method for ill-conditioned problems that have a small residual.

A consequence of the condition squaring effect (2.8) is the fact that while the condition $\kappa(X)\,\varepsilon_M < 1$ is sufficient to ensure that the QR procedure does not break down with a singular computed R-factor, one must impose the much stronger condition $\kappa(X)^2\,\varepsilon_M < 1$ to guarantee that the normal equations method runs to completion. Indeed, merely forming the normal equations may cause valuable information to be lost when X is ill-conditioned, unless the evaluation is done and the results are stored in extended precision arithmetic.

In view of the above comparisons of error and stability properties it is natural to ask why statisticians continue to use the normal equations and why they are generally satisfied with the results. We identify several reasons.

In practical statistical problems the residual vector is usually not very small, so the comparison of the forward error bounds is not strongly in favour of the QR method. Moreover, in many problems the elements of the regression matrix are contaminated by errors of measurement, which are large compared with the rounding errors contemplated by the numerical analyst. If the normal equations are formed and solved in a reasonable precision, the effects of rounding errors will be insignificant compared with the effects of measurement errors. In other words, the problem becomes statistically intractable before it becomes numerically intractable.

To make this assertion precise, let ε_D denote the normwise relative error in the regression matrix. Then, as in the perturbation result (2.9), the data errors alone can induce a perturbation in the regression vector $\underline{\hat{b}}$ of order $\kappa(X)^s \varepsilon_D$, where, roughly, $s = 1$ or 2 according as the residual vector is very small or not. Solution via the normal equations introduces a relative error of order $\kappa(X)^2 \varepsilon_M$, by (2.7). Thus as long as

$$\kappa(X)^{2-s} \varepsilon_M \leqslant \varepsilon_D ,$$

the rounding errors in the normal equations method will play an insignificant role compared with the errors in the regression matrix. For example, if $\varepsilon_M = 10^{-7}$ (as, for example, in IEEE standard arithmetic), $\varepsilon_D = 10^{-3}$ and $\kappa(X) = 10^2$, then $\underline{\hat{b}}$ will have at best one correct figure, because of errors in the data, yet the normal equations method will provide three or more correct figures to the machine problem. However, modifying the example slightly so that $\varepsilon_D = 10^{-5}$ and $\kappa(X) = 10^4$ shows that the normal equations method can fail to solve a meaningful problem.

The conclusion is that if one works to high precision (relative to the accuracy of the data) and takes certain elementary precautions (such as computing estimates of the condition number $\kappa(X)$ (Cline, Moler, Stewart and Wilkinson, 1979)), then one can safely use the normal equations. On the other hand, if one is constructing transportable software which must run on machines with a 32-bit floating point word, then one should use the QR factorisation.

An additional feature which works in favour of the normal equations for statistical problems is that in regression models with a constant term (so that $x_{i1} = 1$ for all i), statisticians often "centre their data" by subtracting the means from the columns of X (Graybill, 1976, p.252; Seber, 1977, p.330). This transformation leads to better conditioned normal equations of order one less (Golub and Styan, 1974). To see this, write

$$X = [\underline{e}, X_2]$$

where $\underline{e} = [1, 1, \cdots, 1]^T$. Subtracting the means from the columns of X_2 is equivalent to forming

$$\tilde{X} = X\begin{bmatrix} 1 & -\bar{\underline{x}}^T \\ \underline{0} & I_{p-1} \end{bmatrix} = [\underline{c}, X_2 - \underline{c}\bar{\underline{x}}^T]$$

where $\bar{\underline{x}}^T = n^{-1}\underline{c}^T X_2$. The new cross-product matrix is

$$\tilde{X}^T\tilde{X} = \begin{bmatrix} n & \underline{0}^T \\ \underline{0} & X_2^T X_2 - n\bar{\underline{x}}\bar{\underline{x}}^T \end{bmatrix},$$

which gives reduced normal equations of order $p-1$ with the coefficient matrix

$$\bar{A} = X_2^T X_2 - n\bar{\underline{x}}\bar{\underline{x}}^T .$$

One can show that if A has Choleski factor

$$R_p = \begin{bmatrix} \sqrt{n} & \underline{r}^T \\ \underline{0} & R_{p-1} \end{bmatrix},$$

then $\bar{A}$ has Choleski factor R_{p-1}. It follows that

$$\kappa(\bar{A}) = \kappa(R_{p-1})^2 \leqslant \kappa(R_p)^2 = \kappa(A) .$$

This potential improvement of the condition can be expected to lead to more accurate computed solutions provided that $\bar{A}$ and the corresponding right-hand side vector are computed using extra precision, or perhaps even in standard precision using formulae of the type (Seber, 1977, pp.331, 333; Chan, Golub and LeVeque, 1983)

$$\bar{a}_{ij} = \sum_{k=1}^{n} (x_{ki} - \bar{x}_i)(x_{kj} - \bar{x}_j) ,$$

where $X_2 = (x_{ij})$, $\bar{\underline{x}} = (\bar{x}_i)$.

There is one case in which the normal equations are undoubtedly to be preferred, even when one must resort to high precision. These problems arise in unbalanced analysis of variance and the analysis of categorical data, where the regression matrix is large and sparse but the normal equations are relatively small and dense. Here, formation of the normal equations will be far more efficient than computation of the QR factorisation, because of intermediate fill-in in the course of computing the R-factor. For example, if $p=5$ and if the rows of X are being processed one at a time, then one may find matrices of the form

$$R_5 = \begin{bmatrix} \times & \times & \times & \times & \times \\ 0 & \times & \times & \times & \times \\ 0 & 0 & \times & \times & \times \\ 0 & 0 & 0 & \times & \times \\ 0 & 0 & 0 & 0 & \times \end{bmatrix} \qquad \underline{x}_6\underline{x}_6^T = \begin{bmatrix} \times & 0 & 0 & \times & 0 \\ 0 & 0 & 0 & 0 & 0 \\ 0 & 0 & 0 & 0 & 0 \\ \times & 0 & 0 & \times & 0 \\ 0 & 0 & 0 & 0 & 0 \end{bmatrix}$$

$$\underline{x}_6^T = (\times \;\; 0 \;\; 0 \;\; \times \;\; 0).$$

In the QR factorisation by Givens rotations the rotation in the (1,6) plane which zeros the (1,6) element has the undesirable effect of "filling in" the rest of the row, which makes subsequent treatment of this row as expensive as if it were a full vector. In contrast, the contribution of the new row $\underline{x}_6^T$ to

$$A = \sum_{i=1}^{n} \underline{x}_i \underline{x}_i^T$$

is inexpensive to compute, as it has only four nonzero elements.

3. PERTURBATION THEORY

We have observed that regression matrices often have errors in their elements. It is natural to attempt to use perturbation theory to assess the effects of these errors on the regression coefficients. The results of such an attempt are generally disappointing, for two reasons. First, the use of triangular and submultiplicative inequalities in the course of deriving the bounds reduces their sharpness. This is not a major concern to numerical analysts, whose errors are typically very small and can suffer some magnification without ill effect. The statistician, on the other hand, with his larger errors must fear that the weakness of the bounds will cause him to declare a good problem intractable.

The second reason why perturbation theory gives disappointing results is that often it is impossible to arrive at a suitable scaling of the problem. For the norm of a regression matrix to represent the sizes of its columns, the columns must be scaled so that they are roughly equal in norm. The same is true of the error in the regression matrix. On the other hand, for a bound on the norm of the vector of regression coefficients to say something meaningful about all the coefficients, the problem must be scaled so that the coefficients are roughly equal. This three way balancing act will in general be unsolvable because any permissible scaling of the regression matrix will scale the error matrix identically and the regression coefficients inversely, as can be seen in the relation (for nonsingular $S \in \mathbb{R}^{p \times p}$)

$$\|(X+E)\underline{b}-\underline{y}\| = \|(XS+ES)S^{-1}\underline{b}-\underline{y}\| . \qquad (3.1)$$

One cure is to produce finer bounds in terms of individual coefficients and columns. A way of doing this is as follows. Suppose X is perturbed in its ith column by a vector $\underline{f}$, so that the error matrix E is given by

$$E = \underline{f}\underline{e}_i^T ,$$

where $\underline{e}_i$ is the ith column of the $p \times p$ identity matrix, and let $\bar{\underline{b}}$ be the corresponding vector of perturbed regression coefficients. Assume that $X+E$ has full rank. Using the expansion

$$(X+E)^+ = X^+ - X^+EX^+ + (X^TX)^{-1}E^T(I-XX^+) + O(\|E\|^2)$$

we have

$$\begin{aligned}\bar{\underline{b}} &= (X+E)^+\underline{y} = \hat{\underline{b}} - X^+E\hat{\underline{b}} + (X^TX)^{-1}E^T(I-XX^+)(\underline{y}-X\hat{\underline{b}}) + O(\|E\|^2)\\ &= \hat{\underline{b}} - X^+\underline{f}\underline{e}_i^T\hat{\underline{b}} + (X^TX)^{-1}\underline{e}_i\underline{f}^T(I-XX^+)\hat{\underline{r}} + O(\|\underline{f}\|^2)\\ &= \hat{\underline{b}} - X^+\underline{f}_1\hat{b}_i + (X^TX)^{-1}\underline{e}_i\underline{f}_2^T\hat{\underline{r}} + O(\|\underline{f}\|^2) ,\end{aligned}$$

where $\underline{f}_1$ and $\underline{f}_2$ are the projections of $\underline{f}$ onto the range of X

and its orthogonal complement respectively. On pre-multiplying by $\underline{e}_j^T$, and using the bound

$$|\underline{e}_j^T(X^TX)^{-1}\underline{e}_i| = |\underline{e}_j^T X^+ X^{+T}\underline{e}_i| \leq \|X^{+T}\underline{e}_j\| \, \|X^{+T}\underline{e}_i\| ,$$

we obtain

$$|\bar{b}_j - \hat{b}_j| \leq \|\underline{x}_j^{(+)}\| \, |\hat{b}_i| \, \|\underline{f}_1\| + \|\underline{x}_j^{(+)}\| \, \|\underline{x}_i^{(+)}\| \, \|\hat{\underline{r}}\| \, \|\underline{f}_2\| + 0(\|\underline{f}\|^2)$$
$$j = 1, \cdots, p, \qquad (3.2)$$

where $\underline{x}_k^{(+)}$ denotes the transpose of the kth *row* of X^+. The interpretation of this bound is clearly much less dependent on the scaling of the problem than is the case for the bound (2.9). In fact, a diagonal scaling $S = \mathrm{diag}(s_i)$ of the form in (3.1) leaves (3.2) unchanged.

The above analysis leads naturally to the introduction of the *collinearity coefficients*

$$\kappa_i = \|\underline{x}_i\| \, \|\underline{x}_i^{(+)}\| , \qquad i = 1, \cdots, p.$$

In addition to playing a key role in the perturbation bound (3.2) the collinearity coefficients have at least two other important properties. First, the reciprocal of κ_i is the smallest relative perturbation in the ith column of X that makes X exactly collinear (that is, rank deficient). This can be shown by using the QR factorisation

$$X = Q\begin{bmatrix} R \\ 0 \end{bmatrix} = Q\begin{bmatrix} R_{11} & \underline{r} \\ \underline{0}^T & r_{pp} \\ 0 & \underline{0} \end{bmatrix}, \qquad R_{11} \in \mathbb{R}^{(p-1)\times(p-1)},$$

where we can assume without loss of generality that the column of interest is the last. Clearly, a perturbation

$$\underline{h} = -Q\begin{bmatrix} \underline{0} \\ r_{pp} \\ \underline{0} \end{bmatrix},$$

to the last column of X makes X collinear, and $\|\underline{h}\| = |r_{pp}|$.

Also it is easily seen that if

$$X + \underline{h}\underline{e}_p^T = Q \begin{bmatrix} R_{11} & \underline{r}+\underline{f} \\ \underline{0}^T & r_{pp}+\rho \\ 0 & \underline{g} \end{bmatrix}, \qquad Q^T\underline{h} = \begin{bmatrix} \underline{f} \\ \rho \\ \underline{g} \end{bmatrix},$$

is collinear, then $\rho = -r_{pp}$ so that $\|\underline{h}\| \geqslant |r_{pp}|$. Thus $|r_{pp}|/\|\underline{x}_p\|$ is the size of the smallest relative perturbation to the last column of X that makes X collinear. But

$$X^+ = [R^+ \; 0]\, Q^+ = \begin{bmatrix} R_{11}^{-1} & -(r_{pp}R_{11})^{-1}\underline{r} & 0 \\ \underline{0}^T & r_{pp}^{-1} & \underline{0}^T \end{bmatrix} Q^T$$

so that $\underline{e}_p^T X^+ = r_{pp}^{-1}\underline{e}_p^T Q^T$, and hence

$$|r_{pp}| = \|\underline{x}_p^{(+)}\|^{-1}.$$

Random errors in the regression matrix X tend to cause a systematic reduction in the size of the regression coefficients when X is ill-conditioned, since such errors tend to increase the size of small or zero singular values. Another use of the collinearity coefficients is to measure the extent of this bias in the regression coefficients; the analysis is fairly lengthy and appears in Stewart (1986).

A desirable property of the numbers κ_i is that they are invariant to diagonal scalings of the columns of X, unlike the standard condition number $\kappa(X)$. Some other interesting properties of the collinearity coefficients are discussed in Stewart (1986).

We mention in passing that Fletcher (1985) attempts to overcome the two drawbacks discussed at the start of this section by using a probabilistic perturbation analysis. This approach may be of particular interest in the context of statistical computations.

4. BENIGN DEGENERACY

The phenomenon of benign degeneracy is best illustrated by an example. The Fisher discriminant function (Graybill, 1976, Section 12.5) is a method for deciding whether a sample vector $\underline{x}$, known to be drawn from one of two populations distributed according to $N(\underline{\mu}_1,\Sigma)$ and $N(\underline{\mu}_2,\Sigma)$ respectively, belongs to the first or to the second. The Fisher discriminant classifies $\underline{x}$ as belonging to the $N(\underline{\mu}_1,\Sigma)$ population if

$$\underline{t}^T\underline{x} > \tfrac{1}{2}\underline{t}^T(\underline{\mu}_1+\underline{\mu}_2)\ ,$$

where

$$\underline{t} = \Sigma^{-1}(\underline{\mu}_1-\underline{\mu}_2)\ ,$$

and to the $N(\underline{\mu}_2,\Sigma)$ population otherwise. As the dispersion matrix Σ approaches singularity the problem of computing the Fisher discriminant becomes increasingly ill-conditioned.

However, the numerical problem does not correspond to an intrinsic statistical problem. To see this, let Σ have the spectral decomposition

$$\Sigma = Q^TDQ\ ,\quad Q^TQ=I\ ,\quad D=\mathrm{diag}(d_i)\ ,$$

and transform to the new coordinate system

$$\underline{x}' = Q\underline{x}-\tfrac{1}{2}Q(\underline{\mu}_1+\underline{\mu}_2)\ ,$$

in which the two populations are distributed according to $N(\underline{\mu}'_1,D)$ and $N(\underline{\mu}'_2,D)$ respectively, where $\underline{\mu}'_1=-\underline{\mu}'_2=\frac{1}{2}Q(\underline{\mu}_1-\underline{\mu}_2)$. The Fisher discriminant declares $\underline{x}'$ to belong to the first population if

$$(\underline{\mu}'_1-\underline{\mu}'_2)^TD^{-1}\underline{x}' > 0\ .$$

In the new coordinate system the singularity of the dispersion matrix corresponds to one or more of the components having zero variance, and in the Fisher discriminant the inverse weights these components infinitely. In other words, if a component has zero variance, then it is sufficient to look at that component alone to determine to which population a sample vector belongs — as is clear intuitively. (If there are several components with zero variance then any single one of these may be considered.)

The question for the numerical analyst is how to evaluate the Fisher discriminant. One possibility is to apply the above-mentioned transformation to diagonalise the problem, after which it is obvious what to do. An alternative is to use the original formula, no matter how ill-conditioned the dispersion matrix. In the unlikely event that one is required to divide by zero, the zero is replaced by a suitable small number. Whether this seemingly risky procedure works is an open question!

REFERENCES

Chan, T.F., Golub, G.H. and LeVeque, R.J. (1983), "Algorithms for computing the sample variance: analysis and recommendations", *Amer. Statist.*, *37*, pp.242-247.

Cline, A.K., Moler, C.B., Stewart, G.W. and Wilkinson, J.H. (1979), "An estimate for the condition number of a matrix", *SIAM J. Numer. Anal.*, *16*, pp.368-375.

Fletcher, R. (1985), "Expected conditioning", *IMA J. Numer. Anal.*, *5*, pp.247-273.

Golub, G.H. (1965), "Numerical methods for solving linear least squares problems", *Numer. Math.*, *7*, pp.206-216.

Golub, G.H. and Styan, G.P.H. (1974), "Some aspects of numerical computations for linear models", *Interface — Proceedings of Computer Science and Statistics*, 7th Annual Symposium on the Interface 1973, pp.189-192.

Golub, G.H. and Van Loan, C.F. (1983), *Matrix Computations*, Johns Hopkins University Press (Baltimore, Maryland).

Golub, G.H. and Wilkinson, J.H. (1966), "Note on the iterative refinement of least squares solution", *Numer. Math.*, *9*, pp.139-148.

Graybill, F.A. (1976), *Theory and Application of the Linear Model*, Duxbury Press (North Scituate, Mass.).

Hammarling, S.J. (1985), "The singular value decomposition in multivariate statistics", *ACM SIGNUM Newsletter*, *20(3)*, pp.2-25.

Seber, G.A.F. (1977), *Linear Regression Analysis*, John Wiley (New York).

Stewart, G.W. (1973), *Introduction to Matrix Computations*, Academic Press (New York).

Stewart, G.W., (1984), "Rank degeneracy", *SIAM J. Sci. Statist. Comput.*, *5*, pp.403-413.

Stewart, G.W. (1986), "Scale invariant measures of the effects of near collinearity", manuscript (submitted for publication).

N.J. Higham
Department of Mathematics
University of Manchester
Manchester M13 9PL
England

G.W. Stewart
Department of Computer Science
University of Maryland
College Park
Maryland 20742
U.S.A.

3 Sparse Matrices

J. K. REID

ABSTRACT

We review the current 'state of the art' in respect of the general-purpose treatment of sparse matrices, emphasizing the developments of the last ten years. The most dramatic of these are the fast analysis of symmetric patterns, improvements to frontal methods and to the Lanczos method for the eigenvalue problem, preconditioning, and multigrid iteration. There have also been many detailed advances and the availability of general-purpose software for direct methods is much enhanced.

1. INTRODUCTION

This paper is a sequel to the survey paper (Reid, 1977), given at the 1976 conference on 'The State of the Art in Numerical Analysis'. The aim here is to concentrate on advances that have taken place in the last ten years, while at the same time providing a survey of the current 'state of the art'. We have tried to avoid any reliance on the earlier paper and other works, with the aim of allowing the reader to come to an appreciation of the present situation without having to do substantial additional reading. However, reading elsewhere will certainly be necessary for an understanding of the fine detail since we have kept to 'broad-brush' descriptions for the sake of brevity. Of course we began preparing this paper by re-reading the earlier survey, and were agreeably surprised to find that little that was said then is seriously dated now, although there are some important omissions. Thus we would not wish to discourage the reader who seeks a little more detail from looking at the 1977 survey. There are also a number of books on sparse matrices available, including those of George and Liu (1981), Østerby and Zlatev (1983), Pissanetsky (1984), and Duff, Erisman and Reid (1986) on direct methods for linear equations, and Parlett (1980) and Cullum and Willoughby (1985) on the Lanczos algorithm.

The treatment of sparse matrices is all about special techniques that exploit the coefficients that are known to be zero, which in many practical problems outnumber the rest by a vast margin. Indeed, without the exploitation of sparsity, many large problems would be totally intractable even on today's supercomputers. Just how many coefficients must be known zeros for it to be worthwhile to exploit them, that is to treat the matrix as sparse, depends on the application and the hardware in use. It is unlikely to be worthwhile unless the zeros are at least a majority, and on modern vector computers they may need to be in quite a substantial majority, say greater than 80%.

It has been traditional to refer to the other coefficients as 'nonzeros' even though they may occasionally have the value zero. Since this can confuse readers and (strictly speaking) is mathematically inconsistent, we here use the term 'entry' instead. Coefficients that are not entries must have the value zero and those that are entries usually have nonzero values.

2. STORAGE SCHEMES AND SIMPLE OPERATIONS

There is no one best storage scheme for a sparse matrix. It all depends on what operations are to be applied and what hardware is to be used. A sparse matrix A may be held

(a) as a full matrix, storing all coefficients, including zeros,

(b) as a band matrix, storing every coefficient a_{ij} for which $(j-i)$ lies in a range $-b_l \leq j-i \leq b_u$,

(c) as a variable-band matrix, storing every coefficient between the first entry in a row and the diagonal and every coefficient between the first entry in a column and the diagonal,

(d) in the 'coordinate' scheme, in which each entry a_{ij} is held as a triple (a_{ij}, i, j) consisting of a numerical value and two indices,

(e) as a linked list, with ready access to the entries of each row or each column, or both,

(f) as a collection of vectors, each holding the indices of the entries of a row or column as a contiguous list, with the numerical values of the entries stored alongside the rows or columns (but not both),

(g) as a sum of matrices, each of which is identically zero except in a few rows and columns and is held as a full submatrix with a list of row and column indices that locate it (known as clique storage), or

(h) as a hypermatrix, that is as a block matrix where each block either is zero, or is held as a full matrix, or is itself held as a hypermatrix.

All these schemes were described in more detail in the earlier survey (Reid, 1977). The advent of vector and parallel hardware has given a boost to full, band, variable-band, and clique storage. Further, the largest order of matrix for which it is never worthwhile to use sparse matrix techniques may be greater on such hardware. Linked lists are not being used very much because they can lead to rows or columns being widely scattered in storage, which is a disadvantage in many memory systems and in particular in paged virtual storage because of the possibility of frequent page faults. The collection of sparse vectors, originally suggested by Gustavson (1972), avoids this difficulty and is widely used despite the fact that to handle fill-ins dynamically it needs some 'elbow room' and occasional garbage collections. For the initial representation of a general sparse matrix, the coordinate scheme is popular because any ordering of the data is permitted. Where the problem arises from a finite-element calculation, clique storage is natural.

It should be noted that many codes use more than one storage form. For instance, the Harwell code MA28 (Duff, 1977) employs the coordinate scheme for initial input and performs a sort to create a collection of sparse vectors. MA27 (Duff and Reid, 1982, 1983) also starts with the coordinate scheme and sorts to a collection of sparse vectors, but then goes on to use clique storage.

An important distinction is between static structures that remain fixed and dynamic structures that can accommodate fill-ins as they occur. Typically a dynamic structure is needed initially, but once the locations of all possible fill-ins have been determined a static structure suffices.

Perhaps the most important operation that is used in sparse matrix work is the addition of a multiple of one sparse vector to another. It occurs, for instance, in Gaussian elimination when a multiple of the pivot row is added to a nonpivot row. If the vector $\underline{y}$ is held in a full-length array and the vector $\underline{x}$ is held in packed form as pairs of values and indices, the operation may be expressed in Fortran thus:

```
      DO 10 K=1,NX
         Y(INDEX(K)) = ALPHA*X(K) + Y(INDEX(K))         (2.1)
   10 CONTINUE
```

Note that the number of operations is proportional to the number of entries in the vector $\underline{x}$, and that there is no searching for relevant entries or work on known zeros. However, there is a serious overhead associated with the indirect addressing. On a conventional scalar computer, each cycle of the loop will typically execute three times slower than comparable full-matrix code. On some vector computers (for example, CRAY-1), the ratio is very much higher since the full-matrix loop vectorizes whereas the other does not. Other computers (for example, CRAY X-MP) have vector hardware for indirect addressing, which allows the loop (2.1) to be vectorized. In essence, this vectorizes the gather operation

```
      DO 20 K=1,NX
         VALUE(K) = Y(INDEX(K))                    (2.2)
   20 CONTINUE
```

and its inverse

```
      DO 30 K=1,NX
         Y(INDEX(K)) = VALUE(K)                    (2.3)
   30 CONTINUE
```

which is called a scatter operation. A sequence of vector operations as in (2.1) may be implemented as a gather, followed by full-matrix loops using gathered values, and finally a scatter operation to store the revised values. Thus, vector hardware for indirect addressing (or gather/scatter operations) restores the ratio between the loop cycle time with full and packed sparse vectors to about the same as it is for a scalar machine.

Where the vector $\underline{y}$ is also in packed form, it is necessary to expand it into a full vector (another scatter operation) and repack it (another gather operation) when the operations are complete. This is preferable to working with two packed vectors, even without vectorized gather/scatter facilities, since otherwise there is much testing and both vectors need to be ordered.

Care has to be taken with any sparse matrix operation if it is to be performed efficiently. An operation that was not mentioned in the 1977 survey because we were not fully aware of the issues at that time is matrix multiplication. The conventional inner-product formulation

$$c_{ij} = \sum_k b_{ik} a_{kj} \tag{2.4}$$

is poor because the inner product between two sparse vectors (row i of B and column j of A) involves at least one action for each entry in each vector, whereas only a few of the entries may form corresponding pairs that actually contribute to c_{ij}.

The outer-product formulation

$$C = \sum_k B_{\cdot k} A_{k \cdot} \; , \tag{2.5}$$

involving column k of B and row k of A, has no such inefficiency. Every pair consisting of an entry of $B_{\cdot k}$ and an entry of $A_{k \cdot}$ makes a contribution to C. The efficient implementation of the form (2.5) depends on B being stored by columns and A by rows, and the formation of C will be easier if the entries of the columns of B are ordered (which allows C to be formed row by row) or the entries of the rows of A are ordered (which allows C to be formed column by column). If both A and B are stored by rows, row i of C may be accumulated as a linear combination of the rows of A by expressing (2.5) in the form

$$C_{i \cdot} = \sum_k b_{ik} A_{k \cdot} \; , \tag{2.6}$$

and there is a similar form for storage by columns.

3. REDUCTION TO BLOCK TRIANGULAR FORM

The situation with respect to algorithms that permute a general sparse matrix to block triangular form

$$PAQ = \begin{bmatrix} B_{11} & & & & \\ B_{21} & B_{22} & & & \\ B_{31} & B_{32} & B_{33} & & \\ \vdots & \vdots & & \ddots & \\ B_{N1} & B_{N2} & B_{N3} & \cdots & B_{NN} \end{bmatrix} , \tag{3.1}$$

is highly satisfactory and was summarized in our earlier survey. Although techniques exist for permuting directly to the form (3.1), they appear to have no advantage over the usual two-stage approach:

(1) permute entries onto the diagonal and

(2) use symmetric permutations thereafter.

For the first stage, a depth-first search (Gustavson, 1976; Duff, 1981a), based on the work of Kuhn (1955), has a worst-case behaviour of $O(n\tau)$ for a matrix of order n with τ entries but on practical problems behaves as $O(n+\tau)$. Duff (1981b) provides Fortran code. The algorithm of Hopcroft and Karp (1973) uses a breadth-first search and has an $O(n^{\frac{1}{2}}\tau)$ bound. So far no code is available that is as efficient as Duff's code except on problems carefully fabricated to demonstrate that $O(n\tau)$ behaviour can occur, though Duff and Wiberg (private communication, 1986) report good results for a variant of the Hopcroft-Karp algorithm. For the second stage, a good algorithm with an $O(n+\tau)$ bound has been provided by Tarjan (1972) and code is available (Duff and Reid, 1978a, 1978b).

4. GAUSSIAN ELIMINATION FOR GENERAL UNSYMMETRIC MATRICES

For the application of Gaussian elimination to general unsymmetric matrices, the pivotal strategy of Markowitz (1957) has stood the test of time well. Let $A^{(k)}$ be the $(n+1-k)\times(n+1-k)$ submatrix that has yet to be factorized at the beginning of the k-th step. Markowitz chooses an entry $a_{ij}^{(k)}$ which minimizes

$$(r_i-1)(c_j-1)\,, \tag{4.1}$$

where r_i is the number of entries in row i of $A^{(k)}$ (the i-th row count) and c_j is the number of entries in column j of $A^{(k)}$ (the j-th column count). It is normally used in conjunction with threshold pivoting, which means that the chosen pivot is restricted by the inequality

$$\left|a_{ij}^{(k)}\right| \geqslant u \max_t \left|a_{tj}^{(k)}\right| \tag{4.2}$$

for a predetermined threshold value u, or the corresponding inequality involving the pivotal row. Condition (4.2) ensures that in any column that is modified during a Gaussian elimination step (that is a column that has an entry in the pivotal row), the maximum entry increases in size by at most the factor $(1+u^{-1})$. If column j is modified p_j times (that is if column j of the factor U of the LU factorization has $1+p_j$ entries), the overall growth in column j is limited to the factor

$$(1+u^{-1})^{p_j} . \tag{4.3}$$

Because p_j is often quite small, this factor can be modest even if u is as small as 0.1 (the value recommended by Duff, 1977), but it is nevertheless often a gross overestimate of the growth. A better bound, based on Hölder's inequality applied to the coefficients of the triangular factors L and U, was suggested by Erisman and Reid (1974) and improved by Barlow (1986). This, too, can seriously overestimate the growth so it has not been a success in practice. We therefore now recommend the direct calculation of the quantity

$$\rho = \max_{i,j,k} \left| a_{ij}^{(k)} \right| . \tag{4.4}$$

Another idea that seemed promising ten years ago, but is little used now, is the generation of loop-free code that is specially tailored for the problem to hand. The principal difficulty is that the generated code may be very voluminous unless the problem is rather small or extremely sparse. For example, Willoughby (1971) reports on a code of length 1.3 Megabytes for a matrix of order 1024 with 15000 entries.

On the other hand, a very simple idea that was certainly known ten years ago, but which was not mentioned in our 1977 survey, has become appreciated. It is to switch to full matrix code when the reduced matrix reaches a certain threshold of density. This can be particularly beneficial when a vector machine is in use. For example, Duff (1984a) found that, for a problem of order 1176, switching once the density of the remaining submatrix exceeds 20% reduces the time for Markowitz analysis by a factor of three.

It remains important to avoid any algorithmic traps that might lead to $O(n^2)$ execution time for a problem of order n, because the efficiency of most sparse matrix computations is such that an $O(n^2)$ component would dominate the total work. A simple example of such a trap is that one might be tempted to use a bubble sort for the row counts, whereas in fact a pocket sort is far more efficient.

One such trap has been exposed by Zlatev (1980). He found a problem with a very large number of entries that had minimum Markowitz cost (4.1) but did not satisfy the threshold inequality (4.2). The original version of the Harwell code MA28 (Duff, 1977) terminates its search immediately it finds an entry that satisfies inequality (4.2) and has a Markowitz cost that cannot be beaten, and for most matrices such termination occurs extremely quickly. However, this was not so for Zlatev's matrix. Furthermore, having rejected a large number of entries at one stage, it retested most of them afresh at the next stage. Zlatev's cure is simple and ruthless: restrict the search to a few of the shortest rows (he recommends three) and take as pivot the best entry found there that satisfies the threshold inequality. Of course, this means that the pivotal strategy is no longer that of Markowitz, but its performance is quite similar. It is now incorporated as an option in MA28.

If a sequence of problems with the same pattern of entries but varying numerical values is to be treated, we may ANALYSE the first matrix and then FACTORIZE the rest using the same pivotal sequence and other stored data such as the positions of potential fill-ins. It is particularly important to monitor the later factorizations since no pivoting is now being performed and they are therefore potentially unstable. Sherman (1978) has suggested including partial pivoting in these factorizations. This can be done with a simple code, but occasionally leads to substantial fill-in. Finally, a single factorization may be used to SOLVE many systems with the same matrix but differing right-hand sides.

Really effective exploitation of vector and parallel hardware for the techniques of this section remains a problem, though much higher computation rates have been achieved with the help of hardware for vectorized indirect addressing and vectorized gather/scatter operations.

5. BAND AND VARIABLE-BAND MATRICES

Band and variable-band methods are based on the fact that, if Gaussian elimination is applied without pivoting, no fill-in takes place to the left of the first entry in a row or above the first entry in a column. This allows the rest of each row of the lower-triangular part of the matrix and the rest of each column of the upper-triangular part to be held as full vectors. Since the inner loop of the resulting code operates on full vectors, good execution speeds on vector processors are obtained.

Of course, the width of the band depends on the ordering of the rows and columns and there is strong demand from users for good automatic ordering techniques. Available algorithms for symmetric permutations of symmetric patterns are based on 'level sets' in the graph that has a node for each variable and an edge (i,j) for each entry a_{ij}. The first level set is a single node and each successive set consists of all the neighbours of the nodes of the previous level set that are not already members of that set or the previous one. A simple example is shown in Figure 5.1, where the level sets are (1), (2,3,4),

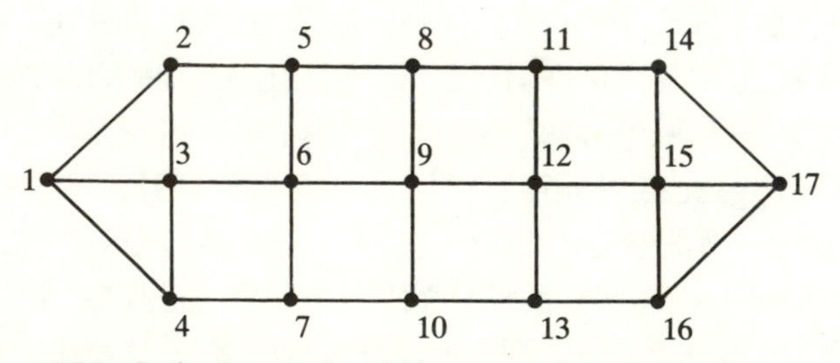

FIG. 5.1 A graph with a good node order

(5,6,7), (8,9,10), (11,12,13), (14,15,16) and (17). If the matrix is permuted to correspond with the level sets taken in order, it has a block triangular form, as illustrated in Figure 5.2. Clearly it is desirable to have a large number of levels with few nodes in each level. This led Gibbs, Poole, and Stockmeyer (1976) to propose seeking a 'pseudo-diameter' which is a pair of nodes such that, if level sets are generated from one, the final level includes the other and restarting from any node in the final set does not yield a greater number of levels. There have been some advances in the practical implementation of this algorithm and good code is provided by Lewis (1982).

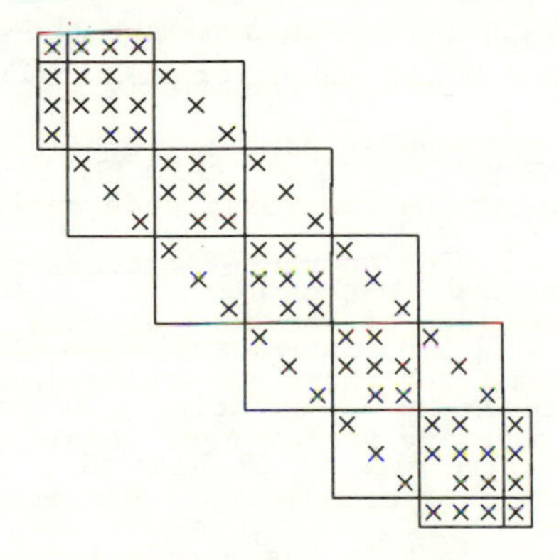

FIG. 5.2 The matrix corresponding to the Fig. 5.1 graph

There remains the problem of choosing the order within each level set. Cuthill and McKee (1969) recommend taking first those nodes which are neighbours of the first node in the previous level set, then those that are neighbours of the second node, and so on. George (1971) found that a worthwhile improvement could be obtained by calculating this ordering but applying it completely reversed. Jennings (1977) explained that the reason for the improvement is that after a Cuthill-McKee ordering there can be no re-entrancy in the profile of leading entries in the rows (the column indices of leading entries form a monotone sequence), but there can be re-entrancy in the profile of trailing entries in the columns. Reversing the order exploits this re-entrancy, as illustrated in Figure 5.3. An alternative for ordering within each level set is to use the strategy of King (1970), which is to order by increasing numbers of neighbours in the next level set. Lewis (1982) provides good code for both these variants.

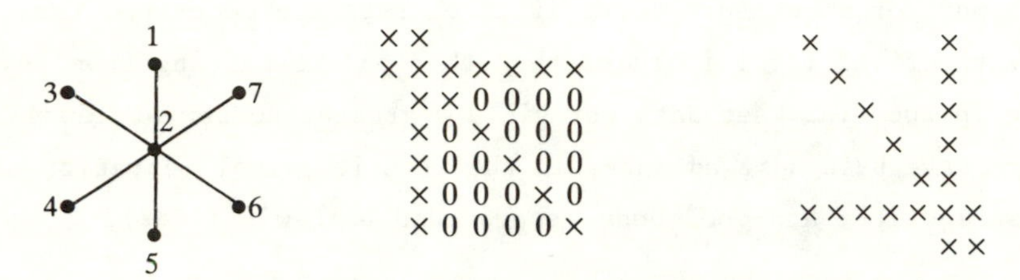

FIG. 5.3 A graph and its associated matrix, ordered by Cuthill-McKee and Reverse Cuthill-McKee

6. THE FRONTAL METHOD FOR SYMMETRIC POSITIVE-DEFINITE MATRICES

The frontal method (Irons, 1970) is in wide use and there have been some exciting developments of it in the last ten years. The method is based on the observation that for a symmetric and positive-definite finite-element problem with matrix

$$A = \sum_k A^{(k)}, \tag{6.1}$$

assembly operations

$$a_{ij} := a_{ij} + a_{ij}^{(k)} \tag{6.2}$$

do not have to be completed for all i and j before elimination operations

$$a_{ij} := a_{ij} - a_{il} a_{ll}^{-1} a_{lj} \tag{6.3}$$

are begun. All that is necessary is that the pivotal row is fully assembled, for otherwise the triple product subtracted in assignment (6.3) will be incorrect.

In the original frontal method, an ordering for the elements is given (that is, for the assembly (6.2)) and each variable is eliminated as soon as its row and column are fully assembled. Intermediate working is performed in a full array called the frontal matrix that corresponds to the submatrix consisting of rows and columns whose assembly has commenced but which have not yet been pivotal.

Automatic ordering methods are as much needed for the frontal method as for band solution and are inherently similar. Recent work by Sloan (1986) is very promising. Like Gibbs *et al* (1976), he commences by determining a pseudo-diameter. He then chooses each variable successively to minimize a weighted sum of the level set number and the resulting front size. He recommends that the weight for the size of the front should be double that for the level number. This heuristic is based on the notion of taking account both of the global information present in the level sets and the local objective of a narrow front-width. The element ordering may be deduced from the node ordering by associating each element with its earliest node in the order. The cost and effectiveness of Sloan's algorithm compare favourably with other algorithms, including those of Lewis (1982) that we mentioned in the previous section.

7. THE FRONTAL METHOD FOR UNSYMMETRIC MATRICES

The frontal method was first extended to unsymmetric matrices by Hood (1976) and the idea has since been developed by Cliffe *et al* (1978) and Duff (1981c, 1983). The frontal matrix itself becomes unsymmetric and need not be square. Two separate index lists are needed to label which rows and which columns are currently in it. Any entry in a fully-summed row and in a fully-summed column may be used as a pivot. This may allow us to find a pivot that satisfies the threshold inequality (4.2). However, this may not be possible for all the fully-summed rows and columns. In the latter case, we simply delay the elimination, which means that the front becomes a little larger than it would otherwise have been. The problem is only temporary because more and more rows and columns become fully summed, which gives more scope for pivoting and eventually the whole frontal matrix is fully summed. Nondefinite symmetric matrices can be treated similarly. In this case, symmetry may be preserved by using a mixture of ordinary diagonal pivots and 2×2 pivots (Bunch and Parlett, 1971; Bunch, 1974).

Duff (1984b) has extended the frontal method to finite-difference equations by treating each matrix row as an unsymmetric finite-element matrix, and his code vectorizes well.

8. MULTIFRONTAL METHOD

It is desirable in the frontal method to keep the size of the front small and that is one of the aims of automatic ordering schemes. However, this may conflict with the aim of reducing the amount of computation. For instance, consider ordering a symmetric positive-definite matrix by the algorithm of minimum degree (Scheme 2 of Tinney and Walker, 1967), where each pivot is chosen from the diagonal entries to minimize the row count (this is actually a variant of the algorithm of Markowitz, 1957). Here many of the early pivotal steps will often involve disjoint sets of variables so the front size is large. However, if the front is ordered so that variables involved in an element or in a pivotal step are grouped, the frontal matrix becomes block diagonal. Great savings can be made by storing the blocks separately, each with its own index list. This is the essence of the multifrontal method. We have many fronts, rather than just one.

Whereas the basic frontal method adds a single element matrix $A^{(k)}$ into the frontal matrix and then eliminates any rows and columns that become fully summed, the multifrontal method adds two or more frontal and element matrices together to make a fresh frontal matrix and then eliminates the rows and columns that become fully summed. Note that the frontal matrices are very like element matrices. Both can be stored as full matrices labelled with an index list. In fact, the frontal matrices are often called generated elements or generalized elements

(Speelpenning, 1978). The sequence of additions can conveniently be represented as a tree whose nodes correspond to the elements and generated elements. The sons of a tree node correspond to the elements and generated elements whose frontal matrices are added to form the frontal matrix of the tree node.

As an example, consider the finite-element problem shown in Figure 8.1. The frontal method is represented by the tree shown in Figure 8.2 and the nested dissection ordering (George, 1973) is represented by the tree shown in Figure 8.3.

1	2	5	6
3	4	7	8

FIG. 8.1
A finite-element problem

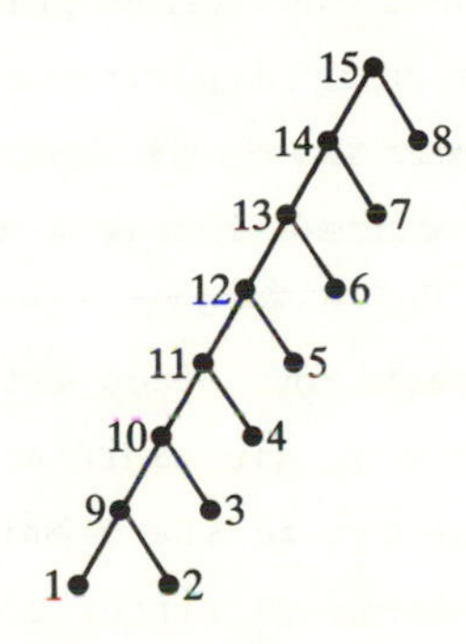

FIG. 8.2
Assembly tree for frontal elimination on the Fig. 8.1 example

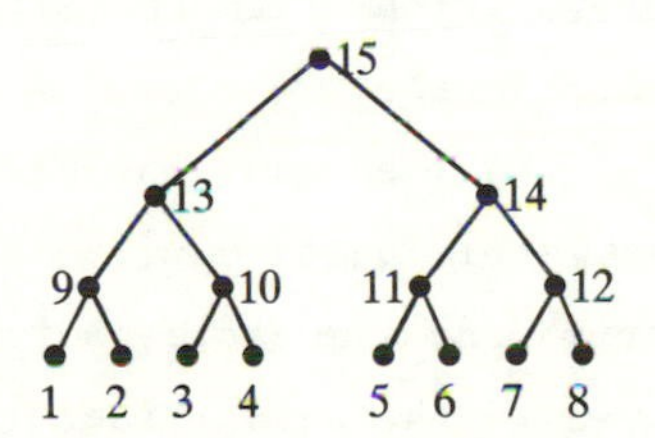

FIG. 8.3
Assembly tree for nested dissection

The whole elimination process may be modelled symbolically by working only with index lists. Corresponding to adding element and generated element matrices, we merge their index lists. Corresponding to eliminations, we remove indices from the list. These symbolic operations are an order of magnitude faster than the corresponding actual operations because (with careful coding) the work is linear rather than quadratic in the lengths of the index lists. Furthermore, the length of a new list cannot exceed the sum of the lengths of the lists that are merged. This is the basis for a very exciting development of the last ten years. It is now possible to ANALYSE the structure of a sparse symmetric matrix without fear of overflowing available storage in a time that is small compared with the time for numerical solution (George and Liu, 1980a, 1980b, 1981). Where the matrix does not arise from a finite-element application, each diagonal entry can be treated as a 1×1 element matrix and each off-diagonal pair can be treated as a 2×2 element matrix. Efficient implementations of the minimum-degree algorithm are available in this way, for example in SPARSPAK (Chu *et al*, 1984).

When the matrix is not positive definite, interchanges can be incorporated into the multifrontal method in just the same way as for the frontal method. For instance, in the Harwell code MA27 (Duff and Reid, 1982, 1983), eliminations are sometimes performed by using 2×2 pivots and are sometimes delayed. The only disadvantage is that an actual factorization may involve slightly more arithmetic and storage than was predicted during the ANALYSE phase.

The idea can also be extended to unsymmetric matrices with a symmetric pattern. The pattern may be analysed just as if the matrix were symmetric, but unsymmetric element matrices are employed in the actual factorization. Interchanges are performed to ensure satisfaction of the threshold inequality (4.2) and eliminations are delayed if necessary. Harwell code MA37 (Duff and Reid, 1984) implements this technique, and it can also treat matrices that do not have a symmetric pattern by embedding the pattern in a symmetric one. However, our experience with seriously unsymmetric patterns is that the amount of fill-in can be far greater than is given by the algorithm of Markowitz, which we considered in Section 4.

9. FAST SYMMETRIC ANALYSE

We explained in Section 8 how a minimum-degree ANALYSE code can be based on multifrontal ideas. Such code cannot fail through lack of storage and is usually much faster in execution than the corresponding numerical factorization for which it prepares data structures. This is highly satisfactory, and should be contrasted with the fact that ANALYSE codes for the general unsymmetric case (see Section 4) typically execute three to seven times more slowly than the subsequent FACTORIZE codes.

This unsatisfactory facet of current codes for unsymmetric cases has spawned alternative approaches. George and Ng (1985) have shown that, if a permutation of the matrix A has the factorization

$$PA = LU \tag{9.1}$$

and if the normal matrix A^TA has the Cholesky factorization

$$A^TA = \bar{U}^T\bar{U} , \tag{9.2}$$

then the patterns of L^T and U are contained in the pattern of $\bar{U}$. This suggests that the pattern of A^TA should be formed and a fast minimum-degree algorithm applied to it to yield the structure of the factorization

$$Q^TA^TAQ = \bar{U}^T\bar{U} , \tag{9.3}$$

where Q is a permutation matrix that is generated by the minimum-degree algorithm. A data structure to accommodate the pattern $\bar{U}^T\bar{U}$ can now be established and Q taken as a predetermined column permutation of A. This leaves P available for row interchanges during numerical factorization, and we can be sure that any resulting fill-ins will always be accommodated. This approach can give very satisfactory results, but the snag is that occasionally the fill-in may be far greater than with the algorithm of Markowitz, for example when A has a full row. The pattern can be reduced (George and Ng, 1984) by an analysis of the pattern of AQ to find the positions of all possible fill-ins that may occur for any P.

10. MULTIGRID ITERATION

Multigrid iteration is a very active and promising technique that has come to the fore during the last ten years. Although it is the subject of the paper of Professor Walsh in this volume, a very brief introduction is given here because we feel that a survey of recent developments in sparse matrix work would be incomplete without it.

The primary application of the multigrid technique is to the solution of partial differential equations, and it is based on the observation that many traditional iterative methods are good at reducing high-frequency components of the error but poor at reducing low-frequency components. Typically, after a few iterations the residual

$$\underline{r}^{(k)} = \underline{b} - A\underline{x}^{(k)} \tag{10.1}$$

is smooth and so can be represented on a coarser grid, say as $\underline{r}_1^{(k)}$. If A_1 is a matrix that represents the same differential operator on the coarse grid as A represents on the given grid, then we may solve the equation

$$A_1 \underline{\delta}_1^{(k)} = \underline{r}_1^{(k)}$$

on the coarse grid and then interpolate $\underline{\delta}_1^{(k)}$ onto the original grid to provide a correction for $\underline{x}^{(k)}$ there. This provides a means of speeding the decay of the smooth components of the error.

The idea can be nested, that is equation (10.2) can itself be solved iteratively with the help of an even coarser grid, etc., and thus very economical overall solution can be obtained. The total amount of work performed on all the coarse grids is often less than for one fine grid iteration, and the overall convergence rate can be very fast and independent of the fineness of the original grid.

Notice that the basic iteration does not necessarily have to be convergent. It has to be able to reduce the high-frequency errors, but does not necessarily have to reduce the low-frequency ones, since these are handled on the coarser grid. It is therefore often called a *smoothing* iteration.

Recently it has been suggested (see, for example, Stüben *et al*, 1984) that the technique can be applied algebraically to problems whose origin is not necessarily restricted to the solution of partial differential equations. It is early yet to judge the efficacy of this approach, but results are encouraging.

11. PRECONDITIONED CONJUGATE GRADIENTS

Another iterative method that has become very prominent in the last ten years is preconditioned conjugate gradients (Meijerink and van der Vorst, 1977). If A is symmetric and positive definite, application of the method of conjugate gradients (Hestenes and Stiefel, 1952) to the problem

$$A\underline{x} = \underline{b} \tag{11.1}$$

is very successful if the range of eigenvalues of A is small, or if all but a few lie within a small range. To see this, suppose $\underline{x}^{(0)}$ is the starting vector. The method automatically yields residuals

$$\underline{r}^{(i)} = \underline{b} - A\underline{x}^{(i)} \tag{11.2}$$

of the form

$$\underline{r}^{(i)} = \underline{r}^{(0)} + \sum_{k=1}^{i} \alpha_k A^k \underline{r}^{(0)} = P_i(A)\,\underline{r}^{(0)}, \tag{11.3}$$

which are such that

$$Q(\underline{r}) = \underline{r}^T A^{-1} \underline{r} \tag{11.4}$$

is minimized. Here P_i is a polynomial of degree i with the property that $P_i(0) = 1$. If A has eigenvalues λ_k, $k=1,2,\cdots,n$, and a corresponding set of orthogonal eigenvectors $\underline{e}_k$, $k=1, 2,\cdots,n$, then the original residual $\underline{r}^{(0)}$ may be expressed in the form

$$\underline{r}^{(0)} = \sum_{k=1}^{n} \beta_k \underline{e}_k \tag{11.5}$$

and the i-th residual is

$$\underline{r}^{(i)} = \sum_{k=1}^{n} \beta_k P_i(\lambda_k)\, \underline{e}_k . \tag{11.6}$$

If there is a polynomial Q_i of degree i with $Q_i(0) = 1$ and

$Q_i(\lambda_k)$, $k = 1, 2, \cdots, n$, all small, the method of conjugate gradients will automatically choose a polynomial P_i that gives a small residual $\underline{r}^{(i)}$ since it minimizes the quadratic form (11.4). There will be such a polynomial if the range of eigenvalues is small, or if all but a few eigenvalues lie in a small range.

Preconditioning seeks to transform the problem (11.1) to one having this property. This is usually done by calculating an approximate factorization

$$A \simeq LL^T, \qquad (11.7)$$

and then working with the system

$$(L^{-1}AL^{-T})\,\underline{y} = L^{-1}\,\underline{b}\ , \qquad (11.8)$$

whose solution is related to $\underline{x}$ by the equation

$$L^T\underline{x} = \underline{y}\ . \qquad (11.9)$$

The matrix L is usually restricted to having a particular sparsity pattern, such as that of the lower-triangular part of A itself. Experience indicates (see, for example, Munksgaard, 1980) that little is gained by working with a more sophisticated choice of sparsity pattern.

The unsymmetric case is much more awkward to treat because the three-term recurrence that is at the heart of the original method is not available. However, it is possible (Vinsome, 1976) to choose a correction to $\underline{x}^{(k)}$ in the space spanned by the recent residuals so that the new residual $\underline{r}^{(k+1)}$ minimizes the quantity

$$\underline{r}^T\underline{r}\,. \qquad (11.10)$$

When combined with good preconditioning so that the number of iterations is small (see, for example, Appleyard *et al*, 1981), this method is very effective.

12. LANCZOS ALGORITHM

The method of Lanzcos (1950) for the symmetric eigenproblem was discussed in some detail in our earlier survey. Given a starting vector $\underline{v}_1$, it provides an automatic way of generating a matrix $V^{(k)}$ whose columns form an orthonormal set that span the space

$$\left\{\underline{v}_1, A\underline{v}_1, A^2\underline{v}_1, \cdots, A^{k-1}\underline{v}_1\right\} \tag{12.1}$$

and that satisfies the equation

$$V^{(k)T} A V^{(k)} = T^{(k)}, \tag{12.2}$$

where $T^{(k)}$ is tridiagonal. The eigenvalues $\lambda_i^{(k)}$ of $T^{(k)}$, particularly those at the ends of the spectrum, provide estimates of the eigenvalues of A and the corresponding eigenvectors $\underline{s}_i^{(k)}$ provide estimates $V^{(k)}\underline{s}_i^{(k)}$ of the eigenvectors of A.

The snag is that roundoff errors lead to nonorthogonal columns. The Lanczos algorithm was analysed by Paige (1976), who showed just how the orthogonality is lost. For an exposition of Paige's principal theorem, we refer the reader to the book of Parlett (1980). Paige's analysis led to the development of methods for maintaining orthogonality by selective orthogonalization, which is practical for large problems, whereas full orthogonalization is not. For a summary of this work, we refer the reader to Simon (1984).

Another approach is to accept the loss of orthogonality and run the algorithm for a large number of iterations, perhaps as many as $3n$ for a problem of order n. Cullum and Willoughby (1981) apply tests to separate good from 'spurious' eigenvalues. Parlett and Reid (1981) track the progress of wanted eigenvalues and accept only those with acceptable error bounds, basing their bounds on the work of Paige.

Eigenvalues near the ends of the spectrum, particularly if they are well-separated from their neighbours, are found quickly. Ericsson and Ruhe (1980) therefore propose transforming the generalized symmetric eigenvalue problem

$$A\underline{x} = \lambda B\underline{x} \tag{12.3}$$

to the form

$$B(A-pB)^{-1} B\underline{x} = (\lambda - p)^{-1} B\underline{x}. \tag{12.4}$$

Eigenvalues of the problem (12.3) that are near p correspond to eigenvalues of the problem (12.4) that are large. Ericsson and Ruhe (1980) adjust p dynamically so that few steps are needed by the Lanczos algorithm to obtain these eigenvalues. A similar transformation has been used by Lewis (private communication, 1985) in a very successful code for large structural problems.

It is remarkable that there has been so much development of a 35-year-old method, enough to justify a two-volume book by Cullum and Willoughby (1985), the second volume of which consists of programs.

13. LEAST-SQUARES PROBLEMS

The most significant development of the last decade for the treatment of sparse least-squares problems is in the effective use of the orthogonal factorization

$$A = QU, \tag{13.1}$$

where A and U are now rectangular. George and Heath (1980) exploit the relation

$$A^T A = (\hat{U}^T \ 0)\ Q^T Q \begin{bmatrix} \hat{U} \\ 0 \end{bmatrix} = \hat{U}^T \hat{U} \tag{13.2}$$

which shows that the Cholesky factors of $A^T A$ are $\hat{U}^T$ and $\hat{U}$. Therefore the sparsity pattern of $\hat{U}$ is contained in the sparsity pattern obtained by an analysis of the pattern of $A^T A$. Furthermore, column permutations of A correspond to symmetric permutations of $A^T A$. George and Heath therefore form the pattern of $A^T A$ and apply an ANALYSE step (Section 9) to it. They apply the permutation it finds to the columns of A, and they use the final pattern it produces as a template for U. U itself is formed into this template row-by-row by using Givens' rotations. This algorithm therefore has the merit of rapid ANALYSE processing, no more fill-in than for the normal equations approach (assuming that the rotations used are not stored), and the satisfactory numerical properties of orthogonal reduction.

It can be extended by partitioning to huge problems for which U cannot be held in main storage (see George, Heath, and Plemmons, 1981). However, it is less satisfactory if there is a need to process further right-hand vectors $\underline{b}$ later since the orthogonal reductions are not stored.

Acknowledgement

I would like to thank Iain Duff and Michael Powell for reading drafts of this paper and making many helpful suggestions.

REFERENCES

Appleyard, J.R., Cheshire, I.M. and Pollard, B.K. (1981), "Special techniques for fully-implicit simulators", Report CSS 112, Harwell Laboratory.

Barlow, J.L. (1986), "A note on monitoring the stability of triangular decomposition of sparse matrices", *SIAM J. Sci. Statist. Comput.*, *7*, pp.166-168.

Bunch, J.R. (1974), "Partial pivoting strategies for symmetric matrices", *SIAM J. Numer. Anal.*, *11*, pp.521-528.

Bunch, J.R. and Parlett, B.N. (1971), "Direct methods for solving symmetric indefinite systems of linear equations", *SIAM J. Numer. Anal.*, *8*, pp.639-655.

Chu, E., George, A., Liu, J. and Ng, E. (1984), "SPARSPAK: Waterloo sparse matrix package user's guide for SPARSPAK-A", Report CS-84-36, University of Waterloo, Canada.

Cliffe, K.A., Jackson, C.P., Rae, J. and Winters, K.H. (1978), "Finite element flow modelling using velocity and pressure variables," Report AERE R9202, HMSO, London.

Cullum, J.K. and Willoughby, R.A. (1981), "Computing eigenvalues of very large symmetric matrices — an implementation of a Lanczos algorithm with no reorthogonalization", *J. Comput. Phys.*, *44*, pp.329-358.

Cullum, J.K. and Willoughby, R.A. (1985), *Lanczos Algorithms for Large Symmetric Eigenvalue Computations. Vol. I: Theory; Vol. II: Programs. Progress in Scientific Computing, Vols. 3 and 4*, Birkhauser (Boston).

Cuthill, E. and McKee, J. (1969), "Reducing the bandwidth of sparse symmetric matrices", in *Proceedings 24th National Conference of the Association for Computing Machinery*, Brandon Press (New Jersey), pp.157-172.

Duff, I.S. (1977), "MA28 — a set of Fortran subroutines for sparse unsymmetric linear equations", Report AERE R8730, HMSO, London.

Duff, I.S. (1981a), "On algorithms for obtaining a maximum transversal", *ACM Trans. Math. Software*, *7*, pp.315-330.

Duff, I.S. (1981b), "Algorithm 575. Permutations for a zero-free diagonal", *ACM Trans. Math. Software*, *7*, pp.387-390.

Duff, I.S. (1981c), "MA32 — A package for solving sparse unsymmetric systems using the frontal method", Report AERE R10079, HMSO, London.

Duff, I.S. (1983), "Enhancements to the MA32 package for solving sparse unsymmetric equations", Report AERE R11009, HMSO, London.

Duff, I.S. (1984a), "The solution of sparse linear systems on the CRAY-1", in *High-speed Computation*, *NATO ASI Series Volume F7*, ed. Kowalik, J.S., Springer-Verlag (Berlin), pp.293-309.

Duff, I.S. (1984b), "Design features of a frontal code for solving sparse unsymmetric linear systems out-of-core", *SIAM J. Sci. Statist. Comput.*, *5*, pp.270-280.

Duff, I.S., Erisman, A.M. and Reid, J.K. (1986), *Direct Methods for Sparse Matrices*, Oxford University Press (London).

Duff, I.S. and Reid, J.K. (1978a), "An implementation of Tarjan's algorithm for the block triangularization of a matrix", *ACM Trans. Math. Software*, *4*, pp.137-147.

Duff, I.S. and Reid, J.K. (1978b), "Algorithm 529. Permutations to block triangular form", *ACM Trans. Math. Software*, *4*, pp.189-192.

Duff, I.S. and Reid, J.K. (1982), "MA27 — A set of Fortran subroutines for solving sparse symmetric sets of linear equations", Report AERE R10533, HMSO, London.

Duff, I.S. and Reid, J.K. (1983), "The multifrontal solution of indefinite sparse symmetric linear equations", *ACM Trans. Math. Software*, *9*, pp.302-325.

Duff, I.S. and Reid, J.K. (1984), "The multifrontal solution of unsymmetric sets of linear systems", *SIAM J. Sci. Statist. Comput.*, *5*, pp.633-641.

Ericsson, T. and Ruhe, A. (1980), "The spectral transformation Lanczos method for the numerical solution of large sparse generalized symmetric eigenvalue problems", *Math. Comp.*, *35*, pp.1251-1268.

Erisman, A.M. and Reid, J.K. (1974), "Monitoring the stability of the triangular factorization of a sparse matrix", *Numer. Math.*, *22*, pp.183-186.

George, A. (1971), *Computer Implementation of the Finite-Element Method*, Report STAN CS-71-208, Ph.D. Thesis, Department of Computer Science, Stanford University.

George, A. (1973), "Nested dissection of a regular finite-element mesh", *SIAM J. Numer. Anal.*, *10*, pp.345-363.

George, A. and Heath, M.T. (1980), "Solution of sparse linear least squares problems using Givens rotations", *Linear Algebra Appl.*, *34*, pp.69-83.

George, A., Heath, M.T. and Plemmons, R.J. (1981), "Solution of large-scale sparse least squares problems using auxiliary storage", *SIAM J. Sci. Statist. Comput.*, *2*, pp.416-429.

George, A. and Liu, J.W.H. (1980a), "A minimal storage implementation of the minimum degree algorithm", *SIAM J. Numer. Anal.*, *17*, pp.282-299.

George, A. and Liu, J.W.H. (1980b), "A fast implementation of the minimum degree algorithm using quotient graphs", *ACM Trans. Math. Software*, *6*, pp.337-358.

George, A. and Liu, J.W.H. (1981), *Computer Solution of Large Sparse Positive Definite Systems*, Prentice-Hall (Englewood Cliffs, New Jersey).

George, A. and Ng, E. (1984), "Symbolic factorization for sparse Gaussian elimination with partial pivoting", Report CS-84-43, University of Waterloo, Canada.

George, A. and Ng, E. (1985), "An implementation of Gaussian elimination with partial pivoting for sparse systems", *SIAM J. Sci. Statist. Comput.*, *6*, pp.390-409.

Gibbs, N.E., Poole, W.G., Jr. and Stockmeyer, P.K. (1976), "An algorithm for reducing the bandwidth and profile of a sparse matrix", *SIAM J. Numer. Anal.*, *13*, pp.236-250.

Gustavson, F.G. (1972), "Some basic techniques for solving sparse systems of linear equations", in *Sparse Matrices and their Applications*, eds. Rose, D.J. and Willoughby, R.A., Plenum Press (New York), pp.41-52.

Gustavson, F.G. (1976), "Finding the block lower-triangular form of a sparse matrix", in *Sparse Matrix Computations*, eds. Bunch, J.R. and Rose, D.J., Academic Press (New York), pp.275-289.

Hestenes, M.R. and Stiefel, E. (1952), "Methods of conjugate gradients for solving linear systems", *J. Res. Nat. Bur. Standards*, *49*, pp.409-436.

Hood, P. (1976), "Frontal solution program for unsymmetric matrices", *Internat. J. Numer. Methods Engrg.*, *10*, pp.379-399.

Hopcroft, J.E. and Karp, R.M. (1973), "An $n^{5/2}$ algorithm for maximum matchings in bipartite graphs", *SIAM J. Comput.*, *2*, pp.225-231.

Irons, B.M. (1970), "A frontal solution program for finite element analysis", *Internat. J. Numer. Methods Engrg.*, *2*, pp.5-32.

Jennings, A. (1977), *Matrix Computation for Engineers and Scientists*, John Wiley and Sons (London).

King, I.P. (1970), "An automatic reordering scheme for simultaneous equations derived from network systems", *Internat. J. Numer. Methods Engrg.*, *2*, pp.523-533.

Kuhn, H.W. (1955), "The Hungarian method for the assignment problem", *Naval Res. Logist. Quart.*, *2*, pp.83-97.

Lanczos, C. (1950), "An iteration method for the solution of the eigenvalue problem of linear differential and integral operators", *J. Res. Nat. Bur. Standards*, *45*, pp.255-282.

Lewis, J.G. (1982), "Implementation of the Gibbs-Poole-Stockmeyer and Gibbs-King algorithms", *ACM Trans. Math. Software*, *8*, pp.180-189 and 190-194.

Markowitz, H.M. (1957), "The elimination form of the inverse and its application to linear programming", *Management Sci.*, *3*, pp.255-269.

Meijerink, J.A. and van der Vorst, H.A. (1977), "An iterative solution method for linear systems of which the coefficient matrix is a symmetric M-matrix", *Math. Comp.*, *31*, pp.148-162.

Munksgaard, N. (1980), "Solving sparse symmetric sets of linear equations by preconditioned conjugate gradients", *ACM Trans. Math. Software*, *6*, pp.206-219.

Østerby, O. and Zlatev, Z. (1983), *Direct Methods for Sparse Matrices*, *Lecture Notes in Computer Science*, *157*, Springer-Verlag (Berlin).

Paige, C.C. (1976), "Error analysis of the Lanczos algorithm for tridiagonalizing a symmetric matrix", *J. Inst. Math. Appl.*, *18*, pp.341-349.

Parlett, B.N. (1980), *The Symmetric Eigenvalue Problem*, Prentice-Hall (Englewood Cliffs, New Jersey).

Parlett, B.N. and Reid, J.K. (1981), "Tracking the progress of the Lanczos algorithm for large symmetric eigenproblems", *IMA J. Numer. Anal.*, *1*, pp.135-155.

Pissanetzky, S. (1984), *Sparse Matrix Technology*, Academic Press (New York).

Reid, J.K. (1977), "Sparse matrices", in *The State of the Art in Numerical Analysis*, ed. Jacobs, D.A.H., Academic Press (London), pp.85-146.

Sherman, A.H. (1978), "Algorithm 533. NSPIV, a Fortran subroutine for sparse Gaussian elimination with partial pivoting", *ACM Trans. Math. Software*, *4*, pp.391-398.

Simon, H.D. (1984), "Analysis of the symmetric Lanczos algorithm with reorthogonalization methods", *Linear Algebra Appl.*, *61*, pp.101-131.

Sloan, S.W. (1986), "An algorithm for profile and wavefront reduction of sparse matrices", *Internat. J. Numer. Methods Engrg.*, *23*, pp.239-251.

Speelpenning, B. (1978), "The generalized element method", Report UIUCDCS-R-78-946, University of Illinois at Urbana-Champaign.

Stüben, K., Trottenberg, U. and Witsch, K. (1984), "Software development based on multigrid techniques", in *PDE Software: Modules, Interfaces and Systems*, eds. Engquist, B. and Smedsaas, T., North-Holland (Amsterdam), pp.241-267.

Tarjan, R.E. (1972), "Depth-first search and linear graph algorithms", *SIAM J. Comput.*, *1*, pp.146-160.

Tinney, W.F. and Walker, J.W. (1967), "Direct solutions of sparse network equations by optimally ordered triangular factorization", *Proc. IEE-E*, *55*, pp.1801-1809.

Vinsome, P.K.W. (1976), "Orthomin, an iterative method for solving sparse sets of simultaneous linear equations", in *Symposium on Numerical Solution of Reservoir Performance*, *SPE 5729*, Soc. of Pet. Eng. of AIME (New York), pp.149-159.

Willoughby, R.A. (1971), "Sparse matrix algorithms and their relation to problem classes and computer architecture", in *Large Sparse Sets of Linear Equations*, ed. Reid, J.K., Academic Press (London), pp.255-277.

Zlatev, Z. (1980), "On some pivotal strategies in Gaussian elimination by sparse technique", *SIAM J. Numer. Anal.*, *17*, pp.18-30.

J.K. Reid
Computer Science and Systems Division
UKAEA
Harwell Laboratory
Didcot
Oxon OX11 ORA
England

4 Multivariate Approximation

CARL DE BOOR

ABSTRACT

The lecture addresses topics in multivariate approximation which have caught the author's interest in the last ten years. These include: the approximation by functions with fewer variables, correct points for polynomial interpolation, the B(ernstein, -ézier, -arycentric)-form for polynomials and its use in understanding smooth piecewise polynomial (pp) functions, approximation order from spaces of pp functions, multivariate B-splines, and surface generation by subdivision.

1. THE SET-UP

The talk concerns the approximation of

$$f : G \subseteq \mathbb{R}^d \to \mathbb{R} ,$$

i.e., of some real-valued function f defined on some domain G in d-dimensional space.

While my first publication dealt with such multivariate (actually, bivariate) approximation, I have concerned myself seriously with multivariate approximation only in the last ten years. This talk reflects some of the experiences I have had during that time.

Sponsored by the United States Army under
Contract No. DAAG29-80-C-0041

The approximating functions are typically polynomials or piecewise polynomials, and just how one describes them will have an effect on one's work with them. Papers on multivariate approximation often sink under the burden of cumbersome notation. As a preventive measure against such a sad fate, I shall follow the "default" convention whereby symbols are left out if they can reasonably be guessed from the context. For example if $\underline{x} = (x(1), \cdots, x(d))$ has just been declared to be a point in $\mathbb{R}^d$, I will feel free to write

$$\sum_i x(i) \quad \text{instead of} \quad \sum_{i=1}^{d} x(i).$$

For polynomials, multi-index notation is standard. With $\underline{x} = (x(1), \cdots, x(d))$ the generic point in $\mathbb{R}^d$, one uses the abbreviation

$$\underline{x}^{\underline{\alpha}} := \prod_i x(i)^{\alpha(i)}, \quad \underline{x} \in \mathbb{R}^d, \quad \underline{\alpha} \in \mathbb{Z}_+^d.$$

The function $\underline{x} \mapsto \underline{x}^{\underline{\alpha}}$ is a monomial of degree $\underline{\alpha}$, or, of (total) degree

$$|\underline{\alpha}| := \sum_i \alpha(i)$$

if only the exponent sum matters. More generally, a polynomial of degree $\leqslant \underline{\alpha}$ is, by definition, any function of the form

$$\underline{x} \mapsto \sum_{\underline{\beta} \leqslant \underline{\alpha}} \underline{x}^{\underline{\beta}} c(\underline{\beta}),$$

with real coefficients $c(\underline{\beta})$. The collection of all such polynomials is denoted by

$$\pi_{\underline{\alpha}} = \pi_{\underline{\alpha}}(\mathbb{R}^d),$$

the collection of all polynomials of total degree at most k by

$$\pi_k = \pi_k(\mathbb{R}^d),$$

and the collection of all polynomials, of whatever degree, by

$$\pi = \pi(\mathbb{R}^d).$$

Many expressions simplify if one makes use of the normalized power function, i.e., the function

$$[\![\;]\!]^{\underline{\alpha}} : \underline{x} \mapsto \underline{x}^{\underline{\alpha}} / \underline{\alpha}! := \prod_i x(i)^{\alpha(i)} / \alpha(i)!.$$

For example, with $\underline{\alpha}, \underline{\xi}, \underline{\upsilon}, \cdots, \underline{\zeta} \in \mathbb{Z}_+^d$, the Multinomial Theorem takes the simple form

$$[\![\underline{x} + \underline{y} + \cdots + \underline{z}]\!]^{\underline{\alpha}} = \sum_{\underline{\xi}+\underline{\upsilon}+\cdots+\underline{\zeta}=\underline{\alpha}} [\![\underline{x}]\!]^{\underline{\xi}} [\![\underline{y}]\!]^{\underline{\upsilon}} \cdots [\![\underline{z}]\!]^{\underline{\zeta}}. \quad (1.1)$$

At times, it pays to give up on the power form altogether. In some contexts, it is very convenient to describe polynomials in terms of the particular homogeneous polynomials

$$\langle Y, \cdot \rangle : \underline{x} \mapsto \prod_{\underline{y} \in Y} \langle \underline{y}, \underline{x} \rangle,$$

with

$$\langle \underline{y}, \underline{x} \rangle := \sum_i y(i)\, x(i)$$

and Y a finite subset of $\mathbb{R}^d$. I will make use of this form in Section 3. A particular instance is the B(ernstein, -ézier)-form, which is the form of choice when dealing with piecewise polynomials on a triangulation (see Section 4).

2. APPROXIMATION BY FUNCTIONS OF FEWER VARIABLES

The simplest approach to multivariate approximation uses tensor products, i.e., linear combinations of functions of the form

$$\underline{x} \mapsto \prod_i g_i(x(i)),$$

each g_i being a univariate function. This neatly avoids dealing with the realities of multivariate functions, but its effectiveness depends on having the information about f correspondingly available in (cartesian) product form, e.g., as function values on a rectangular grid parallel to the coordinate axes. The recent book by Light and Cheney (1986) provides up-to-date material on this practically very important choice of approximating function and certain ready extensions.

The book also deals with the situation when the information about f is not in product form. In that case, tensor product approximants still are attractive since they are composed of univariate functions. The general question of how to approximate a multivariate function by functions of fewer variables has received much attention. A reference with a Numerical Analysis slant is Golomb (1959).

The most remarkable result along this line is

KOLMOGOROV'S THEOREM (Kolmogorov, 1957). There exist $\underline{\lambda} \in]0,1]^d$ and $\Phi \subseteq \mathrm{Lip}_\alpha[0,1] \cap$ strictly monotone, with $\#\Phi = 2d+1$, such that for any $f \in C[0,1]^d$ there is a $g \in C[0,d]$ giving the identity

$$f(\underline{x}) = \sum_{\varphi \in \Phi} g\Bigl(\sum_i \lambda(i)\,\varphi(x(i))\Bigr). \tag{2.1}$$

(Here and below,

$$\#A := \text{the number of elements in } A.)$$

The theorem claims the existence of a set Ψ of $2d+1$ 'universal' maps $\psi : [0,1]^d \to [0,d]$ so that, for each continuous function f on the unit cube $[0,1]^d$, a continuous function g on the interval $[0,d]$ can be found for which

$$f = \sum_{\psi \in \Psi} g \circ \psi .$$

Moreover, each function $\psi \in \Psi$ is of the form

$$\psi(\underline{x}) := \sum_{i=1}^{d} \lambda(i)\ \varphi(x(i)),$$

with each of the $2d+1$ functions $\varphi \in \Phi$ strictly increasing and in $\mathrm{Lip}_\alpha[0,1]$ for some positive α, and $\underline{\lambda}$ some d-vector with positive entries all bounded by 1.

The work of Kolmogorov and his pupil Arnol'd which culminated in this theorem was motivated by Hilbert's Thirteenth Problem which contained (implicitly) the conjecture that not all continuous functions of three variables could be written as superpositions of continuous functions of two variables. The version quoted here reflects further simplifications, chiefly by Lorentz (1962). For a proof and further discussion, see (Lorentz, 1966; pp.168ff).

Practical use of Kolmogorov's Theorem seems elusive since the 'universal' functions $\varphi \in \Phi$ have a fractal 'derivative' (see Section 8), and g need not be smooth even if f is smooth. But it remains a challenge to develop a practical Approximation Theory which can handle approximating functions of the form (2.1). In any case, it suggests a nontraditional form of approximation which is motivated by computational or algorithmic simplicity. Perhaps we have been too accepting of traditional approximation techniques in which we choose the approximating family according to linear degrees of freedom. Perhaps we should consider instead approximating families which are classified by the number of floating-point operations required for their evaluation. Approximation Theory as it now exists has little to offer in this direction, but Computer Science may have something to teach us.

A special case has had much exposure in the times before electronic computers, viz., the approximation by <u>nomographic functions</u>. These are functions of two variables of the specific form

$$\mathbb{R}^2 \to \mathbb{R} : \underline{x} \mapsto g\Big(\varphi(x(1)) + \psi(x(2))\Big).$$

For recent algorithmic work, see von Golitschek (1984).

3. LOSS OF HAAR

The traditional approaches to approximation all start with polynomials, and so will I. Perhaps the greatest change when going to the multivariate set-up is the loss of the Haar property.

To recall, interpolation from a linear space S of functions on some $G \subseteq \mathbb{R}^d$ at a point set $T \subset G$ can be viewed as the task of inverting the restriction map

$$S \to \mathbb{R}^{\#T}: \quad p \longmapsto p_{|T}$$

We are to (re)construct some element $p \in S$ from its prescribed values $p_{|T}$ at the points in T. In these terms, S has the Haar property if every $T \subseteq G$ with $\#T = \dim S$ is correct for S, i.e., is such that

$$S \to \mathbb{R}^{\#T}: p \mapsto p_{|T} \text{ is 1-1 and onto.}$$

In other words, we can interpolate, and uniquely so, from S to any function values given on T.

If $d = 1$, then π_k has the Haar property (for any G with more than k points), but Mairhuber's switching yard argument (cf., e.g., the cover of Lorentz, 1966) shows that this property cannot hold for any S of dimension > 1 on a multidimensional set G.

For the case of polynomials, it is possible to identify various point sets $T \subset \mathbb{R}^d$ which are correct for π_k. A particularly nice example is provided by Chung and Yao (1977) who prove the following: Suppose that V is a finite subset of $\mathbb{R}^d \backslash \underline{0}$, and that $V \cup \underline{0}$ is in general position, which means that π_1 has the Haar property on $V \cup \underline{0}$. This implies that for every $W \subset V$ with $\# W = d$, $\underline{x}_W \in \mathbb{R}^d$ is defined uniquely by the equations

$$1 + \langle \underline{w}, \underline{x}_W \rangle = 0, \quad \underline{w} \in W,$$

since these state that the linear polynomial $1 + \langle \cdot, \underline{x}_W \rangle$ is to vanish on W and take the value 1 at 0. Further, since $1 + \langle \cdot, \underline{x}_W \rangle$ already vanishes on the d points in W, it cannot vanish anywhere on $V \backslash W$, by the Haar property. It follows that the functions

$$l_W : \underline{x} \mapsto \prod_{\underline{v} \in V \backslash W} \frac{1 + \langle \underline{v}, \underline{x} \rangle}{1 + \langle \underline{v}, \underline{x}_W \rangle}$$

are well-defined, are made up of $\# V - d$ linear factors, and satisfy $l_W(\underline{x}_{W'}) = \delta_{WW'}$ (since, for $W' \neq W$, at least one $\underline{v} \in V \backslash W$ is in W', making the corresponding linear factor zero at $\underline{x}_{W'}$); hence in particular $\underline{x}_{W'} \neq \underline{x}_W$ for $W' \neq W$, while

$$\# T = \binom{\# V}{d} = \dim \pi_{\# V - d}(\mathbb{R}^d).$$

This proves

THEOREM (Chung and Yao, 1977). $T := \{\underline{x}_W\}$ is correct for $\pi_{\# V-d}$.

The following result has a different flavour:

THEOREM (Hakopian, 1983). If $T \subset \mathbb{Z}_+^d$ 'contains its shadow', i.e., $\underline{\beta} \leq \underline{\alpha} \in T$ implies $\underline{\beta} \in T$, then T is correct for span $\left([\![\ ]\!]^{\underline{\alpha}}\right)_{\underline{\alpha} \in T}$.

A totally different approach to the correctness problem has been taken by Kergin (1978). He is interested in extending to a multivariate setting H. Whitney's (1957) characterization of functions on some subset T of $\mathbb{R}$ which have extensions to a smooth function on all of $\mathbb{R}$. Since Whitney uses divided differences in an essential way, Kergin looks for a viable generalization of the divided differences. His approach retains the univariate choice of interpolating from π_k at an arbitrary $(k+1)$-set T in $\mathbb{R}^d$ and deals with the many more degrees of freedom available from π_k by enforcing certain mean-value conditions. These conditions are that, for every sufficiently smooth f, the interpolant Pf should be in π_k and, for every $r \leq k$, for every homogeneous polynomial q of degree r, and for every $(r+1)$-subset W of T, there should exist a point in the convex hull of W at which $q(D)f$ and $q(D)Pf$ agree. Here (to be more explicit) $q(D)$ is an r-th order homogeneous constant coefficient differential operator, i.e.,

$$q(D) = \sum_{|\underline{\alpha}|=r} a(\underline{\alpha})\, D^{\underline{\alpha}}$$

for certain coefficients $a(\underline{\alpha})$. Surprisingly, there exists exactly one <u>linear</u> map P with these properties. This linear map can be characterized by the fact that, for every <u>plane wave</u>, i.e., every function f of the form $f := g \circ \underline{\lambda} : \underline{x} \mapsto g(\langle \underline{x}, \underline{\lambda} \rangle)$, P reduces to univariate interpolation at the projected point set $\langle T, \underline{\lambda} \rangle = \{ \langle \underline{t}, \underline{\lambda} \rangle : \underline{t} \in T \} \subset \mathbb{R}$, i.e., $P(g \circ \underline{\lambda}) = (P_{\underline{\lambda} T} g) \circ \underline{\lambda}$, with $P_{\underline{\lambda} T}$ univariate interpolation from $\pi_k(\mathbb{R})$ at $\underline{\lambda} T = \langle T, \underline{\lambda} \rangle$. In other words, for a plane wave f, $Pf(\underline{x})$ is the value at $\langle \underline{x}, \underline{\lambda} \rangle \in \mathbb{R}$ of the univariate polynomial of degree at most k which matches the value $f(\underline{t})$ at $\langle \underline{t}, \underline{\lambda} \rangle$, all $\underline{t} \in T$, where Hermite interpolation is used in case of coincident points. Full understanding of this process (see Micchelli, 1980) led to an understanding of multivariate B-splines, of which more anon.

It seems more promising to give up on polynomials altogether and to choose the interpolating function space S to depend on the point set T at which data are given. The simplest general model has the form

$$\sum_{\underline{t} \in T} \varphi(\cdot - \underline{t})\, c(\underline{t}),$$

with $\varphi : \mathbb{R}^d \to \mathbb{R}$ a function to be chosen 'suitably'. Duchon's <u>thin plate splines</u> (see, e.g., Meinguet, 1979) use

$$\varphi(\underline{x}) := |\underline{x}|^{2m-d} \begin{cases} \ln|\underline{x}|, & n \text{ even} \\ 1, & n \text{ odd}, \end{cases}$$

motivated by a variational argument, while Hardy's <u>multiquadrics</u> correspond to the choice

$$\varphi(\underline{x}) := \sqrt{1 + |\underline{x}|^2}\,.$$

A good source of up-to-date information about such interpolation methods and, in particular, about the question of their correctness, is the recent survey article of Micchelli (1986).

4. THE B-FORM

I now come to a discussion of <u>piecewise polynomial</u> functions, or <u>pp</u> functions for short. I have learned from the people in Computer-Aided Geometric Design that, in dealing with smooth pp functions on some triangulation, it is usually advantageous to write the polynomial pieces in barycentric-Bernstein-Bézier form, or B-form for short. This form relates polynomials to a given simplex. It is hard to appreciate the power and beauty of this form because, even with carefully chosen notation, it looks forbidding at first sight. Still, I want to point out its structure at least.

One starts with a $(d+1)$-subset V of $\mathbb{R}^d$ in general position and considers the barycentric coordinates with respect to it, i.e., the Lagrange polynomials for linear interpolation at V. The typical Lagrange polynomial $\xi_{\underline{v}}$ takes the value 1 at the vertex $\underline{v}$ and vanishes on the facet spanned by $V\backslash\underline{v}$. The B-form for $p\in\pi_k$ employs all possible products of k of these linear polynomials. Explicitly,

$$p := \sum_{|\underline{\alpha}|=k} B_{\underline{\alpha}}\, c(\underline{\alpha}) \tag{4.1}$$

with

$$B_{\underline{\alpha}}(\underline{x}) := |\underline{\alpha}|!\, [\![\underline{\xi}(\underline{x})]\!]^{\underline{\alpha}} = |\underline{\alpha}|! \prod_{\underline{v}\in V} [\![\xi_{\underline{v}}(\underline{x})]\!]^{\alpha(\underline{v})}. \tag{4.2}$$

Here

$$\underline{\xi}(\underline{x}) := \left(\xi_{\underline{v}}(\underline{x})\right)_{\underline{v}\in V} \tag{4.3}$$

is the $(d+1)$-vector containing the barycentric coordinates of $\underline{x}$ with respect to V.

Note that the vector $\underline{\xi}(\underline{x})$ and the multi-index $\underline{\alpha}$ appearing here are conveniently and appropriately indexed by the elements of V (rather than by the numbers $1,2,\cdots,d+1$ or the numbers $0,1,\cdots,d$, which would require an arbitrary indexing of the points in V).

The factor $|\underline{\alpha}|!$ in the definition of the Bernstein basis element $B_{\underline{\alpha}}$ is just right to make $(B_{\underline{\alpha}})_{|\underline{\alpha}|=k}$ a partition of unity. Indeed,

$$\sum_{|\underline{\alpha}|=k} B_{\underline{\alpha}}(\underline{x}) = k! \sum_{|\underline{\alpha}|=k} \prod_{\underline{v}\in V} [\![\xi_{\underline{v}}(\underline{x})]\!]^{\alpha(\underline{v})} = k! [\![\sum_{\underline{v}\in V} \xi_{\underline{v}}(\underline{x})]\!]^{k} = 1,$$

using the Multinomial Theorem (see (1.1)) and the fact that $\Sigma_{\underline{v}} \xi_{\underline{v}} = 1$. The numerical analyst will delight in the alternative formulation of the form,

$$p(\underline{x}) = \langle \underline{\xi}(\underline{x}), E \rangle^k c(\underline{0}) \tag{4.4}$$

which makes use of the <u>shift operator</u> E given by the rule

$$E^{\underline{\beta}} c(\underline{\alpha}) = c(\underline{\alpha} + \underline{\beta}) .$$

More explicitly,

$$\langle \underline{z}, E \rangle = \sum_{\underline{v} \in V} z(\underline{v}) E_{\underline{v}} ,$$

i.e.,

$$\langle \underline{z}, E \rangle c(\underline{\alpha}) = \sum_{\underline{v} \in V} z(\underline{v}) c(\underline{\alpha} + \underline{e}_{\underline{v}})$$

with $\underline{e}_{\underline{v}}$ the $\underline{v}$-unit vector, i.e., $\underline{e}_{\underline{v}}(\underline{w}) = \delta_{\underline{v}\underline{w}}$, $\underline{w} \in V$. This form provides a most convenient starting point for the derivation of efficient algorithms for the evaluation and differentiation of the B-form. For details, see, e.g., Farin (1985) and de Boor (1986).

5. SMOOTH PP FUNCTIONS

The B-form is well suited to pp work since its typical term $B_{\underline{\alpha}}$ vanishes $\alpha(\underline{v})$-fold on the facet spanned by $V \backslash \underline{v}$. This means that the form readily provides information about the behaviour of p at all the bounding faces of the simplex with vertex set V. This is being increasingly exploited in studying the algebraic structure of the space

$$\pi^{\rho}_{k,\Delta}$$

of pp functions of degree $\leqslant k$ on a given triangulation Δ whose pieces join together smoothly to provide a function all of whose derivatives of order $\leqslant \rho$ are continuous.

The problems being studied include: the dimension of such a space, a good basis for such a space, and the approximation power of such a space. For recent results, see Chui and his co-workers, and Schumaker. These results only deal with $d = 2$, and, even for this case, we know relatively little. For example, despite considerable efforts, we still do not know the dimension of the space of continuously differentiable piecewise cubic functions on an arbitrary triangulation in the plane. While we do know that this dimension depends on the quantitative details of the triangulation, we do not know exactly how.

As we understand these problems better and see some of their particular difficulties, we wonder whether $\pi^{\rho}_{k,\Delta}$ is really the right space to study. It now seems that it might be more appropriate to seek out appropriate subspaces, e.g., the subspace spanned by certain compactly supported smooth piecewise polynomials as was done already in Finite Elements. A particularly simple model is provided by approximation from a scale of pp functions.

6. APPROXIMATION FROM A SCALE

Associate with a given function space S the scale (S_h), with

$$S_h := \sigma_h S , \quad (\sigma_h f)(\underline{x}) := f(\underline{x}/h) ,$$

and define the approximation order of S to be

$$\max \{r: \text{ for any smooth } f \ \operatorname{dist}(f, S_h) = O(h^r)\} .$$

This order may well be 0, as it is for $S = \pi_k$. But if S contains functions whose support has diameter δ, then S_h contains functions with supports of diameter $h\delta$, and, for such S, one might hope to obtain closer approximations from S_h as $h \to 0$. Work with specific examples has suggested the following conjectures in case $S \subseteq \pi_{k,\Delta}$:

CONJECTURES: (i) The approximation order of S equals the approximation order of

$$S_{\text{loc}} := \operatorname{span}\{\varphi \in S: \operatorname{supp}\varphi \text{ compact}\} .$$

(ii) S has approximation order $\geqslant 1$ iff S contains a local partition of unity.

(iii) The approximation order is always realized by a good quasi-interpolant.

Here, a map Q into S is a good quasi-interpolant of order r in case it is a linear map which is stable in the sense that, for any f and any $\underline{x} \in G$,

$$|(Qf)(\underline{x})| \leqslant const \sup\{|f(\underline{y})| : \|\underline{y}-\underline{x}\| \leqslant R\}$$

with $const$ and $R < \infty$ independent of f or $\underline{x}$, and which reproduces polynomials of degree $< r$. For example, if Φ is a local and nonnegative partition of unity in S, i.e., $\sup_{\varphi\in\Phi}$ (diam supp φ) $< \infty$, $\varphi \geqslant 0$ for all $\varphi \in \Phi$, and $\sum_{\varphi\in\Phi} \varphi = 1$, then

$$f \mapsto \sum_{\varphi\in\Phi} \varphi f(\underline{\tau}_{\varphi})$$

is a good quasi-interpolant of order 1 (provided, e.g., that $\underline{\tau}_{\varphi} \in \operatorname{supp} \varphi$ for all $\varphi \in \Phi$).

This abstract model can be completely analysed in the very special case when S is spanned by the integer translates of one function φ, i.e.,

$$S := S_{\varphi} := \operatorname{span}\left(\varphi(\cdot-\underline{j})\right)_{\underline{j}\in\mathbb{Z}^d} = \left\{\sum_{\underline{j}\in\mathbb{Z}^d} \varphi(\cdot-\underline{j})\, c(\underline{j}) : c(\underline{j}) \in \mathbb{R}\right\}.$$

For this case, the three conjectures are verified; in particular, Strang and Fix (1973) prove that S has approximation order r iff $\pi_{<r} \subseteq S$.

Already for the slightly more general case when S is the span of integer translates of *several* compactly supported functions, the situation becomes more complicated. A characterization of the approximation order is not yet known for this case, but the somewhat stronger (and practically more interesting) concept of local approximation order can be characterized very simply (de Boor and Jia, 1985): S has local approximation order r iff there exists $\psi \in S_{\mathrm{loc}}$ such that S_{ψ} has approximation order r.

For the general case, even simple questions such as whether a pp space with positive approximation order must contain a compactly supported element have so far remained unanswered.

7. MULTIVARIATE B-SPLINES

The abstract theory of approximation from a scale has found new interest recently because of the advent of multivariate B-splines. These were introduced in 1976 in hopes that they would perform the same service in the study of multivariate smooth pp functions that the B-splines of Schoenberg and Curry provided so nicely for the theory of (univariate) splines.

In retrospect, well, in any case, the idea is simple enough. It involves a body $B \subset \mathbb{R}^n$ and the orthogonal projector $P: \mathbb{R}^n \to \mathbb{R}^d : \underline{x} \mapsto (x(1), \cdots, x(d))$. The map P is used to extend a function φ on $\mathbb{R}^d$ to the function

$$\varphi \circ P : \underline{u} \mapsto \varphi(P\underline{u})$$

on all of $\mathbb{R}^n$. The B-spline M_B is defined as the distribution on $\mathbb{R}^d$ which represents integration over B of the extended function. In formulae:

$$M_B: \varphi \mapsto \int_B \varphi \circ P \quad \text{for all} \quad \varphi \in C_0. \tag{7.1}$$

Here, $C_0 = C_0(\mathbb{R}^d)$ is the collection of all continuous functions on $\mathbb{R}^d$ with compact support. If PB (the projection of the body) is d-dimensional, this can also be written

$$M_B(\varphi) = \int_{\mathbb{R}^d} M_B \varphi = \int_{PB} d\underline{y}\ \varphi(\underline{y}) \int_{B \cap P^{-1}\underline{y}} 1 , \tag{7.2}$$

showing that $M_B(\underline{y}) = \text{vol}_{n-d}\,(B \cap P^{-1}\underline{y})$. This latter formula was the original definition, motivated by a geometric characterization of the (univariate) B-spline due to Curry and Schoenberg (1966) and illustrated, for $n = 3$, $d = 1$, in Figure 7.1.

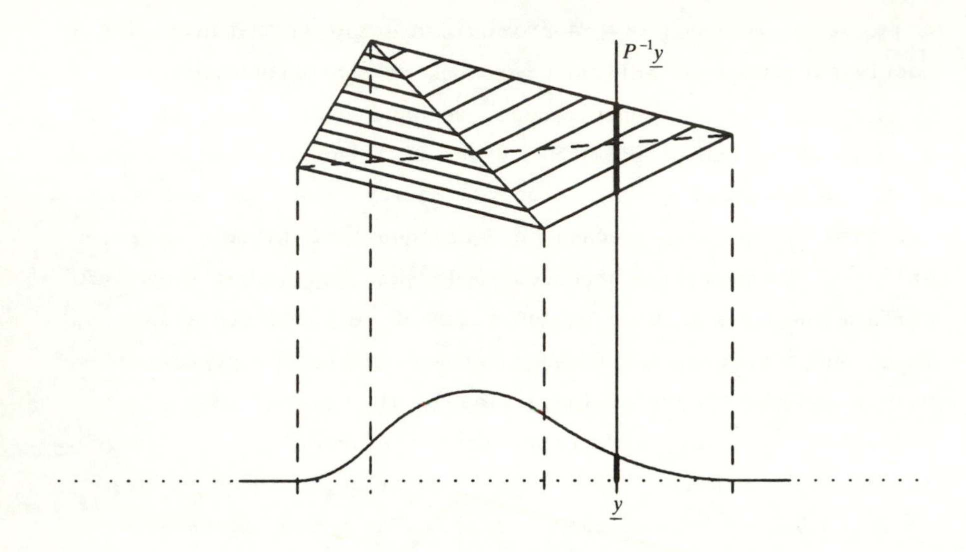

FIG. 7.1 The quadratic B-spline as the "shadow" of a 3-simplex

The value of the B-spline at a point $\underline{y}$ equals the $(n-d)$-dimensional volume of the intersection of the simplex with the hyperplane $P^{-1}\underline{y}$.

It is immediate that M_B has <u>compact support</u>. Further, if B is polyhedral with facets $\{B_i\}$, and if $\underline{z} \in \mathbb{R}^n$, then an application of Stokes' Formula shows that the directional derivative of M_B along $P\underline{z}$ is

$$D_{P\underline{z}} M_B = -\sum_i (\underline{z}, \underline{n}_i) M_{B_i}, \tag{7.3}$$

with $\underline{n}_i$ the outward unit normal to the facet B_i and M_{B_i} the B-spline that is the "shadow" of the "body" B_i. Repeated applications of this differentiation formula show that all derivatives of M_B of order $n-d+1$ must vanish identically away from the projections of the $(d-1)$-dimensional faces of B. Consequently,

$$M_B \in \pi^{\rho}_{n-d,\Delta},$$

with Δ the partition whose partition interfaces are the projections of $(d-1)$-dimensional faces of B, and where ρ is defined by the condition that $n-\rho-2$ equals the largest dimension of a face of B projected entirely into one of the partition interfaces. Thus, in the generic case, we have $\rho=n-d-1$, which is as large as it can possibly be, given that the polynomial degree of M_B is $n-d$.

This surprising smoothness is bought at a price. Since, for a *generic* partition Δ, $\pi_{k,\Delta}^{k-1}$ does not contain any locally supported functions, the partition for M_B must be quite special. Figure 7.2 shows such a partition for a bivariate quadratic simplex spline, i.e., a B-spline that is the "shadow" of a simplex.

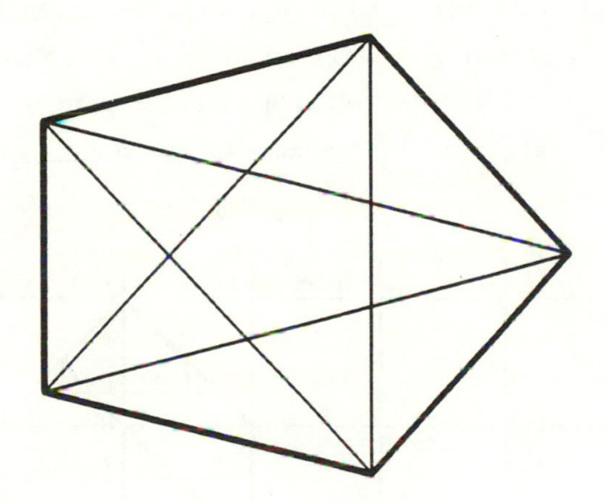

FIG. 7.2 The partition for a bivariate quadratic simplex spline

Thus we cannot expect to obtain B-splines for every partition. At best, we can find B-splines whose partition refines a given one. For the case of simplex splines, such a collection of B-splines of degree k can be constructed rich enough to provide a good quasi-interpolant of order $k+1$. There are even stable recurrence relations, found by Micchelli (1980), for their evaluation. But it seems that their use is computationally quite expensive (see, e.g., Grandine, 1986). It is therefore not likely that simplex splines will be used as a basis for a good subspace of a given smooth pp space of functions. Most likely, translates of a fixed B-spline will find practical employment.

There is a bit more hope for the box splines, i.e., the multivariate B-splines associated with the n-cube $B = [0,1]^n$. For their definition in terms of (7.1), one would allow P to be, more generally, a linear map. Then, with $\underline{\xi}_i := P\underline{e}_i$ the image of the i-th unit vector under P, the box spline can be characterized more explicitly by

$$\int_{\mathbb{R}^d} M(\cdot|\underline{\xi}_1,\cdots,\underline{\xi}_n)\varphi = \int_{[0,1]^n} \varphi\Big(\sum_i \underline{\xi}_i y(i)\Big)d\underline{y}, \quad \varphi \in C_0 .$$

By choosing the $\underline{\xi}_i$ from $\mathbb{Z}^d$ appropriately, the resulting partition can be made to conform to a regular grid; see Figure 7.3 for the supports of two very well known box splines, the Courant element ($d=2$, $n=3$) and the Zwart-Powell element ($d=2$, $n=4$). Further, their evaluation can be accomplished by *subdivision*.

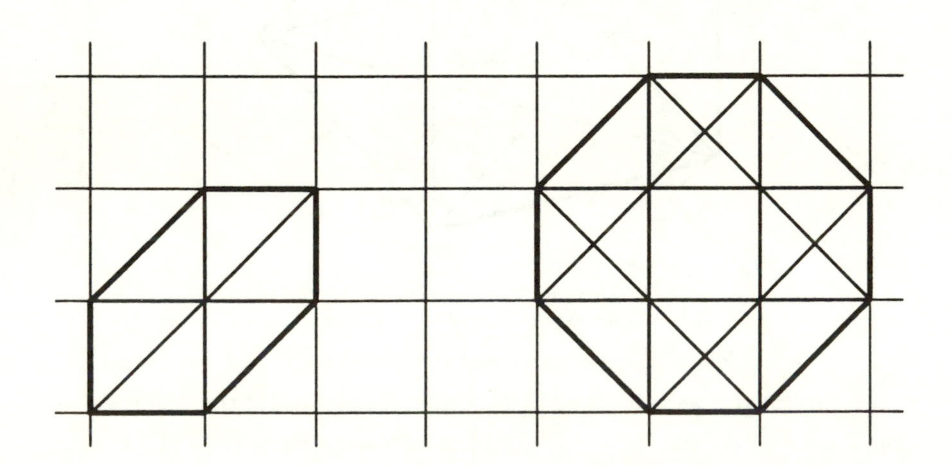

FIG. 7.3 The supports of a linear and a quadratic bivariate box spline

De Boor (1982) gives an introduction to multivariate B-splines. Dahmen and Micchelli (1983) provide a survey of the literature available by 1983 to which they heavily contributed.

Höllig (1986a) gives a more up-to-date introduction, and Höllig (1986b) summarizes what we know about box splines. In addition, I want to stress the beautiful, but more theoretical, developments to which Dahmen and Micchelli were led by their intensive study of box splines (see, e.g., Dahmen and Micchelli, 1985, 1986).

8. SUBDIVISION

I hate to finish on a pessimistic note. I therefore bring up a totally different approach to the generation or approximation of surfaces which comes from Computer Aided Design. I think that this technique has real promise for the generation of "smooth" surfaces which fit to given points in 3-space of more or less arbitrary combinatorial structure. It is at present being used to evaluate linear combinations of box splines (see, e.g., Höllig, 1986a, for the relevant references). But since this idea has not yet been thoroughly studied, I will discuss it only in its original context of curve generation.

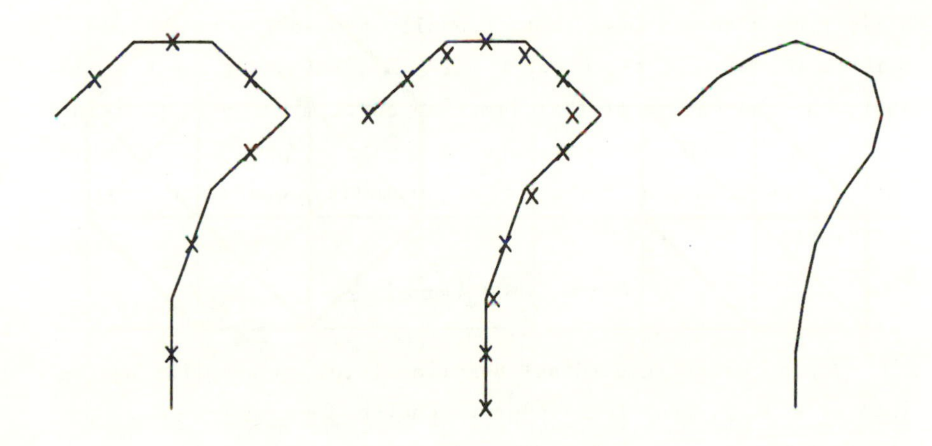

FIG. 8.1 The steps of the simple subdivision algorithm

Here is a very simple version of subdivision, which generalizes Chaikin's algorithm (Chaikin, 1974). Start off with points $\underline{a}_j$ in $\mathbb{R}^d$, where j runs over all of $\mathbb{Z}$, for simplicity. Think of these points as the vertices of a broken line. From the algorithm, one obtains a refined broken line in two steps. In the first step, one introduces the midpoints between neighbouring vertices as new vertices, thus roughly doubling the number of vertices:

$$\underline{b}_{2j} := \underline{a}_j, \quad \underline{b}_{2j+1} := (\underline{a}_j + \underline{a}_{j+1})/2.$$

In the second step, one obtains each vertex of the refined broken line as an average of three neighbouring vertices:

$$\underline{c}_j := \beta \underline{b}_{j-1} + \alpha \underline{b}_j + \gamma \underline{b}_{j+1}, \quad \text{with} \quad \beta + \alpha + \gamma = 1.$$

In fact, for a curve of higher "smoothness", one would repeat this averaging step one or more times, but I will stick with this simple model. Repetition quickly leads to a broken line which, for plotting purposes, is indistinguishable from the limiting curve.

Of course, it is not at all clear *a priori* that there is a limiting curve, though that is easily proved for reasonable choices of the weights, e.g., for $\alpha, \beta, \gamma \geqslant 0$. Nor is it clear just what the nature of that limiting curve might be. Chaikin's algorithm corresponds to the choice $\beta = 0$, $\alpha = \gamma = 1/2$. For this choice, the limiting curve is a parametric quadratic spline curve, viz. the curve

$$t \longmapsto \sum_{\mathbb{Z}} M_3(t-j)\, \underline{a}_j$$

with M_3 a quadratic cardinal B-spline (i.e., a B-spline having integer knots). For the symmetric choice $\beta = \gamma = (1-\alpha)/2$, the limiting curve is a parametric quadratic, resp. cubic, spline curve in case $\alpha = 0$, resp. $1/2$, but for any other choice, the limiting curve appears to be something unmentionable in standard terms, though this is not apparent from the curves themselves;

see, e.g., Figure 8.2.

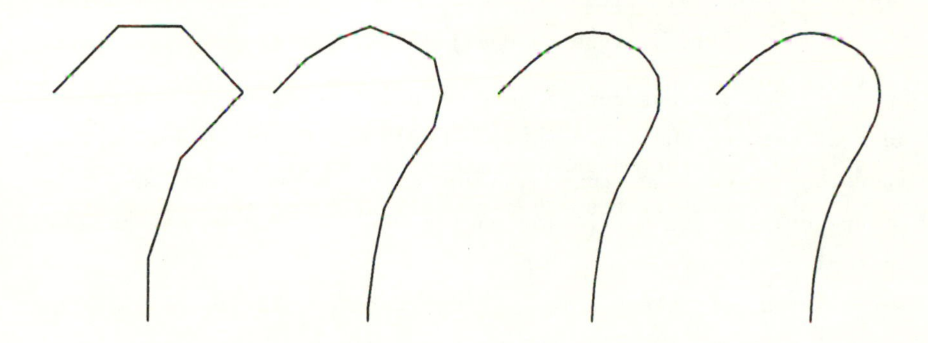

FIG. 8.2 Curve iterates with $\alpha = 1/4$, $\beta = \gamma$

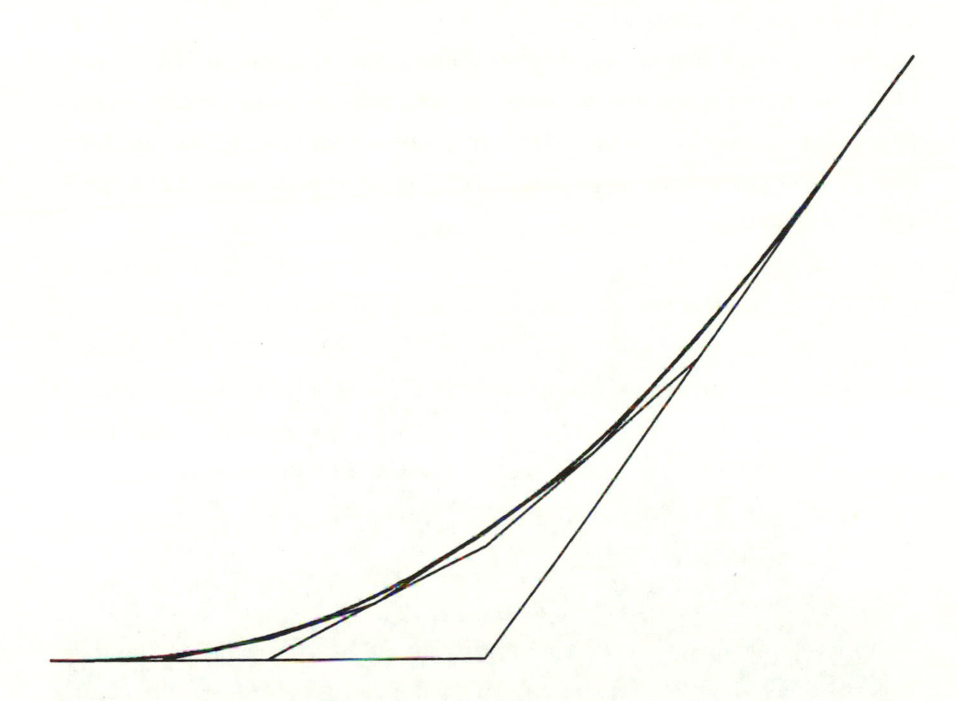

FIG. 8.3 Iterates and limiting curve for one-break broken line

Since the limiting curve depends linearly on the initial data $\underline{a}_j$, it is sufficient for the complete analysis of the process to consider a broken line with just one actual break, e.g., the data

$$\underline{a}_j := (j, j_+), \quad j \in \mathbb{Z}, \tag{8.1}$$

in $\mathbb{R}^2$. The various curves generated will differ only on the segment between the two points $(-1,0)$ and $(1,1)$. Some of the iterates as well as the limiting curve segment are shown in Figure 8.3 for the particular choice $\alpha = 1/4$, and we see nothing unusual.

It is only when we look at derivatives that we realize that something is amiss. For our particular curve, the first component is just $t \mapsto t$, so it makes sense to consider the derivatives of the second component, $t \mapsto y(t)$ say, as an indication of the smoothness of the curve. Since we do not have a formula for $y(t)$, we cannot compute derivatives. But since the iteration is so simple, we can compute $y(t)$ for as many binary fractions t as we care to. That done, we can then compute second divided differences, and these tell the story; see Figure 8.4: the second derivative of y appears to be a fractal.

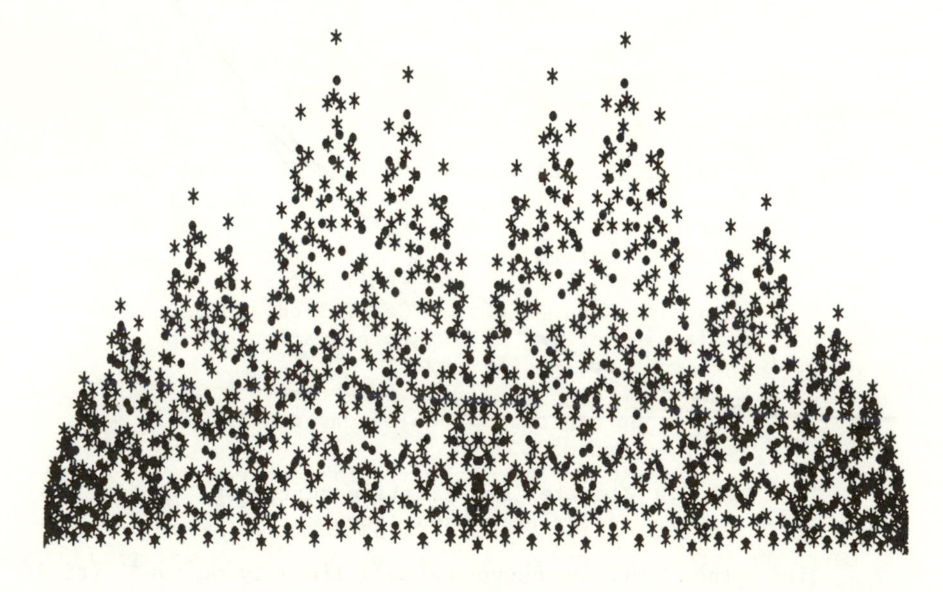

FIG. 8.4 Wisconsin Winter
Second divided differences $[t-h, t, t+h]y$ plotted against t, for $h = 2^{-8}$ (dots), and $h = 2^{-9}$ (stars)

Traditional Approximation Theory views such curves with horror. There is even the seemingly practical objection that it would be difficult to machine curves with such a 'bad' second derivative. But I do not know of experiments which have established such a difficulty nor is it clear *a priori* that there should be any difficulty. On the other hand, the generation of such curves to plotting or machining accuracy is so swift, and their flexibility so great, that this technique is worth exploring in detail. It is this apparent local flexibility that makes subdivision techniques a promising tool for the generation of shape-controlled surfaces of arbitrary combinatorial structure.

For literature, see Catmull and Clark (1978), Doo (1978), and Doo and Sabin (1978). I am indebted to Professor Wanner for the surprising references to work by de Rham (1947, 1953, 1956, 1957, 1959) who, many years ago and, apparently, for pedagogical reasons, investigated a related subdivision algorithm for curves. Professor Wanner referred to it imaginatively as 'the woodcarver's algorithm'.

REFERENCES

de Boor, C. (1982), "Topics in multivariate approximation theory", in *Topics in Numerical Analysis, Lecture Notes in Mathematics 965*, ed. Turner, P.R., Springer-Verlag (Berlin), pp.39-78.

de Boor, C. (1986), "B-form basics", in *Geometric Modeling*, ed. Farin, G., *SIAM Publications* (Philadelphia).

de Boor, C. and Jia, R.-q. (1985), "Controlled approximation and a characterization of the local approximation order", *Proc. Amer. Math. Soc.*, *95*, pp.547-553.

Catmull, E.E. and Clark, J.H. (1978), "Recursively generated B-spline surfaces on arbitrary topological meshes", *Computer Aided Design*, *10*, pp.350-355.

Chaikin, G.M. (1974), "An algorithm for high speed curve generation", *Computer Graphics and Image Processing*, *3*, pp.346-349.

Chung, K.C. and Yao, T.H. (1977), "On lattices admitting unique Lagrange interpolations", *SIAM J. Numer. Anal.*, *14*, pp.735-743.

Curry, H.B. and Schoenberg, I.J. (1966), "Pólya frequency functions IV. The fundamental spline functions and their limits", *J. Analyse Math.*, *17*, pp.71-107.

Dahmen, W. and Micchelli, C.A. (1983), "Recent progress in multivariate splines", in *Approximation Theory IV*, eds. Chui, C.K., Schumaker, L.L. and Ward, J.D., Academic Press (New York), pp.27-121.

Dahmen, W. and Micchelli, C.A. (1985), "On the solution of certain systems of partial difference equations and linear dependence of translates of box splines", *Trans. Amer. Math. Soc.*, *292*, pp.305-320.

Dahmen, W. and Micchelli, C.A. (1986), "On the number of solutions to linear diophantine equations and multivariate splines", manuscript.

Doo, D.W.H. (1978), "A subdivision algorithm for smoothing down irregularly shaped polyhedrons", in *Proceedings: Interactive Techniques in Computer Aided Design*, IEEE Computer Society (Los Alamitos), pp.157-165.

Doo, D. and Sabin, M.A. (1978), "Behaviour of recursive division surfaces near extraordinary points", *Computer Aided Design*, *10*, pp.356-360.

Farin, G. (1985), "Triangular Bernstein-Bézier patches", CAGD Report 11-1-85, Department of Mathematics, University of Utah.

von Golitschek, M. (1984), "Shortest path algorithms for the approximation by nomographic functions", in *Approximation Theory and Functional Analysis*, *ISNM 65*, eds. Butzer, P.L., Stens, R.L. and Sz.-Nagy, B., Birkhäuser Verlag (Basel), pp.281-301.

Golomb, M. (1959), "Approximation by functions of fewer variables", in *On Numerical Approximation*, ed. Langer, R.E., The University of Wisconsin Press (Madison), pp.275-327.

Grandine, T. (1986), "The computational cost of simplex spline functions", MRC Report 2926, University of Wisconsin, Madison.

Hakopian, H. (1983), "Integral remainder formula of the tensor product interpolation", *Bull. Polish Acad. Sci. Math.*, *31*, pp.267-272.

Höllig, K. (1986a), "Box splines", in *Approximation Theory V*, eds. Chui, C.K., Schumaker, L.L. and Ward, J.D., Academic Press (New York).

Höllig, K. (1986b), "Multivariate splines", in *Approximation Theory*, *Proc. Sympos. Appl. Math.*, *36*, ed. de Boor, C., American Mathematical Society (Providence, R.I.), pp.103-127.

Kergin, P. (1978), *Interpolation of C^k Functions*, Ph.D. Thesis, University of Toronto, Canada: published as "A natural interpolation of C^k functions", *J. Approx. Theory*, *29* (1980), pp.278-293.

Kolmogorov, A.N. (1957), "On the representation of continuous functions of many variables by superposition of continuous functions of one variable and addition", *Dokl. Akad. Nauk. SSSR*, *114*, pp.953-956.

Light, W.A. and Cheney, E.W. (1986), *Approximation Theory in Tensor Product Spaces*, *Lecture Notes in Mathematics*, *1169*, Springer-Verlag (Berlin).

Lorentz, G.G. (1962), "Metric entropy, widths, and superpositions of functions", *Amer. Math. Monthly*, *69*, pp.469-485.

Lorentz, G.G. (1966), *Approximation of Functions*, Holt, Rinehart and Winston (New York).

Meinguet, J. (1979), "Multivariate interpolation at arbitrary points made simple", *Z. Angew. Math. Phys.*, *30*, pp.292-304.

Micchelli, C.A. (1980), "A constructive approach to Kergin interpolation in $\mathbb{R}^k$: multivariate B-splines and Lagrange interpolation", *Rocky Mountain J. Math.*, *10*, pp.485-497.

Micchelli, C.A. (1986), "Algebraic aspects of interpolation", in *Approximation Theory*, *Proc. Sympos. Appl. Math.*, *36*, ed. de Boor, C., American Mathematical Society (Providence, R.I.), pp.81-102.

de Rham, G. (1947), "Un peu de mathématique à propos d'une courbe plane", *Elemente der Math.*, *2*, pp.73-76 and 89-97 (*Collected Works*, pp.678-689).

de Rham, G. (1953), "Sur certaines équations fonctionnelles", in *Ecole Polytechnique de l'Université de Lausanne*, *Centenaire, 1853-1953*, Ecole Polytechnique (Lausanne), pp.95-97 (*Collected Works*, pp.690-695).

de Rham, G. (1956), "Sur une courbe plane", *J. Math. Pures. Appl.*, *35*, pp.25-42 (*Collected Works*, pp.696-713).

de Rham, G. (1957), "Sur quelques courbes définies par des équations fonctionnelles", *Univ. e Politec. Torino Rend. Sem. Mat.*, *16*, pp.101-113 (*Collected Works*, pp.714-727).

de Rham, G. (1959), "Sur les courbes limites de polygones obtenus par trisection", *Enseignement Math.*, *5*, pp.29-43 (*Collected Works*, pp.728-743).

Strang, G. and Fix, G. (1973), "A Fourier analysis of the finite element variational method", in *Constructive Aspects of Functional Analysis*, ed. Geymonat, G., C.I.M.E., pp.793-840.

Whitney, H. (1957), "On functions with bounded n-th differences", *J. Math. Pures Appl.*, *36*, pp.67-95.

Carl de Boor
Mathematics Research Center
610 Walnut Street
Madison
Wisconsin 53705
U.S.A.

5 Data Approximation by Splines in One and Two Independent Variables

M. G. COX

ABSTRACT

This paper reviews developments in data approximation by univariate and bivariate splines since the mid-1970s. The emphasis is algorithmic and, because of the degree of generality afforded by their use, gives particular attention to B-splines as a basis for splines. Two important advances are highlighted: (i) the use and advantages of methods other than the traditional minimum norm solution in determining least squares bivariate spline approximants to scattered data sets, (ii) the value of curved knot lines, in terms of which tensor product splines have the potential to handle a much wider range of practical problems. Other issues discussed include approximation norms, constrained approximation and the determination of statistical information associated with least squares spline approximants.

I. INTRODUCTION

This paper reviews methods for the solution of data approximation problems in which univariate and bivariate splines are used as approximating functions. Emphasis is placed on algorithms that in the last ten years have proved to be of practical value, and mention is made of concepts that are likely to lead to useful new algorithms. All the methods discussed make use of a B-spline basis to represent the spline. Such a basis permits splines of any order, uniform and nonuniform knot sets, simple and multiple knots, and various orders of continuity. Moreover, it provides a ready ability to incorporate boundary information, generally good numerical conditioning (at least for low orders of spline), and straightforward extensions to periodic and parametric splines. Univariate splines are represented in terms of a univariate B-spline basis and bivariate splines in terms of a tensor product of two univariate B-spline bases.

Some knowledge of B-splines is assumed. See Cox (1982b) for basic concepts and notation and de Boor (1978) for a valuable guide to the use of B-splines. Important properties of B-splines are as follows. Each B-spline is a "bell-shaped" function and has compact support: a B-spline of order n (degree $n-1$) is nonzero only over an interval spanning $n+1$ knots. It is strictly positive within the open interval spanned by these knots. As a result of this compact support, the system of linear algebraic equations defining the coefficients of a spline interpolant or approximant has special structure. In the nonperiodic case the associated matrix has band form and in the periodic case it is a bordered band matrix. As a consequence of their usual normalization the B-splines sum to unity for each value of their argument in a defining interval. The B-spline coefficients in a spline interpolant or approximant s tend to mimic the function values from which s was obtained. The value of s at a point is a convex combination of the B-spline coefficients; all but n (n^2 in the bivariate case) of the multipliers in the combination are zero.

The B-splines satisfy a three-term recurrence relation (Cox, 1972; de Boor, 1972) which is unconditionally stable regardless of the distribution or multiplicity of the knots. The relative errors of the B-splines evaluated by recurrence in floating-point arithmetic satisfy a bound which grows only linearly with n (Cox, 1972). The B-spline derivatives also satisfy a three-term recurrence relation (de Boor, 1972). At the endpoints of an interval (when suitable exterior knots are chosen — see Section 2.1) all derivatives computed by recurrence also satisfy a similar error bound, as do derivatives of order $n-2$ at the knots (Cox, 1975b). These results permit stable implementations of boundary conditions in spline approximation (Cox, 1978) and enable inequality constraints in shape-preserving spline approximation to be imposed (Cox and Jones, 1987).

Spline approximation methods are now employed in such a wide variety of application areas that it is impossible to enumerate them here. At NPL alone they have been used in problems ranging from the design of musical instruments to the analysis of plant growth to the representation of aircraft flight profiles. Spline approximation overlaps with several other mathematical and computational areas, both in the sense that spline approximation methods are used in these areas and in that methods from other numerical analysis disciplines are valuable in solving spline approximation problems. Cases in the first category are finite element calculations (in which the set of finite elements essentially forms a spline or piecewise polynomial) and computer aided design (CAD) (particularly as a means of modelling engineering components). Included in the second category are spline interpolation (a useful approach to obtaining spline approximants, although it is normally recommended only when the prescribed data is accurate), linear algebra (of key importance in analysing and solving the linear equations to which many spline approximation problems can be reduced) and mathematical optimization (in its theoretical concepts that are valuable in characterizing splines that in some sense are optimal, and in providing algorithms for determining good approximants, e.g. by enabling suitable knot sets to be computed).

2. B-SPLINES FOR INTERPOLATION AND APPROXIMATION

This section provides background material for the subsequent discussion.

2.1 *Exterior Knots*

In order to define a full set of B-spline basis functions in a univariate spline approximation problem it is necessary to introduce additional knots that are "exterior" to the interval of interest. The choice of these knots does not affect s within this interval, but it influences the numerical conditioning of the problem and the values of the first few and the last few B-spline coefficients in the B-spline representation of s. Although there are circumstances in which it is important to make other choices, the use of coincident end knots (i.e. additional knots placed at the endpoints of the interval) is often advantageous. For these knots the spectral condition numbers of the matrices associated with a wide range of practical least-squares spline approximation problems were found to be significantly smaller (Cox, 1975b) than those for other (reasonable) choices. Moreover, Kozak (1980) has established theoretically that the condition number, with respect to the maximum norm, associated with the matrices occurring in spline interpolation problems is minimal for this choice. A further desirable consequence is that the imposition of boundary conditions is facilitated (see Section 6.1).

2.2 *Conditions for Solution*

Given a set of m distinct data points $\{(x_i, f_i)\}_1^m$ and a prescribed set of N (interior) knots $\{\lambda_j\}_1^N$ (where N is assumed to be chosen such that the dimension $q\,(=N+n)$ of the resulting spline space equals m), the univariate spline interpolation problem is to determine the coefficients $\{c_j\}_1^q$ in the B-spline representation of s such that

$$s(x_i) = f_i, \qquad i = 1, \cdots, m. \tag{2.2.1}$$

In matrix form, (2.2.1) may be expressed as $A\underline{c} = \underline{f}$, where $a_{i,j}$ is the value of the j-th B-spline at x_i, and $\underline{c}$ and $\underline{f}$ are respectively vectors holding $\{c_j\}$ and $\{f_j\}$. Schoenberg and Whitney (1953) provided necessary and sufficient conditions for the existence and uniqueness of s: they are that a unique data point whose abscissa lies strictly within the support of the B-spline can be associated with each B-spline in the basis. These conditions can rapidly be tested within a spline interpolation algorithm before attempting to form and solve the system $A\underline{c} = \underline{f}$.

In the case of univariate least-squares spline approximation, knots are chosen such that their number N results in $q \leqslant m'$, the number of distinct data abscissae. It is required to calculate $\underline{c}$ such that the system (2.2.1), which is now overdetermined, is satisfied in the least-squares sense. To determine whether the solution is unique, the Schoenberg-Whitney criterion may be extended in the following way. The solution is unique if and only if there exists a subset of the distinct data abscissae such that these conditions hold for this subset. Cox and Hayes (1973) provide an algorithm for determining whether such a subset exists.

For bivariate spline interpolation and least-squares problems in which data is defined on a rectangular mesh, the above conditions extend in a natural way through the concept of separability (see Sections 2.5 and 2.6). For problems in which the data is scattered irregularly there is no analogue of the Schoenberg-Whitney conditions. Indeed, tensor-product spline interpolants cannot normally be obtained for such data. In the least-squares case, it is important to use linear algebraic methods that are able to detect nonuniqueness. Nonuniqueness is manifested as rank deficiency in the rectangular matrix A (see Section 3).

2.3 *Univariate Spline Interpolation*

As a consequence of the compact support of the B-splines, the matrix A is banded, of bandwidth n. Moreover, A is totally positive, i.e. all its minors are nonnegative (de Boor and DeVore, 1985): a linear system with such a coefficient matrix can be solved quite safely without pivoting (de Boor and Pinkus, 1977). An algorithm for solving this system of equations, which exploits the structure of A, is outlined by Cox (1982b). Because pivoting is unnecessary the elimination can be carried out *in situ*, and thus no additional array space to accommodate fill-in is required.

A backward error analysis (Cox, 1975a) of the complete problem (i.e. the formation and solution of the defining equations) shows that the computed solution represents an exact interpolant for a neighbouring point set. The bound derived for the departure of this set from the given point set is independent of the condition number of A.

2.4 *Univariate Least-Squares Spline Approximation*

In the case of univariate least-squares spline approximation, the matrix A is rectangular, of dimensions $m \times q$, again of bandwidth n. The corresponding overdetermined system can be solved as follows. First, form the normal equations $A^T A \underline{c} = A^T \underline{f}$ ($A^T A$ has bandwidth $2n-1$). Second, determine the Cholesky decomposition,

$$A^T A = R^T R \ , \qquad (2.4.1)$$

where R, the Cholesky factor, is a band upper triangular matrix of order q and bandwidth n. Finally, solve the resulting lower and upper triangular systems, $R^T \underline{\theta} = A^T \underline{f}$ and $R\underline{c} = \underline{\theta}$. An algorithm based on this approach is given by de Boor (1978). Alternatively, and more reliably, solution can be effected by orthogonal triangularization. With this approach orthogonal transformations are applied to A to reduce it to triangular form R (which is mathematically equal to the Cholesky factor provided in the normal-equations approach). The same transformations are applied to $\underline{f}$ to produce a vector $\underline{\theta}$ (again mathematically the same vector as before), and the upper triangular system $R\underline{c} = \underline{\theta}$ is then solved. Again, full account can be taken of the structure of A. Details are given by Cox (1981b). The method described there is based on the use of Givens rotations to effect the orthogonal transformations. Reid (1967) gives a method which uses Householder transformations. The former approach permits successive rows of A and $\underline{f}$ to be formed and processed, and requires minimal memory. Since their introduction (Gentleman, 1973, 1974) for general matrices and their specialization to band (Cox and Hayes, 1973; Cox, 1981b) and bordered band systems (Cox, 1981b), Givens-based methods seem to be becoming "standard" for a wide class of linear least-squares problems in general and least-squares spline approximation problems in particular. A backward error analysis (Cox, 1975b) shows that the computed vector $\underset{\sim}{c}$ corresponds to a spline approximant that is exact for a neighbouring point set. This time the bound is (inevitably) dependent on the condition number of A.

2.5 *Bivariate Spline Interpolation*

Let the bivariate spline $s(x,y)$ based on a given set of N_x interior x-knots and a given set of N_y interior y-knots be of order n_x in x and order n_y in y. Let $q_x = N_x + n_x$, $q_y = N_y + n_y$ and $q = q_x q_y$. Given data values

$$F = \{f_{i,j}: i = 1,\cdots,m_x;\ j = 1,\cdots,m_y\}$$

specified at the vertices of an $m_x \times m_y$ rectangular mesh

$$\{x,y: x \in \{x_1,\cdots,x_{m_x}\},\ y \in \{y_1,\cdots,y_{m_y}\}\},$$

the problem is to construct a bivariate spline interpolant $s(x,y)$ such that

$$s(x_i,y_j) = f_{i,j},\quad i = 1,\cdots,m_x,\quad j = 1,\cdots,m_y.$$

If s is expressed in terms of a basis formed from the tensor product of a univariate B-spline basis in x and a univariate B-spline basis in y, and if $q_x = m_x$ and $q_y = m_y$, a system of $m_x m_y$ linear algebraic equations can be constructed in the q coefficients $C = \{c_{i,j}\}$ of the basis functions, which defines the coefficients if an extension of the Schoenberg-Whitney condition is satisfied. However, the work involved in solving this system is enormously reduced if advantage is taken of separability. Thus the defining equations can be written as the matrix equation $A_x C A_y^T = F$, and therefore solved in the following two stages, through an intermediate $m_x \times m_y$ matrix B:

$$A_x B = F,\quad A_y C^T = B^T. \tag{2.5.1}$$

The matrix A_x is of order m_x, of bandwidth n_x, and contains the B-spline basis functions in x evaluated at the x-meshlines, and A_y is the corresponding $m_y \times m_y$ matrix for y. Thus each of the above two stages corresponds to the solution of a univariate spline interpolation problem having m_y or m_x "right-hand sides". De Boor (1979) and Cox (1980) describe efficient methods for implementing such "tensor-product" schemes. The use of these methods, if account is taken of the band structure of A_x and A_y, gives an execution time that for a fixed order of spline is $O(m_x m_y)$.

2.6 *Bivariate Least-Squares Spline Approximation: Data on Mesh*

We distinguish between two cases: data on a mesh and arbitrarily-scattered data. In the first case the problem separates, as for interpolation. The problem is a special case of least-squares approximation by a tensor product of two sets of general basis functions, which has been analysed by Greville (1961). It reduces to the least-squares solutions of overdetermined systems of the form (2.5.1), where now A_x is $m_x \times q_x$ of bandwidth n_x, A_y is $m_y \times q_y$ of bandwidth n_y, C is $q_x \times q_y$, and F is $m_x \times m_y$.

As for interpolation, for fixed orders of spline these systems can be solved in $O(m_x m_y)$ time. If no account is taken of structure, $O(q m_x m_y)$ time is required. For a problem of quite modest size the corresponding computation times could be 10 seconds and 1 hour.

2.7 *Bivariate Least-Squares Spline Approximation: Scattered Data*

There appears to be greater concentration in the literature on the development of algorithms for scattered data interpolation than for approximation. The reason may be due either to the fact that it is easier to characterize the solution to an interpolation problem (for example, no decisions have to be made in respect of choice of approximation norm or value of a smoothing parameter), or to a belief that interpolation problems occur more frequently in practice. The former reason is indeed valid but the second, we maintain, is untenable. In our opinion most sets of scattered data are the result of measurement and therefore subject to error; hence approximation methods are often more appropriate. In any particular circumstance, unless there are overriding reasons to the contrary, an approximant should be sought that reproduces the data to the accuracy warranted by the data. (Similar comments apply also to univariate interpolation.)

When the data is scattered the problem does not separate as in the case of data on a mesh. The resulting formulation is a relatively large matrix problem. However, the matrix A is still banded, with bandwidth $n_y q_x - q_x + n_x$ (see Cox, 1982a), but there are only $n_x n_y$ nonzero elements in each row. It is not straightforward to take significant account of this fine structure. In forming the Cholesky factor R of $A^T A$, most of the structure within the band is lost. Some examples are given by Cox (1982a).

For some of these problems iterative methods, such as the Lanczos-based algorithm LSQR of Paige and Saunders (1982), may be competitive. For a class of model problems, Cox (1982a) shows that iterative methods have smaller memory requirements for bivariate problems and are more efficient in both space and time for higher-dimensional problems. An important consideration, however, in many fields of application is that iterative methods are unable readily to provide statistical information (see Section 7) associated with the solution (although LSQR delivers crude estimates of the variances of the coefficients).

3. RANK DEFICIENT PROBLEMS

3.1 *Sources of Rank Deficiency*

In univariate spline approximation problems, satisfaction of the Schoenberg-Whitney conditions ensures that the resulting matrix A has full rank and therefore that the spline approximant is unique. However, this result applies to exact arithmetic. It is straightforward to construct (not particularly extreme) counterexamples (see Cox, 1982b) in which floating-point underflow causes a mathematically full-rank matrix to be rank deficient on the computer. Thus the methods used to solve the defining equations should be designed to cope satisfactorily with such circumstances or, in practice, with ill-conditioned full-rank matrices. The problem is more critical in two independent variables particularly because, as stated, there is no counterpart of the Schoenberg-Whitney conditions, except for special positions of the data points. Moreover, rank deficient matrices arise very frequently in the bivariate case; indeed it may be argued that they are the usual case. Consider a scatter of data points in the (x,y)-plane. A typical approach to determining a bivariate spline approximant based on these points would be as follows. First, contain the points by a rectangle with sides parallel to the axes. Lay down a rectangular knot mesh, taking into account any prior knowledge of the behaviour of the underlying function. Obtain a bivariate spline approximant of some order based on this knot mesh. In the light of the results obtained, modify the mesh as appropriate and repeat the computation. In practice scattered data sets often form a roughly elliptical cloud of points in the (x,y)-plane. Thus, unless the knot mesh is sufficiently coarse, there will be panels within the mesh which contain no or very few points. Corner and edge panels are likely candidates. A corner panel containing no data gives rise to a zero column in A, and several adjacent panels in which the local data density is critically low compared with the knot density gives rise to a subset of columns of A which are linearly dependent.

3.2 *Choice of Solution*

If A is rank deficient, the least-squares problem has an infinity of solutions and it is therefore necessary to select a particular one. The minimum norm solution is often suggested in linear algebra texts. This solution is unique and minimizes the 2-norm of the solution vector. An algorithm for obtaining it in the case of bicubic spline approximation (i.e. $n_x = n_y = 4$) to scattered data is given by Hayes and Halliday (1974). It uses Householder transformations with column interchanges to permit reliable determination of the rank of A and thence to compute the minimum norm solution. An alternative solution is the basic or reduced-parameter solution (Rosen, 1964). The number of non-zero components of this solution is at most the rank of A (k say).

In order to compute the minimum norm solution, or indeed any particular solution to a rank deficient problem, either the singular value decomposition or orthogonal transformations with column pivoting (as in the Hayes and Halliday, 1974, approach) should be used for stability. Unfortunately, both of these methods destroy the structure of A. Therefore, in order to preserve structure it is necessary to use an algorithm which sacrifices the guarantee of stability by avoiding column interchanges. Such an algorithm is potentially less stable, but we believe that serious instabilities will arise (for a reasonable implementation) only in the case of pathologically constructed examples. Our belief is based on computational experience with the following strategy, which was first implemented in Mark 5 of the NAG Library in 1975. A similar approach has been advocated by Heath (1982).

3.3 *Outline of Algorithm*

First, A is reduced to triangular form R by pre-multiplying by orthogonal transformations. Rank deficiency in A will be manifested mathematically as $q-k$ zero rows of $(R, \underline{\theta})$. For example, the transformations may give the structures

```
XXXXXXXXXX    X
 XXXXXXXXX    X
  XXXXXXXX    X
   XXXXXXX    X
    XXXXXX    X
     00000    0
      XXXX    X
       000    0
        XX    X
         X    X

     R        θ
```

where X denotes a nonzero element. Consider now the solution of the system $R\underline{c} = \underline{\theta}$. Backsubstitution can be carried out in the usual way, except that any element of $\underline{c}$ that corresponds to a zero diagonal element of R will be indeterminate. Specific choices for such elements will yield a particular least-squares solution. The choice of zero for these values yields the basic solution. The minimum norm solution may be obtained by forming and solving a further, smaller, least-squares problem in $q-k$ variables (Heath, 1982). Other solutions can be determined by updating $(R,\underline{\theta})$ to take account of appropriately constructed additional rows — termed resolving constraints (Gentleman, 1973).

In finite-precision arithmetic a similar strategy can be adopted, except that nonzero values may appear in rows of R where mathematically only zeros would appear. The following approach may be used. Examine each row of R in turn, starting with the first. If its diagonal element is smaller in magnitude than a specified threshold (which may vary from column to column), replace the element by zero and carry out a sequence of plane rotations between this row and subsequent rows to annihilate the remaining elements in the row. The resulting modified R is now in a form to which the above approach can be applied.

3.4 *Disadvantage of the Minimum Norm and Basic Solutions*

The basic and minimum norm solutions share an unfortunate property (except in the full-rank case) in respect of determining spline approximants. Suppose an approximant $s(x,y)$ to data

$\{(x_i, y_i, f_i)\}_1^m$ is computed. Then, unless the corresponding matrix A has full rank, the approximating spline for the data $\{(x_i, y_i, f_i + K)\}_1^m$ will not be $s(x,y)+K$, where K is any nonzero constant. To demonstrate this, consider a set of data and choice of knotlines such that every panel contains a large number of points except for one corner panel which is empty. The empty panel results in one of the bivariate basis functions having no data within its support and thus the corresponding column of A will contain only zeros (cf Section 3.1). The resulting R will have rank $q-1$, with one zero row. Upon backsubstitution, assuming without loss of generality that the zero row is the first row of R, all elements of C but the last-determined one will be uniquely defined. The element corresponding to the zero row is in general arbitrary, but for both the minimum norm and basic solutions the value of $c_{1,1}$ will necessarily be zero. The addition of K to all the function values will increase all $c_{i,j}$ by K except for $c_{1,1}$ which will be unchanged. In particular, both surfaces so obtained will take the value zero at the corresponding vertex of the rectangle, but over all panels but the corner one they will differ by the constant K. In other words, translating the given function values by a constant distance does not modify the approximant by the same amount throughout the complete rectangle. Thus the basic and minimum norm solutions are not geometrically invariant. We maintain that such invariance is important, because the shape or form of the solution should not depend on the origin of the axis system, which is artificial, but only on the intrinsic properties of the function represented by the data.

3.5 *Geometrically Invariant Solution*

A different form of solution is proposed by Cox (1982b). To determine it a special set of resolving constraints is appended to the system $R\underline{c} = \underline{\theta}$. These constraints are chosen to be minimal in number in order to define a unique solution and, moreover, to furnish a geometrically invariant solution. Each constraint can be described in terms of a discrete Laplacian-like operator, here represented in its simplest form when the spacing between knots is uniform:

$$c_{i-1,j} + c_{i+1,j} - 4c_{i,j} + c_{i,j-1} + c_{i,j+1} = 0.$$

The way in which the constraint is applied is as follows. Consider the two-dimensional array C expressed as a long vector. Each column of R corresponds to an element of this vector. If a diagonal element of R is zero, a constraint is constructed with the operator "centred" on the corresponding element of C, with the remaining elements chosen to correspond in position with the four neighbouring elements of C. (The constraint needs to be modified, however, when the relevant element lies on the edge of the C-array.) The resulting solution has the required geometric invariance property and moreover is also invariant with respect to a planar shift of the values of the dependent variable (Cox, 1982b). Algorithms based on this approach have proved useful in providing sensible solutions in cases where the data contains one or more "holes": the resulting approximant provides a reasonable "bridge" across the sparse region, following the trend of neighbouring data, whereas the basic and minimum norm solutions may introduce a spurious depression or peak there.

Figure 3.5.1a shows a smooth "test" function,

$$f(x,y) = \exp(-x^2 - 2y^2)$$

defined over $S = [0,1] \times [0,1]$. Data was generated by sampling f at 1000 points distributed randomly over $S - O$, where O is the circle with radius 1/4 and centre (1/2, 1/2) (see Fig. 3.5.1b). Least-squares bicubic spline approximants based on 9 uniformly spaced x-knots and 9 uniformly spaced y-knots were computed using both the minimum norm (Fig. 3.5.1c) and the geometrically invariant criterion (Fig. 3.5.1d). The data was then modified by subtracting 2 from every function value, and approximants based on both criteria were again computed. Figure 3.5.1e shows the new minimum norm solution. The geometrically invariant solution was, apart from the shift of 2, identical to that in Figure 3.5.1d.

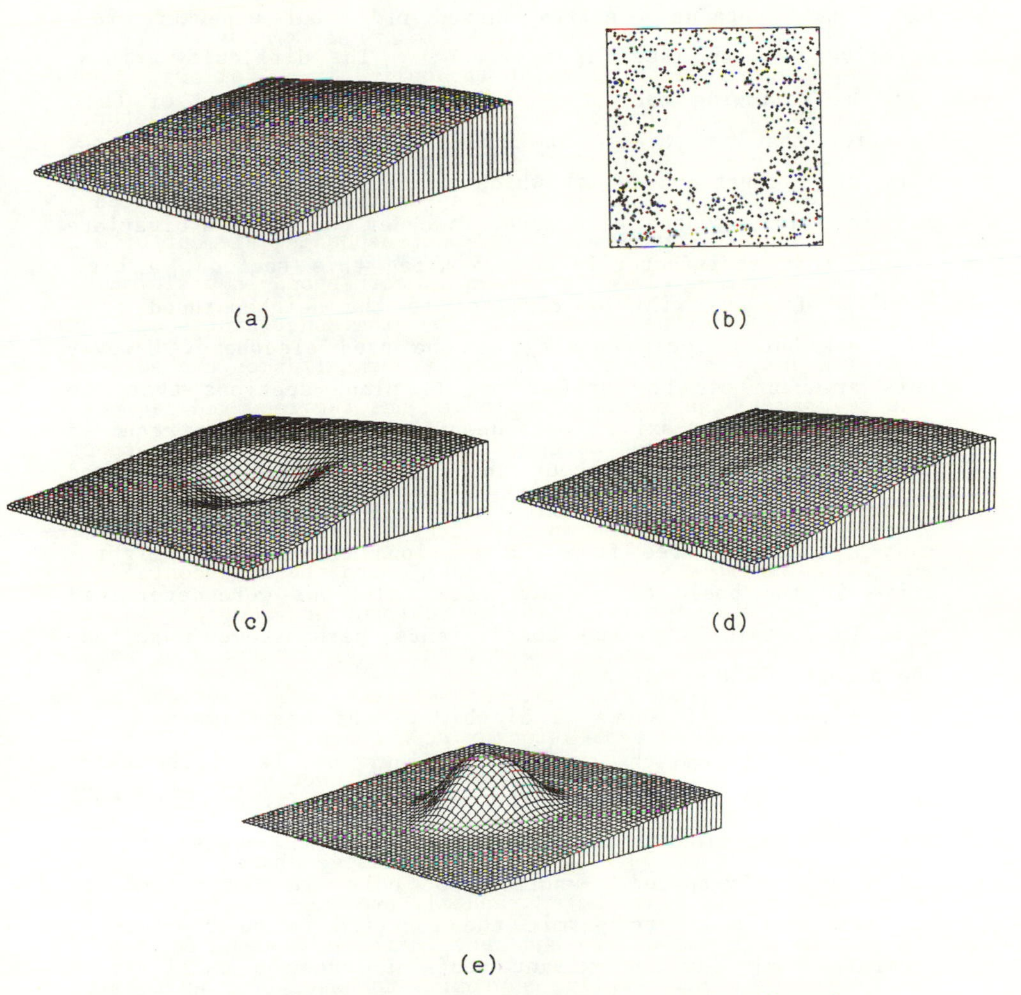

FIG. 3.5.1 (a) The test function $f(x,y) = \exp(-x^2 - 2y^2)$, (b) the 1000 sampled data points, (c) a bicubic spline approximant s to f based on 9 uniformly spaced x-knots and 9 uniformly spaced y-knots using the minimum norm criterion, (d) as (c) but using the geometrically invariant criterion, (e) as (c) but after subtracting 2 from all function values (and drawn on a different vertical scale)

4. CURVED KNOT LINES

4.1 *A Difficulty with Tensor Product Splines*

Tensor product splines cannot cope efficiently with data sampled from certain types of function. An example is a function that contains a narrow curved ridge but elsewhere is relatively smooth (see Figure 4.1.1a). The difficulty arises for the following reason. Consider a plane section of the function that is parallel to the x-axis. This will take the form of a function (of x) which contains a sharp narrow peak but is fairly smooth elsewhere. In order to obtain a bivariate spline that satisfactorily approximates this section, a high density of knots will be required in the neighbourhood of the peak but a lower density can be used elsewhere. However, this argument can be applied to all plane sections that are parallel to the x-axis. Consequently knots will be required at a fine spacing throughout the region. Although the method described in Section 3.5 would hopefully deliver an approximant s that was free from the spurious extrema that might arise if the basic or minimum norm solutions were determined, s would contain very many coefficients, perhaps even more than the number of data points m.

Figure 4.1.1b shows a bicubic spline approximant to points sampled from the function in Figure 4.1.1a at the vertices of a 50 by 50 mesh. The approximant (which has 169 B-spline coefficients) is based on 9 uniformly-spaced x-knots and 9 uniformly-spaced y-knots. Not only are these numbers of knots too small to permit the function to be reproduced accurately but the approximant contains unwanted oscillations. An approximant based on 19 uniformly spaced x-knots and 19 uniformly-spaced y-knots (529 B-spline coefficients) is identical to graphical accuracy to the original function.

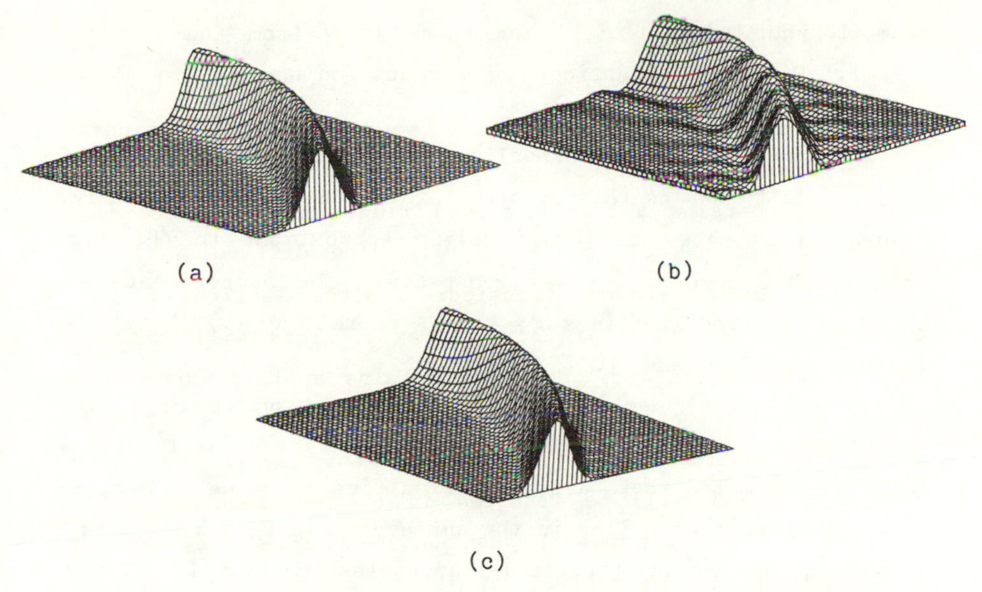

FIG. 4.1.1 (a) A function f containing a narrow curved ridge that cannot efficiently be approximated by a conventional tensor product spline, (b) a bicubic spline approximant to f based on 9 uniformly-spaced x-knots and 9 uniformly-spaced y-knots, (c) as (b) but based on 5 curved knot lines, a constant x-distance apart

4.2 *Value of Curved Knot Lines*

The concept of curved knot lines was introduced by Hayes (1974) in order to handle such problems efficiently. Normally, a tensor product spline is defined, as in Sections 2.5 to 2.7, in terms of a rectangular mesh of knotlines, one set parallel to the y-axis and the other to the x-axis. The B-spline basis functions in x at any point (x_0, y_0) take values that depend only on x_0 and the x-knotlines, and the functions in y depend only on y_0 and the y-knotlines. In the case of curved knot lines the dependence is stronger. The i-th x-knotline is a curve of the form $x = \lambda_i(y)$ and the j-th y-knotline is of the form $y = \mu_j(x)$. The mesh is required to be topologically equivalent to a rectangular mesh with the exception that neighbouring lines are permitted to merge or coalesce. Consider the point (x_0, y_0) in terms of this mesh. Here the B-splines in x are computed from the knot values $\lambda_i(y_0)$ and those in y from the values $\mu_j(x_0)$. With this exception the computation proceeds as for a rectangular mesh of knotlines.

A choice of curved knotlines which follow the ridge of Figure 4.1.1a enables this problem to be handled satisfactorily. Figure 4.1.1c shows a bicubic spline approximant to the data from which Figure 4.1.1a was generated. The approximant is based on 5 curved knotlines $x = \lambda_i(y)$, where $\lambda_{i+1}(y) - \lambda_i(y)$ is independent of i and y.

Hayes (1974) demonstrates a further advantage of curved knot lines. A choice of coalescing knot lines enables functions to be approximated that have discontinuities in value or derivatives along part of a line in the surface. Figure 4.2.1 shows bicubic splines having three and four coalescing knot lines, that contain respectively (a) a "fading" discontinuity in gradient and (b) a discontinuity in value over part of the surface.

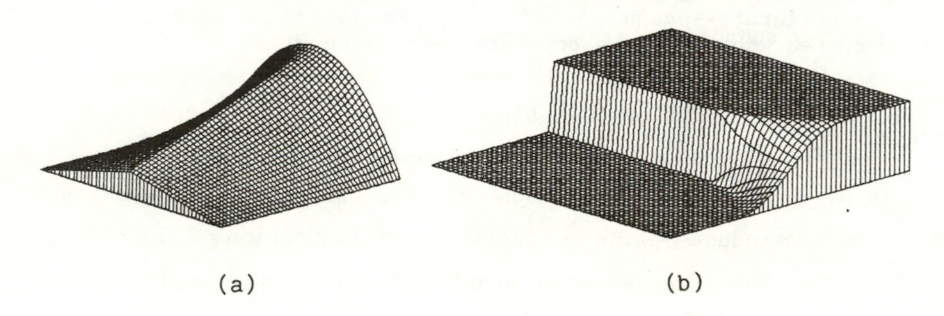

FIG. 4.2.1 Bicubic splines having coalescing curved knot lines that contain (a) a "fading" discontinuity in gradient and (b) a discontinuity in value over part of the surface

5. OTHER APPROXIMATION NORMS

We have not mentioned so far spline approximation in norms other than ℓ_2. Certainly less attention is given to other norms in the literature, at least from the practical viewpoint, possibly for the following reasons. First, least squares is the traditional approach to data approximation, and there are often sound reasons for employing it in particular applications. Second, because of the availability of software that takes advantage of the structure of the associated matrices, the least-squares solution can usually be determined more rapidly and with less computer memory. Third, uncertainty estimates (under appropriate statistical assumptions) can readily be formed (see Section 7).

Apart from ℓ_2, the two most popular norms are ℓ_1 and the maximum (Chebyshev) norm. The ℓ_1 norm, for which the sum of the moduli of the residuals is minimized, has useful properties (see Barrodale, 1968), the most important of which is the tendency to ignore "wild" points in the data, the presence of which may adversely affect an ℓ_2 solution. The maximum norm, which minimizes the residual of maximum magnitude and which is the statistically correct choice when data errors follow a rectangular distribution, is suitable when the data corresponds to correctly rounded values of some function.

Algorithms for general linear ℓ_1 and maximum norm discrete approximation problems are reviewed by Hayes (1987) in his paper on facilities in the NAG Library for data and function approximation. These algorithms may be applied to several of the spline approximation problems considered in earlier sections. Since they were designed for general (full) matrices (which arise from general basis functions) they do not exploit structure and so execution times and memory requirements may be much greater than necessary in spline calculations (although many problems are so small that this issue is academic). For ℓ_1 approximation, methods based on linear programming formulations (Barrodale and Roberts, 1978) or orthogonal transformations (Bartels, Conn and Sinclair, 1978) are used. The former approach is potentially less stable, but considerable experience with it has indicated no difficulties. For the maximum norm, the algorithm of Barrodale and Phillips (1975) is available.

6. CONSTRAINTS

There are fewer publications and indeed available algorithms for constrained spline approximation than for constrained spline interpolation (compare the comments in Section 2.7). We concentrate only on the former class of problems here.

There are two distinct types of constraint that can be imposed, namely equality and inequality constraints. The first type occurs when s (and possibly its derivatives) is required to take particular values at specified points (termed point constraints). In the bivariate case, s or a derivative of some order may also be required to have a plane section that is a specifed univariate spline (a line constraint). The second type includes conditions such as positivity, monotonicity and convexity. It may be required to impose such conditions either globally, i.e. throughout the complete region of interest, or locally, i.e. over specified parts of the range. The B-spline representation is a useful tool for expressing many constraints of the above types in a form that permits efficient algorithms to be constructed.

6.1 *Equality Constraints*

Point and line constraints can be imposed by expressing them, through the use of the recurrence relations for the values and derivatives of B-splines, as linear combinations of the coefficients in the B-spline representation of s. For the ℓ_2 norm the resulting formulation is a linear least-squares problem with linear equality constraints. There are several reliable approaches to the solution of such problems. The method described by Hayes and Halliday (1974) for bicubic spline approximation to scattered data permits point and line constraints to be included. The algorithm checks for self-consistency of the constraints, which must be specified in their B-spline form. A general class of such methods is described by Björck (1984). Unfortunately, due to stability considerations, much of the structure in the problem is often lost using these approaches.

The recent work of Van Loan (1985) shows how the method of weighting may be used in a stable way. In terms of its ability to preserve structure, the approach is as effective as algorithms for unconstrained problems, so this method may prove to be a useful tool in constructing new constrained approximation algorithms. For the ℓ_1 and maximum norms the methods mentioned in Section 5 may be applied, since they permit the inclusion of general linear constraints.

The imposition of boundary conditions on a spline interpolant or approximant is a commonly occurring problem that gives rise to equality constraints. An important class of boundary conditions can be handled straightforwardly by using coincident end knots (Section 2.1), and taking advantage of the properties of the B-spline basis defined on these knots. For univariate problems in which s and its first few derivatives are to take specified values at the endpoints of the interval of interest the problem decomposes. The required spline can be written as $s = s_1 + s_2$, where s_1 can be determined from the boundary conditions alone, and s_2 by computing a spline approximant to the modified function values $\tilde{f}_i = f_i - s_1(x_i)$. The B-spline coefficients of s_1 are obtained by solving triangular systems of equations whose elements can be determined stably by recurrence. For details see Cox (1978), where methods for several other types of boundary conditions are also discussed.

In the bivariate case the analogue of boundary conditions is specified functional forms for s (and its normal derivatives) at the boundary of the rectangular region. It is required to compute a bivariate spline approximant whose plane sections (and derivative plane sections) at the boundary are identical to the given functions. There are similarities here to the blending-function methods introduced by Coons (1964) and since used widely by the CAD fraternity. A method with a considerable degree of generality for solving such approximation problems is given by Anthony and Cox (1987).

6.2 *Inequality Constraints*

For inequality constrained approximation problems the properties of the B-splines can again be used to advantage. In particular, as a result of the convex combination property, bounds on s or one of its derivatives can be replaced by a finite set of linear inequality constraints in the B-spline coefficients of s. In general, the replacing constraints are stronger than the original ones, but in some cases they are equivalent. Specifically, if it is required that $s^{(r)} \geqslant 0$ for some r, the replacing constraints are both necessary and sufficient if $r = n-2$, but otherwise sufficient only (Cox and Jones, 1987). Included in the former class are nonnegative linear splines, monotonic quadratic splines and convex cubic splines, all of which are of considerable practical importance. Moreover, if $r = n-2$, the constraint equations can be determined in an unconditionally stable way (cf Section 1). If $r \neq n-2$, the spline so computed is not in general optimal but nevertheless is often "good" (and of course feasible).

Hanson (1979) discusses the computation of constrained least-squares spline approximants, giving an iterative method that ensures necessity and sufficiency, and Cox and Jones (1987) describe an approach in which the ℓ_1 norm is employed.

7. STATISTICS

There are many approximation problems in which it is valuable to assess the extent to which the errors in the given function values influence the approximant. In this respect the area of discussion overlaps with that of regression analysis. The point we wish to make here is that, in the case of the ℓ_2 norm, useful statistical information is readily obtainable from the Cholesky factor R.

Let "var" denote variance and "cov" covariance. Suppose that $\underline{g}$ and $\underline{h}$ are arbitrary prescribed vectors in the space that contains $\underline{c}$, the vector of coefficients. Then (assuming full rank for A) simple statistical theory (e.g. Kendall and Stuart, 1968) gives

$$\mathrm{var}(\underline{g}^T\underline{c}) = \mathrm{cov}(\underline{g}^T\underline{c}, \underline{g}^T\underline{c})$$

and

$$\mathrm{cov}(\underline{g}^T\underline{c}, \underline{h}^T\underline{c}) = \sigma^2\underline{g}^T(A^TA)^{-1}\underline{h},$$

where σ^2 is the variance of the errors in the data $\{f_i\}$. The use of (2.4.1) yields

$$\begin{aligned}\mathrm{cov}(\underline{g}^T\underline{c}, \underline{h}^T\underline{c}) &= \sigma^2\underline{g}^T(R^TR)^{-1}\underline{h}\\ &= \sigma^2(R^{-T}\underline{g})^T(R^{-T}\underline{h})\\ &= \sigma^2\underline{u}^T\underline{v},\end{aligned}$$

where $\underline{u}$ and $\underline{v}$ are the solutions to the triangular systems

$$R^T\underline{u} = \underline{g}, \qquad R^T\underline{v} = \underline{h}.$$

For univariate spline approximation, R is banded. Thus the determination of an individual covariance requires the solution of two band triangular systems and the formation of an inner product. For a variance just one band triangular solution is required. Matrix inversion is unnecessary and the additional storage requirement over the approximation algorithm itself is just two q-element vectors. The information so obtained is valuable in providing variances of s and its derivatives and integrals (each of which is a linear combination of the B-spline coefficients) at any point in the region of interest. In the case of bivariate spline approximation over a rectangular mesh, advantage can again be taken of separability in computing the statistics (Cox, 1981a).

8. ITEMS NOT COVERED

The field addressed in this paper is broad, despite the fact that, with a few exceptions, algorithms for computing with splines only started appearing in the late 1960s. A previous IMA state-of-the-art paper (Cox, 1977) endeavoured to cover both data and function approximation and was not restricted to spline approximating functions. In this survey I have omitted several important areas of research within which significant progress has been made over the last ten years. Among these areas are:

Smoothing splines: functions which establish a compromise between the smoothness of the approximant and its proximity to the data (de Boor, 1978). The use of cross-validation to determine automatically the optimal mix is important (Hutchinson and de Hoog, 1985).

Splines other than polynomial: for special purposes periodic splines and parametric splines — hinted at in Section 1 — (Dierckx, 1982), trigonometric splines (Lyche and Winther, 1979) and exponential splines — splines in tension — (Pruess, 1978) provide useful capabilities.

Nontensor product splines: a disadvantage of tensor product splines has already been pointed out in Section 4.1 of this paper. Approaches based on subdivisions of the region which do not correspond topologically to a rectangular knot mesh have been studied intensively in recent years. See de Boor (1987) for a recent survey.

Splines defined over nonplanar regions: splines on the sphere or torus or other bodies have value in various applications. The sphere is a particularly important case for modelling various physical phenomena over the surface of the globe (Dierckx, 1984).

Adaptive knot placement: the ability to place the knots automatically in a spline approximation problem in an optimal or at least "good" way is of great practical significance, and has captured the imagination of a number of workers. See Cox (1982b) for a summary. Very much work needs to be done to develop algorithms that will provide good results for a wide class of data sets.

9. FUTURE WORK

The following topics I believe constitute fruitful areas for research. Effective algorithms in these areas will serve an important practical need and be widely used.

Curved knot lines: the original initiative of Hayes (1974) in introducing the concept of curved knot lines has yet to be developed into the form of generally usable algorithms. Some work has been done, however, by Hartley (1980) and Ferguson (1986). The concept is potentially a very powerful one. One reason why it has not been developed further is the difficulty of designing a suitable user interface.

Adaptive knot placement: as mentioned in Section 8 improved knot placement algorithms are a prime requirement. Existing algorithms include ones by de Boor and Rice (1968), Powell (1970) and de Boor (1974). Recently a number of promising ideas have informally been mooted by workers in the area. A significant amount of development work and comparative assessment needs to be undertaken.

Constrained spline approximation: further work in this area, possibly to extend the approaches proposed by Hanson (1979) and Cox and Jones (1987), will permit the user to provide a wide class of qualitative information about his problem in addition to the raw data. Good interactive facilities will feature here.

Nontensor product spline algorithms: this is an exciting area in which a great deal of excellent theoretical work has been carried out (de Boor, 1987). The development of the mathematics to provide viable algorithms will be a major objective in the next decade.

Acknowledgements

Gerald Anthony made valuable comments on the manuscript. Sally Jones helped with the preparation of the figures.

REFERENCES

Anthony, G.T. and Cox, M.G. (1987), "The National Physical Laboratory's Data Approximation Subroutine Library", in *Algorithms for Approximation*, eds. Mason, J.C. and Cox, M.G., Oxford University Press (Oxford), to be published.

Barrodale, I. (1968), "L_1 approximation and the analysis of data", *Applied Statist.* (*J. Roy. Statist. Soc. Ser. C*), *17*, pp.51-57.

Barrodale, I. and Phillips, C. (1975), "Solution of an overdetermined system of linear equations in the Chebyshev norm (Algorithm 495)", *ACM Trans. Math. Software*, *1*, pp.264-270.

Barrodale, I. and Roberts, F.D.K. (1978), "An efficient algorithm for discrete ℓ_1 linear approximation with linear constraints", *SIAM J. Numer. Anal.*, *15*, pp.603-611.

Bartels, R.H., Conn, A.R. and Sinclair, J.W. (1978), "Minimization techniques for piecewise differentiable functions: the ℓ_1 solution to an overdetermined linear system", *SIAM J. Numer. Anal.*, *15*, pp.224-241.

Björck, Å. (1984), "A general updating algorithm for constrained linear least squares problems", *SIAM J. Sci. Statist. Comput.*, *5*, pp.394-402.

de Boor, C. (1972), "On calculating with B-splines", *J. Approx. Theory*, *6*, pp.50-62.

de Boor, C. (1974), "Good approximation by splines with variable knots II", in *Numerical Solution of Differential Equations, Lecture Notes in Mathematics 363*, ed. Watson, G.A., Springer-Verlag (Berlin), pp.12-20.

de Boor, C. (1978), *A Practical Guide to Splines*, Springer-Verlag (New York).

de Boor, C. (1979), "Efficient computer manipulation of tensor products", *ACM Trans. Math. Software*, *5*, pp.173-182.

de Boor, C. (1987), Multivariate approximation", in *The State of the Art in Numerical Analysis*, this volume, pp.87-109.

de Boor, C. and DeVore, R. (1985), "A geometric proof of total positivity for spline interpolation", *Math. Comp.*, *45*, pp.497-504.

de Boor, C. and Pinkus, A. (1977), "Backward error analysis for totally positive linear systems", *Numer. Math.*, *27*, pp.485-490.

de Boor, C. and Rice, J.R. (1968), "Least squares cubic spline approximation II — variable knots", Report No. CSD TR 21, Purdue University.

Coons, S.A. (1964), "Surfaces for computer-aided design of space figures," ESL Memorandum 9442-M-139, MIT, Cambridge, USA.

Cox, M.G. (1972), "The numerical evaluation of B-splines", *J. Inst. Math. Appl.*, *10*, pp.134-149.

Cox, M.G. (1975a), "An algorithm for spline interpolation", *J. Inst. Math. Appl.*, *15*, pp.95-108.

Cox, M.G. (1975b), *Numerical Methods for the Interpolation and Approximation of Data by Spline Functions*, Ph.D. Thesis, City University, London.

Cox, M.G. (1977), "A survey of numerical methods for data and function approximation", in *The State of the Art in Numerical Analysis*, ed. Jacobs, D.A.H., Academic Press (London), pp.627-668.

Cox, M.G. (1978), "The incorporation of boundary conditions in spline approximation problems", in *Numerical Analysis, Lecture Notes in Mathematics 630*, ed. Watson, G.A., Springer-Verlag (Berlin), pp.51-63.

Cox, M.G. (1980), "Versatile parameter lists for scientific library routines", in *Production and Assessment of Numerical Software*, eds. Hennell, M.A. and Delves, L.M., Academic Press (London), pp.249-262.

Cox, M.G. (1981a), "Parameters and statistics of separable linear models", Oxford, Gatlingberg VIII conference lecture, unpublished.

Cox, M.G. (1981b), "The least squares solution of overdetermined linear equations having band or augmented band structure", *IMA J. Numer. Anal.*, *1*, pp.3-22.

Cox, M.G. (1982a), "Direct versus iterative methods of solution for multivariate spline-fitting problems", *IMA J. Numer. Anal.*, *2*, pp.73-81.

Cox, M.G. (1982b), "Practical spline approximation", in *Topics in Numerical Analysis, Lecture Notes in Mathematics 965*, ed. Turner, P.R., Springer-Verlag (Berlin), pp.79-112.

Cox, M.G. and Hayes, J.G. (1973), "Curve fitting: a guide and suite of algorithms for the non-specialist user", NAC Report No. 26, National Physical Laboratory, Teddington.

Cox, M.G. and Jones, Helen M. (1987), "Shape-preserving spline approximation in the ℓ_1 norm", in *Algorithms for Approximation*, eds. Mason, J.C. and Cox, M.G., Oxford University Press (Oxford), to be published.

Dierckx, P. (1982), "Algorithms for smoothing data with periodic and parametric splines", *Comput. Graph. Image Proc.*, *20*, pp.171-184.

Dierckx, P. (1984), "Algorithms for smoothing data on the sphere with tensor product splines", *Computing*, *32*, pp.319-342.

Ferguson, D.R. (1986), "Construction of curves and surfaces using numerical optimization techniques", *Comput. Aided Design*, *18*, pp.15-21.

Gentleman, W.M. (1973), "Least squares computations by Givens transformations without square roots", *J. Inst. Math. Appl.*, *12*, pp.329-336.

Gentleman, W.M. (1974), "Basic procedures for large, sparse or weighted linear least squares problems", *Applied Statist. (J. Roy. Statist. Soc. Ser. C)*, *23*, pp.448-454.

Greville, T.N.E. (1961), "Note on fitting of functions of several independent variables", *J. Soc. Indust. Appl. Math.*, *9*, pp.109-115.

Hanson, R.J. (1979), "Constrained least squares curve fitting to discrete data using B-splines — A users guide", Report No. SAND 78-1291, Sandia Laboratories.

Hartley, P. (1980), "On using curved knot lines", in *Approximation Theory III*, ed. Cheney, E.W., Academic Press (New York), pp.485-490.

Hayes, J.G. (1974), "New shapes from bi-cubic splines", in *Proc. CAD 74*, IPC Business Press (Guildford), fiche 36G/37A; also available as NAC Report No. 58, National Physical Laboratory, Teddington (September, 1974).

Hayes, J.G. (1987), "NAG algorithms for the approximation of functions and data", in *Algorithms for Approximation*, eds. Mason, J.C. and Cox, M.G., Oxford University Press (Oxford), to be published.

Hayes, J.G. and Halliday, J. (1974), "The least-squares fitting of cubic spline surfaces to general data sets," *J. Inst. Math. Appl.*, *14*, pp.89-103.

Heath, M.T. (1982), "Some extensions of an algorithm for sparse linear least squares problems", *SIAM J. Statist. Comput.*, *3*, pp.223-237.

Hutchinson, M.F. and de Hoog, F.R. (1985), "Smoothing noisy data with spline functions", *Numer. Math.*, *47*, pp.99-106.

Kendall, M.G. and Stuart, A. (1968), *The Advanced Theory of Statistics*, *Volume II*, Charles Griffin (London).

Kozak, J. (1980), "On the choice of the exterior knots in the B-spline basis for a spline basis for a spline space", MRC Report 2148, University of Wisconsin.

Lyche, T. and Winther, R. (1979), "A stable recurrence relation for trigonometric B-splines", *J. Approx. Theory*, *25*, pp.266-279.

Paige, C.C. and Saunders, M.A. (1982), "LSQR: an algorithm for sparse linear equations and sparse least squares", *ACM Trans. Math. Software*, *8*, pp.43-71.

Powell, M.J.D. (1970), "Curve fitting by splines in one variable", in *Numerical Approximation to Functions and Data*, ed. Hayes, J.G., Athlone Press (London), pp.65-83.

Pruess, S. (1978), "An algorithm for computing smoothing splines in tension", *Computing*, *19*, pp.365-373.

Reid, J.K. (1967), "A note on the least squares solution of a band system of linear equations by Householder reductions", *Comput. J.*, *10*, pp.188-189.

Rosen, J.B. (1964), "Minimum and basic solutions to singular linear systems", *SIAM J. Appl. Math.*, *12*, pp.156-162.

Schoenberg, I.J. and Whitney, Anne (1953), "On Polya frequency functions III", *Trans. Amer. Math. Soc.*, *74*, pp.246-259.

Van Loan, C. (1985), "On the method of weighting for equality-constrained least-squares problems", *SIAM J. Numer. Anal.*, *22*, pp.851-864.

M.G. Cox
Division of Information Technology & Computing
National Physical Laboratory
Teddington
Middlesex TW11 0LW,
England

6 Methods for Best Approximation and Regression Problems

G. A. WATSON

ABSTRACT

This paper is primarily concerned with methods for certain classes of discrete data fitting problems using criteria other than least squares. Recent developments in the calculation of best nonlinear ℓ_1 and ℓ_∞ approximations are considered, as well as methods for robust regression, in particular the use of the Huber M-estimator. The problem of best Chebyshev approximation to continuous functions on an N-dimensional continuum is also addressed, and some recent methods for obtaining solutions are described. Finally, an introduction is given to some ways of solving data fitting problems in which there are errors in all the problem variables.

1. INTRODUCTION

Most of the problems considered in this paper fall into the category of best approximation problems, which may be stated in general terms as follows: given a point g and a set M in a normed linear space S, $\|\cdot\|$, find a point of M of minimum distance from g. In practice, a particular subset M of S is usually identified through an appropriate parameterization, so that the elements of M have the form $F(\underline{a})$, where $F:\mathbb{R}^n \to M$. If F is linear in $\underline{a}$ then it is well known that a best approximation exists, otherwise in general it may not.

Most best approximation problems can be subdivided into *discrete* problems, where $S=\mathbb{R}^m$, and *continuous* problems, where $S=C(X)$, with X a compact set; attention has mainly been focussed on problems set in these spaces normed with the associated ℓ_p norms, in particular the cases $p = 1, 2, \infty$. Important examples of discrete best approximation problems arise from data fitting or regression problems, and these are of particular interest here as much has happened over the last 10 years. When the normed linear space is strictly convex, and $F(\underline{a})$ is continuously differentiable, there has always been an obvious link between methods for the appropriate approximation problems and algorithms for unconstrained optimization. This is particularly true of nonlinear least squares problems, a class of problems which is not considered here as it has its own eponymous place elsewhere in these proceedings. For linear discrete ℓ_1 and ℓ_∞ problems, the simplex method of linear programming is an important tool, and good methods are well-known and widely available. For the ℓ_∞ problem, recent developments have involved the idea of permitting more than one exchange of variables per step (Hopper and Powell, 1977, and Armstrong and Kung, 1980); for the ℓ_1 problem there is some evidence that improvement can be gained by using a normalized steepest edge test (which is scale invariant) for choosing the variable to enter the basis: see Bloomfield and Steiger (1980). Perhaps the major influence on the construction of algorithms for both linear and nonlinear discrete ℓ_1 and ℓ_∞ problems in the past decade has been the attention which problems such as these have attracted from numerical analysts researching into optimization technques; this is to a large extent associated with the development of optimization methods which no longer require traditional smoothness assumptions. Examples in the linear case are the projected gradient methods of Bartels, Conn and Charalambous (1978), and Bartels, Conn and Sinclair (1978), and the penalty algorithm of Bartels and Joe (1983). A very comprehensive account of the various algorithms available for those linear problems and the relationship between them is contained in Osborne (1985).

Of particular interest here are nonlinear discrete problems. It is convenient to introduce these in Section 2 as part of a wider class of problems, for there are useful criteria for fitting regression models which involve convex functions which are not necessarily norms. Details are given in Section 3 of the development of methods for problems with the important ℓ_1 and ℓ_∞ norms. Section 4 is concerned with robust regression, and in particular with the use of the Huber *M*-estimator. The construction of algorithms for regression problems using this criterion is still at a relatively early stage, and only methods for linear problems are considered here.

Section 5 is devoted to an account of recent methods for continuous Chebyshev approximation problems. This is the only example of a continuous problem which is included, a fact which is judged to reflect the more limited interest in algorithms for such problems. Satisfactory methods for ℓ_1 problems in $C[a,b]$ have also been developed (for example, Glashoff and Schultz, 1979, and Watson, 1981, 1982) but the Chebyshev problem is treated in some detail because of its acknowledged role as a central problem in approximation theory.

Finally in Section 6 another class of regression problems is introduced. Problems of this class differ from those considered earlier in that they are based on modifying the usual model equations to take account of the situation when there are errors in all the variables of the underlying physical problem. There are different ways in which this can be done, and the intention here is merely to exemplify some approaches in a best approximation context. This is an area where there is still considerable scope for the development of methods.

It is inevitable that the material included in a paper of this kind should be limited not only by space, but also by the interests and knowledge of its author. Before proceeding, then, a word about what else is *not* included: approximation in the complex plane is not included; the incorporation of constraints on problems is not considered; also the treatment of special classes of approximating functions is excluded. Spline functions are of course well represented elsewhere. Another class of approximating functions which is usually regarded as important is that of rational functions: an algorithm for discrete ℓ_1 approximation by generalized rational functions is described by Watson (1984a); Remes-type methods for rational Chebyshev approximation on $C[a,b]$ have been developed by Belogus and Liron (1978), Kaufman, Leeming and Taylor (1978, 1980), and Breuer (1986), and a generalized differential correction algorithm for generalized rational functions is given by Cheney and Powell (1987).

2. NONLINEAR REGRESSION

A typical nonlinear regression problem may be described as follows: given observations y_i at points x_i (which need not be scalars), $i = 1,2,\cdots,m$, and a model $F(x,\underline{a})$ which depends nonlinearly on the parameter vector $\underline{a} \in \mathbb{R}^n$, $n<m$, determine values of the parameters so that

$$y_i \approx F(x_i,\underline{a}), \qquad i = 1,2,\cdots,m . \tag{2.1}$$

Because of observation errors, it is not normally possible to force equality to hold in (2.1) and the usual approach is to introduce perturbations f_i into the y_i values so that the model equations

$$f_i + y_i = F(x_i,\underline{a}) , \qquad i = 1,2,\cdots,m, \tag{2.2}$$

are satisfied exactly. The vector $\underline{a}$ is then chosen so that the f_i are small in some sense, for example through the solution of the problem

$$\text{find } \underline{a} \in \mathbb{R}^n \text{ to minimize } \psi(\underline{f}(\underline{a})) , \tag{2.3}$$

where $\underline{f}$ has ith component f_i (and the dependence on $\underline{a}$ is shown explicitly), and $\psi : \mathbb{R}^m \to \mathbb{R}$ is given. There are many possible choices of ψ, and usually it is a convex function. Then, if F (and therefore $\underline{f}$) is a continuously differentiable function of $\underline{a}$, it is straightforward to give first order necessary conditions for $\underline{a}^*$ to solve (2.3), and this leads to the following definition.

DEFINITION. Let $A = \nabla \underline{f} \in \mathbb{R}^{m \times n}$. Then $\underline{a}^*$ is a stationary point of (2.3) if

$$\underline{0} \in S^* = \{A^{*T}\underline{v} : \underline{v} \in \partial\psi(\underline{f}^*)\} , \tag{2.4}$$

where the superscript * denotes evaluation at $\underline{a}^*$, and ∂ denotes the subdifferential.

Most methods for finding $\underline{a}^*$ satisfying (2.4) (or equivalent definitions) are iterative: for example Gauss-Newton-type methods use an affine model of $\underline{f}$ in (2.3) and therefore consist of the sequence of subproblems

$$\left.\begin{array}{ll} \text{find } \underline{d}^k \text{ to minimize } & \psi(\underline{f}^k + A^k\underline{d}) \\ \text{set } \quad \underline{a}^{k+1} = \underline{a}^k + \gamma\underline{d}^k & \end{array}\right\} , \tag{2.5}$$

where $\underline{f}^k$ and A^k are evaluated at the current point $\underline{a}^k$, and γ is a suitably chosen step length. Rules for choosing γ can be given which make this into a satisfactory method for many problems; however algorithms based on (2.5) lack good generally provable global convergence properties, and an improved subproblem is

$$\left.\begin{array}{l} \text{find } \underline{d}^k \text{ to minimize } \psi(\underline{f}^k + A^k\underline{d}) \\ \text{subject to } \|\underline{d}\| \leq \delta^k , \\ \text{set } \quad \underline{a}^{k+1} = \underline{a}^k + \gamma\underline{d}^k , \ \gamma = 0 \text{ or } 1 \end{array}\right\} , \tag{2.6}$$

where δ^k is a positive parameter. This is the basis for Levenberg-Marquardt or trust region methods. If the choice $\gamma = 1$ is unsuitable, then the trust region radius δ^k must be reduced, and this can be done in such a way that the resulting method has safe global convergence properties. Model algorithms are given by Watson (1980), Osborne (1981), and Madsen (1985). It is possible for the final rate of convergence of these methods to be second order, and a sufficient condition is that the limit point $\underline{a}^*$ is a strongly unique local minimum (Cromme, 1978, and Madsen, 1985): an equivalent statement is that

$$\underline{0} \in \mathrm{int}(S^*). \tag{2.7}$$

When ψ is differentiable at $\underline{f}^*$, then $\underline{S}^*$ is a singleton and so (2.7) does not hold. When ψ is not differentiable at $\underline{a}^*$, then (2.7) is satisfied occasionally. However, the situation is complicated by the fact that this condition is stronger than is necessary for second order convergence: necessary and sufficient conditions are given by Jittorntrum and Osborne (1980) when ψ is a polyhedral norm. When the rate of convergence is not second order, it can be unacceptably slow, and to improve matters, it is necessary to incorporate second derivative information: this can for example be done by the replacement of the objective function of (2.6) by

$$\psi(\underline{f}^k + A^k\underline{d}) + \tfrac{1}{2}\,\underline{d}^T B^k \underline{d}\ , \tag{2.8}$$

where B^k is an $n \times n$ symmetric matrix chosen in a certain way. A model (trust region) algorithm for the case where ψ is the polyhedral convex function

$$\psi(f) = \max_{1 \leqslant i \leqslant l} (\underline{f}^T\underline{h}_i + b_i)\ , \tag{2.9}$$

where $\underline{h}_i$ and b_i are given, $i = 1, 2, \cdots, l$, is presented by Fletcher (1982). A line search algorithm for a similar class of problems is given by Womersley and Fletcher (1986), based on replacing the nonsmooth part of (2.8) by linear constraints. Conditions on the sequence of matrices $\{B^k\}$ for superlinear convergence of methods based on (2.8) are given by Powell (1983), Powell and Yuan (1984), and Womersley (1985): when there are no derivative discontinuities, these follow from the theory for unconstrained optimization given by Dennis and Moré (1974).

An important class of functions $\psi(\underline{f})$ is given by

$$\psi(\underline{f}) = \|\underline{f}\| ,$$

and the most commonly used norm is the least squares norm. However, of value also are other ℓ_p norms, for example the ℓ_1 and ℓ_∞ norms: from a data fitting point of view, the former is of particular interest because it gives minimum weight to wild points or gross errors in the data.

3. METHODS FOR NONLINEAR DISCRETE ℓ_1 AND ℓ_∞ PROBLEMS

For the case where

$$\psi(\underline{f}) = \sum_{i=1}^{m} |f_i| , \tag{3.1}$$

the subdifferential is given by

$$\partial\psi(\underline{f}) = \left\{\underline{v} \in \mathbb{R}^m : |v_i| \leq 1 , \; f_i = 0 ; \; v_i = \text{sign}(f_i) , \; f_i \neq 0\right\} . \tag{3.2}$$

Define

$$\left.\begin{aligned} Z^f &= \{i : f_i = 0\} \\ \theta_i^f &= \text{sign}(f_i) , \; i \notin Z^f \end{aligned}\right\} , \tag{3.3}$$

and let Z^* denote Z^{f^*} etc. Then the condition for a stationary point $\underline{a}^*$ may be written

$$Z^* = \{i : f_i(\underline{a}^*) = 0\} , \tag{3.4}$$

$$\sum_{i \in Z^*} v_i \nabla f_i^* + \sum_{i \notin Z^*} \theta_i^* \nabla f_i^* = \underline{0} , \tag{3.5}$$

where

$$|v_i| \leq 1 , \; i \in Z^* . \tag{3.6}$$

When A has full rank, it is well known that a solution to (2.5) provided by standard methods will have (assuming non-degeneracy) exactly n zero components of $\underline{f}^k + A^k\underline{d}^k$. If $|Z^*| = n$, then it follows that locally the Gauss-Newton method with $\gamma = 1$ corresponds to the application of Newton's method to the system of equations

$$f_i(\underline{a}) = 0\,, \quad i \in Z^*\,, \tag{3.7}$$

and a second order convergence rate is possible. However, for many problems $|Z^*| < n$, and so $\underline{a}^*$ is not determined by (3.7) alone. Most modern methods for (2.3), (3.1) fall into one of the following two categories.

A Single Phase Methods

These methods involve the solution of a sequence of problems for the step (or search direction) $\underline{d}^k$, and incorporate an active set strategy for updating the current approximation $\hat{Z}^k$ (which may differ from Z^k) to the index set Z^*. Most methods are based on the subproblem

$$\left.\begin{array}{ll} \text{minimize} & \underline{g}^{kT}\underline{d} + \tfrac{1}{2}\underline{d}^T B^k \underline{d} \\ \text{subject to} & f_i^k + \underline{d}^T \nabla f_i^k = 0\,, \; i \in \hat{Z}^k \end{array}\right\}, \tag{3.8}$$

where $\{\underline{g}^k = \Sigma\theta_i^k \nabla f_i^k\,;\ i \notin \hat{Z}^k\}$, with $\underline{d}$ possibly restricted to a trust region. This differs from the use of (2.8) essentially in that the zero components of $\underline{f}^k + A^k\underline{d}^k$ are predetermined, so that the nonsmooth part of (2.8) can be removed. Therefore the methods are normally locally equivalent.

Define the Lagrangian function

$$\mathcal{L}(\underline{v}, \underline{a}) = \sum_{i \in \hat{Z}^k} v_i f_i + \sum_{i \notin \hat{Z}^k} \theta_i f_i\,.$$

Then (3.5) may be written

$$\nabla \mathcal{L}(\underline{v}, \underline{a}^*) = \underline{0}\,. \tag{3.9}$$

Examination of the Kuhn-Tucker necessary (and sufficient) conditions for the above subproblems shows that for both methods, provided that $B^k = \nabla^2 \mathcal{L}^k$ $(= \nabla^2 \mathcal{L}(\underline{v}^k, a^k))$ and $\hat{Z}^k = Z^*$, then $\underline{d}^k$ is just the Newton step in $\underline{a}$ at the point $(\underline{v}^k, \underline{a}^k)$ for the solution of (3.7), (3.9). An advantage of using (3.8) is that the constraints may be eliminated and, instead of working with $\nabla^2\mathcal{L}^k$, it is possible to work with the reduced Hessian matrix $W^{kT}\nabla^2\mathcal{L}^k W^k$, where W^k is a matrix with orthonormal columns spanning the null space of the vectors $\{\nabla f_i^k : i \in \hat{Z}^k\}$. The reduced Hessian matrix is normally positive definite at $\underline{a}^*$, and so is readily approximated using quasi-Newton techniques. Line search methods based on (3.8) are given by Murray and Overton (1981), and Bartels and Conn (1982): the treatment of degeneracy in these methods is considered by Busovača (1984). Trust region algorithms are given by Fletcher (1981, 1985).

B Two-Phase or Hybrid Methods

Methods of this type avoid the need for an active set strategy by attempting as a first phase to identify Z^*. The local convergence properties of the first phase method are therefore not so important, and for example (2.6) may be used. The zero components of $\underline{f}^k + A^k\underline{d}^k$ together with additional information from the characterization conditions for (2.6) may be used to help identify Z^*: if $|Z^*| = n$, then it is usually appropriate to continue to iterate with (2.6), because provided that the step restriction condition is eventually inactive, a second order rate of convergence may be expected; if $|Z^*| < n$, then Newton's method (or an approximate Newton method) may be applied directly to (3.7), (3.9): the Jacobian matrix of this system is

$$J = \left[\begin{array}{c|c} \nabla^2\mathcal{L} & H \\ \hline H^T & 0 \end{array}\right]$$

where $H = [\nabla f_i^T : i \in Z^*]$.

A method due to McLean and Watson (1980) uses the exact Jacobian matrix. A more sophisticated method due to Hald and Madsen (1985) features quasi-Newton approximations to $\nabla^2\mathcal{L}^k$ (using either the Powell-symmetric-Broyden or BFGS update), these being built up during the first phase; it allows several switches between phases (usually there are few in practice) and has good proven local (superlinear) and global convergence properties.

Precisely analogous methods can be developed for the discrete Chebyshev problem, when

$$\psi(f) = \max_{1 \leqslant i \leqslant m} |f_i| = \|\underline{f}\| .$$

Define

$$E^f = \{i\colon |f_i| = \|\underline{f}\|\} ,$$
$$\theta_i^f = \operatorname{sign}(f_i) , \quad i \in E^f .$$

Then $\underline{a}^*$ is a stationary point if

$$E^* = \{i\colon |f_i| = \|\underline{f}^*\|\} , \qquad (3.10)$$

$$\sum_{i \in E^*} \lambda_i \nabla f_i^* = \underline{0} , \qquad (3.11)$$

where

$$\lambda_i \theta_i^* \geqslant 0 , \quad i \in E^* , \qquad (3.12)$$

and the numbers λ_i are not all zero. A necessary condition for a second order convergence rate of the Gauss-Newton method is that $|E^*| \geqslant n+1$.

To obtain a subproblem analogous to (3.8) one can argue as follows. Introduce an additional variable h, and define $h^k = \|\underline{f}^k\|$ at each iteration. If $\underline{d}$ and p represent the increments in $\underline{a}$ and h respectively, then the problem corresponding to (3.8) is

$$\left.\begin{array}{ll} \text{minimize} & p + \tfrac{1}{2}\underline{d}^T B^k \underline{d} \\ \text{subject to} & \theta_i^k f_i^k - h^k + \theta_i^k \underline{d}^T \nabla f_i^k - p = 0 , \quad i \in \hat{E}^k \end{array}\right\} . \qquad (3.13)$$

If

$$B^k = \sum_{i \in \hat{E}^k} \lambda_i^k \nabla^2 f_i^k ,$$

and if $\hat{E}^k = E^*$, it is easily seen that $\underline{d}^k$ is just the Newton step in $\underline{a}$ at the point $(\underline{a}^k, h^k, \underline{\lambda}^k)$ for the solution of the first order conditions for a stationary point. Line search methods based on (3.13) are given by Conn (1979), Murray and Overton (1980), and Womersley and Fletcher (1986). A related method but based on inequality constraints (thus avoiding the choice of $\hat{E}^k$) is given by Han (1981). Two phase methods are developed in Watson (1979) and Hald and Madsen (1979, 1981).

It is interesting that two-phase (or hybrid) methods are currently popular for ℓ_2 problems, with combined Gauss-Newton and BFGS methods prominent. However, whether or not this is the most suitable approach for these nondifferentiable cases remains to be established.

4. ROBUST REGRESSION

There is clearly a desire among (some) statisticians to move away from more traditional criteria for fitting data towards the use of robust estimators (see, for example, Huber, 1981, and Rocke *et al*, 1982). The appropriate 'loss functions' have the form

$$\psi(\underline{f}) = \sum_{i=1}^{m} \rho(f_i) \tag{4.1}$$

and are not necessarily norms (or even convex). Since insensitivity to large errors is a primary requirement, the ℓ_1 norm is an important special case of (4.1). Another case which is becoming increasingly prominent is Huber's *M*-estimator. defined by

$$\rho(f_i) = \begin{cases} \frac{1}{2}f_i^2, & |f_i| \leq c, \\ c|f_i| - \frac{1}{2}c^2, & |f_i| > c, \end{cases} \tag{4.2}$$

where c is a scale factor to be estimated from the data. This makes (4.1) a convex function, and corresponds to the maximum likelihood estimate for a perturbed normal distribution (Huber, 1977). Additional scaling parameters may be incorporated, but to avoid clutter these will not be included. Methods for this problem in the nonlinear case are still fairly rare, and only the linear case will be considered here, where

$$\underline{f} = A\underline{a} - \underline{y}\,, \tag{4.3}$$

with A a constant $m \times n$ matrix assumed to have full rank n, and $\underline{y} \in \mathbb{R}^m$ a constant vector.

When $\rho(\cdot)$ has the form (4.2) and c can be fixed beforehand, there are 3 standard ways to solve (4.1). Let $\underline{v} \in \mathbb{R}^m$ be defined by

$$v_i = \rho'(f_i)\,, \qquad i = 1, 2, \cdots, m\,.$$

Then a necessary and sufficient condition for $\underline{a}^*$ to solve (4.1) in this case is that

$$A^T \underline{v}^* = \underline{0}\,,$$

and this system of equations for $\underline{a}^*$ may be solved by the iteration

$$\left.\begin{aligned} A^T D^k A \underline{d}^k &= -A^T \underline{v}^k \\ \underline{a}^{k+1} &= \underline{a}^k + \gamma \underline{d}^k \end{aligned}\right\}\,, \tag{4.4}$$

with the following choices of D^k and γ:

(1) $D^k = I\,, \quad \gamma = 1\,.$

This is the modified residual method (or Huber-Dutter or Huber method) which has guaranteed linear convergence. Notice that no refactorization of the matrix in (4.4) is needed.

(2) $D^k = \mathrm{diag}\left\{\rho'(f_i^k)/f_i^k\,; \quad i = 1, 2, \cdots, m\right\}, \ \gamma = 1.$

This is the method known as iteratively reweighted least squares. Again linear convergence is guaranteed.

(3) $D^k = \mathrm{diag}\left\{\rho''(f_i^k); \quad i = 1, 2, \cdots, m\right\}.$

This corresponds (formally) to Newton's method. A line search procedure is necessary for global convergence.

The relative effectiveness of these methods is examined by Holland and Welsch (1977).

More realistically, c is not known in advance, but may be required to satisfy a specified condition. Let p denote the number of indices such that $|f_i| > c$ at the solution. Then for example c may be required to be such that p is a particular fraction of the data. The so-called Huber proposal 2 requires that

$$\frac{1}{(m-n)}\Big(\sum_{i:|f_i|\leq c} f_i^2 + pc^2\Big) = kc^2 , \tag{4.5}$$

for some positive number k (the Huber proposal 1 is that c can be selected beforehand by intuition). Another possibility that has been suggested is to choose c so that

$$c = \text{med}\{|f_i|\} .$$

Modifications of Newton's method so that (4.5) is also satisfied have been implemented by Ekblom (1985), and Shanno and Rocke (1986).

A different way of looking at the problem exploits the following observation of Huber (1973): 'The search for the M-estimator is the search for the correct partition'. For given c, let $\underline{a}^*$ minimize (4.1) and let the index set

$$\nu = \{i : |f_i^*| \leq c\}$$

be assumed known. Then the problem may be written as the minimization of

$$\psi(\underline{a}) = \tfrac{1}{2}\sum_{i\in\nu} f_i^2 + c\sum_{i\notin\nu}(|f_i| - \tfrac{1}{2}c) . \tag{4.6}$$

Now, if (4.6) is least, then using (4.3)

$$\nabla\psi(\underline{a}) = \sum_{i\in\nu} f_i\underline{\alpha}_i + c\sum_{i\notin\nu}\theta_i\underline{\alpha}_i = \underline{0} ,$$

where $\underline{\alpha}_i^T$ denotes the ith row of A, and $\theta_i = \text{sign}(f_i)$, $i \notin \nu$. Thus

$$\Big(\sum_{i\in\nu}\underline{\alpha}_i\underline{\alpha}_i^T\Big)\underline{a} = \sum_{i\in\nu} y_i\underline{\alpha}_i - c\sum_{i\notin\nu}\theta_i\underline{\alpha}_i , \tag{4.7}$$

so that $\underline{a}^*$ is uniquely defined if the $n\times n$ matrix on the left hand side is nonsingular: because A is assumed to have rank n, there always exists a solution with $|\nu| \geq n$, and so this is always possible. It follows from (4.7) that $\underline{a}^*$ is an *affine function* of c on any interval of c on which the classification of residuals does not change. In fact, $\underline{a}^*(c)$ is a continuous piecewise linear function for all $c \geq 0$ (Clark, 1981). In addition, the minimization of the ℓ_1 norm of $\underline{f}$ corresponds to the case $c = 0$, with $\nu = \{i: |f_i^*| = 0\}$, and the minimization of the ℓ_2 norm of $\underline{f}$ corresponds to the cases $c \geq c_0 = \max\{|f_i^*|: i = 1, 2, \cdots, m\}$ with $\nu = \{1, 2, \cdots, m\}$. In other words the partitions of the residuals and $\underline{a}^*$ can be easily determined in these extreme cases. These observations suggest that a procedure based on continuation with respect to c could be applied to the M-estimation problem until some target value of c is reached (which need not be known in advance). Methods of this type are given by Clark (1981), Boncelet and Dickinson (1984), and Clark and Osborne (1986). In contrast to the iterative methods discussed earlier, these algorithms are provably finite, and can be given in the following form:

(1) solve the ℓ_1 (or ℓ_2) problem to determine an initial partition,

(2) increase (decrease) c to find the next critical point where there is a change in ν,

(3) update ν,

(4) continue until c reaches a target value.

In the event of degeneracy (more than one index changing simultaneously), the trajectory can be lost, and it may be necessary to recompute the correct partition. A finite algorithm for doing this is given by Clark (1981). Some interesting examples of unusual occurrences related to degeneracy are illustrated by Clark (1985).

5. CHEBYSHEV APPROXIMATION OF CONTINUOUS FUNCTIONS

Let X be a compact set in $\mathbb{R}^N$, and let $C(X)$ be normed with the Chebyshev norm

$$\|r(\cdot)\| = \max_{\underline{x} \in X} |r(\underline{x})| .$$

Then of interest here is the problem

$$\text{find } \underline{a} \in \mathbb{R}^n \text{ to minimize } \|f(\cdot, \underline{a})\| , \qquad (5.1)$$

where $f: X \times \mathbb{R}^n \to \mathbb{R}$ is continuous. Techniques and analyses for this problem are naturally closely related to those for the discrete case considered in Section 3: of course if X is finite, there is nothing new, so it will be assumed that X is a continuum, for example the Cartesian product of closed intervals $[\alpha_1, \beta_1] \times [\alpha_2, \beta_2] \times \cdots \times [\alpha_N, \beta_N]$. In addition it will be assumed that f is a continuously differentiable function of its parameters. As usual, the first order necessary conditions for a local solution provide the definition of a stationary point: $\underline{a}^*$ is a stationary point if there is a set $E = \{\underline{x}_1, \underline{x}_2, \cdots, \underline{x}_t\} \subset X$ with $t \leqslant n+1$ such that

$$|f(\underline{x}_i, \underline{a}^*)| = \|f(\cdot, \underline{a}^*)\| , \quad i = 1, 2, \cdots, t , \qquad (5.2)$$

and λ_i, $i = 1, 2, \cdots, t$, not all zero, such that

$$\left.\begin{aligned} &\sum_{i=1}^{t} \lambda_i \nabla f(\underline{x}_i, \underline{a}^*) = \underline{0} , \\ &\lambda_i \operatorname{sign}(f(\underline{x}_i, \underline{a}^*)) \geqslant 0 , \quad i = 1, 2, \cdots, t \end{aligned}\right\} .$$

For linear problems, f may be expressed in the form

$$f(\underline{x}, \underline{a}) = \sum_{i=1}^{n} a_i \phi_i(\underline{x}) - y(\underline{x}) ,$$

and if $N = 1$ and the functions $\{\phi_i(\underline{x})\}$ form a Chebyshev set on X, then $t = n+1$, and the above definition may be interpreted in the form of the familiar alternating set characterization result. In addition the well-known methods of Remes may be successfully applied. For more general linear (and nonlinear) problems, the number of points of X where the norm is attained at a solution is frequently less than $(n+1)$, and it is necessary to look beyond the traditional methods because $\underline{a}^*$ is not determined by the solution of a system of equations of the form (5.2).

An approach which can be effective for general problems is to consider two phases. The first phase is typically a solution of a discretization of the original problem, with the aim of identifying t, and obtaining good approximations to $\underline{a}^*$ and $\underline{\lambda}$. For nonlinear problems, the appropriate methods of Section 3 may be used; for linear problems, the simplex method of linear programming may be applied to the dual problem in the usual way. However, it is important to realise that while the use of a fine mesh over X may lead to accurate information, it may not only be computationally expensive but may also give ill-conditioning. Consider for example the linear case: then if $t<n+1$ in (5.2), because there will be $(n+1)$ points where the (discrete) Chebyshev norm is attained at the solution to the discrete problem, adjacent points will occur in the optimal dual basis. A method based on a numerically stable simplex algorithm and successive grid refinement is given by Hettich (1986) for one- and two-dimensional linear problems: solutions are obtained successfully for problems with n up to 37 and 32,761 grid points.

Assuming that, in particular, t and $\theta_i = \text{sign}(f(\underline{x}_i, \underline{a}^*))$, $i = 1, 2, \cdots, t$ have been correctly determined, the second phase is the solution of the system of nonlinear equations

$$f(\underline{x}_i, \underline{a}) = \theta_i h, \quad i = 1, 2, \cdots, t,$$

$$\sum_{i=1}^{t} \lambda_i \nabla f(\underline{x}_i, \underline{a}) = \underline{0},$$

$$\sum_{i=1}^{t} \lambda_i \theta_i = 1,$$

together with zero derivative conditions arising because the points $\underline{x}_i$ are local maxima or minima of f. There are therefore effectively $Nt+t+n+1$ equations in the same number of unknowns which may be solved by, for example, Newton's method. Failure to converge, or convergence to a point which can be identified as nonstationary, requires re-entry into Phase 1 with a refined mesh. Two-phase methods of this type are given by Watson (1976), Hettich (1976), and Gustafson (1983).

There are some unsatisfactory features of two-phase methods for this class of problems. Firstly, it can be difficult to get correct information at the end of Phase 1, particularly when $N>1$, without resort to a very fine mesh; secondly, there is no easy way to monitor progress in the second phase, and for example to ensure that limit points $h^*, \underline{a}^*$ of the iteration also satisfy

$$\|f(\cdot,\underline{a}^*)\| = h^* ; \tag{5.3}$$

thirdly, the application of Newton's method requires second derivatives to be calculated, and without modification may break down. For these reasons, it is of interest to investigate the possibility of a more robust second phase procedure, capable of converging from poorer starting points, even with an incorrect value of t, or incorrect positional information on the extrema.

At the current approximation $\underline{a}^k$, define $M(\underline{a}^k)$ to be the set of local maxima of $|f(\underline{x},\underline{a}^k)|$ in X, and assume that $|M(\underline{a}^k)| = t^k < \infty$, and for $\underline{a} \in \text{Nbhd}(\underline{a}^k)$, each $\underline{x}_i \in M(\underline{a})$, $i=1,2,\cdots,t^k$, is a differentiable function of $\underline{a}$. Then locally

$$\|f(\cdot,\underline{a})\| = \max_{1\leqslant i\leqslant t^k} \theta_i^k f(\underline{x}_i(\underline{a}),\underline{a})$$

where

$$\theta_i^k = \text{sign}(f(\underline{x}_i,\underline{a}^k)), \quad i=1,2,\cdots,t^k .$$

For $\underline{a} \in \text{Nbhd}(\underline{a}^k)$, let

$$g_i(\underline{a}) = \theta_i^k f(\underline{x}_i(\underline{a}),\underline{a}), \quad i=1,2,\cdots,t^k ,$$

and define

$$q^k(\underline{d}) = \max_{1 \leq i \leq t^k} \left\{ g_i(\underline{a}^k) + \underline{d}^T \nabla g_i(\underline{a}^k) \right\} + \tfrac{1}{2} \underline{d}^T B^k \underline{d}\,, \qquad (5.4)$$

where B^k is an $n \times n$ symmetric matrix. The minimization of $q^k(\underline{d})$ is equivalent to the solution of a quadratic programming problem: let $\underline{d}^k$ minimize $q^k(\underline{d})$. Then, if B^k is positive semi-definite, it is readily shown by examination of the Kuhn-Tucker conditions that (1) $\underline{d}^k = \underline{0}$ implies that $\underline{a}^k$ is a stationary point of (5.1), (2) $\underline{d}^k \neq \underline{0}$ is a descent direction for $\|f(\cdot, \underline{a})\|$ at $\underline{a}^k$. Line search and trust region algorithms based on (5.4) are given by Jónasson and Watson (1982), and Jónasson (1985). If the choice $B^k = G(\underline{a}^k, \underline{\lambda}^k)$ is made, where

$$G(\underline{a}, \underline{\lambda}) = \sum_{i=1}^{t^k} \lambda_i \nabla^2 g_i(\underline{a})\,, \qquad (5.5)$$

with $\underline{\lambda}^k$ an approximation to $\underline{\lambda}$, if $t^k = t$, and if $\theta_i^k = \theta_i^*$, $i = 1, 2, \cdots, t$, then $\underline{d}^k$ is the Newton step for the solution of the system of equations defined above, and therefore a satisfactory local convergence result may be established. (The right hand side of (5.5) may of course be regarded as the Hessian matrix of a Lagrangian function.) If $G(\underline{a}^k, \underline{\lambda}^k)$ is indefinite, then a suitable choice of B^k is

$$B^k = G(\underline{a}^k, \underline{\lambda}^k) + \tau^k I\,,$$

where $\tau^k > 0$ is a suitably chosen scalar. However, because $G(\underline{a}, \underline{\lambda})$ may not be positive definite at a stationary point, the algorithms of Jónasson (1985) eventually allow $\tau^k = 0$ to be used anyway: under mild conditions it may be shown that, if $\underline{a}^k \in \text{Nbhd}(\underline{a}^*)$, then $\underline{d}^k$ remains a descent direction at $\underline{a}^k$. It is interesting that, if in a neighbourhood of $\underline{a}^*$, $\underline{a}^{k+1} = \underline{a}^k + \underline{d}^k$ is possible when $B^k = G(\underline{a}^k, \underline{\lambda}^k)$, then the application of the method to linear Chebyshev set problems gives a procedure precisely equivalent to the second algorithm of Remes.

Examples of up to 3-dimensional linear and nonlinear

problems are solved in Jónasson (1985). The methods can be used as single phase methods, but it is generally more efficient to use the solution of a discretized problem as an initial approximation: for nonlinear problems, this can readily be achieved by redefining the method so that $M(\underline{a}^k)$ is the set of local maxima on the discrete set. The main disadvantage of the approach is the fact that eventually at each iteration $M(\underline{a}^k)$ (as originally defined) is required, and this is not a finite problem. However it may be argued that software for this is necessary anyway, even if only to check that (5.3) holds. A general method for this problem in an N-dimensional region is given in Jónasson (1985).

6. REGRESSION PROBLEMS WITH ERRORS IN ALL VARIABLES

A basic assumption in the use of the model equations (2.2) is that all errors are in the dependent variables (y_i values) and the independent variables (x_i values) are error free, or have negligible error. In many situations, however, this is an oversimplification and use of the usual methods may lead to bias in the estimated parameter and variance values. The more general errors-in-variables problem has a long history, but it is only recently that serious attention has been given to the provision of satisfactory numerical methods. As is to be expected, the sum of squared errors criterion has attracted most interest: attention will be directed here to the use of more general criteria.

Assume first that F is a linear function of $\underline{a}$ so that (2.1) may be expressed as

$$\underline{y} \approx A\underline{a}\,, \tag{6.1}$$

where the ith row of A depends only on x_i, $i=1,2,\cdots,m$. Then the effect of inaccuracies in the values of both x_i and y_i may be taken into account by introducing perturbations into both $\underline{y}$ and A and satisfying the model equations

$$\underline{y}+\underline{f} = (A+E)\underline{a}\,, \tag{6.2}$$

for small $\underline{f}$ and E. For example, if $\|\cdot\|$ is a norm on $m\times(n+1)$ matrices, this leads to the problem of calculating $\underline{a}$, E and $\underline{f}$ to

$$\text{minimize } \|E:\underline{f}\| \text{ subject to (6.2).} \tag{6.3}$$

This is known as the *total approximation problem*: the name comes from the use of the term 'total least squares' by Golub and Van Loan (1980) for the special case of (6.3) with the Euclidean norm.

Define the class of *separable norms* on $m\times(n+1)$ matrices as norms which have the property that, if $\|\cdot\|_R$ and $\|\cdot\|_D$ are norms on $\mathbb{R}^m$ and $\mathbb{R}^{n+1}$ respectively, then, for any $\underline{u}\in\mathbb{R}^m$ and $\underline{v}\in\mathbb{R}^{n+1}$,

$$\|\underline{u}\underline{v}^T\| = \|\underline{u}\|_R\,\|\underline{v}\|_D^*,$$

$$\|\underline{u}\underline{v}^T\|^* = \|\underline{u}\|_R^*\,\|\underline{v}\|_D,$$

where the superscript * denotes the dual norm. Examples of separable norms are the ℓ_p norms

$$\|M\| = \left(\sum_{i=1}^{m}\sum_{j=1}^{n+1}|m_{ij}|^p\right)^{1/p}, \quad 1\leqslant p<\infty,$$

where $\|\cdot\|_R=\|\cdot\|_p$ and $\|\cdot\|_D=\|\cdot\|_q$, with $p^{-1}+q^{-1}=1$, and operator norms

$$\|M\| = \max_{\|\underline{v}\|_D\leqslant 1}\|M\underline{v}\|_R.$$

The total approximation problem in separable norms has a close relationship with the following problem:

$$\text{minimize } \|[A:\underline{y}]\underline{v}\|_R \text{ subject to } \|\underline{v}\|_D=1. \tag{6.4}$$

Let $\underline{v}$ solve (6.4) with $v_{n+1}\neq 0$. Then it is shown in Osborne and Watson (1985) that

(1) $\underline{a}$ d fined by $\underline{v}=\tau\begin{bmatrix}\underline{a}\\-1\end{bmatrix}$ solves (6.3),

(2) the minimizing matrix $[E:\underline{f}]$ is the rank 1 matrix $[E:\underline{f}]=-[A:\underline{y}]\,\underline{v}\underline{w}^T$, where $\underline{w}\in\partial\|\underline{v}\|_D$.

Because neither (6.3) nor (6.4) is a convex problem, both problems may have a number of stationary points: these points are also related through the above results.

There are some generalizations of (6.3) which can be treated in a straightforward manner. Firstly it is possible to pre- and post-multiply $[E : \underline{f}]$ by positive diagonal weighting matrices: the effect on (6.4) is to transform $[A : \underline{y}]$ similarly. Secondly it is possible to take account of exact columns of A by omitting the corresponding components of $\underline{v}$ from the normalization condition of (6.4).

Algorithms for (6.4) when $\|\cdot\|_R = \|\cdot\|_p$ and $\|\cdot\|_D = \|\cdot\|_q$ with $1 < p, q < \infty$ are given in Watson (1984b) (see also Watson, 1985). A finite algorithm for the total ℓ_1 problem ($\|\cdot\|_R = \|\cdot\|_1$, $\|\cdot\|_D = \|\cdot\|_\infty$) is given by Osborne and Watson (1985).

In addition to the restriction that F be linear, a limitation of the model equations (6.2) is that no account is taken of the interrelationship of the elements of E. If perturbations are introduced directly into both x_i and y_i values, then from (2.1), the model equations are

$$y_i + f_i = F(x_i + e_i, \underline{a}), \quad i = 1, 2, \cdots, m, \tag{6.5}$$

and it seems appropriate to minimize some norm of the vector $(\underline{f}^T, \underline{e}^T)$ subject to satisfaction of (6.5). There is clearly much structure in (6.5) which can be exploited, and good methods should aim to do this. So far, it appears that only calculations involving the (weighted) least squares norm have been attempted: this gives the generalized least squares or orthogonal distance regression problem. However numerical methods based on (6.5) could well prove a burgeoning industry in the future.

REFERENCES

Armstrong, R.D. and Kung, D.S. (1980), "A dual method for discrete Chebyshev curve fitting", *Math. Programming*, *19*, pp.186-199.

Bartels, R.H. and Conn, A.R. (1982), "An approach to nonlinear ℓ_1 data fitting", in *Numerical Analysis, Cocoyoc 1981, Lecture Notes in Mathematics 909*, ed. Hennart, J.P., Springer-Verlag (Berlin), pp.48-58.

Bartels, R.H., Conn, A.R. and Charalambous, C. (1978), "On Cline's direct method for solving overdetermined linear equations in the ℓ_∞ sense", *SIAM J. Numer. Anal.*, *15*, pp.255-270.

Bartels, R.H., Conn, A.R. and Sinclair, J.W. (1978), "Minimization techniques for piecewise differentiable functions: the ℓ_1 solution to an overdetermined linear system", *SIAM J. Numer. Anal.*, *15*, pp.224-241.

Bartels, R.H. and Joe, B. (1983), "An exact penalty method for constrained, discrete, linear ℓ_∞ data fitting", *SIAM J. Sci. Statist. Comput.*, *4*, pp.69-84.

Belogus, D. and Liron, N. (1978), "DCR2: An improved algorithm for ℓ_∞ rational approximation on intervals", *Numer. Math.*, *31*, pp.17-29.

Bloomfield, P. and Steiger, W.L. (1980), "Least absolute deviations curve fitting", *SIAM J. Sci. Statist. Comput.*, *1*, pp.290-301.

Boncelet, C.G. Jr. and Dickinson, B.W. (1984), "A variant of Huber robust regression", *SIAM J. Sci. Statist. Comput.*, *5*, pp.720-734.

Breuer, P.T. (1986), "A new method for real uniform rational approximation", in *Algorithms for the Approximation of Functions and Data, Shrivenham 1985*, eds. Cox, M.G. and Mason, J.C., Oxford University Press (Oxford), to be published.

Busovača, S. (1984), *Handling Degeneracy in a Nonlinear ℓ_1 Algorithm*, Ph.D. Thesis, University of Waterloo.

Cheney, E.W. and Powell, M.J.D. (1987), "The differential correction algorithm for generalized rational functions", *Constr. Approx.* (to be published).

Clark, D.I. (1981), *Finite Algorithms for Linear Optimization Problems*, Ph.D. Thesis, Australian National University (Canberra).

Clark, D.I. (1985), "The mathematical structure of Huber's *M*-estimator", *SIAM J. Sci. Statist. Comput.*, *6*, pp.209-219.

Clark, D.I. and Osborne, M.R. (1986), "Finite algorithms for Huber's *M*-estimator", *SIAM J. Sci. Statist. Comput.*, *7*, pp.72-85.

Conn, A.R. (1979), "An efficient second order method to solve the (constrained) minimax problem", Report CORR 79-5, University of Waterloo.

Cromme, L. (1978), "Strong uniqueness: a far reaching criterion for the convergence analysis of iterative procedures", *Numer. Math.*. *29*, pp.179-193.

Dennis, J.E. and Moré, J.J. (1974), "A characterization of superlinear convergence and its applications to quasi-Newton methods", *Math. Comp.*, *28*, pp.549-560.

Ekblom, H. (1985), "A new algorithm for the Huber estimator in linear models", Department of Mathematics Report 1985-4, University of Luleå.

Fletcher, R. (1981), "Numerical experiments with an exact ℓ_1 penalty function method", in *Nonlinear Programming 4*, eds. Mangasarian, O.L., Meyer, R.R. and Robinson, S.M., Academic Press (New York), pp.99-129.

Fletcher, R. (1982), "A model algorithm for composite non-differentiable optimization problems", *Math. Programming Stud.*, *17*, pp.67-76.

Fletcher, R. (1985), "An ℓ_1 penalty method for nonlinear constraints", in *Numerical Optimization 1984*, eds. Boggs, P.T. Byrd, R.H. and Schnabel, R.B., SIAM Publications (Philadelphia), pp.26-40.

Glashoff, K. and Schultz, R. (1979), "Uber die genaue Berechnung von besten ℓ_1-Approximierenden", *J. Approx. Theory*, *25*, pp.280-293.

Golub, G.H. and Van Loan, C.F. (1980), "An analysis of the total least squares problem", *SIAM J. Numer. Anal.*, *17*, pp.883-893.

Gustafson, S.-Å. (1983), "A three-phase algorithm for semi-infinite programs", in *Semi-infinite Programming and Applications*, *Lecture Notes in Economics and Mathematical Systems 215*, eds. Fiacco, A.V. and Kortanek, K.O., Springer-Verlag (Berlin), pp.138-157.

Hald, J. and Madsen, K. (1979), "A 2-stage algorithm for minimax optimization", in *International Symposium on Systems Optimization and Analysis*, *Lecture Notes in Control and Information Sciences*, *14*, eds. Bensoussan, A. and Lions, J., Springer-Verlag (Berlin), pp.225-239.

Hald, J. and Madsen, K. (1981), "Combined LP and quasi-Newton methods for minimax optimization", *Math. Programming*, *20*, pp.49-62.

Hald, J. and Madsen, K. (1985), "Combined LP and quasi-Newton methods for nonlinear ℓ_1 optimization", *SIAM J. Numer. Anal.*, *22*, pp.68-80.

Han, S.-P. (1981), "Variable metric methods for minimizing a class of nondifferentiable functions", *Math. Programming*, *20*, pp.1-13.

Hettich, R. (1976), "A Newton method for nonlinear Chebyshev approximation", in *Approximation Theory*, *Lecture Notes in Mathematics*, *556*, eds. Schaback, R. and Scherer, K., Springer-Verlag (Berlin), pp.222-236.

Hettich, R. (1986), "An implementation of a discretization method for semi-infinite programming", *Math. Programming*, *34*, pp.354-361.

Holland, P.W. and Welsch, R.E. (1977), "Robust regression using iteratively reweighted least squares", *Comm. Statist.*, *A6*, pp.813-827.

Hopper, M.J. and Powell, M.J.D. (1977), "A technique that gains speed and accuracy in the minimax solution of overdetermined linear equations", in *Mathematical Software III*, ed. Rice, J.R., Academic Press (New York), pp.15-33.

Huber, P.J. (1973), "Robust regression: asymptotics, conjectures and Monte Carlo", *Ann. Statist.*, *1*, pp.799-821.

Huber, P.J. (1977), *Robust Statistical Procedures, CBMS Regional Conference Series in Applied Mathematics*, *27*, SIAM Publications (Philadelphia).

Huber, P.J. (1981), *Robust Statistics*, Wiley (New York).

Jittorntrum, K. and Osborne, M.R. (1980), "Strong uniqueness and second order convergence in nonlinear discrete approximation", *Numer. Math.*, *34*, pp.439-455.

Jónasson, K. (1985), *Numerical Methods for Continuous Chebyshev Approximation Problems*, Ph.D. Thesis, University of Dundee.

Jónasson, K. and Watson, G.A. (1982), "A Lagrangian method for multivariate continuous Chebyshev approximation problems", in *Multivariate Approximation Theory II*, *ISNM 61*, eds. Schempp, W. and Zeller, K., Birkhauser Verlag (Basel), pp.211-221.

Kaufman, E.H., Leeming, D.J. and Taylor, G.D. (1978), "A combined Remes-differential correction algorithm for rational approximation", *Math. Comp.*, *32*, pp.233-242.

Kaufman, E.H., Leeming, D.J. and Taylor, G.D. (1980), "A combined Remes-differential correction algorithm for rational approximation: experimental results", *Comput. Math. Appl.*, *6*, pp.155-160.

Madsen, K. (1985), "Minimization of nonlinear approximation functions", Report No. NI-85-01, Danmarks Tekniske Højskole.

McLean, R.A. and Watson, G.A. (1980), "Numerical methods for nonlinear discrete ℓ_1 approximation problems", in *Numerical Methods of Approximation Theory*, *ISNM 52*, eds. Collatz, L., Meinardus, G. and Werner, H., Birkhauser Verlag (Basel), pp.169-183.

Murray, W. and Overton, M.L. (1980), "A projected Lagrangian algorithm for nonlinear minimax optimization", *SIAM J. Sci. Statist. Comput.*, *1*, pp.345-370.

Murray, W. and Overton, M.L. (1981), "A projected Lagrangian algorithm for nonlinear ℓ_1 optimization", *SIAM J. Sci. Statist. Comput*, *2*, pp.207-224.

Osborne, M.R. (1981), "Algorithms for nonlinear discrete approximation", in *The Numerical Solution of Nonlinear Problems*, eds. Baker, C.T.H. and Phillips, C., Clarendon Press (Oxford), pp.270-286.

Osborne, M.R. (1985), *Finite Algorithms in Optimization and Data Analysis*, Wiley (Chichester).

Osborne, M.R. and Watson, G.A. (1985), "An analysis of the total approximation problem in separable norms, and an algorithm for the total ℓ_1 problem", *SIAM J. Sci. Statist. Comput.*, *6*, pp.410-424.

Powell, M.J.D. (1983), "General algorithms for discrete non-linear approximation calculations", in *Approximation Theory IV*, eds. Chui, C.K., Schumaker, L.L. and Ward, J.D., Academic Press (New York), pp.187-218.

Powell, M.J.D. and Yuan, Y. (1984), "Conditions for superlinear convergence in ℓ_1 and ℓ_∞ solutions of overdetermined non-linear equations", *IMA J. Numer. Anal.*, *4*, pp.241-251.

Rocke, D.M., Downs, G.W. and Rocke, A.J. (1982), "Are robust estimators really necessary?", *Technometrics*, *24*, pp.95-101.

Shanno, D.F. and Rocke, D.M. (1986), "Numerical methods for robust regression: linear models", *SIAM J. Sci. Statist. Comput.*, *7*, pp.86-97.

Watson, G.A. (1976), "A method for calculating best nonlinear Chebyshev approximations", *J. Inst. Math. Appl.*, *18*, pp.351-360.

Watson, G.A. (1979), "The minimax solution of an overdetermined system of nonlinear equations", *J. Inst. Math. Appl.*, *23*, pp.167-180.

Watson, G.A. (1980), "An algorithm for a class of nonlinearly constrained nondifferentiable optimization problems", in *Constructive Methods of Finite Nonlinear Optimization, ISNM 55*, eds. Collatz, L., Meinardus, G. and Wetterling, W., Birkhauser Verlag (Basel), pp.197-211.

Watson, G.A. (1981), "An algorithm for linear ℓ_1 approximation of continuous functions", *IMA J. Numer. Anal.*, *1*, pp.157-167.

Watson, G.A. (1982), "A globally convergent method for (constrained) nonlinear continuous ℓ_1 approximation problems", in *Numerical Methods of Approximation Theory, ISNM 59*, eds. Collatz, L., Meinardus, G. and Werner. H., Birkhauser Verlag (Basel), pp.233-243.

Watson, G.A. (1984a), "Discrete ℓ_1 approximation by rational functions", *IMA J. Numer. Anal.*, *4*, pp.275-288.

Watson, G.A. (1984b), "The numerical solution of total ℓ_p approximation problems", in *Numerical Analysis Dundee 1983. Lecture Notes in Mathematics 1066*, ed. Griffiths, D.F., Springer-Verlag (Berlin), pp.221-238.

Watson, G.A. (1985), "On a class of algorithms for total approximation", *J. Approx. Theory*, *45*, pp.219-231.

Womersley, R.S. (1985), "Local properties of algorithms for minimizing nonsmooth composite functions", *Math. Programming*, *32*, pp.69-89.

Womersley, R.S. and Fletcher, R. (1986), "An algorithm for composite nonsmooth optimization problems", *J. Optim. Theory Appl.*, *48*, pp.493-523.

(for name and address, see over)

G.A. Watson
Department of Mathematical Sciences
University of Dundee
Dundee DD1 4HN
Scotland

7 Branch Cuts for Complex Elementary Functions or Much Ado About Nothing's Sign Bit

W. KAHAN

ABSTRACT

Zero has a usable sign bit on some computers, but not on others. This accident of computer arithmetic influences the definition and use of familiar complex elementary functions like $\sqrt{}$, arctan and arccosh whose domains are the whole complex plane with a slit or two drawn in it. The Principal Values of those functions are defined in terms of the logarithm function from which they inherit discontinuities across the slit(s). These discontinuities are crucial for applications to conformal maps with corners. The behaviour of those functions on their slits can be read off immediately from defining Principal Expressions introduced in this paper for use by analysts. Also introduced herein are programs that implement the functions fairly accurately despite roundoff and other numerical exigencies. Except at logarithmic branch points, those functions can all be continuous up to and onto their boundary slits when zero has a sign that behaves as specified by IEEE standards for floating-point arithmetic; but those functions must be discontinuous on one side of each slit when zero is unsigned. Thus does the sign of zero lay down a trail from computer hardware through programming language compilers, run-time support libraries and applications programmers to, finally, mathematical analysts.

1. INTRODUCTION

Conventions dictate the ways nine familiar multiple-valued complex elementary functions, namely

$$\sqrt{},\ \ell n,\ \text{arcsin},\ \text{arccos},\ \text{arctan},\ \text{arcsinh},\ \text{arccosh},\ \text{arctanh},\ z^w,$$

shall be represented by single-valued functions called "Principal Values". These single-valued functions are defined and analytic throughout the complex plane except for discontinuities across certain straight lines called "slits" so situated as to maximize the reign of continuity, conserving as many as possible of the properties of these functions' familiar real restrictions to apt segments of the real axis. There can be no dispute about where to put the slits; their locations are deducible. However, Principal Values have too often been left ambiguous on the slits, causing confusion and controversy insofar as computer programmers have had to agree upon their definitions. This paper's thesis is that most of that ambiguity can and should be resolved; however, on computers that conform to the IEEE standards 754 and p854 for floating-point arithemetic the ambiguity should not be eliminated entirely because, paradoxically, what is left of it usually makes programs work better.

What has to be ambiguous is the sign of zero. In the past, most people and computers would assign no sign to zero except under duress, and then they would treat the sign as + rather than −. For example, the real function

$$\begin{aligned} \mathrm{signum}(x) &:= +1 \quad \text{if} \quad x > 0 \,, \\ &:= \;\; 0 \quad \text{if} \quad x = 0 \,, \\ &:= -1 \quad \text{if} \quad x < 0 \,, \end{aligned}$$

illustrates the traditional noncommittal attitude toward zero's sign, whereas the Fortran function

$$\begin{aligned} \mathrm{sign}(1.0, x) &:= +1.0 \quad \text{if} \quad x \geqslant 0 \,, \\ &:= -1.0 \quad \text{if} \quad x < 0 \,, \end{aligned}$$

must behave as if zero had a + sign in order that this function and its first argument have the same magnitude. Just as $\mathrm{sign}(1.0, x)$ is continuous at $x = 0+$, i.e. as x approaches zero from the right, so can each principal value above be continuous as its slit is reached from one side but not from the other. Sides can be chosen in a consistent way among all the elementary complex functions, as they have been chosen for the implementations built into the Hewlett-Packard hp-15C calculator that will be used to illustrate this approach.

The IEEE standards 754 and p854 take a different approach. They prescribe representations for both $+0$ and -0 but do not distinguish between them during ordinary arithmetic operations, so the ambiguity is benign. Rather than think of $+0$ and -0 as distinct numerical values, think of their sign bit as an auxiliary variable that conveys one bit of information (or misinformation) about any numerical variable that takes on 0 as its value. Usually this information is irrelevant; the value of $3+x$ is no different for $x := +0$ than for $x := -0$, and the same goes for the functions signum(x) and sign(y, x) mentioned above. However, a few extraordinary arithmetic operations *are* affected by zero's sign; for example $1/(+0) = +\infty$ but $1/(-0) = -\infty$. To retain its usefulness, the sign bit must propagate through certain arithmetic operations according to rules derived from continuity considerations; for instance $(-3)(+0) = -0$, $(-0)/(-5) = +0$, $(-0)-(+0) = -0$, etc. These rules are specified in the IEEE standards along with the one rule that had to be chosen arbitrarily:

$s - s := +0$ for every string s representing a finite real number.

Consequently when $t = s$, but $0 \neq t \neq \infty$, then $s-t$ and $t-s$ both produce $+0$ instead of opposite signs. (That is why, in IEEE style arithmetic, $s-t$ and $-(t-s)$ are numerically equal but not necessarily indistinguishable.) Implementations of elementary transcendental functions like sin(z) and tan(z) and their inverses and hyperbolic analogs, though not specified by the IEEE standards, are expected to follow similar rules; if $f(0) = 0 < f'(0)$, then the implementation of $f(z)$ is expected to reproduce the sign of z as well as its value at $z = \pm 0$. That does happen in several libraries of elementary transcendental functions; for instance, it happens on the Motorola 68881 Floating-Point Coprocessor, on Apple computers in their Standard Apple Numerical Environment, in Intel's Common Elementary Function Libraries for the i8087 and i80287 floating-point coprocessors, in analogous libraries now supplied with the Sun III, with the ELXSI 6400 and with the IBM PC/RT, and in the *C* Math Library currently distributed with 4.3 BSD UNIX for machines that conform to IEEE 754. With a few unintentional exceptions, it happens also on the hp-71B hand-held computer, whose arithmetic was designed to conform to IEEE p854.

If a programmer does not find these rules helpful, or if he does not know about them, he can ignore them and, as has been necessary in the past, insert explicit tests for zero in his program wherever he must cope with a discontinuity at zero. On the other hand, if the standards' rules happen to produce the desired results without such tests, the tests may be omitted leaving the programs simpler in appearance though perhaps more subtle. This is just what happens to programs that implement or use the elementary functions named above, as will become evident below.

2. WHERE TO PUT THE SLITS

Each of our nine elementary complex functions $f(z)$ has a slit or slits that bound a region, called the *principal domain*, inside which $f(z)$ has a *principal value* that is single valued and analytic (representable locally by power series), though it must be discontinuous across the slit(s). That principal value is an extension, with maximal principal domain, of a real elementary function $f(x)$ analytic at every interior point of its domain, which is a segment of the real x-axis. To conserve the power series' validity, points strictly inside that segment must also lie strictly inside the principal domain; therefore the slit(s) cannot intersect the segment's interior. Let $z^* = x - iy$ denote the complex conjugate of $z = x + iy$; the power series for $f(x)$ satisfy the identity $f(z^*) = f(z)^*$ within some complex neighbourhood of the segment's interior, so the identity should persevere throughout the principal domain's interior too. Consequently complex conjugation must map the slit(s) to itself/themselves. The slit(s) of an *odd* function $f(z) = -f(-z)$

must be invariant under reflection in the origin $z = 0$. Finally, the slit(s) must begin and end at *branch-points*: these are singularities around which some branch of the function cannot be represented by a Taylor nor Laurent series expansion. A slit can end at a branch point at infinity.

Consequently the slit for $\sqrt{}$, ℓn and z^w turns out to be the negative real axis. Then the slits for arcsin, arccos and arctanh turn out to be those parts of the real axis not between -1 and $+1$; similarly those parts of the imaginary axis not between $-i$ and $+i$ serve as slits for arctan and arcsinh. The slit for arccosh, the only slit with a finite branch-point (-1) inside it, must be drawn along the real axis where $z \leqslant +1$. None of this is controversial, although a few other writers have at times drawn the slits elsewhere either for a special purpose or by mistake; other tastes can be accommodated by substitutions sometimes so simple as writing, say, $\ell n(-1) - \ell n(-1/z)$ in place of $\ell n(z)$ to draw its slit along (and just under) the positive real axis instead of the negative real axis.

3. WHY DO SLITS MATTER ?

A computer program that includes complex arithmetic operations must be a product of a deductive process. One stage in that process might have been a model formulated in terms of analytic expressions that constrain physically meaningful variables without telling explicitly how to compute them. From those expressions somebody had to deduce other complex analytic expressions that the computer will evaluate to solve the given physical problem. The deductive process entails transformations among which some may resemble algebraic manipulations of real expressions, but with a crucial difference:

> Certain transformations, generally valid for real expressions, are valid for complex expressions only while their variables remain within suitable regions in the complex plane.

Moreover, those regions of validity can depend disconcertingly upon the computer that will be used to evaluate the expressions in question. For example, simplifying the expression $\sqrt{(z/(z-1))}\sqrt{(1/(z-1))}$ to $\sqrt{(z)}/(z-1)$ seems legitimate in so far as they both describe the same complex function, one that is continuous everywhere except for a pole at $z=1$ and a jump-discontinuity along the negative real axis $z<0$. And when those two expressions are evaluated upon a variety of computers including the ELXSI 6400, the Sun III, the IBM PC/RT, the IBM PC/AT, PC/XT and PC using i80287 or i8087, and the hp-71B, they agree *everywhere* within a rounding error or two. But when the same expressions are evaluated upon a different collection of computers including CRAYs, the IBM 370 family, the DEC VAX line, and the hp-15C, those expressions take opposite signs along the negative real axis! An experience like this could undermine one's faith in some computers.

What deserves to be undermined is blind faith in the power of Algebra. We should not believe that the equivalence class of expressions that all describe the same complex analytic function can be recognized by algebraic means alone, not even if relatively uncomplicated expressions are the only ones considered. To locate the domain upon which two analytic expressions take equal values generally requires a combination of algebraic, analytical and topological techniques. The paradigm is familiar to complex analysts, but it will be summarized here for the sake of other readers, using the two expressions given above for concrete illustration.

<u>How to decide where two analytic expressions describe the same function</u>.

1. Locate the singularities of each constituent subexpression of the given expressions.

The singularities of an analytic function are the boundary points of its domain of analyticity. These will consist of poles, branch-points and slits in this paper; but more generally they would include certain contours of integration, boundaries of regions of convergence, etc. In general, singularities can be hard to find; in our examples the singularities are obviously the pole at $z=1$, the branch-point $z=0$, and respective slits $0<z<1$, $z<1$ and $z<0$ whereon the quantities under square root signs are negative real.

2. Taken together, the singularities partition the complex plane into a collection of disjoint connected components. Inside each such component locate a *small continuum* upon which the equivalence of the given two expressions can be decided; that decision is valid throughout the component's interior.

The "small continuum" might be a small disk inside which both expressions are represented by the same Taylor series; or it could be a curvilinear arc within which both expressions take values that can be proved equal by the laws of real algebra. Other possibilities exist; some will be suggested by whatever motivated the attempt to prove that the given expressions are equivalent. In our example, the two expressions are easily proven equal on that part of the real axis where $z>1$, which happens to lie inside the one connected component into which the slits along the rest of the real axis divide the complex plane. Therefore the two expressions must be equivalent everywhere in the complex plane except possibly for real $z \leqslant 1$.

3. The singularities constitute loci in the plane upon which the processes in steps 1 and 2 above can be repeated, finally leaving isolated singular points to be handled individually. End of paradigm.

In our example, the slit along $z<1$ is partitioned into two connected components by the branch-point at $z=0$. Each component has to be handled separately. Whether the two expressions are equivalent on a component must depend upon the definition of complex $\sqrt{z}$ on its slit where $z<0$; there diverse computers appear to disagree. That is what this paper is about.

More generally, programmers who compose complex analytic expressions out of the nine elementary functions listed at this paper's beginning will have to verify whether their expressions deliver the functions that they intend to compute. In principle, that verification could proceed without prior agreements about the functions' values on their slits if instead analysts and programmers were obliged to supply an explicit expression to handle every boundary situation as they intend. Such a policy seems inconsiderate (not to say unconscionable) considering how hard some singularities are to find, and how easy to overlook; but that policy is not entirely heartless since verifying correctness along a boundary costs the intellect nearly as much as writing down a statement of intent about that boundary. The trouble with those statements is that they generally contain inequalities and tests and diverse cases, and as they accumulate they burden proofs and programs with a dangerously enlarged capture cross-section for errors. And almost all of those statements become superfluous in programs after we agree upon reasonable definitions for the functions in question on their slits.

For instance, in our example above we had to discover whether the two expressions agreed on an interval $0 < z < 1$ that lies strictly inside the domain of the desired function's analyticity, not on its boundary. That interval turns out to be a *removable singularity*, and it does remove itself from all the computers mentioned above because they evaluate both expressions correctly on that interval; diverse computers disagree only on the boundary where the desired function is discontinuous. Perhaps that's just luck. (Unlucky examples do exist and one will be presented later.) Let us accept good luck with gratitude whenever it simplifies our programs.

Complex analytic expressions that involve slits and other singularities are intrinsically complicated, and they get more complicated when rounding errors are taken into account. Our objective cannot be to make complicated things simple but rather, by choosing reasonable values for our nine elementary functions on their slits, to make them no worse than necessary.

4. PRINCIPAL VALUES ON THE SLITS, IEEE STYLE

Since all the slits in question lie on either the real or the imaginary axis, every point z on a slit is represented in at least two ways, at least once with a $+0$ and at least once with a -0 for whichever of the real and imaginary parts of z vanishes. Benignly, ambiguity in z at a discontinuity of $f(z)$ permits $f(z)$ to be defined formally continuously, except possibly at the ends of some slits, by continuation from inside the principal domain. This continuity goes beyond mere formality. By analytic continuation, the domain of each of our nine elementary functions $f(z)$ extends until it fills out a *Riemann Surface*; think of this surface as a multiple covering wrapped like a bandage around the *Riemann Sphere* and mapped onto it continuously by f. To construct f's principal domain, cut the bandage along the slit(s) and discard all but one layer covering the sphere. That layer is a *closed* surface mapped by f continuously onto a subset of the sphere. The shadow of that layer projected down upon the sphere is the principal domain; it consists of the whole sphere, but with slit(s) covered twice. That is why we wish to represent slits ambiguously.

Here are some illustrative examples, the first of a real function that is recommended for any implementation of IEEE standard 754 or p854.

$$\left.\begin{array}{l} \text{copysign}(x,\, y) := \pm x \text{ where the sign bit is that of } y\text{, so} \\ \text{copysign}(1,+0) = +1 = \lim \text{copysign}(1,\, y) \quad \text{at} \quad y = 0+\text{, and} \\ \text{copysign}(1,-0) = -1 = \lim \text{copysign}(1,\, y) \quad \text{at} \quad y = 0-\,. \end{array}\right\} \quad (4.1)$$

$$\left.\begin{array}{l} \sqrt{(-1+i0)} = +0+i = \lim \sqrt{(-1+iy)} \quad \text{at} \quad y = 0+\,; \\ \sqrt{(-1-i0)} = +0-i = \lim \sqrt{(-1+iy)} \quad \text{at} \quad y = 0-\,. \end{array}\right\} \quad (4.2)$$

Consequently, $\sqrt{(z^*)} = \sqrt{(z)}^*$ for every z, and $\sqrt{(1/z)} = 1/\sqrt{(z)}$ too. These identities persist within roundoff provided the programs used for square root and reciprocal are those, supplied in this paper, that would have been chosen anyway for their efficiency and accuracy.

$$\left.\begin{aligned} \arccos(2 + i0) &= +0 - i \operatorname{arccosh}(2) \\ &= \lim \arccos(2 + iy) \text{ at } y = 0+ ; \\ \arccos(2 - i0) &= +0 + i \operatorname{arccosh}(2) \\ &= \lim \arccos(2 + iy) \text{ at } y = 0- . \end{aligned}\right\} \quad (4.3)$$

An implementation of arccos that preserves full accuracy in the imaginary part of $\arccos(2 + iy)$ when $|y|$ is very tiny can be expected to get its sign right when $y = \pm 0$ too without extra tests in the code; such a program is supplied later in this paper.

But the foregoing examples make it all seem too simple. The next example presents a more balanced picture.

Let function $a(x) := \sqrt{(x^2 - 1)}$ for real x with $x^2 \geqslant 1$, and let $b(x) := a(x)$ for real $x \geqslant 1$; note that $b(x)$ is not yet defined when $x \leqslant -1$. The principal values of the complex extensions of a and b following the principles enunciated above turn out to be

$$a(z) = \sqrt{(z^2 - 1)} \qquad = a(-z), \text{ and}$$
$$b(z) = \sqrt{(z - 1)}\sqrt{(z + 1)} = -b(-z) .$$

Both a and b are defined throughout the complex plane and both have a slit on the real axis running from -1 to $+1$, but a has another slit that runs along the entire imaginary axis separating the right half-plane where $a = b$ from the left half-plane where $a = -b$. The functions are different because generally

$$\begin{aligned} \sqrt{(\xi)}\sqrt{(\eta)} &= \sqrt{(\xi\eta)} \quad \text{when} \quad |\arg(\xi) + \arg(\eta)| < \pi , \\ &= -\sqrt{(\xi\eta)} \quad \text{when} \quad |\arg(\xi) + \arg(\eta)| > \pi , \\ &= \pm\sqrt{(\xi\eta)} \quad \text{(hard to say which) when} \quad \xi\eta \leqslant 0 . \end{aligned}$$

Both functions a and b are continuous up to and onto ambiguous boundary points in IEEE style arithmetic, as described above, only if that arithmetic is implemented carefully; in particular, the expression $z + 1$ should not be replaced by the ostensibly equivalent $z + (1 + i0)$ lest the sign of zero in the imaginary part of z be reversed wrongly. (Generally, mixed-mode arithmetic combining real and complex variables should be performed directly, not by first coercing the real to complex, lest the sign of zero be rendered uninformative; the same goes for combinations of pure imaginary quantities with complex variables. And doing arithmetic directly this way saves execution time that would otherwise be squandered manipulating zeros.) When z is near ± 1 the expression $a(z)$ nearly vanishes and loses its relative accuracy to roundoff. Although this loss could be avoided by rewriting $a(z) := \sqrt{((z-1)\,(z+1))}$, doing so would obscure the discontinuity on the imaginary axis in a cloud of roundoff which obliterates $\mathrm{Re}(z)$ whenever it is very tiny compared with 1 as well as when it is ± 0.

Also obscure is what happens at the ends of some slits. Take for example $\ell n(z) = \ell n(\rho) + i\theta$, where $\rho = |z|$ and $\theta = \arg(z)$ are the polar coordinates of $z = x + iy$ and satisfy

$$x = \rho\cos\theta\,, \quad y = \rho\sin\theta\,, \quad \rho \geqslant 0 \text{ and } -\pi \leqslant \theta \leqslant \pi\,.$$

Evidently $\rho := +\sqrt{(x^2 + y^2)}$, and when $0 < \rho < +\infty$ then

$$\theta := 2\arctan\left(y/(\rho + x)\right) \quad \text{if } x \geqslant 0\,, \quad \text{or}$$
$$:= 2\arctan\left((\rho - x)/y\right) \quad \text{if } x \leqslant 0\,.$$

At the end of the slit where $z = x = y = \rho = 0$ (and $\ell n(\rho) = -\infty$) the value of θ may seem arbitrary, but in fact it must cohere with other almost arbitrary choices concerning division by zero and arithmetic with infinity. A reasonable choice is to interpose the reassignment

$$\text{if } \rho = 0 \text{ then } x := \mathrm{copysign}(1,\, x)$$

between the computations of ρ and θ above. More about that later.

The foregoing examples provide an unsettling glimpse of the complexities that have daunted implementers of compilers and run-time libraries who would otherwise extend to complex arithmetic the facilities they have supplied for real floating-point computation. These complexities are attributable to failures, in complex floating-point arithmetic, of familiar relationships like algebraic identities that we have come to take for granted in the arena of real variables. Three classes of failures can be discerned:

(i) The domain of an analytic expression can enclose singularities that have no counterparts inside the domain of its real restriction. That is why, for example, $\sqrt{(z^2-1)} \neq \sqrt{(z-1)}\sqrt{(z+1)}$.

(ii) Rounding errors can obscure the singularities. That is why, for example, $\sqrt{(z^2-1)} = \sqrt{((z-1)(z+1))}$ fails so badly when either $z^2 = 1$ very nearly or when $z^2 < 0$ very nearly. To avoid this problem, the programmer may have to decompose complex arithmetic expressions into separate computations of real and imaginary parts, thereby forgoing some of the advantages of a compact notation.

(iii) Careless handling can turn infinity or the sign of zero into misinformation that subsequently disappears leaving behind only a plausible but incorrect result. That is why compilers must not transform $z-1$ into $z-(1+i0)$, as we have seen above, nor $-(-x-x^2)$ into $x+x^2$, as we shall see below, lest a subsequent logarithm or square root produce a nonzero imaginary part whose sign is opposite to what was intended.

The first two classes are hazards to all kinds of arithmetic; only the third kind of failure is peculiar to IEEE style arithmetic with its signed zero. Yet all three kinds must be linked together esoterically because the third kind is not usually found in an applications program unless that program suffers also from the second kind. The link is fragile, easily broken if the rational operations or elementary functions, from which applications programs are composed, contain either of the last two kinds of failures. Therefore, implementers of compilers and run-time libraries bear a heavy burden of attention to detail if applications programmers are to realize the full benefit of the IEEE style of complex arithmetic. That benefit deserves some discussion here if only to reassure implementers that their assiduity will be appreciated.

The first benefit that users of IEEE style complex arithmetic notice is that familiar identities tend to be preserved more often than when other styles of arithmetic are used. The mechanism that preserves identities can be revealed by an investigation of an analytic function $f(z)$ whose domain is slit along some segment of the real or imaginary axis; say the real axis. When $z = x + iy$ crosses the slit, $f(z)$ jumps discontinuously as y reverses sign although $f(z)$ is continuous as z approaches one side of the slit or the other. Consequently the two limits

$$f(x + i0) := \lim f(x+iy) \text{ as } y \to 0+ \text{ and}$$
$$f(x - i0) := \lim f(x+iy) \text{ as } y \to 0-$$

both exist, but they are different when x has a real value inside the slit. Ideally, a subroutine $F(z)$ programmed to compute $f(z)$ should match these values; $F(x \pm i0) = f(x \pm i0)$ respectively should be satisfied within a small tolerance for roundoff. This normally happens in IEEE style arithmetic as a by-product of whatever steps have been taken to ensure that $F(x + iy) = f(x + iy)$, within a similarly small tolerance, for all sufficiently small but nonzero $|y|$. To generate a discontinuity, the subroutine F must contain constructions similar to copysign$(\cdots, y)$ or $\arctan(1/y)$ possibly with "y" replaced by some other expression that either vanishes or tends to infinity as $y \to 0$. That expression cannot normally be a sum or difference like $\arctan(y-1) + \pi/4$ or $\exp(y) - 1$ that vanishes by cancellation, because roundoff can give such expressions values (typically 0) that have the wrong sign when $|y|$ is tiny enough. Instead, to preserve accuracy when $|y|$ is tiny, that expression must normally be a real product or quotient involving a power of y or $\sin(y)$ or some other built-in function that vanishes with y and therefore should inherit its sign at $y = \pm 0$. Thus does careful implementation of compiler and library combine with careful applications programming to yield correct behaviour on and near the slit. And if two such carefully programmed subroutines $F(z)$, though based upon different formulas, agree within roundoff everywhere near the slit, then the foregoing reasoning implies that normally they have to agree on the slit too; this is the way IEEE style arithmetic preserves identities like $\sqrt{(z^*)} = (\sqrt{z})^*$

and $\sqrt{(1/z)} = 1/\sqrt{z}$ that would have to fail on slits if zero had no sign.

Of course, applications programmers generally have things more important than the preservation of identities on their minds. Figure 1 shows a more typical and realistic example. Here $f(z) := 1 + z^2 + z\sqrt{(1+z^2)} + \ell n(z^2 + z\sqrt{(1+z^2)})$, and we construe the equation $\zeta := f(z)$ as a conformal map, from the plane of $z = x + iy$ to the plane of $\zeta = \xi + i\eta$, that maps the right half-plane $x \geqslant +0$ onto the space occupied by a liquid that is forced by high pressure to jet into a slot. The walls of the slot, where $\xi < 0$ and $\eta = \pm\pi$, should be the images of those parts of the imaginary axis $z^2 < -1$ lying beyond $\pm i$. The free surfaces of the jet, curving forward from $\zeta = \pm i\pi$ and then back to $\zeta = -\infty \pm i\pi/2$, should be the image of that segment of the imaginary axis $-1 < z^2 < 0$ between $\pm i$.

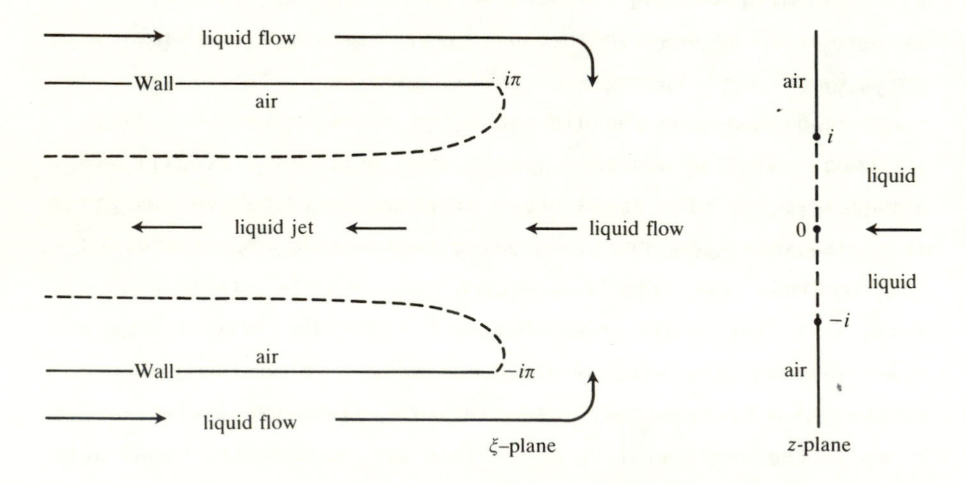

FIG. 1 Conformal map $\zeta := f(z)$ of half-plane to jet with free boundary

The picture of $f(z)$ should be symmetrical about the real axis because $f(z^*) = f(z)^*$. As z runs up the imaginary axis, with $x = +0$ and y running from $-\infty$ through -1 toward -0 and then from $+0$ through $+1$ toward $+\infty$, its image $\zeta = f(z)$ should run from left to right along the lower wall and back along the lower free boundary of the jet, then from left to right along the jet's upper free boundary and back along the upper wall. This is just what happens when $f(z)$ is plotted from a one-line program on the hp-71B calculator, which implements the proposed IEEE standard p854. But when $f(z)$ is programmed onto the hp-15C, whose zero is unsigned, the lower wall disappears. Its pre-image, the lower part of the imaginary axis where $z/i < -1$, is mapped during the computation of $f(z)$ into the slit that belongs to $\sqrt{}$ and ℓn; the upper part $z/i > 1$ gets mapped onto the same slit. For lack of a signed zero, that slit gets attached to a side that is right for the upper wall but wrong for the lower wall, thereby throwing the pre-image of the lower wall away into a tiny segment of the upper wall. To put the lower wall back, x must be increased from 0 to a tiny positive value while y runs from $-\infty$ to -1. (How tiny should x be? That's a nontrivial question.)

The misbehaviour revealed in the foregoing example $f(z)$ may appear to be deserved because $f(z)$ has slits on the imaginary axis $z^2 < -1$ beyond $\pm i$. Should mapping a slit to the wrong place be blamed upon the discontinuity there rather than upon arithmetic with an unsigned zero? No. Arithmetic with an unsigned zero can also cause other programs to misbehave similarly at places where the functions being implemented are otherwise well behaved. For example consider $c(z) := z - i\sqrt{(iz+1)}\sqrt{(iz-1)}$, whose slit lies in the imaginary axis $-1 < z^2 < 0$ between $\pm i$. Now $\zeta := c(z)$ maps the slit z plane onto the ζ plane outside the circle $|\zeta| \geqslant 1$; vertical lines in the z plane map to stream lines in the vertical flow of a fluid around the circle. Implementing $c(z)$, the programmer notices that he can reduce two expensive square roots to one by rewriting

$$c(z) := z + \sqrt{(z^2+1)}\ \mathrm{copysign}(1, \mathrm{Re}(z)).$$

The two expressions for $c(z)$ match everywhere in IEEE style arithmetic; but when zero has only one sign, say $+$, the second expression maps the lower part of the imaginary axis, where $z/i < -1$, into the inside instead of the outside of the circle, although $c(z)$ should be continuous there.

The ease with which IEEE style arithmetic handled the important singularities near $z=\pm i$ in the examples above should not be allowed to persuade the reader that all singularities can be dispatched so easily. The singularities $f(0)$ and $f(\infty)$ and the overflows near $z=\infty$ would have to be handled in the usual ways if they did not lie so far off the left-hand side of the picture that nobody cares. Another kind of singularity that did not matter here, but might matter elsewhere, insinuated weasel words like "not usually", "tends to be" and "normally" into the earlier discussion of sums and differences that normally vanish by cancellation. Sums and differences can vanish without cancellation if they combine terms that have already vanished; an example is $h(x):=x+x^2$ when $x=0$. Evaluating $h(\pm 0)$ in IEEE style real arithemetic yields $+0$ instead of ± 0 respectively, losing the sign of zero. $h(x)$ has other troubles: it signals Underflow when x is very tiny, suffers inaccuracy when x is very near -1, and becomes Invalid at $x=-\infty$. Simply rewriting $h(x):=x(1+x)$ dispels all these troubles, but is slightly less accurate for very tiny $|x|$ than is $h(x):=-(-x-x^2)$, which preserves accuracy and the sign of zero for all tiny real x. Complex arithmetic complicates this situation. Both expressions $z+z^2$ and $z(1+z)$ produce zeros with the wrong sign for $\mathrm{Im}(h(z))$ on various segments of the real z-axis; to get the correct sign and better accuracy requires an expression like

$$h(x+iy) := x(1+x)-y^2+2iy(x+0.5)$$

regardless of arithmetic style. For similar reasons, the expression for $f(z)$ used above for the conformal map would have to be rewritten if the interesting part of its domain were the left instead of right half-plane.

IEEE style complex arithmetic appears to burden the implementers of compilers and run-time libraries with a host of complicated details that need rarely bother the user if they are dispatched properly; and then familiar identities will persist, despite roundoff, more often than in other styles of arithmetic. This thought would comfort us more if the aberrations were easier to uncover. Locating potential aberrations remains an onerous task for an applications programmer, regardless of the style of arithmetic; however that style can affect the locus of aberration fundamentally. In IEEE style arithmetic, a programmed implementation of a complex analytic function can take aberrant boundary values, different from what would be produced by continuation from the interior, because of roundoff or similar phenomena. In arithmetic without a signed zero, such an aberration can be caused as well by an unfortunate choice of analytic expression, though the programmer has implemented it faithfully. The fact that an analytic expression determines the values of an analytic function correctly inside its domain is no reason to expect the boundary values to be determined correctly too when zero is unsigned.

5. PRINCIPAL VALUES ON THE SLITS, hp–15C STYLE

Of course, the hp-15C is not the only machine with an unsigned zero; a DEC VAX 11 model is similar but lacks so far a careful software implementation of some of the functions under discussion – in time that lack will be remedied. Many other machines, the IBM 370 series among them, have a signed zero in their hardware but no provision for propagating its sign in a coherent and useful way, so they are customarily programmed as if zero were unsigned. All these machines discourage attempts to distinguish one side of a slit from the other on the slit itself.

What we have to do is attach each slit to one of its sides in accordance with some reasonable rule, thereby obtaining a principal value which is continuous up to the slit from that side but not from the other. In other words, we have to assign a sign to zero on each slit and then compute the same principal value as would have been computed using IEEE-style arithmetic. The assignment cannot be arbitrary; for instance we cannot change sides in the middle of the slit lest a gratuitous singularity be insinuated by the change. On the other hand, some degree of arbitrariness is obligatory. For instance, the two functions

$$b(z) := \sqrt{(z-1)}\,\sqrt{(z+1)} \quad \text{and} \quad -b(-z)$$

are indistinguishable everywhere except in the slit $-1 < z < 1$ across which they are discontinuous, but in hp-15C style arithmetic one function must be continuous onto the top of the slit and the other onto the bottom. Evidently no general rule attaching a slit to one of its sides can depend solely upon the slit's shape nor solely upon the function's values off the slit. And yet, paradoxically, the hp-15C appears to follow just such a rule, namely

Counter-Clockwise Continuity (CCC):

> Attach each slit to whichever side is approached when the finite branch-point at its end is circled counter-clockwise.

Thus when z is real and negative CCC defines $\sqrt{z} = i\sqrt{|z|}$ and $\ell n(z) = \ell n|z| + i\pi$. Actually CCC is merely a mnemonic summary of the implications, for the nine functions that are the subject of this note, of the following more general convention applicable also to $b(z)$ above, as CCC is not.

The Principal Expression:

> Assign to each elementary function in question not merely a Principal Value but also a *Principal Expression* in terms of $\ell n(z)$ and $\sqrt{z}$, using the simplest formula that manifests its behaviour at finite branch-points without gratuitous singularities elsewhere.

What makes this convention effective is a canonical association between the archetypal branch-points of $\ell n(z)$ and $\sqrt{z}$ on the one hand, and on the other any isolated branch-point at the end of a slit belonging to any other elementary function. For example,

$\operatorname{arcsin}(z) = \pi/2 - (\text{power series in } 1-z)\sqrt{(1-z)}$ for z near 1,
$\operatorname{arccosh}(z) = \ln(2z) - (\text{power series in } 1/z)$ when $|z|$ is huge,
$\operatorname{arctanh}(z) = -0.5\ln(1-z) + (\text{power series in } 1-z)$ for z near 1.

In each case the power series is determined uniquely. In general, if β is a finite branch-point at the end of a slit belonging to one of our nine functions $f(z)$, and if the function is analytic inside some circular disk $|z-\beta| < \rho$ except on the slit, then $f(z)$ can be represented inside that slit disk by one of the formulas

$$f(z) = P(z-\beta) + p(z-\beta)\sqrt{((z-\beta)/c)}\ , \text{ or}$$
$$f(z) = P(z-\beta) + p(z-\beta)\ln((z-\beta)/c)\ , \text{ or}$$
$$f(z) = P(\text{some nonintegral power of } \sqrt{((z-\beta)/c)})\ ,$$

where $c = \lim(\beta - z)/|\beta - z|$ as $z \to \beta$ along the slit, so $|c| = 1$ and $(z-\beta)/c < 0$ in the slit, and $P(t)$ and $p(t)$ are representable by power series around $t = 0$. Given β and f and its slit, c and P and p are *canonical* (determined uniquely). Formulas slightly more general than these, but still essentially unique, cope with more general elementary functions or with isolated branch-points at ∞.

The dominant terms of these canonical formulas provide approximations useful near branch-points, and are therefore precious to analysts and programmers who have to exploit or compensate for singularities, so these formulas should not be violated unnecessarily on the slits. Programs that handle singularities are complicated enough without the additional burden of treating specially those slits that need no special care so long as programs remain as valid on the slits as off them near their ends. Then programmers can predict from Principal Expressions how their programs will behave on slits. The Principal Expressions for all nine of our elementary functions are determined by convention and tabulated nearby. For other functions the choice of Principal Expression is forced by the choice of slits except when a slit contains just two singularities, both finite branch points at its ends. In the exceptional case the Principal Expression tells which side of that slit is attached to it. For instance, the Fortran programmer can define the

```
COMPLEX FUNCTION B(Z) = CSQRT(Z - 1.0)*CSQRT(Z + 1.0)
```

when he wishes to attach its slit to its upper side, and invoke $-B(-Z)$ when he wishes to attach the slit to its lower side. Another example has two definitions

$$\mathrm{arccot}(z) := \arctan(1/z) \quad \text{and} \quad \mathrm{arccot}(z) := \pi/2 - \arctan(z)$$

that are both widely used though they differ by π in the left half-plane. The first has one slit on the imaginary axis $-1 < z^2 < 0$ between $z = \pm i$. The second has two slits on the imaginary axis $z^2 < -1$ beyond $z = \pm i$. But $\arctan(1/z)$ is not a Principal Expression for $\mathrm{arccot}(z)$ because it has a gratuitous singularity at $z = 0$ where its slit changes sides. A correct Principal Expression for the first definition of $\mathrm{arccot}(z)$ is either $i\,\ell\mathrm{n}((z-i)/(z+i))/2$ or $\ell\mathrm{n}((z+i)/(z-i))/(2i)$ according to whether its slit be attached respectively to the left half-plane or to the right; except on the slit, these Principal Expressions are equal and satisfy $\mathrm{arccot}(-z) = -\mathrm{arccot}(z)$. Whichever one be chosen, the other is $-\mathrm{arccot}(-z)$. Similarly for $\pm\mathrm{arccoth}(\pm z) := \ell\mathrm{n}((z+1)/(z-1))/2$.

Table 1

Conventional Principal Expressions for Elementary Functions:

$-\pi \leqslant \arg(z) \leqslant \pi$; and $-\pi < \arg(z)$ if 0 has just one sign.

$$\ell\mathrm{n}(z) := \ell\mathrm{n}(|z|) + i \arg(z)$$

$$z^w := \exp(w\,\ell\mathrm{n}(z)) \qquad (\text{and } z^0 = 1 \,,\; 0^w = 0 \text{ if } \mathrm{Re}(w) > 0)$$

$$\surd(z) := z^{1/2}$$

$$\mathrm{arctanh}(z) := (\ell\mathrm{n}(1+z) - \ell\mathrm{n}(1-z))/2 \qquad = -\mathrm{arctanh}(-z)$$

$$\arctan(z) := \mathrm{arctanh}(iz)/i \qquad = -\arctan(-z)$$

$$\mathrm{arcsinh}(z) := \ell\mathrm{n}(z + \surd(1+z^2)) \qquad = -\mathrm{arcsinh}(-z)$$

$$\arcsin(z) := \mathrm{arcsinh}(iz)/i \qquad = -\arcsin(-z)$$

$$\arccos(z) := 2\,\ell\mathrm{n}(\surd((1+z)/2) + i\surd((1-z)/2))/i \;=\; \pi/2 - \arcsin(z)$$

$$\mathrm{arccosh}(z) := 2\,\ell\mathrm{n}(\surd((z+1)/2) + \surd((z-1)/2))$$

In general the definitions of Principal Expressions can and should be honoured in all styles of arithmetic, though they must be implemented carefully if they are to survive roundoff. Careful implementations of our nine elementary functions will be presented later in this paper. But some familiar identities satisfied in IEEE style arithmetic must be violated when 0 is unsigned no matter how the slits be attached. For instance, no elementary function f in the table except arctan and arcsinh can satisfy $f(z^*) = f(z)^*$ when z lies in a slit in the real axis. Similarly,

$$\ell n(1/z) = -\ell n(z) \quad \text{and} \quad \sqrt{(1/z)} = 1/\sqrt{(z)}$$

must be violated at $z = -1$ and therefore everywhere in the slit $z < 0$. Other familiar identities violated only in a slit include

$\operatorname{arctanh}(z) = \ell n((1+z)/(1-z))/2$, violated when $z > 1$,
$\arctan(z) = i\,\ell n((i+z)/(i-z))/2$, violated when $iz < -1$, and
$\arccos(z) = 2\arctan(\sqrt{((1-z)/(1+z))})$, violated when $z < -1$.

Other writers have put forward different formulas as definitions for our nine elementary functions. Comparing various definitions, and choosing among them, is a tedious business prone to error. Some ostensibly different definitions, like

$$\operatorname{arccosh}(z) = \ell n(z + \sqrt{(z-1)}\sqrt{(z+1)}),$$

give the same results as ours. Some are quite wrong, as are

$$\operatorname{arccosh}(z) = \ell n(z + \sqrt{(z^2-1)}) \quad \text{and} \quad \arccos(z) = \ell n(z + \sqrt{(z^2-1)})/i,$$

because their slits are in the wrong places. Some are different on only part of a slit, as is

$$\operatorname{arccosh}(z) = -\ell n(z - \sqrt{(z-1)}\sqrt{(z+1)})$$

which is continuous from below that part of the slit where $z < -1$ and therefore violates the canonical formula around infinity. Some are very close to ours; for instance, a proposal to introduce complex functions into APL recommended the formula

$$\operatorname{arccosh}(z) = \ell n(z + (z+1)\sqrt{((z-1)/(z+1))})$$

which yields the same principal value as our formula except for a gratuitous removable singularity at $z = -1$. The same proposal advocated

$$\arctan(z) = -i\,\ell n((1+iz)\surd(1/(z^2+1)))$$

because its range matches that of $\arcsin(z)$, though no reason was given why the ranges should match (but see below), and because it was alleged that the CCC rule should be reversed around a branch point at which the function is infinite, though doing so would introduce anomalies in the relation between ℓn and $\surd$, thereby vitiating the formula being advocated. Another well-known formula

$$\arctan(z) = i\ \ell n(\surd((i+z)/(i-z)))$$

is continuous one way around one branch-point and the opposite way around the other, thereby violating $\arctan(-z) = -\arctan(z)$ on the slits. Our formula given earlier, which is equivalent to

$$\arctan(z) = i(\ell n(1-iz) - \ell n(1+iz))/2,$$

follows the CCC rule and seems simplest, but it does violate two cherished formulas

$$\arcsin(z) = \arctan(z/\surd(1-z^2)) \text{ and}$$
$$\arccos(z) = 2\arctan(\surd((1-z)/(1+z)))$$

on the slit. These formulas are satisfied almost everywhere by the APL proposal's definition of arctan mentioned above, the exceptions $\arccos(-1)$ and $\arcsin(\pm 1)$ arising because, like zero, $1/0$ has no sign and therefore $\arctan(1/0)$ has to be either undefined or chosen arbitrarily from $\{\pm\pi/2\}$. Rather than debate the merits of cherished formulas satisfied everywhere except at some finite branch-points versus canonical formulas satisfied around every finite branch-point, we choose what seem to be the more perspicuous definitions. For similar reasons, our formula above for arctanh seems preferable to the APL proposal's

$$\text{arctanh}(z) = \ell n((1+z)\surd(1/(1-z^2))).$$

Regardless of whether our Principal Expressions really are preferable to someone else's, and regardless of the style of arithmetic, good reasons exist to seek universal agreement upon a set of Principal Expressions to define Principal Values for familiar elementary functions. The first to benefit from such an agreement would be analysts, who would suffer less confusion when reading each other's results. More importantly, programmers would make fewer mistakes, and find them sooner, when implementing conformal maps from complex analytic expressions. Although those benefits might follow from any kind of agreement, Principal Expressions offer the further advantage that they introduce no unnecessary singularities. That advantage goes beyond mere parsimony, because control of singularities is the essence of the subject.

Programs that involve singularities are especially difficult to debug because so many programmers tend to think more like algebraists than like analysts or geometers. Unaccustomed to manipulating inequalities, they have trouble locating the slits that are implicit in complex expressions that contain any of our nine elementary functions. Instead, too many programmers are inclined to test complex expressions in the same way as they often test real expressions, by evaluating them at a handful of trial arguments to see whether the results agree with prior expectations. Because this test strategy usually works for real analytic expressions, programmers mostly ignore warnings that it is unreliable; what else should we expect in a society where drunk driving is still regarded widely as a mere *peccadillo*? But this strategy is truly a dangerous way to test complex analytic expressions of conformal maps with corners because those maps are notoroious for mapping tiny regions into huge ones. When a tiny region like that is missed by a scattering of trial arguments, the test can be quite deceptive. The next example illustrates the point.

Let $g(z) := 2\,\mathrm{arccosh}(1+2z/3) - \mathrm{arccosh}(5/3-(8/3)/(z+4))$, and construe the equation $\zeta := g(z)$ as a conformal map of the z-plane, slit along the negative real axis $z < 0$, onto a slotted strip in the plane of $\zeta = \xi + i\eta$. The strip lies where $|\eta| \leqslant 2\pi$, and the slot within it lies where $\xi < 0$ and $|\eta| < \pi$. The boundary of the slotted strip is the image of both sides of the slit in IEEE style arithmetic; with an unsigned zero the slit maps onto only that part of the boundary in the upper half-plane.

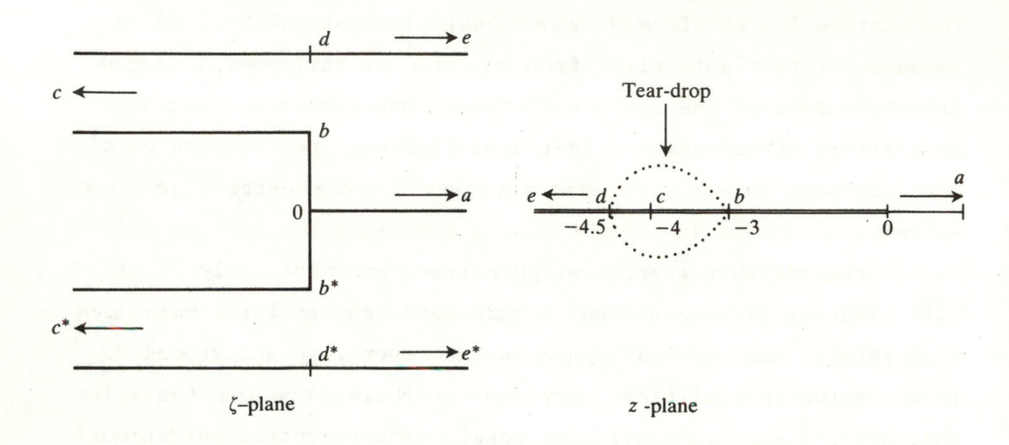

FIG. 2 Conformal map $\zeta := g(z)$ of slit plane to slotted strip

The cost of computing $g(z)$ comes mostly from two logarithms entailed by two calls upon arccosh. Two logarithms can be reduced to one by means of a page or so of algebraic manipulation starting from the Principal Expression tabulated for arccosh above; the result is a *proof* that

$$g(z) = 2\,\ell\mathrm{n}\,(\sqrt{((z+4)/3)}(\sqrt{(z+3)} + \sqrt{z})^2/(2\sqrt{(z+3)} + \sqrt{z}))\,.$$

Without Principal Expressions, one might resort instead to formulas like

$$\mathrm{Arccosh}(z) = \ell\mathrm{n}(z \pm \sqrt{(z^2-1)}) + 2ik\pi \quad \text{for } k = 0, \pm 1, \pm 2, \cdots$$

or to identities like

$$\mathrm{Arccosh}(z) \pm \mathrm{Arccosh}(\zeta) = \mathrm{Arccosh}(z\zeta \pm \sqrt{((z^2-1)(\zeta^2-1))})\ ,$$

with results that are hard to predict. A possible outcome is the expression

$$q(z) := 2\,\mathrm{arccosh}(2(z+3)\sqrt{((z+3)/(27(z+4)))})$$

which matches the desired $g(z)$ everywhere in the z-plane except in a small tear-drop shaped region situated symmetrically about the segment $-4.5 < z < -3$ on the real axis. The tear-drop's boundary is the locus in the plane of $z = x + iy$ whereon the argument of arccosh in $q(z)$ takes values on the slit between 0 and -1; the boundary's equation is

$$y^2 + (x+3)^2 (2x+9)/(2x+5) = 0 \quad \text{for} \quad -4.5 \leqslant x \leqslant -3.$$

Whereas $\zeta = g(z)$ maps the tear-drop onto two half-strips in the left-half of the ζ-plane, $\zeta = q(z)$ maps the tear-drop into two half-strips in the right half-plane. Indeed, $q(z) = -g(z)$ in the tear-drop except, if zero is unsigned, $q(z) = -g(z)^*$ for $-4.5 < z < -4$. Is it likely that a few trial evaluations will reveal the difference between $q(z)$ and $g(z)$?

The examples presented in this paper may give the impression that an analyst will benefit far less than a programmer from Principal Expressions because their benefits seem meagre unless slits run along straight lines. Moreover a signed zero seems useless except when slits lie in the real and imaginary axes. True; but not the whole truth. Despite that applications of elementary functions frequently relocate their slits to non-standard places, the functions so constructed have to be communicated to humans and to computers in terms of combinations of the standard elementary functions with which we are all acquainted. For instance, let $e(z)$ be an analytic extension of $\arcsin(z)$ from the upper half-plane across its slits $z^2 > 1$ into the lower half-plane, where we relocate the slits to run down from ± 1 along some paths to $-i\infty$. Can $e(z)$ be expressed in terms of $\arcsin(z)$? Yes. In the upper half-plane or between the new slits, $e(z) := \arcsin(z)$. Elsewhere we define $s :=$ copysign(1, Im(z)) and calculate

$$e(z) := s \arcsin(z) + \text{copysign}((1-s)\pi/2,\ \text{Re}(z)),$$

which is continuous across the old slits in IEEE style arithmetic. If 0 is unsigned, the last expression must be replaced by something somewhat more complicated.

Readers who recoil from tedious labour may rather acquiesce to all the foregoing assertions than verify any of them personally, despite that such assertions are notoriously rife with mistakes. Yet, lest the pleasures of analysis be eschewn altogether, the writer tenders some simple exercises for the reader's amusement; in each group the object is to discover the whole domain, including boundary, wherein one expression equals another.

Exercises: Where are Two Expressions in the Same Group Equal ?

Group 1: $\sqrt{(z^2-1)}$, $\sqrt{(z-1)}\sqrt{(z+1)}$, $-\sqrt{(1-z)}\sqrt{(-1-z)}$, $i\sqrt{(1-z)}\sqrt{(1+z)}$.

Group 2: $\sqrt{(z-1)}/\sqrt{z}$, $\sqrt{(1-1/z)}$, $\sqrt{(z^2-z)}/z$, $\sqrt{(x(x-1)-y^2+2iy(x-1/2))}/z$.

Group 3: $\sqrt{(z)}/\sqrt{(z-1)}$, $\sqrt{(z/(z-1))}$.

Group 4: $2\,\mathrm{arctanh}(z)$, $\ell n((1+z)/(1-z))$, $\mathrm{arcsinh}(2z/(1-z^2))$.

Group 5: $\cos(n \arccos(z))$, $\cosh(n\,\mathrm{arccosh}(z))$, for integers n.

Group 6: $\arctan(z)+\arctan(1/z)$, $\pi/2$, $-\pi/2$.

Group 7: $\mathrm{arccosh}(z)$, $\mathrm{arccosh}(2z^2-1)/2$, $2\,\mathrm{arcsinh}(\sqrt{((z-1)/2)})$, $i\arccos(z)$.

Group 8: $\mathrm{arccosh}(z)-\mathrm{arccosh}(-z)$, $i\pi$, $-i\pi$.

The answers may depend upon whether arithmetic is performed in hp-15C style or in IEEE style, the difference appearing only when a slit lies in the real or the imaginary axis.

6. SUMMARY

Two different styles of arithmetic induce two different mental attitudes towards the connection between analytic expressions and analytic functions.

IEEE style arithmetic encourages the extension by continuity of every complex analytic function from the interior of its domain to the boundary, including both sides of slits that are distinguishable with the aid of a signed ± 0. Consequently, two expressions that represent the same function everywhere inside its domain are likely to match everywhere on the boundary too; most exceptions are correlated with roundoff problems.

Arithmetic with an unsigned 0 permits continuous extension

to one side of a slit but not to both. Consequently, two expressions that represent the same function everywhere inside its domain often take different values on the boundary. Choosing among such expressions is tantamount to choosing among boundary values for what is otherwise the same function. Our nine elementary functions are among those defined by Principal Expressions determined along with their Principal Values by convention. Other complex functions have to be defined on and inside boundaries by apt compositions of Principal Expressions, or else by *ad hoc* assignments on boundaries.

Regardless of the style of arithmetic, analytic expressions provide at best a statement of intent, at worst wishful thinking about complex analytic functions. Implementations faithful to the expressions despite roundoff and over/underflow must overcome nontrivial technical challenges.

7. IMPLEMENTATION NOTES

Six inverse trigonometric and hyperbolic functions are defined in terms of ℓn and $\sqrt{}$ by Principal Expressions tabulated above in such a way as might appear to provide one-line programs to compute those functions in, say, Fortran. Unfortunately, roundoff can cause such programs to lose their relative accuracy near their zeros or poles; and overflow can occur for large arguments even though the desired function has an unexceptionable value. Programs to compute complex elementary functions robustly and fairly accurately are surprisingly complicated, so much so as to justify supplying them in this paper. Actually, we supply algorithms that can be converted into programs on various machines by being adapted to the peculiarities of diverse programming languages and computing environments.

Certain *Environmental Constants* that characterize important attributes of computer arithmetic may be specified precisely when that arithmetic conforms to IEEE 754 or p854; otherwise they might be slightly vague:

Ω := Overflow threshold = Nextafter($+\infty$, 0)

ε := Roundoff threshold = 1.0 − Nextafter (1.0, 0)

λ := Underflow threshold = $4(1-\varepsilon)/\Omega$ in IEEE 754

Smallest possible no. = Nextafter(0.0, 1) = $2\varepsilon\lambda$ in IEEE 754.

Here *Nextafter* is a function specified in the appendix to IEEE 754; it perturbs its first argument by one *ulp* (one *U*nit in its *L*ast *P*lace) towards the second. That appendix also includes *copysign*, which was described early in this paper, and two functions *scalb* and *logb* that will be used later. Let β be the arithmetic's *radix*, 2 for IEEE 754, or 2 or 10 for p854. For any floating-point x and integer N, scalb(x, N) := $\beta^N x$ computed without first computing β^N, so Over/Underflow is signalled only if the final value deserves it. Logb(NaN) is NaN, which stands for "Not a Number" and is produced by invalid operations like 0/0, 0∞, ∞/∞ and $\infty-\infty$; logb($\pm\infty$) := $+\infty$; logb(0) := $-\infty$ with Divide-by-Zero signalled; and if $\lambda \leqslant |x| < \infty$ then logb(x) is an integer such that $1 \leqslant |\text{scalb}(x, -\text{logb}(x))| < \beta$. The same may be true when $0 < |x| < \lambda$, but early implementations may instead yield logb(x) := logb(λ) in that case. Like the procedures *ldexp* and *frexp* in the *C* library, scalb and logb are practically indispensible for scaling and for computing logarithms and exponentials.

Certain details, particularly those that pertain to ∞ and NaN, are peculiar to IEEE style arithmetic. Otherwise the algorithms presented here for various complex elementary transcendental functions, though designed for IEEE style arithmetic, can be used with other reasonably rounded binary floating-point arithmetics to get comparable results. Our algorithms assume either that zero always has a + sign, or else that its sign obeys the rules specified by IEEE 754 and p854. Those standards also specify rules for $+\infty$ and $-\infty$ and for NaN. Predicates like $x = y$, $x \leqslant y$ and $x < y$ are all *false*; but $x \neq y$ and $x \not> y$ are *true* when either or both of x and y are NaN.

Algebraic operations upon a NaN reproduce it. Both infinities and NaNs can be produced by our algorithms, and both will be accepted as inputs to them.

The IEEE standards prescribe responses to five kinds of *exceptions*:

Invalid Operation, *Overflow*, *Divide-by-Zero*,
Underflow, *Inexact*.

Each kind has its *flag*, to be raised to signal that its kind of exception has occurred; each kind produces a *default result*, respectively

NaN, $+\infty$, $+\infty$, gradual underflow, rounded result.

Gradual underflow approximates any value between $\pm\lambda$ with an error smaller than $\epsilon\lambda$ instead of flushing it to zero. Neither this feature nor flags figure as much as they could and should in our algorithms. In environments that conform fully to IEEE 754, as does the Standard Apple Numerical Environment (SANE) on Apple computers, robust exception-handling complicates programs much less than ours have been complicated by our desire to provide algorithms adaptable also to machines that do *not* conform to the standards. Most of our algorithms can be adapted to such machines by merely excising references to features that those machines do not support. For instance, a statement like "If $x = \infty$ then $\cdots$" will be deleted for machines that have no infinity; however, some obvious precaution against division by zero may have to be inserted elsewhere instead. Machines that flush underflows to zero instead of underflowing gradually may produce less accurate results when they approach the underflow thresholds $\pm\lambda$.

Our algorithms would be simpler, some much simpler, if every arithmetic operation accepted and produced intermediate results of wider range and precision than our algorithms are normally expected to accept or produce. Such a situation arises when the transcendental functions are intended for a higher-level language like Fortran that supports only Single- and Double- precision variables, but the implementer has access to another wider format like IEEE 754's *Extended* format. That is implemented in floating-point coprocessor chips such as the Intel *i*8087 and *i*80287 used in the IBM PC, PC/XT and PC/AT, the Motorola 68881 used in a host of 68000-based workstations, the Western Electric 32106, and also in Apple's SANE. But no such Extended format is provided by the National Semiconductor 32081 used in the IBM PC/RT, nor by the Weitek 1164/1165 chips used in the Sun III among others, nor by the NCUBE multiprocessor array, nor by Fairchild's *Clipper*; for their sakes we use devious formulas to preserve accuracy and avoid spurious overflows.

In the programs below, $\beta, \rho, \theta, s, t, u, v, x, y, \xi$ and η denote real variables; $w := u + iv$, $z := x + iy$ and $\zeta := \xi + i\eta$ denote complex variables; and a star denotes not multiplication but complex conjugation: $z^* = x - iy$. Mixed-mode arithmetic upon one real and one complex variable is presumed *NOT* to be performed by coercing the real to complex, but rather in a way that avoids unnecessary hazards like 0∞ or $\infty - \infty$ by avoiding unnecessary real operations:

$$\beta + z := (\beta + x) + iy\,, \quad \beta z := \beta x + i\beta y, \quad z/\beta := x/\beta + iy/\beta\,; \quad \text{but}$$

$$\beta/z := \beta/(x + (y/x)y) - i(y/x)(\beta/(x + (y/x)y)) \quad \text{if} \quad |y| \leqslant |x|,$$

$$:= (x/y)(\beta/(y + (x/y)x)) - i\beta/(y + (x/y)x) \quad \text{if} \quad |x| \leqslant |y|,$$

with due attention to spurious over/underflows and zeros and infinities.

Ideally, the operators Re and Im, that select the *Re*al and *Im*aginary parts respectively, should be interpreted in a way that avoids unnecessary computation of the unwanted part whenever possible. For instance, $\mathrm{Re}(wz)$ should be evaluated by computing only $ux - vy$, without evaluating $\mathrm{Im}(wz)$ too. Besides saving time, this policy avoids spurious exceptions like over/underflow that might afflict only the unwanted part.

Note too, to conserve ± 0, that $-z$ is not $0 - z$ though they be equal arithmetically; and similarly $w - z$ is the same as $-z + w$ but not $-(z - w)$. Multiplication or division by $i = \sqrt{-1}$

should be accomplished not by actual multiplication but rather by swaps and sign reversal; $iz := -y + ix$. In a similar way, an expression that is syntactically pure imaginary with an unsigned zero for its real part should be handled in a way that avoids both unnecessary arithmetic and unnecessary hazards. For instance,

$$i\beta + z := x + i(\beta + y)\ , \quad (i\beta)z := i(\beta z),$$
$$z/(i\beta) := -i(z/\beta)\ , \quad (i\beta)/z := i(\beta/z).$$

In languages where a construction like CMPLX(x,y) is used to create the complex value $z := x + iy$, the expression CMPLX(0,β) should be treated as $i\beta$, whereas CMPLX(+0,β) and CMPLX(−0,β) should be treated as intentional attempts by the programmer to introduce an appropriately signed zero into the calculation. Of course, both attempts will produce the same CMPLX(+0,β) on a machine whose only zero is +0.

8. COMPLEX ZEROS AND INFINITIES

All four zeros $\pm 0 \pm i0$ are arithmetically equal. Whether all complex infinities should be arithmetically equal is a topological question. When dealing with complex algebraic (not transcendental) functions, the most convenient topology is that of the Riemann sphere with its unique point at infinity. A *metric* (distance function) that induces that topology is the *Chordal Metric* :

$$\begin{aligned} \text{Chord}(z,\zeta) &:= |z-\zeta|/\sqrt{((1+|z|^2)(1+|\zeta|^2))} \\ &\qquad\qquad \text{if } |z| < \infty \text{ and } |\zeta| < \infty, \\ &:= \text{Chord}(1/z\ ,\ 1/\zeta) \quad \text{if } z \neq 0 \text{ and } \zeta \neq 0; \\ &\leqslant \text{Chord}(0,\infty) := \text{Chord}(\infty,0) := 1\ . \end{aligned}$$

In this topology, every algebraic function is a continuous (though perhaps multi-valued) map of the sphere to itself. So are our nine elementary functions $f(z)$. Only a function discontinuous at infinity can be affected by its multiplicity of representations there; an important instance is the equality case $f(z) = z$. To combat ambiguity at infinity a programmer can map all its representations upon one of them, namely real

$+\infty$, by invoking the function

$$\begin{aligned}\mathrm{PROJ}(x+iy) &:= x+iy \quad \text{if} \quad |x| \neq \infty \quad \text{and} \quad |y| \neq \infty\,,\\ &:= +\infty + i\ \mathrm{copysign}(0,y) \quad \text{otherwise,}\end{aligned}$$

before performing any operation discontinuous at infinity. Of course, PROJ is just the identity function on machines that lack a way to represent ∞.

The topology of the Riemann sphere is inappropriate for functions like e^z that have an essential singularity at infinity. Instead, different representations of infinity are customarily associated with different paths that tend to infinity in some asymptotic way, justifying assertions like

$$\exp(-\infty + iy) = 0 \quad \text{and} \quad |\exp(+\infty + iy)| = \infty \quad \text{for all finite } y.$$

For example, "$\infty + i2$" could represent a path asymptotically parallel to the positive real axis and 2 units above it; "$\infty + i\infty$" would have to represent a path parallel to that traced by $\exp(\beta + i\theta)$ as $\beta \to +\infty$ for some fixed but unknown θ strictly between 0 and $\pi/2$. Unfortunately, programming languages like Fortran represent complex variables by pairs of reals in such a way as allows at most nine asymptotic directions (θ) to be represented by two real variables of which at least one is $\pm\infty$. Those directions are

θ:	$\pm\pi$	$-3\pi/4$	$-\pi/2$	$-\pi/4$	± 0
z:	$-\infty \pm i\beta$	$-\infty - i\infty$	$\beta - i\infty$	$+\infty \pm i\infty$	$+\infty \pm i\beta$
θ:	$\pi/4$	$\pi/2$	$3\pi/4$		NaN
z:	$+\infty + i\infty$	$\beta + i\infty$	$-\infty + i\infty$		$\mathrm{NaN} \pm i\infty$ or $\pm\infty \pm i\,\mathrm{NaN}$.

(Here β stands for any finite real number.)

These complex infinities z are the only ones available. By default, in the absence of some contrivance programmed explicitly to cope with other asymptotic directions, every infinite complex result, especially of multiplication and division, has to be approximated by something chosen from the available complex infinities z in a fashion resembling the way real numbers are rounded to the ones representable in floating-point. That default rounding, while fully satisfactory in the topology of the Riemann sphere, can approximate arbitrary asymptotic directions at best crudely,

Crudely, but not quite arbitrarily. The approximations should be predictable and consistent with reasonable expectations; in particular, it seems reasonable to expect

$$wz = \exp(\ln(w) + \ln(z)) \quad \text{and} \quad w/z = \exp(\ln(w) - \ln(z))$$

to hold within an allowance for roundoff even for infinite or zero products and quotients. These relations imply $|wz| = |w||z|$ and $|w/z| = |w|/|z|$ at 0 and ∞, equations that can be satisfied exactly; another implication is that

$$\arg(wz) = \arg(w) + \arg(z) \bmod 2\pi \quad \text{and}$$
$$\arg(w/z) = \arg(w) - \arg(z) \bmod 2\pi$$

have to be approximated within the set of ten values available for $\arg(\zeta)$ when ζ is zero or infinite. Those values turn out to be:

$$\arg(+0 \pm i0) = \arg(+\infty \pm i\beta) = \pm 0 \quad \text{for all finite } \beta,$$
$$\arg(+\infty \pm i\infty) = \pm\pi/4,$$
$$\arg(\beta \pm i\infty) = \pm\pi/2 \quad \text{for all finite } \beta,$$
$$\arg(-\infty \pm i\infty) = \pm 3\pi/4,$$
$$\arg(-0 \pm i0) = \arg(-\infty \pm i\beta) = \pm\pi \quad \text{for all finite } \beta;$$

$\arg(\text{NaN} + i\,\text{Anything})$ and $\arg(\text{Anything} + i\,\text{NaN})$ are both NaN. Thus, any coherent scheme for computing complex products, quotients and logarithms at zero and infinity can be regarded as a scheme that rounds $\arg(\zeta)$ into one of the ten values above when ζ is zero or infinite. To be acceptable, such a scheme should not add much to the cost of complex multiplication and division. The procedure *Box* that follows seems tolerable.

9. THE PROCEDURES

Box supplants the explicit calculation of arg during multiplication and division. It is followed by procedures and auxiliary procedures that calculate the Principal Expressions of the Elementary Functions of Table 1, and algorithms for CTANH and CTAN are given too. Several real special functions are used by these procedures; indeed the only complex auxiliary function that occurs during the computation of the inverse trigonometric and hyperbolic functions is CSQRT. It is assumed that the radix of the computer arithmetic is 2.

••• To compute $x+iy = z := \text{Box}(\zeta) = \text{Box}(\xi+i\eta)$.

CBOX($\xi+i\eta$): ••• Defined *only* for zero and infinite arguments.

If $\xi=0$ and $\eta=0$ then $z :=$ copysign$(1,\xi)+i\eta$

else if $|\xi|=\infty$

then { if $|\eta|=\infty$

then $z :=$ copysign$(1,\xi)+ i\,$copysign$(1,\eta)$

else $z :=$ copysign$(1,\xi)+ i\eta/\xi$ }

else if $|\eta|=\infty$ then $z := \xi/\eta+i$ copysign$(1,\eta)$

else $z := (0+i0)/0$; ••• Invalid use.

Return z; end CBOX.

••• To compute $\rho := |z| = |x+iy| = \sqrt{(x^2+y^2)}$.

ABS $(x+iy)$: ••• = Fortran's CABS(Z) = C's hypot(x,y).

••• The obvious formula can produce errors bigger than one

••• ulp, and could over/underflow spuriously. Not so for

••• what follows.

Constants $r2 := \sqrt{2}$, $r2p1 := 1+\sqrt{2}$, $t2p1 := 1+\sqrt{2}-r2p1$;

••• These constants must be correctly rounded to work-

••• ing precision; consequently $r2p1+t2p1 = 1+\sqrt{2}$

••• to double that precision.

Save invalid flag; ••• This suppresses spurious Invalid

••• Operation signals from NaN comparison or $\infty-\infty$;

••• but spurious inexact signals *can* be generated by

••• this program.

$x := |x|$; $y := |y|$; $s := 0.0$;

If $x<y$ then swap x and y; ••• so $x \geqslant y \geqslant 0$ if not NaN.

```
If  y=∞ then x := y ;
t := x−y ;
If  x≠∞ and t≠x then
    { ••• executed if x≠∞, y≠∞ and y is not negligible.
    Save Underflow flag;
    If  t > y
        then ••• when  2 < x/y < 2/ε,
            { s := x/y ;  s := s + √(1 + s²) }
        else ••• when  1 ⩽ x/y ⩽ 2 ,
            { s := t/y ;  t := (2+s) s ;
              s := ((t2p1 + t/(r2 + √(2+t))) + s) + r2p1 } ;
    s := y/s ••• Harmless Gradual Underflow can occur here.
    Restore  Underflow flag;
    } ;
Restore  Invalid flag ; ••• Only if deserved can Overflow
    ••• happen now.
Return  x + s ;   end   ABS.
```

```
•••To compute  θ := arg(z) = arg(x+iy).
ARG (x+iy) : ••• = Fortran's  ATAN2(y,x) .
    If  x = 0  and  y = 0  then  x := copysign(1,x) ;
    If  |x| = ∞  or  |y| = ∞  then  z := CBOX(z) ;
                                        ••• leaves signs unchanged.
    If  |y| > |x|  then  θ := copysign(π/2, y) − arctan(x/y)
    else if  x < 0 then  θ := copysign(π, y) + arctan(y/x)
                   else  θ := arctan(y/x) ;
    Suppress any Underflow signal unless  |θ| < 0.125, say;
    ••• Better accuracy may be obtained by further case
    ••• reduction and use of identities like
    ••• arctan(y/x) = π/4 + arctan((y−x)/(y+x)).
    Return  θ ;   end   ARG.
```

```
••• To compute  x + iy = z := ζ² = (ξ + iη)².
CSQUARE(ξ + iη) :
    x := (ξ − η)(ξ + η) ;  ••• Not  ξ² − η² .
    y := ξη + ξη ;         ••• ONE multiply, one add.
    ••• If a spurious NaN is created by overflow it gets
    ••• removed thus:
    If  x ≠ x  then
                  { if  |y| = ∞  then  x := copysign(0, ξ)
                    else if  |η| = ∞  then  x := −∞
                    else if  |ξ| = ∞  then  x := ∞ }
    else if  y ≠ y  and  |x| = ∞  then  y := copysign(0, y) ;
    Return  (x + iy) ;  end  CSQUARE .
```

```
••• To compute  ρ := |(x + iy)/2^k|²  scaled to avoid Over/Underflow.
CSSQS(x + iy) : ••• = ρ + ik,  with an integer  k .
     Integer  k ;
     k := 0 ;
     Save and reset the Over/Underflow flags ;
     ρ := x² + y² ; ••• Multiply twice and add.
     If  (ρ ≠ ρ  or  ρ = ∞)  and  (|x| = ∞  or  |y| = ∞)  then  ρ := ∞
     else if  { the Overflow  flag was just raised,  or
                the Underflow flag was just raised and  ρ < λ/ε }
        then  { k := logb(max(|x| , |y|)) ;
                ρ := scalb(x, −k)² + scalb(y, −k)² } ;
     Restore the Over/Underflow flags :
     Return (ρ + ik) ;  end  CSSQS .
```

```
••• To compute  ξ + iη = ζ := √z = √(x + iy).
CSQRT(x + iy):
    Real  ρ ; Integer  k ;
    ρ + ik := CSSQS(x + iy) ;
    ••• Sum-of-Squares Scaled :  see above .
    If  x = x  then  ρ := scalb(|x| , -k) + √ρ ;
    If  k  is odd  then    k := (k - 1)/2
                 else  { k := k/2 - 1 ; ρ := ρ + ρ } ;
    ρ := scalb(√ρ , k) ;
    ••• = √((|x + iy| + |x|)/2)  without over/underflow.
    ξ := ρ ;  η := y ;
    If  ρ ≠ 0  then
        { if  |η| ≠ ∞  then  { η := (η/ρ)/2 ;
              if  η  underflowed, signal it } ;
          if  x < 0    then  { ξ := |η| ;
               η := copysign(ρ, y) }
        } ;
    Return (ξ + iη);
    •••
    ••• This program seems to handle all cases correctly :
    ••• √(-β ± i0) = +0 ± i√(β)   for all  β ≥ 0.
    ••• √(x ± i∞) = +∞ ± i∞  for all  x,  finite,
    •••   infinite or  NaN, and if  x  is  NaN  then
    •••   "Invalid Comparison" is signalled too.
    ••• For all  finite  β ,
    •••    √(NaN + iβ) , √(β + iNaN)  and  √(NaN + iNaN)
    •••    are all  NaN + iNaN;
    •••    √(+∞ ± iβ) = +∞ ± i0 ;
    •••    √(+∞ ± iNaN) = +∞ + iNaN ;
    •••    √(-∞ ± iβ) = +0 ± i∞;
    •••    √(-∞ ± iNaN) = NaN ± i∞ .

    End  CSQRT
```

••• To compute $\xi + i\eta = \zeta := \ell n(2^J z) = \ell n(2^J (x+iy))$ (integer J).
CLOGS($x + iy$, J) : ••• For use with $J \neq 0$ only when $|x+iy|$
    ••• is huge. This program is particularly helpful for
    ••• inverse trigonometric and hyperbolic functions that
    ••• behave like $\ell n(2z)$ for huge $|z|$. This program uses
    ••• a subprogram $\ell n1p(x) := \ell n(1+x)$ presumed to be
    ••• available with full relative accuracy for all tiny
    ••• real x. Such a program exists in various math.
    ••• libraries, included that for 4.3 BSD Unix, Intel's
    ••• CEL and Apple's SANE. The accuracy of $\ell n1p$
    ••• influences the choice of thresholds $T0$, $T1$ and $T2$.
    Constants $T0 := 1/\sqrt{2}$; $T1 := 5/4$; $T2 := 3$; $\ell n2 := \ell n(2)$;
    Real ρ ; Integer k ;
    $\rho + ik :=$ CSSQS $(x+iy)$; ••• $= |(x+iy)/2^k|^2 + ik$; see above.
    $\beta := \max(|x|, |y|)$; $\theta := \min(|x|, |y|)$;
    If $k=0$ and $T0 < \beta$ and ($\beta \leqslant T1$ or $\rho < T2$)
        then $\rho := \ell n1p((\beta-1)(\beta+1)+\theta^2)/2$
        else $\rho := \ell n(\rho)/2 + (k+J)\,\ell n2$;
    $\theta :=$ ARG$(x+iy)$;
    Return $(\rho + i\theta)$; end CLOGS .

••• To compute $\xi + i\eta = \zeta := \ell n(z) = \ell n(x+iy)$.
CLOG(z) := CLOGS(z, 0).

••• To compute $\xi + i\eta = \zeta := \arccos(z) = \arccos(x+iy)$.
CACOS(z) : ••• Based upon formulas :
    ••• $\xi := 2\arctan(\mathrm{Re}(\sqrt{(1-z)})/\mathrm{Re}(\sqrt{(1+z)}))$;
        ••• Suppress any Divide-by-Zero signal when $z \leqslant -1$.
    ••• $\eta := \mathrm{arcsinh}(\mathrm{Im}(\sqrt{(1+z)}^* \sqrt{(1-z)}))$;
    Return $(\xi + i\eta)$; end CACOS .

```
••• To compute  ξ + iη = ζ := arccosh(z) = arccosh(x + iy) .
CACOSH(z) : ••• Based upon formulas :
    ••• ξ := arcsinh(Re(√(z − 1)* √(z + 1))) ;
    ••• η := 2 arctan(Im(√(z − 1))/Re(√(z + 1))) ;
        •••Suppress any Divide-by-Zero signal when z ≤ −1 .
    Return (ξ + iη) ; end CACOSH .

••• To compute  ξ + iη = ζ := arcsin(z) = arcsin(x + iy) .
CASIN(x + iy) : •••Based upon formulas :
    ••• ξ := arctan(x/Re(√(1 − z) √(1 + z))) ;
        ••• Suppress any Divide-by-Zero signal when z ≤ −1 .
    ••• η := arcsinh(Im(√(1 − z)* √(1 + z))) ;
    Return (ξ + iη) ; end CASIN .

••• To compute  ξ + iη = ζ := arcsinh(z) = arcsinh(x + iy) .
CASINH(z) := −i CASIN(iz) .

••• To compute  ξ + iη = ζ := arctanh(z) = arctanh(x + iy) .
CATANH(x + iy):
    Constants θ := √(Ω)/4,  ρ := 1/θ ;
    β := copysign(1, x) ; z := βz* ; ••• Copes with unsigned 0.
    If x > θ or |y| > θ ••• To avoid overflow.
        then { η := copysign(π/2, y) ; ξ := Re(1/(x + iy)) }
    else if x = 1
        then { ξ := ℓn(√(√(4 + y²))/√(|y| + ρ)) ;
               η := copysign(π/2 + arctan((|y| + ρ)/2), y)/2}
    else ••• Normal case. Using ℓn1p(u) := ℓn(1 + u)
        ••• accurately even if u is tiny.
            { ξ := ℓn1p(4x/((1 − x)² + (|y| + ρ)²))/4 ;
              η := arg((1 − x)(1 + x) − (|y| + ρ)² + 2iy)/2 }
        ••• All cases appear to be handled correctly .
    Return (βζ*) ; end CATANH .
```

••• To compute $\xi + i\eta = \zeta := \arctan(z) = \arctan(x+iy)$.

CATAN(z) := $-i$CATANH(iz).

••• To compute $x + iy = z := \tanh(\zeta) = \tanh(\xi + i\eta)$.

```
CTANH(ξ + iη) :
    If |ξ| > arcsinh(Ω)/4 ••• Avoid overflow.
        then z := copysign(1,ξ) + i copysign(0,η)
        else {
            t := tan(η) ; ••• Suppress any Divide-by-Zero
            ••• signal here.
            β := 1 + t² ; ••• = 1/cos²η .
            s := sinh(ξ) ;
            ρ := √(1 + s²); ••• = cosh ξ .
            if |t| = ∞
                then z := ρ/s + i/t ••• May signal if s = 0.
                else z := (βρs + it)/(1 + βs²)
            } ;
    Return z ; end CTANH .
```

••• To compute $x + iy = z := \tan(\zeta) = \tan(\xi + i\eta)$.

CTAN(ζ) := $-i$ CTANH($i\zeta$).

10. THE EXPONENTIAL FUNCTION z^w, AND 0^0

The function z^w has two very different definitions. One is recursive and applicable only when w is an integer:

$$z^0 = 1 \quad \text{and} \quad z^{(w+1)} = z^w z \quad \text{whenever } z^w \text{ exists.}$$

The second definition is analytic:

$$z^w := \lim_{\zeta \to z} \exp(w \, \ell n(\zeta)) ,$$

provided the limit exists using the principal value and domain of $\ell n(\zeta)$. The limit process is necessary to cope smoothly with $z = 0$. Since the recursive definition makes sense when z is a number or a square matrix or a nonlinear map of some domain into itself, regardless of whether $\ell n(z)$ exists, the fact that both definitions coincide when w is an integer and $\ell n(z)$ exists must be a nontrivial theorem. The fact that both definitions agree that $z^0 = 1$ for every z is doubly significant because programmers who have implemented z^w on computers have so often decreed 0^0 to be a capital offence.

I can only speculate on why 0^0 might be feared. Perhaps fear is induced by the singularity that z^w possesses at $z = w = 0$; if both z and w are compelled to approach 0 but allowed to do so independently along any paths, then paths may be chosen on which z^w holds fast to any preassigned value whatsoever. Assuming for the sake of argument (because it is generally not so) that neither z nor w could be exactly zero but must instead be approximately zero because of roundoff or underflow, the expression 0^0 would have to be treated as if it really ought to have been $(\text{roundoff})^{\text{roundoff}}$, which generally defies estimation.

To draw conclusions based upon something better than fear or speculation, we need estimates for certain costs and benefits. Setting $z^0 := 1$ without exception confers the benefit of adherence to simply stated rules; but it introduces some risk that we might unwittingly accept 1 for 0^0 instead of an unknown but preferred value ζ^v with tiny ζ and v. That added risk should be judged in the light of the greater and unavoidable risk that

z^w might unwittingly be accepted when z and w are both nonzero but tiny and quite wrong because of roundoff. In other words, only on those extremely rare occasions when a program of unknown reliability betrays its inaccuracy by a chance encounter with 0^0 will we benefit from outlawing 0^0. But outlawing 0^0 incurs the cost of departing from a simple rule; it imposes upon those programmers who prefer to take $z^0 = 1$ for granted, regardless of whether $z = 0$, the extra burden of remembering to insert extra code to cope with a rare eventuality.

There are two situations in which programmers are fully entitled to take $0^0 = 1$ for granted. The first arises in languages like Fortran and Pascal that distinguish variables of type INTEGER from floating-point variables of type REAL and COMPLEX. Suppose that M is of type INTEGER but w has a floating-point type; then z^M can be distinguished from z^w, and particularly z^0 from $z^{0.0}$, because they call upon different subroutines from a library of intrinsic functions. Since roundoff cannot possibly obscure the value of an exponent M of type INTEGER in the way it might obscure the value of a floating-point variable w that happens to vanish, there is no reason to doubt that $z^0 = 1$ for every z regardless of one's fears about $0.0^{0.0}$. Therefore, in every language in which M can be declared of INTEGER type, the exponential function z^M must be consistent with its recursive definition even if computed, at least when $|M|$ is huge, with the aid of logarithms; in short,

$$\text{when } M = 0 \text{ then } z^M = 1 \text{ regardless of } z.$$

A second situation in which programmers might presume that $0.0^{0.0} = 1$ arises frequently. Consider two expressions $z := z(\xi)$ and $w := w(\xi)$ that depend upon some variable ξ, and suppose that $z(\beta) = w(\beta) = 0$ and that z and w are analytic functions of ξ in some open neighbourhood of $\xi = \beta$. This means that $z(\xi)$ and $w(\xi)$ can be expanded in Taylor series in powers of $\xi - \beta$ valid near $\xi = \beta$, and both series begin with positive powers of $\xi - \beta$. Then we find that $z \to 0$ and $w \to 0$ and $z^w = \exp(w\,\ell n(z)) \to 1$ as

$\xi \to \beta$ regardless of the branch chosen for ℓn. Since this phenomenon occurs for *all* pairs of analytic expressions z and w, it is very common.

In the light of the foregoing considerations, $0.0^{0.0} = 0^0 = 1$ seems to be the only reasonable choice; similar considerations imply $\infty^{0.0} = \infty^0 = 1$ too. Some other exponential expressions involving infinite operands require further thought. For instance, 1.0^∞ is clearly an invalid operation, but $1^\infty = 1$ might be acceptable. Somewhat less clear are the signs of results like

$$(\pm 0.5)^\infty = 0^\infty = (\pm 2)^{-\infty} = (\pm\infty)^{-\infty} = 0\,, \text{ and}$$
$$(\pm 0.5)^{-\infty},\ 0^{-\infty},\ (\pm 2)^\infty,\ (\pm\infty)^\infty,\ \text{all } \pm\infty\,.$$

It is possible to argue that all these results should be assigned + signs in real arithmetic on any North American computer; since all sufficiently big floating-point numbers on such machines are even integers, taking the limit makes ∞ an even integer too. Whether equally fulgent reasoning can be applied to complex arithmetic remains to be seen. And whether $0^{-\infty} = \infty$ should signal "Division by Zero", as 0^{-1} and $1/0$ must, seems to be a matter of taste until we realize that no signal is needed for $0^{-\infty}$ because "Division by Zero" is a misnomer imposed for historical reasons in place of the more appropriate phrase

"an infinite result produced exactly from finite operands".

When z is neither zero nor infinite, and when w is not an integer, the complex function z^w could be assigned a multiplicity of values; they are arranged around a circle if w is real, or otherwise along an Archimedean spiral in the complex plane. What distinguishes the Principal Value defined above from all others is that its logarithm has minimum magnitude; this definition is conventional. Respectable accuracy can be difficult to achieve when either $|w|$ or $|w\ \ell n(z)|$ is big, requiring extraordinarily careful calculation of $\ell n(z)$, but that is a story to unfold elsewhere.

Acknowledgements

I am indebted to Prof. Paul Penfield Jr. of M.I.T. for a conversation that illuminated some of the reasons behind the differences between his APL proposal and the complex arithmetic implemented on the hp-15C by Dr J. Tanzini, then at Hewlett-Packard. The author's own work has been supported in part also by grants from the U.S. Department of Energy, the Office of Naval Research, and the Air Force Office of Scientific Research.

This paper is an extension of, and completely supersedes, an earlier version that appeared in September 1982 as report PAM-105 of the Center for Pure and Applied Mathematics at the University of California at Berkeley. That version was prepared as an exercise on an APPLE text formatter slightly modified using the PASCAL editor and Colin McMaster's SCRIPT. The author thanks his friends at APPLE for that opportunity.

BIBLIOGRAPHIC NOTES

Penfield's proposal "Principal Values and Branch Cuts in Complex APL" appeared in *APL Quote Quad* vol. 12, no. 1, Sept. 1981.

The complex functions implemented in the hp-15C are described in Section 3 of the *Hewlett-Packard* HP-15C *Advanced Functions Handbook*, Aug. 1982, part no. 00015-90011. The formulas that tell where that calculator puts the branch cuts were first published in an article "Scientific Pocket Calculator Extends Range of Built-In Functions" by Eric Evett, Paul McClellan and Joe Tanzini in the *Hewlett-Packard Journal* of May 1983, vol. 34, no. 5, pp.25-35. More about that calculator, plus a formula for computing arccosh accurately, may be found in my paper "Mathematics Written in Sand", pp.12-26 in the Statistical Computing Section of the *Proceedings* of the Joint Statistical Meetings of the American Statistical Association etc. held in Toronto in August 1983. The conformal map onto a slotted strip is adapted from that paper.

The ANSI/IEEE standard 754-1985 for Binary Floating-Point Arithmetic is available as stock number SH10116 from the IEEE Service Center, 445 Hoes Lane, Piscataway, NJ 08854: telephone (201) 981-0060. A more readable exposition of 754 and the proposed Binary and Decimal Standard p854 was published in pp.86-100 of the Aug. 1984 issue of the IEEE magazine *MICRO*: to obtain a reprint from the IEEE, cite document number 0272-1732/84/0800-0086. Early versions of 754, now superseded, plus some supporting materials have appeared in the March 1981 and January 1980 issues of the IEEE magazine *Computer*, and in a special issue, October 1979, of the ACM *SIGNUM Newsletter*. Implementations of IEEE 754 abound, ranging in size and speed from the ELXSI 6400 to the Apple II and Macintosh. The Standard Apple Numerical Environment (SANE) is now the most thorough implementation, and is documented in the *Apple Numerics Manual* published in 1986 by Addison-Wesley, Reading, Mass.

The Intel i8087 and i80287 floating-point coprocessor chips were designed to conform to an early draft of IEEE 754; they very nearly conform to the present standard. Though widely used in the IBM PC, PC/XT and PC/AT, they are not yet well supported by software in that realm. A fine library of elementary functions for them, real ones coded by Steve Baumel, complex by Dr Phil Faillace, comes with Intel's Fortran for its 286/310 and 286/330 computers running under both Xenix and RMX86 operating systems. That library's algorithms are much like ours above. The real functions are documented in Intel's *80287 Support Library Reference Manual* (1983), order no. 122129. Real functions similar to those, and almost as accurate, are implemented on the Motorola 68881 and documented in the *MC68881 Floating-Point Coprocessor User's Manual* (1985, preliminary edition), order no. MC68881UM/AD. I do not yet have public documentation for analogous libraries running on the ELXSI 6400 (programmed by Peter Tang), on the National Semiconductor 32081 floating-point slave processor chip, and on the IBM PC/RT. The latter two machines' libraries are very much like the *C* Math Library for IEEE 754-conforming machines programmed mostly by Dr Kwok-Choi Ng and now distributed with 4.3 BSD UNIX by the University of California at Berkeley; that library is intended ultimately to be distributed independently of Berkeley UNIX.

The hp-71B is currently the only implementation in Decimal arithmetic of p854; that hand-held computer is the subject of the July 1984 issue of the *Hewlett-Packard Journal*, vol. 35, no. 17. Many of the complex elementary functions, plus PROJ, have been implemented in the hp-71B's Math Pac, HP 82480A; but its implementers were compelled by limitations upon time and space to acquiesce to a few compromises that I wish they could have avoided. For instance, users of that machine have to write Z*Z instead of Z∧2 to compute z^2, and (−IMPT(Z), REPT(Z)) instead of (0,1)∗Z to compute iz, if they wish to conserve the sign of zero.

Some of the ideas that lead to canonical formulas around branch-points are explained in pp.276-286 of volume III of A.I. Markushevich's *Theory of Functions of a Complex Variable* translated by R.I. Silverman, 1967, Prentice-Hall, N.J. The conformal map from the right half-plane to a liquid jet was adapted, with corrections, from pp.122-5 of *Theory of Functions as Applied to Engineering Problems*, edited by Rothe, Ollendorf and Pohlhausen, translated by Herzenberg in 1933, reprinted in 1961 by Dover, N.Y. Another Dover reprint is the *Handbook of Mathematical Functions with Formulas, Graphs, and Mathematical Tables*, edited by M. Abramowitz and Irene Stegun, issued originally in 1964 as no. 55 in the U.S. National Bureau of Standards Applied Math. Series. Its Chapter 4 locates the slits for all nine elementary functions considered here, but its formulas 4.4.37-9 for complex Arcsin, Arccos, and Arctan are noncommittal on the slits and generally vulnerable to roundoff; and it lacks a formula for complex Arccosh. During the Handbook's ninth reprinting its definition of arccot(z) changed from $\pi/2-\arctan(z)$ to $\arctan(1/z)$. Finally, H. Kober's *Dictionary of Conformal Representations* contains pictures of many useful conformal maps; this too was reprinted by Dover, in 1957.

W. Kahan
Elect. Eng. and Computer Science
and Mathematics Departments
University of California
Berkeley
California 94720
U.S.A.

8 Recent Developments in Linear and Quadratic Programming

R. FLETCHER

ABSTRACT

This paper describes recent developments in linear programming, including the ellipsoid algorithm, Karmarkar's algorithm, new strategies for updating LU factors in the simplex method, and methods with guaranteed termination in the presence of degeneracy and round-off errors. Various new algorithms for quadratic programming are discussed, and the choice of matrix factorizations and their updates is considered. The use of l_1 penalty functions in linear and quadratic programming is also mentioned briefly.

1. INTRODUCTION

The casual student of mathematical programming might regard linear programming (LP) as a well established branch of the subject in which few recent significant advances would be expected. To a lesser extent the same might be thought true of quadratic programming (QP). In fact the past decade has been remarkable in the large number of highly interesting developments that have occurred in both these fields, and some of these are reviewed in this paper. New advances elsewhere, particularly in SLP (sequential linear programming) and SQP methods for nonlinear and nonsmooth programming, make it essential to strive for the ultimate efficiency and reliability in LP and QP and much recent research has been devoted to this end.

Arguably the most notable event has been the discovery of

polynomial time algorithms for linear programming, for this has not only answered a long standing theoretical question but has also raised the possibility of developing algorithms that are more efficient than the simplex method. This subject, in particular the ellipsoid algorithm of Khachiyan and a recent algorithm of Karmarkar, is considered in Section 2. In the context of matrix updating formulae, another development has been to show that it is possible to update LU factors explicitly in a stable manner when column or row interchanges are made. This is described in Section 3 and is also of some interest because it has widely been considered that it could not be done. In Section 4, some work is described which I regard as being important, regarding the development of methods and codes for LP which have guaranteed performance, even when degeneracy and round-off errors are taken into account. An overview of recent developments in quadratic programming is given in Section 5, both in regard to the choice of overall method and to the development of new matrix updating strategies. New advances have been made in both of these areas and the current state of the art is very fluid.

Before embarking on these discoveries, however, I shall first make another point which I regard as being important. This is that it may not be best to solve LP (or QP) problems at all, but rather to solve what might be called l_1LP (or l_1QP) problems. For example consider the simple form of an LP problem

$$\text{minimize} \quad \underline{c}^T \underline{x}$$
$$\text{subject to} \quad A^T \underline{x} \geqslant \underline{b}.$$

This can be transformed into the l_1LP problem

$$\text{minimize} \quad \nu \underline{c}^T \underline{x} + \sum_i \max(0,\ b_i - \underline{a}_i^T \underline{x})$$

in which a weighted combination of the objective function and an l_1 sum of constraint violations is calculated. If the LP problem has a solution and $\nu > 0$ is sufficiently small then the solutions of these problems are identical. Methods for l_1LP can be determined that are very similar to those for LP. Advantages of the l_1LP formulation are that the Phase 1 / Phase 2 approach to LP is avoided and less time is required to solve the problem. The reason for this is that in l_1LP the search for a feasible point is biased towards minimizing the objective function. Another feature of l_1LP is that advantage can always be taken of an estimate of the correct active set, even if this does not give rise to a feasible vertex. The only difficulty is the need to supply the weighting parameter ν. In practice however a little trial and error (Fletcher, 1985b) soon enables a suitable value to be determined. A similar transformation is possible for QP problems, and similar comments apply. Apart from the above advantages, l_1LP and l_1QP subproblems also occur in their own right within some methods for nonlinear and nonsmooth programming.

2. POLYNOMIAL TIME ALGORITHMS

In many ways one of the most interesting developments of the last decade has been that of polynomial time algorithms, which have given a new impetus and excitement to research into linear programming. The *ellipsoid algorithm*, attributed to Khachiyan (1979) but based on the work of a number of Russian authors, first attracted widespread interest. A later more efficient polynomial algorithm of Karmarkar (1984), claimed controversially to be many more times faster than the simplex method, has also encouraged research into radically different practical alternatives to the simplex method. Both these developments attracted wide publicity and were even reported in the national press, an event that is possibly without precedence in the history of numerical analysis! Yet the cautious research worker will in my opinion need more convincing evidence, and access to definitive methods or developed software, before coming to a favourable appraisal.

The reason for seeking an alternative to the simplex method is that in the worst case it can perform remarkably badly, and the number of iterations can increase exponentially in the number of variables n. This is despite the fact that in practice the algorithm is generally rated as being reasonably effective and reliable. This "exponential time" behaviour was pointed out by Klee and Minty (1972) and later made explicit in an example due to Chvatal. Thus the KMC problem of order n is

$$\text{maximize} \quad \sum_j 10^{j-1} x_j ,$$

$$\text{subject to} \quad x_i + 2 \sum_{j>i} 10^{j-i} x_j \leqslant 10^{2n-2i}, \quad i = 1,2,\ldots,n,$$

$$\underline{x} \geqslant \underline{0} .$$

It is possibly more illuminating to write down this problem for a particular value of n, for example the case $n=5$ is

$$\begin{array}{ll} \text{maximize} & x_1 + 10\,x_2 + 100\,x_3 + 1000\,x_4 + 10000\,x_5 \\ \text{subject to} & x_1 + 20\,x_2 + 200\,x_3 + 2000\,x_4 + 20000\,x_5 \leqslant 10^8 \\ & \qquad x_2 + 20\,x_3 + 200\,x_4 + 2000\,x_5 \leqslant 10^6 \\ & \qquad\qquad x_3 + 20\,x_4 + 200\,x_5 \leqslant 10^4 \\ & \qquad\qquad\qquad x_4 + 20\,x_5 \leqslant 10^2 \\ & \qquad\qquad\qquad\qquad x_5 \leqslant 10^0 \\ & \underline{x} \geqslant \underline{0} . \end{array}$$

For the general problem the feasible region has 2^n vertices, which is not in itself unusual. However starting from $\underline{x}=\underline{0}$ (all slacks basic) the simplex method in exact arithmetic visits all 2^n vertices, hence the exponential time solution. In practice, in floating point arithmetic, Williams (1981) indicates that standard codes often do considerably better on these problems than might be expected. However this is not our experience, which may be due to the fact that Williams also includes some upper bounds in the formulation which can become active and so perturb the simplex sequence of iterates.

Khachiyan (1979) first showed that it is possible to solve LP problems in polynomial time. However his algorithm permitted some obvious improvements and the version described by Wolfe (1980) is presented here. The basic method relates to finding a feasible point of the system of inequalities $A^T\underline{x} \leqslant \underline{b}$ where A is an $n \times m$ matrix. The algorithm is iterative and holds a current point $\underline{x}_c$ which is the centre of an ellipsoid

$$E(\underline{x}_c, J_c) = \{\underline{x}: \|J_c^{-T}(\underline{x}-\underline{x}_c)\|_2 \leqslant 1\}$$

that contains a feasible point, if one exists. Such an ellipsoid is readily constructed initially (Gacs and Lovasz, 1979). A constraint $\underline{a}^T\underline{x} \leqslant \beta$ is found which is violated by $\underline{x}_c$. For the part of the ellipsoid which is cut off by this constraint, a new ellipsoid $E(\underline{x}_+, J_+)$ is constructed which contains it, having smallest volume. This process is repeated until convergence. Letting

$$d = \frac{\underline{a}^T\underline{x}_c - \beta}{\|J_c\underline{a}\|_2} \ ,$$

the parameters of the new ellipsoid are

$$\underline{x}_+ = \underline{x}_c - \frac{(1+nd)\, J_c^T J_c \underline{a}}{(1+n)\|J_c\underline{a}\|_2}$$

and

$$J_+ = \left(\frac{1-d^2}{1-n^{-2}}\right)^{\frac{1}{2}} \left[I - \left\{1 - \left(\frac{(n-1)(1-d)}{(n+1)(1+d)}\right)^{\frac{1}{2}}\right\} \frac{J_c\underline{a}(J_c\underline{a})^T}{\|J_c\underline{a}\|_2^2}\right] J_c \, .$$

It seems to be advantageous to choose a substantially violated inequality, but not necessarily the constraint that maximizes d. If $d > 1$ then there is no feasible point. Each iteration of the method reduces the volume of the ellipse by the factor

$$\left(\frac{1-d^2}{1-n^{-2}}\right)^{n/2} \left(\frac{(n-1)(1-d)}{(n+1)(1+d)}\right)^{\frac{1}{2}} < e^{-1/2n} \qquad (2.1)$$

so the volume converges linearly to zero with rate at least $e^{-1/2n}$. If the part of the feasible region contained in the original ellipsoid has positive volume, V_f say, and the volume of the initial ellipsoid is V_0, then the algorithm must terminate in at most $p = [p'] + 1$ iterations, where

$$V_f = \left(e^{-1/2n}\right)^{p'} V_0$$

which implies that $p = O(n)$.

If the feasible region has zero volume (for example if the problem essentially contains equations) and the constraints are relaxed by an amount 2^{-L}, then it can be shown (Gacs and Lovasz, 1979) that the volume of the feasible set is at least 2^{-nL}. Thus to this accuracy, and following the same argument, the algorithm will find a solution in $p = O(n^2 L)$ iterations. By relating L to the number of bits required to determine the problem, the solvability of the unperturbed system can also be determined, but this is more of interest to workers in computational complexity. Thus the number of iterations required to solve the problem is *polynomial* in n, and this is the major advance of Khachiyan's work. However not even Khachiyan himself has claimed that this algorithm provides a practical alternative to the simplex method. For example to improve the accuracy of each component in the solution by a factor of 10, it can be expected that the volume of the current ellipsoid must be reduced by a factor 10^{-n}. Using the above estimate, in k steps the volume will be reduced by a factor $e^{-k/2n}$, which gives $k = 4.6n^2$ as the expected number of iterations. The simplex method would normally perform much faster than this. Since both methods require $O(n^2)$ operations per iteration, the ellipsoid algorithm is not competitive in practice.

As presented the ellipsoid algorithm is a method for finding feasible points. If it is required to solve an LP problem, minimizing $\underline{c}^T\underline{x}$ subject to the same constraints, then there are at least two ways in which the ellipsoid algorithm can be used. One is to find a feasible point, $\underline{x}_c$ say, as above, and then cut out this point with the constraint $\underline{c}^T(\underline{x}-\underline{x}_c)<0$, re-solving to find another feasible point. This is repeated until the algorithm converges. An alternative approach is to convert the LP into the primal-dual system

$$\underline{c}^T\underline{x} = \underline{b}^T\underline{\lambda}, \quad A^T\underline{x} \leqslant \underline{b}, \quad \underline{x} \geqslant \underline{0}, \quad A\underline{\lambda} \geqslant \underline{c}, \quad \underline{\lambda} \geqslant \underline{0}$$

and solve this using the ellipsoid algorithm. Some other references for the ellipsoid algorithm are given by Wolfe (1980).

A more recent polynomial time algorithm with a faster rate of convergence is given by Karmarkar (1984). This algorithm is claimed to be substantially faster than the simplex method, a claim received with enthusiasm by some members of the mathematical programming community, and scepticism by others. Unfortunately the details on which the claim is made have not been publicised, and there is the lack of a definitive method and software. Already many researchers are deriving similar algorithms: however to revise our ideas of what is the standard algorithm for LP requires a thorough comparison that shows a distinct overall improvement on a wide range of large scale LP test problems. In my opinion this has not yet been done, despite some interesting attempts. Whilst it is not unlikely in the future that a polynomial time algorithm will be adopted as the standard, that time is not yet here, and it is by no means certain even that such a method will be a development of the Karmarkar algorithm.

Nonetheless Karmarkar's ideas are of substantial interest and it is appropriate to summarize them here. In its basic form the method solves a very particular form of an LP problem, namely

$$\begin{aligned} &\text{minimize} \quad && f \equiv \underline{c}^T\underline{x}, \quad \underline{x} \in \mathbb{R}^n \\ &\text{subject to} \quad && A\underline{x} = \underline{0}, \quad \underline{e}^T\underline{x} = 1, \quad \underline{x} \geqslant \underline{0}, \end{aligned} \tag{2.2}$$

where $\underline{e} = (1, 1, \cdots, 1)^T$. The constraints $\underline{e}^T\underline{x} = 1$ and $\underline{x} \geq \underline{0}$ define a regular simplex in $\mathbb{R}^n$, and the feasible region is the intersection of this set with the homogeneous constraint $A\underline{x} = \underline{0}$. It is assumed in the first instance that $A\underline{e} = \underline{0}$, which implies that the initial point $\underline{x}_c = \underline{e}/n$ (the centre of the simplex) is feasible. It is also assumed that the solution value of the LP is zero, so that $f^* \equiv \underline{c}^T\underline{x}^* = 0$. Conceptually the first iteration of the method is as follows. Let S_r be the sphere

$$\{\underline{x}: \|\underline{x}-\underline{x}_c\|_2 \leq r,\ \underline{e}^T\underline{x} = 1\},$$

where r is the largest value that keeps S_r inscribed within the simplex. A new point $\underline{x}_+$ is determined by minimizing f on the set $S_r \cap \{\underline{x}: A\underline{x} = \underline{0}\}$. This is a relatively straightforward calculation which is considered in more detail below. Let S_R be the concentric sphere that circumscribes the simplex, so that $R/r = n-1$, and let $\underline{x}_R$ minimize f on the set $S_R \cap \{\underline{x}: A\underline{x} = \underline{0}\}$. Because f is linear, it follows that $(f_c - f_R) = (n-1)(f_c - f_+)$. But the set $S_R \cap \{\underline{x}: A\underline{x} = \underline{0}\}$ contains the feasible region of the LP. Hence $f_R \leq f^* = 0$ and so

$$f_+ \leq (n-2)\, f_c/(n-1) < e^{-1/(n-1)} f_c. \qquad (2.3)$$

Comparing this with expression (2.1) for the ellipsoid algorithm, it can be seen that if this reduction were obtained on every iteration, then $O(nL)$ iterations would be required to get f within a tolerance 2^{-L} of f^*, and the algorithm would be polynomial.

After the first iteration, the current point $\underline{x}_c$ is no longer the centre of the simplex, but can be made so by transforming the problem. Karmarkar uses a *projective transformation*

$$\underline{x}' = D^{-1}\underline{x}/\underline{e}^T D^{-1}\underline{x},$$

where $D = \mathrm{diag}((\underline{x}_c)_i)$, which transforms the simplex into itself and makes $\underline{x}'_c$ the centre of the simplex. Therefore subsequent iterations can proceed as above, except that linearity in f is not preserved by the transformation, so the linear approximation

$\underline{c}^T D \underline{x}'$ is used. Karmarkar shows by means of a potential function that a similar result to (2.3) holds good and so the polynomial behaviour follows.

The subproblem which must be solved in the transformed variables $\underline{x}'$ is

$$\text{minimize} \quad \underline{c}^T D \underline{x}'$$
$$\text{subject to} \quad \underline{x}' \in S', \quad AD\underline{x}' = \underline{0},$$

where S' is the sphere $\{\underline{x} : \|\underline{x} - \underline{x}_c\|_2 \leq r,\ \underline{e}^T \underline{x} = 1\}$. It is not difficult to show that the solution to this problem is on the line through $\underline{x}'_c$ with direction $\underline{d}'$, which is defined by

$$\text{minimize} \quad \tfrac{1}{2} \underline{d}'^T \underline{d}'$$
$$\text{subject to} \quad -\underline{c}^T D \underline{d}' = 1, \quad AD\underline{d}' = \underline{0}, \quad \underline{e}^T \underline{d}' = 0.$$

Introducing multipliers λ, $\lambda\underline{\pi}$ and $\lambda\varphi$ respectively, the first order conditions give

$$\underline{d}' + \lambda(D\underline{c} - DA^T \underline{\pi} - \varphi \underline{e}) = \underline{0}$$

and by normalizing $\underline{d}'$ suitably, it follows that

$$\underline{d}' = -D\underline{c} + DA^T \underline{\pi} + \varphi \underline{e}. \tag{2.4}$$

Using $AD\underline{e} = A\underline{x}_c = \underline{0}$, the solution separates into $\varphi = \underline{c}^T \underline{x}_c / n$ and

$$AD^2 A^T \underline{\pi} = AD^2 \underline{c}. \tag{2.5}$$

Thus $\underline{\pi}$ is defined by the normal equations of a weighted least squares problem. Well understood methods exist for its solution, and it is the dominant cost in an iteration of Karmarkar's method.

The next iterate in the method is defined by

$$\underline{x}'_+ = \underline{x}'_c + \alpha' \underline{d}'$$

where α' is a suitably chosen step length; its choice is not resolved with any certainty by Karmarkar. When transformed back into the original coordinate system, the new vector of variables is

$$\underline{x}_+ = D\underline{x}'_+ / \underline{e}^T D\underline{x}'_+ .$$

Using $D\underline{e} = \underline{x}_c$, $\underline{e}^T\underline{x}_c = 1$, $\underline{x}'_c = \underline{e}/n$ and (2.4), it follows that

$$\underline{e}^T D(\underline{x}'_c + \alpha'\underline{d}') = (1 + n\alpha'(\varphi - \mu_k))/n = (\gamma^{-1})/n$$

say, where $\mu_k = \underline{x}_c^T D(\underline{c} - A^T\underline{\pi})$, and hence $\underline{x}_+$ can be rearranged as

$$\underline{x}_+ = \underline{x}_c + \gamma n\alpha'(\mu_k \underline{x}_c - D^2(\underline{c} - A^T\underline{\pi})). \tag{2.6}$$

Thus Karmarkar's algorithm can also be regarded as a line search method in the original variables.

Karmarkar does suggest some modifications to his algorithm in order to solve a more general form of LP problem, and in particular to avoid the assumption that $f^* = 0$. Modifications with a similar aim are given by Gay (1985) and Lustig (1985). Karmarkar's idea has also engendered a number of papers proposing other interior point methods, based for example on different types of transformation: these include contributions by Iri and Imai (1985), Neumaier (1986), Sonnevend (1986) and Strang (1985). However a particularly perceptive paper is that of Gill *et al.* (1986) who derive a relationship between Karmarkar's method and the logarithmic barrier function, normally thought of as the basis of a somewhat out-of-date interior point method for nonlinear programming. For an LP problem in standard form (minimize $\underline{c}^T\underline{x}$ subject to $A\underline{x} = \underline{b}$, $\underline{x} \geqslant \underline{0}$) the bounds can be handled in a log barrier function, giving the problem

$$\begin{aligned} &\text{minimize} \quad \varphi(\underline{x}) \equiv \underline{c}^T\underline{x} - \mu \sum_i \ln x_i \\ &\text{subject to} \quad A\underline{x} = \underline{b}. \end{aligned} \tag{2.7}$$

If this transformation is applied to the LP format (2.2) used by Karmarkar, and the iteration formula for Newton's method from an initial point $\underline{x}_c$ is written down, then

$$\underline{\nabla}\varphi = \underline{c} - \mu D^{-1}\underline{e}, \quad \nabla^2\varphi = \mu D^{-2}$$

and the search direction $\underline{d}$ is defined by solving the QP problem

$$\text{minimize} \quad \tfrac{1}{2}\mu \underline{d}^T D^{-2} \underline{d} + (\underline{c} - \mu D^{-1}\underline{e})^T \underline{d}$$

$$\text{subject to} \quad A\underline{d} = \underline{0}, \quad \underline{e}^T\underline{d} = 0.$$

Ignoring the constraint $\underline{e}^T\underline{d} = 0$, an explicit expression for $\underline{d}$ (e.g. Fletcher, 1981) is

$$\underline{d} = \underline{x}_c - \mu^{-1}D^2(\underline{c} - A^T\underline{\pi})$$

(using $D\underline{e} = \underline{x}_c$, $A\underline{x}_c = \underline{0}$ and (2.5)). The penalty parameter μ in (2.7) is chosen to have the value $\mu_k = \underline{x}_c^T D(\underline{c} - A^T\underline{\pi})$ that occurs in (2.6). It follows from $\underline{e}^T\underline{x}_c = 1$ and $D\underline{e} = \underline{x}_c$ that $\underline{e}^T\underline{d} = 0$ is nonetheless satisfied. Hence incorporating a step length β, the next point is defined by

$$\underline{x}_+ = \underline{x}_c + \beta\mu^{-1}(\mu\underline{x}_c - D^2(\underline{c} - A^T\underline{\pi})).$$

Comparison with (2.6) shows that the search directions in both methods are parallel and the iterates are identical if the step lengths satisfy $\beta\mu^{-1} = \gamma n\alpha'$. Thus Karmarkar's method is seen to be just a special case of the log barrier function applied to the LP problem.

It would seem to follow *a forteriori* therefore that, if good results can be obtained by Karmarkar's method, then it should be possible to do at least as well by using the log barrier function. Gill *et al.* (1986) investigate this possibility, solving (2.5) by means of a preconditioned conjugate gradient method. (Another possibility would be to use sparse QR factors of DA^T, modified to treat dense columns specially.) On the basis of a number of large scale tests, they report that the method can be competitive with the simplex method on *some* problems, particularly those with a favourable sparsity structure for QR. However the overall performance of the simplex method is better, and the interior point algorithms have difficulties with degenerate problems, and in making use of a good initial estimate of the solution. Gill *et al.* are not able to reproduce the significant improvements reported by Karmarkar. Clearly more research is needed before a final evaluation of the merit of interior point methods in LP can be made.

3. UPDATING LU FACTORS

In the simplex method a representation of a nonsingular $n \times n$ matrix A is required which enables inverse operations to be carried out effectively. On each iteration of the method one column of A is changed and the representation is updated. Many different representations have been suggested but the one most commonly preferred in other circumstances, namely the factors $PA = LU$ computed by Gaussian elimination with partial pivoting, has not been used because it was widely regarded that these factors could not be updated in an adequately stable way. However a new algorithm proposed by Fletcher and Matthews (1984) has caused this attitude to be revised, and this method is now the best one to use with dense LP codes.

To explain the idea, let factors $PA = LU$ of A be available, and remove a column from A giving $PA' = LU'$, where A' and U' are $n \times (n-1)$ matrices and U' has a step on the diagonal. The first stage of the method is to restore U' to upper triangular form, giving factors $P'A' = L^*U^*$, and this part of the method is new. The second stage is to add a new column to A' giving

$$P'A^* = P'[A' : \underline{a}] = L^*[U^* : \underline{u}] ,$$

and $\underline{u}$ is readily computed by solving $L^*\underline{u} = P'\underline{a}$ by forward substitution, which gives the final updated factors.

The main interest then is in how to restore U' to upper triangular form. Consider the (worst) case when column 1 of A is removed so that U' is upper Hessenberg. Let L have columns $\underline{l}_1, \underline{l}_2, \cdots, \underline{l}_n$ and U' have rows $\underline{u}_1, \underline{u}_2, \cdots, \underline{u}_n$ so that

$$PA' = LU' = \sum_i \underline{l}_i \underline{u}_i .$$

The first step of the method is to make $u'_{21} = 0$. One possibility is to write the first two terms in the sum as

$$\underline{l}_1\underline{u}_1 + \underline{l}_2\underline{u}_2 = \begin{bmatrix} \underline{l}_1 & \underline{l}_2 \end{bmatrix} \begin{bmatrix} \underline{u}_1 \\ \underline{u}_2 \end{bmatrix} = \begin{bmatrix} \underline{l}_1 & \underline{l}_2 \end{bmatrix} B\, B^{-1} \begin{bmatrix} \underline{u}_1 \\ \underline{u}_2 \end{bmatrix}, \tag{3.1}$$

where B is a 2×2 matrix, and define

$$\begin{bmatrix} \underline{l}_1^+ & \underline{l}_2^+ \end{bmatrix} = \begin{bmatrix} \underline{l}_1 & \underline{l}_2 \end{bmatrix} B, \qquad \begin{bmatrix} \underline{u}_1^+ \\ \underline{u}_2^+ \end{bmatrix} = B^{-1} \begin{bmatrix} \underline{u}_1 \\ \underline{u}_2 \end{bmatrix}.$$

To eliminate u'_{21} define $r = u'_{21}/u'_{11}$ and

$$B^{-1} = \begin{bmatrix} 1 & 0 \\ -r & 1 \end{bmatrix}, \qquad B = \begin{bmatrix} 1 & 0 \\ r & 1 \end{bmatrix}. \tag{3.2}$$

Then $\underline{l}_1, \underline{l}_2, \underline{u}_1, \underline{u}_2$ are replaced by $\underline{l}_1^+, \underline{l}_2^+, \underline{u}_1^+, \underline{u}_2^+$ respectively. The process is repeated to eliminate u'_{32}, u'_{43}, $\cdots$ and so on, until an upper triangular matrix is obtained. This method is essentially that of Bennet (1965). Unfortunately it fails when $u'_{11} = 0$ and is numerically unstable because r can be arbitrarily large.

To circumvent this difficulty, when $|u'_{21}| > |u'_{11}|$ one might consider first interchanging $\underline{u}_1$ and $\underline{u}_2$, so that

$$\begin{bmatrix} \underline{u}_1^+ \\ \underline{u}_2^+ \end{bmatrix} = B^{-1} P \begin{bmatrix} \underline{u}_1 \\ \underline{u}_2 \end{bmatrix} = \begin{bmatrix} 1 & 0 \\ -r^{-1} & 1 \end{bmatrix} \begin{bmatrix} 0 & 1 \\ 1 & 0 \end{bmatrix} \begin{bmatrix} \underline{u}_1 \\ \underline{u}_2 \end{bmatrix}.$$

Unfortunately the corresponding operations on L, namely $[\underline{l}_1^+, \underline{l}_2^+] = [\underline{l}_1, \underline{l}_2]\, PB$, destroy the lower triangular structure. However Bartels and Golub (1969) suggest a method in which the elementary operations on L are stored in a file (in condensed form) and only U' is updated, giving a representation like

$$PA' = [LP_1B_1P_2B_2 \cdots]\, U^*.$$

This is an example of a *product-form method* and forms the basis of some efficient sparse LP software (Reid, 1975). Another popular product-form method is that of Forrest and Tomlin (1972). A feature of this type of method is that when a sequence of simplex updates is carried out, the file grows longer after each update, and operating with it becomes progressively more expensive. Hence new LU factors must be computed periodically (*reinversion*).

The new method of Fletcher and Matthews is able to update explicit LU factors without the need for a product form, and with reasonable stability properties. When r in (3.2) is small in a certain sense, then the update based on (3.2) is used to eliminate u'_{21}. In the difficult case that u'_{21} is relatively large, a different method of elimination is used. Here the idea is to interchange *rows* 1 and 2 of L, making a corresponding change to P to preserve equality in $PA' = LU'$. An operation of the form BB^{-1} is introduced as in (3.1), but this time B is defined by

$$B = \begin{bmatrix} 0 & 1 \\ 1 & -l_{21} \end{bmatrix} \begin{bmatrix} 1 & 0 \\ s & 1 \end{bmatrix}, \quad B^{-1} = \begin{bmatrix} 1 & 0 \\ -s & 1 \end{bmatrix} \begin{bmatrix} l_{21} & 1 \\ 1 & 0 \end{bmatrix} \tag{3.3}$$

where $s = u'_{11}/\Delta$ and $\Delta = l_{21}u'_{11} + u'_{21}$. The purpose of the left hand matrix in B is to return L to lower triangular form, and the right hand matrix in B^{-1} gives the corresponding change to U'. This operation replaces the elements u'_{11} and u'_{21} by Δ and u'_{11} respectively. The left hand matrix in B^{-1} then eliminates the new element in the u_{21} position as above.

In exact arithmetic it is easy to see that one of the above types of operation (based on either (3.2) or (3.3)) must be able to eliminate u'_{21} (if $u'_{21} \neq 0$ then r and Δ cannot both be zero). When round-off errors are present Fletcher and Matthews choose the operation on the basis of a test which minimizes a bound on growth in the LU factors. Subsequent steps to eliminate u'_{32}, $u'_{43}, \cdots$ are made in a similar way. The cost of the double operation (3.3) is twice that of (3.2), and if p is the column of A

that is removed then a total of $c(n-p)^2 + O(n)$ flops are required to reduce U' to upper triangular form, where $1 \leqslant c \leqslant 2$. This is the same order of magnitude as the cost of the forward substitution ($\frac{1}{2}n^2$ flops) required for the second stage. Essentially the same algorithm can also be used to update A when rows as well as columns may be replaced. This problem arises for example in some QP algorithms (see Section 5).

As regards numerical stability, experiments in which the updates are repeated many times show excellent round-off error control, even for ill-conditioned matrices and numerically singular matrices. In pathological cases we are familiar with growth of 2^n in the LU factors when using Gaussian elimination with partial pivoting. The same is true of the Bartels-Golub method. A similar type of result holds for the Fletcher-Matthews algorithm but the rate of growth in the worst case is more severe. Powell (1987) has produced worst case examples in which the growth is compounded on successive updates with the method. However the situation can be monitored, and factors can be recomputed if large growth is detetected, because explicit factors are available. This is not the case for product form algorithms. On the other hand, as with Gaussian elimination, it may be that these pathological situations are so unlikely that it is not worth including special code to detect them in practice.

For dense LP the Fletcher-Matthews algorithm is currently the most attractive one for maintaining an invertible representation of the basis matrix. For sparse LP, the same may not be true because considerable effort has gone into increasing the efficiency of product form methods, and these are currently regarded as the most suitable in this case. If the Fletcher-Matthews algorithm is to be used it will be particularly important to detect zeroes caused by cancellation in the updating process and so avoid unneccessary fill in. A promising approach might be:

(i) Start with Markowitz factors to try to minimize fill in;
(ii) Use Fletcher-Matthews updates in sparse form;
(iii) Reinvert if the file length has increased by some factor.

The hope would be that the time between reinversions would be longer with the Fletcher-Matthews update as against using a product form method. The Fletcher-Matthews update is very convenient for coding in BLAS (Basic Linear Algebra Subroutines) and can readily take advantage of any available optimized code for these routines.

4. RELIABLE LINEAR PROGRAMMING

As well as being important in their own right, LP and QP codes are being used increasingly often to solve subproblems that arise in nonlinear or nonsmooth programming. It is therefore of particular importance that these codes do not get 'stuck' in the subproblem. Some codes can perform unreliably when round-off errors at a near-degenerate vertex cause the method to cycle and fail to make progress. Many techniques exist for handling degeneracy which are guaranteed to work if the arithmetic is exact (although they are not always used in codes) and some attempt is often made to control the effects of round-off errors, for example by the use of tolerances. A recent development has been an attempt (Fletcher, 1985a) to derive an LP code which is *guaranteed* to terminate with an indicated solution, even in the presence of both degeneracy and round-off error. The idea is very recent and has by no means yet become accepted by the mathematical programming community, but I regard the subject as being of some importance and worthy of serious attention.

The idea is based on a recursive method for handling degeneracy by the repeated use of duality which is similar to a method of Balinski and Gomory (1963). The particular feature that makes this method suitable for round-off error control is that there is a *cost function* directly available at any level of recursion. Small changes to any cost function which are negligible at any given level of precision are taken as an indication of degeneracy and handled accordingly. When this degeneracy is resolved it is guaranteed that the cost function can be improved or optimality recognised. Hence every iteration at any level of recursion always guarantees to make progress.

The method is a Phase 1/Phase 2 method and it suffices to describe Phase 1 which aims to find a feasible vertex of a system of inequalities $\underline{r} \equiv A^T\underline{x} - \underline{b} \geqslant \underline{0}$ where A is an $n \times m$ matrix of rank n. This can be partitioned as

$$\begin{bmatrix} \underline{r}_B \\ \underline{r}_N \end{bmatrix} = \begin{bmatrix} A_B^T \\ A_N^T \end{bmatrix} \underline{x} - \begin{bmatrix} \underline{b}_B \\ \underline{b}_N \end{bmatrix} \geqslant \underline{0},$$

where B (basic residuals) represent *active* constraints and N (nonbasic residuals) are *inactive* constraints, the active constraints being chosen so that A_B is square and nonsingular. The current vertex $\hat{\underline{x}}$ is therefore defined by $A_B^T\hat{\underline{x}} = \underline{b}_B$ (that is by $\hat{\underline{r}}_B = \underline{0}$). If $\underline{x}$ is eliminated by virtue of $\underline{x} = \hat{\underline{x}} + A_B^{-T}\underline{r}_B$, the problem can be expressed locally in terms of $\underline{r}_B$ as

$$\begin{aligned} \underline{r}_N \equiv \hat{\underline{r}}_N + \hat{A}^T\underline{r}_B &\geqslant \underline{0} \\ \underline{r}_B &\geqslant \underline{0} \end{aligned} \qquad (4.1)$$

where $\hat{\underline{A}} = A_B^{-1}A_N$. The vector $\hat{\underline{r}}_N = A_N^T\hat{\underline{x}} - \underline{b}_N$ represents the current values of the inactive constraint residuals which fall into the following categories

$$\hat{r}_j \ (j \in N) \begin{cases} < 0 & \text{violated} \\ > 0 & \\ = 0 & \end{cases} \quad \begin{matrix} \left.\begin{matrix} <0, >0 \end{matrix}\right\} \text{nondegenerate} \\ \left.\begin{matrix} >0, =0 \end{matrix}\right\} \text{nonviolated} \\ =0 \ \text{degenerate.} \end{matrix}$$

The method starts by finding the most violated inactive constraint residual, $\hat{r}_q$ say, and considers the LP

$$\begin{aligned} &\text{maximize} \quad \hat{r}_q + \underline{\hat{a}}_q^T \underline{r}_B \\ &\text{subject to} \quad \hat{r}_j + \underline{\hat{a}}_j^T \underline{r}_B \geq 0, \qquad j : \hat{r}_j \geq 0, \\ &\qquad\qquad\qquad \underline{r}_B \geq \underline{0}, \end{aligned} \tag{4.2}$$

where $\underline{\hat{a}}_j$ refers to column j of $\hat{A}$, in which the constraints are the currently nonviolated constraints in (4.1). The aim is not so much to solve the LP, although this possibility does arise, but is mainly to make $\hat{r}_q$ nonnegative whilst keeping feasibility for the nonviolated constraints. As usual in LP there is an *optimality test* to identify some $p \in B$ for which $\hat{a}_{pq} > 0$. If so $\hat{r}_q$ can be increased by increasing r_p in a *line search*, which is terminated by a previously inactive constraint, q' say, becoming active. The iteration is completed by interchanging $p \leftrightarrow q'$ and updating the tableau. Iterations of this type are continued until:

(i) The infeasibility is removed, that is $\hat{r}_q = 0$, in which case the method returns to search for further violated residuals and terminates if there are none; or

(ii) Optimality is detected, that is no suitable $p \in B$ exists, which implies that there is no feasible point; or

(iii) Degeneracy is detected, that is a step of zero length is taken in the line search because a degenerate inactive constraint blocks any progress.

In case (iii) an outline of the measures that are taken is the following. The problem is first *localized* by ignoring nondegenerate inactive constraints (the set $N - Z_1$ below), since the aim is to find whether or not any ascent is possible from the current vertex. After introducing multipliers $\underline{\lambda}_N$ and $\underline{\lambda}_B$ for the constraints in (4.1), the *dual* of the *localized* system is taken, giving rise to the system of inequalities

$$\underline{\lambda}_B \equiv -\underline{\hat{a}}_q - \hat{A}\underline{\lambda}_N \geq \underline{0}$$

$$\lambda_j \begin{cases} \geq 0, & j \in Z_1 \\ = 0, & j \in N - Z_1 \end{cases}$$

where

$$Z_1 = \{j:\ j \in N,\quad \hat{r}_j = 0\}$$

is the current set of degenerate constraints. This is closely related to (4.1), the main difference being that $\hat{A}$ is no longer transposed, thus switching the status of B and N. Also there are changes in sign and some of the variables are fixed at zero. Notice that the *residuals* in the dual are minus the *costs* $\hat{a}_q$ in the primal, and inequality p is violated in the dual because $\hat{a}_{pq} > 0$ in the primal optimality test. The next step is to set up a dual LP, analogous to (4.2), to reduce $\hat{a}_{pq}$ to zero. Iterations in the dual LP change B and N but these changes only involve indices corresponding to zero valued constraints in the primal, so do not change any values of $\hat{r}_j$ for nondegenerate constraints. These iterations are carried out until one of the following situations arises (c.f. (i) – (iii) above):

(i) $\hat{a}_{pq}$ becomes zero: this implies that a negative cost has been removed in the primal, so the method returns to the primal to reassess the optimality test; or

(ii) $-\hat{a}_{pq}$ is maximized in the dual but cannot be increased to zero: this implies that the degeneracy block in the primal has been removed, so the method returns to the primal to carry out the line search; or

(iii) Degeneracy in the dual is detected: in this case the dual problem is localized as described above, and a further dual is taken of the localized problem. This results in a system of inequalities to be solved which is handled in the same way as above.

It can be seen that a recursive method is set up with a stack of problems at different levels, Level 1 = Primal, Level 2 = Dual, Level 3 = Dual dual, etc. ... Each level has its own *cost function* corresponding to some infeasibility that is being reduced (in Phase 2, the Level 1 cost function is $\underline{c}^T\underline{x}$, the usual LP cost function). In exact arithmetic a termination proof for the algorithm can be deduced, consequent on the following facts. Because each dualizing step removes at least one variable from the problem, there is a bound $L_{max} \leq \min(2m+1, 2n)$ on the highest level number. Also, each iteration always makes progress in that *either* a reduction in some cost function is guaranteed *or* the number of infeasibilities in some problem is reduced. Finally the number of vertices in any one problem is finite.

For inexact arithmetic, a similar termination proof follows if it is ensured that all iterations always make progress. Thus any iteration is regarded as being degenerate if a small step is taken in the line search which does not reduce the current cost function due to round-off error. The fact about the number of vertices being finite is replaced in the proof by the fact that the set of floating point numbers on a computer is finite. Some other precautions also have to be taken, for example nonviolated constraints must not be allowed to become violated due to round-off error.

Despite being recursive, the method is easy to code, mainly because the *same tableau* is used at each level. Thus only a single index needs to be stacked when recursion takes place, and the overheads are minimal. Standard matrix handling techniques can be used such as those described in Section 3. The method extends to solving l_1LP problems (Fletcher, 1985a) and this circumvents the criticism of the above method that minimizing the maximum violation is less efficient than minimizing an l_1 sum of violations.

5. AN OVERVIEW OF QUADRATIC PROGRAMMING

This section reviews developments over the last decade in solving the quadratic programming problem, concentrating on the form

$$\begin{aligned} &\text{minimize} \quad \tfrac{1}{2}\underline{x}^T G\underline{x} + \underline{g}^T\underline{x}\ , \qquad \underline{x} \in \mathbb{R}^n\ , \\ &\text{subject to} \quad A^T\underline{x} \geq \underline{b}\ , \end{aligned} \tag{5.1}$$

for simplicity of exposition. The first part of the section considers the different overall methods that are regarded as important in the current state of the art, and discusses their strengths and weaknesses. The second part looks at the matrix updating formulae that are required to make the methods perform as efficiently as possible. Currently no one approach is regarded as being uniformly best and the subject is still one of active research interest. To describe the methods, some simple concepts from the theory of quadratic programming (e.g. Fletcher, 1981) will be assumed. The set of *active constraints* is written as $\mathcal{A} = \{i : \underline{a}_i^T \underline{x} = b_i\}$ where $\underline{a}_i$ refers to column i of the matrix A. The *Kuhn-Tucker* (*KT*) *necessary conditions* for a solution to (5.1) are that there exist $\underline{x}$ and $\underline{\lambda}$ such that $A^T\underline{x} \geqslant \underline{b}$ (primal feasibility), $\underline{\lambda} \geqslant \underline{0}$ (dual feasibility) and

$$A\underline{\lambda} = G\underline{x} + \underline{g}. \tag{5.2}$$

In addition for each i, either $\underline{a}_i^T\underline{x} = b_i$ or $\lambda_i = 0$ must hold (complementary slackness). An *equality constraint subproblem* is denoted by

$$P(\mathcal{A}): \quad \text{minimize} \quad \tfrac{1}{2}\underline{x}^T G\underline{x} + \underline{g}^T\underline{x}$$
$$\text{subject to} \quad \underline{a}_i^T\underline{x} = b_i, \quad i \in \mathcal{A}.$$

Ten years ago there were two popular methods for solving QP problems, namely *active set methods* and methods based on the *linear complementarity problem* (*LCP*). To some extent the same is true today although there is more awareness of the different possibilities within these categories. The standard *primal active set method* (e.g. Fletcher, 1971) is still an important technique. In it a sequence of vectors $\underline{x}$ and $\underline{\lambda}$ is calculated which satisfy the KT conditions except for dual feasibility. The initial $\underline{x}$ is any feasible vertex and every $\underline{\lambda}$ is calculated to satisfy (5.2). A major iteration of the method consists of the following steps:

(i) Pick some $q \in \mathcal{A}$ such that $\lambda_q < 0$;

(ii) Move towards the solution of $P(A\backslash q)$;

(iii) If an inactive primal constraint becomes active, add it to A and go to (ii).

The major iteration is completed when the solution of $P(A\backslash q)$ is reached in step (ii). Major iterations are continued until dual feasibility ($\underline{\lambda} \geqslant \underline{0}$) is recognised in step (i). Although the method looks for and removes dual infeasibilities, it is not necessarily the case that a dual variable which is made feasible will stay feasible. The termination property of the algorithm is assured by the fact that each iteration is guaranteed to reduce the objective function. Primal feasibility is maintained throughout.

Good features of the algorithm are that it is a general method that allows G to be an indefinite matrix, in which case it will usually find a local minimizer of the QP problem. It tends to approach the correct active set by starting from a vertex, which often includes many bound constraints whose structure can be used advantageously. For convex QP, *multiple drops* are possible in step (i) and the evidence is that this can reduce the number of major iterations (Goldfarb, 1986), although it is less clear whether the overall savings outweigh the extra complexity and loss of generality. Disadvantages of the method are that degeneracy is possible as in primal LP, and special measures are required to counteract this and prevent cycling. In this respect it is possible to adapt the ideas of Section 4 to this context, but there are some additional difficulties when G is indefinite so the best approach is not yet clear. The effects of dual degeneracy also have to be taken into account. Another point is that in starting from a feasible vertex, a Phase 1 process is required. Both these features detract from the apparent simplicity of a primal active set method code. An alternative to the Phase 1 / Phase 2 approach is to pose the problem as an l_1QP problem as described in Section 1. This approach also enables the method to take advantage of a good estimate of $\mathcal{A}$ even when solving $P(\mathcal{A})$ gives rise to an infeasible point (the primal active set method for QP is deficient in this respect). A primal active set method for the l_1QP problem can be derived in a similar way (Fletcher, 1985b).

A more recent method for convex QP is suggested by Goldfarb and Idnani (1983) and can be interpreted as a *dual active set method*. In it a sequence of vectors $\underline{x}$ and $\underline{\lambda}$ is calculated which satisfy the KT conditions except for primal feasibility. Initially $\underline{x} = -G^{-1}\underline{g}$ is the unconstrained minimizer of the quadratic function, $\mathcal{A} = \emptyset$, and $\underline{\lambda} = \underline{0}$ is a vertex in the dual space. A major iteration of the method consists of the following steps:

(i) Pick some q such that $\underline{a}_q^T\underline{x} < b_q$;

(ii) If $\underline{a}_q \in \text{span}\{\underline{a}_i : i \in \mathcal{A}\}$ then drop from $\mathcal{A}$ the index of a constraint function that will then become positive in step (iii);

(iii) Move towards the solution of $P(\mathcal{A} \cup q)$;

(iv) If any $\lambda_i \downarrow 0$ then remove i from $\mathcal{A}$ and go to (iii).

The major iteration is completed when the solution of $P(\mathcal{A} \cup q)$ is reached in step (iii). Major iterations are continued until primal feasibility $(A^T\underline{x} \geqslant \underline{b})$ is recognised in step (i). If the QP problem that is dual to (5.1) is set up (e.g. Fletcher, 1981) then the Goldfarb and Idnani method is equivalent to the primal active set method being applied to this dual problem, and it is instructive to compare the properties of these primal and dual active set methods with this in mind. Although the Goldfarb and Idnani method looks for and removes primal infeasibilities, it is not necessarily the case that a primal constraint stays feasible once it is made feasible. The termination property of the method is assured by the fact that each iteration increases both the primal and dual objective functions. Dual feasibility is maintained throughout.

Good features of the method are that it can readily take advantage of the availability of Choleski factors of the matrix G such as can occur in some SQP methods for nonlinear programming. The method is most effective when there are only a few active constraints at the solution, and is also able to take advantage of a good estimate of A. Degeneracy of the dual constraints $\underline{\lambda} \geqslant \underline{0}$ is not possible. The only difficulty caused by degeneracy in the primal constraints arises near the solution and it is claimed that this can readily be handled by the use of tolerances. Adverse features of the method include its lack of generality, in that the matrix G must be positive definite. In this respect, it is also the case that difficulties will arise when G is ill-conditioned, for example large elements and hence large round-off errors can arise in the vector $\underline{x} = -G^{-1}\underline{g}$. However Powell (1985) claims that with care these difficulties can be circumvented. The method also does not take very much advantage of the structure of primal constraints which are simple bounds.

It is possible to develop other what might be called *primal-dual* methods for convex QP which allow both primal and dual infeasibilities, and take steps similar to those in the methods above (e.g. Goldfarb and Idnani, 1981). However the well known transformation of a QP problem to an LCP can also be regarded as giving rise to a number of primal-dual methods. If bounds $\underline{x} \geqslant \underline{0}$ are included in the statement of the problem (5.1), the LCP transformation introduces multipliers $\underline{\pi}$ for the bounds, and slack variables $\underline{r} = A^T\underline{x} - \underline{b}$ for the general constraints, so that the KT conditions may be written

$$\begin{bmatrix} \underline{\pi} \\ \underline{r} \end{bmatrix} - \begin{bmatrix} G & -A \\ A^T & 0 \end{bmatrix} \begin{bmatrix} \underline{x} \\ \underline{\lambda} \end{bmatrix} = \begin{bmatrix} \underline{g} \\ -\underline{b} \end{bmatrix} \tag{5.3}$$

$$\underline{\pi}^T\underline{x} = 0 , \quad \underline{\lambda}^T\underline{r} = 0 , \quad \underline{\pi}, \underline{r}, \underline{x}, \underline{\lambda} \geqslant \underline{0} .$$

The *principal pivoting method* (e.g. Cottle and Dantzig, 1968) removes infeasibilities in this system in a similar way to that described in Section 4 above. It differs from the active set methods in that once a variable becomes feasible, then it is not subsequently allowed to go infeasible. Another method is that of Lemke (1965) which is claimed not to require special treatment of degeneracy, and can be regarded as solving the QP problem in a parametric fashion (Fletcher, 1981). Moreover the first order conditions for the l_1QP problem can also be given in a similar form to (5.3), but with the addition of upper bounds on the variables $\underline{x}$ and $\underline{\lambda}$ (Fletcher, 1985b). Thus methods for solving the LCP can readily be extended to solve the l_1QP problem.

Currently none of the above methods for QP or l_1QP stands out as being uniformly best and each type of method has both its advantages and disadvantages. The subject is still one of active interest to researchers and the state of the art can be expected to advance over the next few years.

The rest of this section considers the processes in matrix algebra that are required to implement the above methods. For convenience, A subsequently denotes the matrix with columns $\{\underline{a}_i : i \in \mathcal{A}\}$. To solve the equality constraint subproblem $P(\mathcal{A})$ requires the matrices H and T defined by

$$\begin{bmatrix} G & -A \\ -A^T & 0 \end{bmatrix}^{-1} = \begin{bmatrix} H & -T \\ -T^T & U \end{bmatrix} \tag{5.4}$$

(Fletcher, 1981). In a *null space method*, matrices Y and Z are defined by

$$[Y : Z] = [A : V]^{-T}$$

where V is any matrix for which $[A : V]$ is nonsingular. Because $A^T Z = 0$, the columns of Z span the null space of A^T. This method stores and updates an invertible representation of $[A : V]$ and also the Choleski factors of the reduced Hessian matrix, that is

$$Z^T G Z = LL^T .$$

The matrices H and T are then implicitly determined by

$$H = ZL^{-T} L^{-1} Z^T$$
$$T = (I - HG)Y . \tag{5.5}$$

(Of course these matrices are not multiplied out but are used to indicate how vector products with H and T are formed.) Various methods exist (Fletcher, 1981) for choosing V and hence Z, but the following are the most popular. If the QR factors of A are

$$A = \begin{bmatrix} Q_1 : Q_2 \end{bmatrix} \begin{bmatrix} R \\ 0 \end{bmatrix} = Q_1 R$$

then $V = Z = Q_2$ may be chosen so as to give an orthogonal basis for the null space (Gill and Murray, 1978). This choice minimizes a bound on the l_2 norm of the reduced Hessian matrix, and is therefore a safe choice from the point of view of numerical stability. In the variable elimination method, the choice $V^T = [0 : I]$ is essentially made and it gives a highly efficient method. In particular, considerable savings from the structure of active bounds can be gained by using the format

$$[A : V] = \begin{bmatrix} A_1 & & \\ A_2 & I & \\ A_3 & & I \end{bmatrix} \begin{matrix} \} \text{ Elimination variables} \\ \} \text{ Bound variables} \\ \} \text{ Free variables} \end{matrix}$$

and the factorization $PA_1 = LU$, which is updated where necessary using the Fletcher-Matthews method of Section 3 for the row and column interchanges that are required (Matthews, 1984).

For convex QP problems an alternative approach is the *range space method*. Such a method requires the Hessian matrix to be positive definite and makes use of Choleski factors $G = LL^T$. Furthermore it employs either explicitly or implicitly the QR factors defined by

$$L^{-1}A = [Q_1 : Q_2] \begin{bmatrix} R \\ 0 \end{bmatrix} = Q_1 R ,$$

and H and T in (5.4) are determined from

$$T = L^{-T} Q_1 R^{-T}$$

$$H = L^{-T}(I - Q_1 Q_1^T) L^{-1} .$$

Gill, Gould *et al*. (1984) give one such method in which Q_1 (but not Q_2) and R are updated explicitly. Goldfarb and Idnani (1983) give a related method in which they introduce the matrix $S = L^{-T}Q$, and they store and update S and R. Clearly the formulae above allow $L^{-T}Q_1$ to be replaced by S_1: it may be that the method could be improved by only storing and updating S_1 and R. Powell (1985) compares the Goldfarb and Idnani representation with a direct QR representation, and concludes that, whilst the former may lack numerical stability in certain pathological cases, it is entirely adequate for all practical purposes. A further matrix scheme might be called the *low storage method* because it just updates R, replacing Q_1 by the product $L^{-1}AR^{-1}$ in the above formulae. This transformation squares the condition number of the problem (e.g. Björck, 1985) and so is undesirable as it stands. However it may be that iterative refinement could be used to advantage in this context, along the lines given by Björck. Range space methods are of most use when the Choleski factor of G is given *a priori* and does not have to be calculated. The operation counts for these methods are most favourable when there are few constraints, and the matrix operations do not allow much advantage to be taken of simple bounds. Therefore these methods are more appropriate for use with the Goldfarb and Idnani dual active set method.

For linear complementarity problems the commonly used matrix method is that of pivoting in the tableau derived from (5.3). Thus any of the standard methods for LP problems could be used and the most up to date choice would be from the methods described in Section 3. For large sparse QP problems this would be one way to proceed. For combining sparse matrix techniques with an active set method, the best possibility is probably to use the variable elimination technique and a null space method. However it is unlikely that much sparsity would be present in the Choleski factor of the reduced Hessian matrix Z^TGZ, so this factor would be stored in full, which would limit the applicability of such a technique to problems of at most a few hundred free variables (number of columns of Z). Another interesting technique for sparse problems is the *Schur complement update* which enables one to avoid directly updating some representation of a matrix when the direct update would cause serious fill in. This and other sparse matrix techniques are reviewed by Gill, Murray *et al.* (1984). For dense QP it is very difficult to recommend a single best choice of method. It depends not only on the type of problem, but also on the relative weight given to storage and time costs, and to numerical stability. My current preference for general use would be some sort of null space method, but there may be a case for having more than one routine available in a library.

Finally a brief mention of some special forms of QP problems is given. The most important of these is the *least squares QP problem* in which the objective function is the sum of squares of a linear function, $\frac{1}{2}(B\underline{x}-\underline{c})^T(B\underline{x}-\underline{c})$ say, where B is a $p \times n$ matrix and $\underline{c} \in \mathbb{R}^p$. It is possible to use a standard QP method by writing $G = B^TB$ and $\underline{g} = -B^T\underline{c}$. However it is well known in linear least squares that to explicitly form the matrix B^TB squares the condition number of the problem and often causes unnecessary loss of accuracy. The same can be true here and the same remedy of using some form of QR factorization also applies. A null space method could be obtained by updating QR factors of the matrix BZ where Z spans the null space of A as described above. That is to say,

$$BZ = P \begin{bmatrix} U \\ 0 \end{bmatrix} = P_1 U$$

is defined where P is $p \times p$ orthogonal, and U is $(n-m) \times (n-m)$ upper triangular, m being the number of active constraints. It follows that $U^T U$ is the Choleski factorization of $Z^T GZ$ and can be used in (5.5) to solve the null space equations. An algorithm of this type is given by Schittkowski and Stoer (1979) in which Z is also defined by QR factors of A. Alternatively, if $B = Q_1 L^T$ is the QR factorization of B itself, then L is the Choleski factor of G and could therefore be used in any of the range space methods described above. Other special cases of QP problems are *least distance problems* and *bounds-only problems* and special methods for such problems are reviewed by Fletcher (1981).

REFERENCES

Balinski, M.L. and Gomory, R.E. (1963), "A mutual primal-dual simplex method", in *Recent Advances in Mathematical Programming*, eds. Graves, R.L. and Wolfe, P., McGraw-Hill (New York), pp.17-26.

Bartels, R.H. and Golub, G.H. (1969), "The simplex method of linear programming using LU decomposition", *Comm. A.C.M.*, *12*, pp.266-268.

Bennett, J.M. (1965), "Triangular factors of modified matrices", *Numer. Math.*, *7*, pp.217-221.

Björck, A. (1985), "Stability analysis of the method of semi-normal-equations for linear least squares problems", Report LiTH-MAT-R-1985-08, Linköping University, Sweden.

Cottle, R.W. and Dantzig, G.B. (1968), "Complementary pivot theory of mathematical programming", *Linear Algebra Appl.*, *1*, pp.103-125.

Fletcher, R. (1971), "A general quadratic programming algorithm", *J. Inst. Math. Appl.*, *7*, pp.76-91.

Fletcher, R. (1981), *Practical Methods of Optimization*, *Volume 2*, *Constrained Optimization*, John Wiley (Chichester).

Fletcher, R. (1985a), "Degeneracy in the presence of round-off errors", Report NA/79, University of Dundee.

Fletcher, R. (1985b), "An l_1 penalty method for nonlinear constraints", in *Numerical Optimization 1984*, eds. Boggs, P.T., Byrd, R.H. and Schnabel, R.B., SIAM Publications (Philadelphia), pp.26-40.

Fletcher, R. and Matthews, S.P.J. (1984), "Stable modification of explicit LU factors for simplex updates", *Math. Programming*, *30*, pp.267-284.

Forrest, J.J.H. and Tomlin, J.A. (1972), "Updated triangular factors of the basis to maintain sparsity in the product form simplex method", *Math. Programming*, *2*, pp.263-278.

Gacs, P. and Lovasz, L. (1979), "Khachian's algorithm for linear programming", Report CS 750, Stanford University.

Gay, D.M. (1985), "A variant of Karmarkar's linear programming algorithm for problems in standard form", Numerical Analysis Manuscript 85-10, AT & T Bell Labs., Murray Hill, NJ.

Gill, P.E. and Murray, W. (1978), "Numerically stable methods for quadratic programming", *Math. Programming*, *14*, pp.349-372.

Gill, P.E., Gould, N.I.M., Murray, W., Saunders, M.A. and Wright, M.H. (1984), "A weighted Gram-Schmidt method for convex quadratic programming", *Math. Programming*, *30*, pp.176-195.

Gill, P.E., Murray, W., Saunders, M.A. and Wright, M.H. (1984), "Sparse matrix methods in optimization", *SIAM J. Sci. Statist. Comput.*, *5*, pp.562-589.

Gill, P.E., Murray, W., Saunders, M.A., Tomlin, J.A. and Wright, M.H. (1986), "On projected Newton barrier methods for linear programming and an equivalence to Karmarkar's projective method", *Math. Programming*, *36*, pp.183-209.

Goldfarb, D. (1986), "Strategies for constraint deletion in active set algorithms", in *Numerical Analysis 1985*, eds. Griffiths, D.F. and Watson, G.A., Pitman (London), pp.66-81.

Goldfarb, D. and Idnani, A. (1981), "Dual and primal-dual methods for solving strictly convex quadratic programs", in *Numerical Analysis*, *Proceedings Cocoyoc*, *Mexico 1981*, *Lecture Notes in Mathematics 909*, ed. Hennart, J.P., Springer-Verlag (Berlin), pp.226-239.

Goldfarb, D. and Idnani, A. (1983), "A numerically stable dual method for solving strictly convex quadratic programs", *Math. Programming*, *27*, pp.1-33.

Iri, M. and Imai, H. (1985), "A multiplicative penalty function method for linear programming — Another 'new and fast' algorithm", in *Proceedings of the 6th Mathematical Programming Symposium in Japan*, Tokyo 1985, pp.97-120.

Karmarkar, N. (1984), "A new polynomial-time algorithm for linear programming", *Combinatorica*, *4*, pp.373-395.

Khachiyan, L.G. (1979), "A polynomial algorithm in linear programming", *Dokl. Akad. Nauk. SSSR*, *244*, pp.1093-1096 (translated as *Soviet Math. Dokl.*, *20*, pp.191-194).

Klee, V. and Minty, G.J. (1972), "How good is the simplex algorithm?" in *Inequalities III*, ed. Shisha, O., Academic Press (New York), pp.159-175.

Lemke, C.E. (1965), "Bimatrix equilibrium points and mathematical programming", *Management Sci.*, *11*, pp.681-689.

Lustig, I.J. (1985), "A practical approach to Karmarkar's algorithm", Report SOL 85-5, Stanford University.

Matthews, S.P.J. (1984), *Matrix Algebra and Other Aspects of Linear and Quadratic Programming*, Ph.D. Thesis, University of Dundee.

Neumaier, A. (1986), "The solution of linear programming problems by weighted least squares", working paper, Inst. Ang. Math., University of Freiburg.

Powell, M.J.D. (1985), "On the quadratic programming algorithm of Goldfarb and Idnani", *Math. Programming Stud.*, *25*, pp.46-61.

Powell, M.J.D. (1987), "On error growth in the Bartels-Golub and Fletcher-Matthews algorithms for updating matrix factorizations", *Linear Algebra Appl.* (to be published).

Reid, J.K. (1975), "A sparsity-exploiting version of the Bartels-Golub decomposition for linear programming bases", Report CSS 20, AERE Harwell.

Schittkowski, K. and Stoer, J. (1979), "A factorization method for the solution of constrained linear least squares problems allowing subsequent data changes", *Numer. Math.*, *31*, pp.431-463.

Sonnevend, Gy. (1986), "An 'analytical centre' for polyhedrons and new classes of global algorithms for linear (smooth, convex) programming", in *Proc. 12th IFIP Conference on System Modelling and Optimization, Budapest 1985, Lecture Notes in Control and Information Sciences*, *84*, eds. Thoma, M. and Wyner, A., Springer-Verlag (Berlin), pp.866-875.

Strang, G. (1985), "Karmarkar's algorithm in a nutshell", *SIAM News*, *18*, *No.6*.

Williams, A.C. (1981), "Computational experience with ellipsoid algorithms for linear programming", *Math. Programming Soc. COAL Newsletter*, *No.5*, pp.37-48.

Wolfe, P. (1980), "The ellipsoid algorithm", *Optima (Math. Programming Soc. Newsletter)*, *No.1*, pp.1-5.

R. Fletcher
Department of Mathematical Sciences
The University
Dundee DD1 4HN
Scotland

9 Solving Systems of Nonlinear Equations by Tensor Methods

ROBERT B. SCHNABEL and PAUL D. FRANK

ABSTRACT

Tensor methods are a class of general purpose methods for solving systems of nonlinear equations. They are especially intended to efficiently solve problems where the Jacobian matrix at the solution is singular or ill-conditioned, while remaining at least as efficient as standard methods on nonsingular problems. Their distinguishing feature is that they base each iteration on a quadratic model of the nonlinear function. The model has a simple second order term that allows it to interpolate more information about the nonlinear function than standard, linear model based methods, without significantly increasing the cost of forming, storing, or solving the model.

This paper summarizes two types of tensor methods, derivative tensor methods that calculate an analytic or finite difference Jacobian at each iteration, and secant tensor methods that avoid Jacobian evaluations. Both are shown to require no more function or derivative information per iteration, and hardly more storage or arithmetic operations per iteration, than standard linear model based methods. Computational results are presented that indicate that both tensor methods are consistently at least as reliable as the corresponding linear model based methods, and are significantly more efficient, both on nonsingular and on singular test problems.

Research supported by NSF grant DCR-8403483 and ARO contract DAAG 29-84-K-0140

1. INTRODUCTION

This paper summarizes a recently developed class of methods, called tensor methods, for solving the nonlinear equations problem

$$\text{given } \underline{F}: \mathbb{R}^n \to \mathbb{R}^n, \quad \text{find } \underline{x}_* \in \mathbb{R}^n \text{ such that } \underline{F}(\underline{x}_*) = \underline{0} \tag{1.1}$$

where it is assumed that $\underline{F}(\underline{x})$ is at least once continuously differentiable. Tensor methods are especially intended to efficiently solve problems where the Jacobian matrix of $\underline{F}$ at $\underline{x}_*$, $\underline{F}'(\underline{x}_*) \in \mathbb{R}^{n \times n}$, is singular or ill-conditioned. They also are intended to be at least as efficient as standard methods on problems where $\underline{F}'(\underline{x}_*)$ is nonsingular. Their distinguishing feature is that they base each iteration on a quadratic model of $\underline{F}(\underline{x})$ whose second order term has a simple form.

Systems of nonlinear equations arise frequently in many practical applications including equilibrium calculations, curve tracing problems, and as subproblems in solving nonlinear systems of differential equations. In many important situations, $\underline{F}'(\underline{x}_*)$ is singular or ill-conditioned. For example, in some stiff systems of ordinary differential equations the Jacobian of the associated system of nonlinear equations is nearly singular for all $\underline{x}$. The calculation of turning points in curve tracing problems and the solution of over-parameterized data fitting problems are other common situations that lead to singular systems of equations. In all these cases, it is important to notice that the (near) rank deficiency in the derivative matrix usually is small. This is the case in which our methods are intended to improve upon standard methods.

Standard methods for solving (1.1) base each iteration upon a linear model $\underline{M}(\underline{x})$ of $\underline{F}(\underline{x})$ around the current iterate $\underline{x}_c \in \mathbb{R}^n$,

$$\underline{M}(\underline{x}_c + \underline{d}) = \underline{F}(\underline{x}_c) + J_c \underline{d} \tag{1.2}$$

where $\underline{d} \in \mathbb{R}^n$ and $J_c \in \mathbb{R}^{n \times n}$. These methods can be divided into two classes: derivative methods, where J_c is the current

Jacobian matrix $\underline{F}'(\underline{x}_c)$ or a finite difference approximation to it, and secant methods, where J_c is a secant (quasi-Newton) approximation to the Jacobian. For a general description of these methods, see e.g. Dennis and Schnabel (1983).

When the analytic Jacobian is available, the linear model (1.2) becomes

$$\underline{M}(\underline{x}_c + \underline{d}) = \underline{F}(\underline{x}_c) + \underline{F}'(\underline{x}_c)\,\underline{d}\,. \tag{1.3}$$

The standard method for nonlinear equations, Newton's method, consists of setting the next iterate $\underline{x}_+$ to the value of $\underline{x}_c + \underline{d}$ that solves (1.3),

$$\underline{x}_+ = \underline{x}_c - \underline{F}'(\underline{x}_c)^{-1}\,\underline{F}(\underline{x}_c). \tag{1.4}$$

If $\underline{F}'(\underline{x}_c)$ is Lipschitz continuous in a neighbourhood containing the root $\underline{x}_*$ and $\underline{F}'(\underline{x}_*)$ is nonsingular, then the sequence of iterates produced by (1.4) converges locally and q-quadratically to $\underline{x}_*$. This means that there exist $\delta > 0$ and $c \geqslant 0$ such that the sequence of iterates $\{\underline{x}_k\}$ produced by Newton's method obeys

$$\|\underline{x}_{k+1} - \underline{x}_*\| \leqslant c\,\|\underline{x}_k - \underline{x}_*\|^2$$

if $\|\underline{x}_0 - \underline{x}_*\| \leqslant \delta$. In practice, local q-quadratic convergence means eventual fast convergence.

Newton's method usually is not quickly locally convergent, however, if $\underline{F}'(\underline{x}_*)$ is singular. For example when applied to one equation in one unknown $(n = 1)$ where $f'(x_*) = 0$ but $f''(x_*) \neq 0$, Newton's method is locally q-linearly convergent with constant converging to $\frac{1}{2}$, meaning that the sequence of iterates $\{x_k\}$ obeys

$$|x_{k+1} - x_*| = c_k\,|x_k - x_*|\,, \qquad \lim_{k\to\infty} c_k = \tfrac{1}{2}$$

if $|x_0 - x_*|$ is sufficiently small. For systems of equations, the situation is more complex and has been analyzed by many authors, including Decker and Kelley (1980a, 1980b, 1982), Decker, Keller and Kelley (1983), Griewank (1980a, 1980b, 1985), Griewank and Osborne (1981, 1983), Keller (1970), Kelley and Suresh (1983), Rall (1966), and Reddien (1978, 1979). In summary, their papers show that, from many starting points, Newton's method for systems of equations also is locally q-linearly convergent with constant converging to $\frac{1}{2}$, although for some problems with starting points arbitrarily close to $\underline{x}_*$, (1.4) may be undefined or lead further away from the solution (see e.g. Griewank and Osborne, 1983). In practice, Newton's method usually exhibits local linear convergence with constant $\approx\frac{1}{2}$ on singular problems, much slower convergence than one would like.

When analytic derivatives are unavailable and function evaluation is expensive, (1.1) generally is solved by a secant method. These methods attempt, as much as possible, to solve (1.1) using only the function values at the iterates. The model (1.2) still is used but the matrix J_c is generated from these function values and may be a very rough approximation to $\underline{F}'(\underline{x}_c)$. In the most commonly used secant method for systems of equations, namely Broyden's method, the Jacobian approximation J_c is chosen to be the smallest change to the previous Jacobian approximation which causes the new linear model $\underline{M}(\underline{x})$ to interpolate the value of $\underline{F}(\underline{x})$ at the previous iterate. This results in a rank one change to the Jacobian approximation at each iteration. (The details are given in Section 3.1.) The initial Jacobian approximation is made by finite differences, and sometimes it is necessary to reset J_c to a finite difference approximation at subsequent iterations.

The sequence of iterates produced by Broyden's method converges locally and q-superlinearly to $\underline{x}_*$ as long as $\underline{F}'(\underline{x}_c)$ is Lipschitz continuous in a neighbourhood containing the root $\underline{x}_*$ and $\underline{F}'(\underline{x}_*)$ is nonsingular (Broyden, Dennis and Moré, 1973). This means that there exist $\delta>0$ and $\tau>0$ such that the sequence of iterates $\{\underline{x}_k\}$ obeys

$$\lim_{k\to\infty} \|\underline{x}_{k+1}-\underline{x}_*\| / \|\underline{x}_k-\underline{x}_*\| = 0$$

if $\|\underline{x}_0-\underline{x}_*\| \leq \delta$ and $\|J_0-\underline{F}'(\underline{x}_0)\| \leq \tau$. In practice, secant methods are quickly convergent on nonsingular problems, and while they usually require more iterations than Newton's method, they usually require fewer function evaluations than a finite difference implementation of Newton's method.

However secant methods, like Newton's method, are slowly convergent on problems where $\underline{F}'(\underline{x}_*)$ is singular. For example on one variable problems with $f'(x_*)=0$ but $f''(x_*)\neq 0$, the secant method is locally q-linearly convergent with constant converging to 0.618, a slightly slower rate than Newton's method. For multiple variable problems with rank $(\underline{F}'(\underline{x}_*))=n-1$, Decker and Kelley (1985) have shown that this same rate of convergence is obtained by Broyden's method from certain starting points. As in the case of Newton's method, this slow linear convergence usually is observed in practice, making quicker methods desirable.

Several papers, for example Decker and Kelley (1982), Decker, Keller and Kelley (1983), Griewank (1980a, 1985), Kelley (1985), and Kelley and Suresh (1983), propose methods that are rapidly convergent on some singular problems. Many of these methods are related to the one dimensional acceleration technique of taking j times the Newton step if one has a root of multiplicity j. Some other methods explicitly calculate and use higher derivative information in null space directions. To our knowledge, no computational experience with a complete method of this type has been published, and it is not clear how amenable these techqniues are to solving general systems of nonlinear equations when it is unknown *a priori* whether $\underline{F}'(\underline{x}_*)$ is singular or not.

The major aim of tensor methods is to provide general purpose methods that have rapid convergence even when $\underline{F}'(\underline{x}_*)$ is singular. In addition, the methods should not experience any special difficulty when J_c is singular or ill-conditioned, while methods based on (1.2) must be modified in this case.

Tensor methods are based on expanding the linear model (1.2) of $\underline{F}(\underline{x})$ around $\underline{x}_c$ to the quadratic model

$$\underline{M}_T(\underline{x}_c + \underline{d}) = \underline{F}(\underline{x}_c) + J_c\underline{d} + \tfrac{1}{2}T_c\underline{d}\underline{d} \tag{1.5}$$

where $T_c \in \mathbb{R}^{n\times n\times n}$ and J_c is $\underline{F}'(\underline{x}_c)$ or a secant approximation to it. The three dimensional object T_c often is referred to as a tensor; hence we call (1.5) a *tensor model*, and methods based upon (1.5) *tensor methods*. The term $T_c\underline{d}\underline{d}$ is defined by $(T_c\underline{d}\underline{d})[i] = \underline{d}^T H_i \underline{d}$, where H_i is the ith horizontal face of T_c. Thus the model $\underline{M}_T(\underline{x}_c + \underline{d})$ is the n-vector of quadratic models of the component functions of $\underline{F}(\underline{x})$,

$$(\underline{M}_T(\underline{x}_c + \underline{d}))[i] = f_i + \underline{g}_i^T\underline{d} + \tfrac{1}{2}\underline{d}^T H_i \underline{d}, \quad i = 1,2,\cdots,n,$$

where $f_i = \underline{F}(\underline{x}_c)[i]$, $\underline{g}_i^T = \text{row } i$ of $\underline{F}'(\underline{x}_c)$ or an approximation to it, and H_i is (an approximation to) the Hessian matrix of the ith component function of $\underline{F}(\underline{x})$. Notice that we are denoting components of a vector $\underline{v} \in \mathbb{R}^n$ by $\underline{v}[i] \in \mathbb{R}$.

The obvious choice of T_c in (1.5) is the tensor $\underline{F}''(\underline{x}_c)$ of second partial derivatives of $\underline{F}$ at $\underline{x}_c$; if J_c is $\underline{F}'(\underline{x}_c)$, this makes (1.5) the first three terms of the Taylor series expansion of $\underline{F}(\underline{x}_c + \underline{d})$ around $\underline{x}_c$. Several serious disadvantages, however, make (1.5) with $T_c = \underline{F}''(\underline{x}_c)$ unacceptable for algorithmic use. First, the n^3 second partial derivatives of $\underline{F}$ at $\underline{x}_c$ would have to be computed at each iteration. Second, the model would take more than $n^3/2$ locations to store as compared to the n^2 locations for the standard model. Third, to find a root of the model, at each iteration one would have to solve a system of n quadratic equations in n unknowns, which for $n > 1$ requires an iterative procedure. Finally, the model might not have a real root.

To use a model of form (1.5) and avoid these disadvantages, our tensor methods use a very restricted form of T_c. In particular, our tensor methods require no additional derivative or function information; the additional costs of forming and solving the tensor model are small compared to the $O(n^3)$ arithmetic cost per iteration of standard methods; and the additional storage required for our tensor models is small compared to the n^2 storage required for the Jacobian. The remainder of this paper describes how we utilize the tensor term T_c in the model (1.5) and what benefits we obtain from its inclusion. In Section 2 we summarize the use of a tensor model in derivative methods for nonlinear equations, and our computational experience with this method. Section 3 similarly presents the use of and computational experience with a tensor model in secant methods for nonlinear equations. In Section 4 we briefly comment on extensions of tensor methods to nonlinear least squares and to unconstrained optimization. More details on this research can be found in Frank (1984), Schnabel and Frank (1984), and an upcoming paper by Frank and Schnabel.

2. DERIVATIVE TENSOR METHODS

Derivative tensor methods base an algorithm for solving systems of nonlinear equations on a model of the form

$$\underline{M}_T(\underline{x}_c + \underline{d}) = \underline{F}(\underline{x}_c) + \underline{F}'(\underline{x}_c)\,\underline{d} + \tfrac{1}{2}\, T_c \underline{d}\underline{d}, \qquad (2.1)$$

where it is assumed that $\underline{F}'(\underline{x}_c)$ either is supplied analytically or is calculated by finite differences. Their aim is to choose $T_c \in \mathbb{R}^{n\times n\times n}$ so that the model (2.1) is hardly more expensive to form, store, or solve than the standard model (1.3), while still leading to an algorithm that requires fewer function evaluations than standard methods to solve difficult problems.

2.1 *Forming the Tensor Model*

The first step in deriving a method based on (2.1) is to choose the second order term T_c. We do not use any second derivative information in constructing T_c. Instead, we construct the second order term in (2.1) by asking the model to interpolate additional values of the function $\underline{F}(\underline{x})$ that have already been computed by the algorithm. In particular, we ask the model to satisfy

$$\underline{F}(\underline{x}_{-k}) = \underline{F}(\underline{x}_c) + \underline{F}'(\underline{x}_c)\,\underline{s}_k + \tfrac{1}{2}T_c\underline{s}_k\underline{s}_k, \quad k = 1,2,\cdots,p, \tag{2.2a}$$

where

$$\underline{s}_k = \underline{x}_{-k} - \underline{x}_c, \quad k = 1,2,\cdots,p, \tag{2.2b}$$

and $\underline{x}_{-1},\cdots,\underline{x}_{-p}$ are some set of p past iterates that need not be consecutive.

For the equations (2.2) to be consistent, the past points $\{\underline{x}_{-k}\}$ must be selected so that the set of directions $\{\underline{s}_k\}$ is linearly independent. In fact, we enforce a far more restrictive condition. We always set $\underline{x}_{-1}$ to the most recent iterate. We then include each remaining past iterate in the set of points to be interpolated if the step from it to $\underline{x}_c$ makes an angle of at least θ degrees with the subspace spanned by the steps to the already selected more recent iterates. Here θ is some fixed angle between 20 and 45 degrees. In addition, we consider at most $\sqrt{n}$ past iterates. The bounds $\sqrt{n}$ and 20 – 45 degrees have been shown by computational experience to be reasonable. This procedure for selecting past iterates to interpolate is implemented easily using a modified Gram – Schmidt algorithm, and requires about n^2 multiplications and additions.

The equations (2.2) are a set of $np \leq n^{1.5}$ linear equations in the n^3 unknowns comprising T_c. Thus T_c is underdetermined, so we follow the standard and successful practice in secant methods for nonlinear equations and optimization (see e.g., Dennis and Schnabel, 1979) and choose T_c to be the solution to

$$\underset{T_c \in \mathbb{R}^{n\times n\times n}}{\text{minimize}} \quad \|T_c\|_F \tag{2.3}$$

$$\text{subject to} \quad T_c\underline{s}_k\underline{s}_k = \underline{t}_k = 2(\underline{F}(\underline{x}_{-k}) - \underline{F}(\underline{x}_c) - \underline{F}'(\underline{x}_c)\,\underline{s}_k), \quad k = 1,2,\cdots,p,$$

where $\|\cdot\|_F$ is the Frobenius norm. If we denote by $\underline{u}\underline{v}\underline{w}$ the rank one tensor whose ith horizontal face is the rank one matrix $\underline{u}[i](\underline{v}\underline{w}^T)$, then the solution to (2.3) is shown by Schnabel and

Frank (1984) to be

$$T_c = \sum_{k=1}^{p} \underline{a}_k \underline{s}_k \underline{s}_k \tag{2.4}$$

where

$$(\underline{a}_1[i], \cdots, \underline{a}_p[i])^T = M^{-1} * (\underline{t}_1[i], \ldots, \underline{t}_p[i])^T \,, \quad i = 1,2,\ldots,n \,,$$

and $M \in \mathbb{R}^{p \times p}$ is the positive definite matrix defined by

$$M[i,j] = (\underline{s}_i^T \underline{s}_j)^2 \,, \quad 1 \leq i,j \leq p.$$

Substituting (2.4) into the tensor model (2.1) gives

$$\underline{M}_T(\underline{x}_c + \underline{d}) = \underline{F}(\underline{x}_c) + \underline{F}'(\underline{x}_c)\underline{d} + \tfrac{1}{2} \sum_{k=1}^{p} \underline{a}_k (\underline{d}^T \underline{s}_k)^2 \,. \tag{2.5}$$

The simple form of the second order term in (2.5) is the key to being able to efficiently form, store, and solve the tensor model. Since $p \leq \sqrt{n}$, the additional storage for the entire method, including values of $\underline{x}_{-k}$ and $\underline{F}(\underline{x}_{-k})$, is at most $4\sqrt{n}$ n-vectors, compared to the n^2 storage for $\underline{F}'(\underline{x}_c)$. The dominant cost of forming the tensor model by the above procedure is at most $n^{2.5}$ multiplications and additions per iteration, small compared to the at least $n^3/3$ multiplications and additions per iteration required by standard methods.

2.2 *Solving the Derivative Tensor Model*

To base an efficient algorithm on the tensor model (2.5), we need to find efficiently a root of this model, that is a $\underline{d} \in \mathbb{R}^n$ for which

$$\underline{M}_T(\underline{x}_c + \underline{d}) = \underline{F}(\underline{x}_c) + \underline{F}'(\underline{x}_c)\underline{d} + \tfrac{1}{2} \sum_{k=1}^{p} \underline{a}_k (\underline{d}^T \underline{s}_k)^2 = \underline{0} \,. \tag{2.6}$$

In some cases, the tensor model may have no root; it is then appropriate to choose $\underline{d}$ to minimize the tensor model in some norm. We choose the ℓ_2 norm, so that the general problem we wish to solve is to choose $\underline{d} \in \mathbb{R}^n$ to minimize $\|\underline{M}_T(\underline{x}_c + \underline{d})\|_2$.

The basic idea behind efficiently solving (2.6) is that, since $\underline{M}_T$ is quadratic only on the p dimensional subspace spanned by $\{\underline{s}_k\}$, and is linear on the orthogonal complement to this subspace, it may be possible to satisfy (2.6) by solving a system of p quadratic equations in p unknowns plus a system of $n-p$ linear equations in $n-p$ unknowns. This is accomplished by a procedure given in Schnabel and Frank (1984). This procedure first makes an orthogonal transformation of the variable space to $\hat{\underline{d}} = Q^T\underline{d}$, so that all n equations are quadratic only in the last p components of $\hat{\underline{d}}$, $\hat{\underline{d}}_2 \in \mathbb{R}^p$, and are linear in the first $n-p$ components, $\hat{\underline{d}}_1 \in \mathbb{R}^{n-p}$. It then makes an orthogonal transformation of the equations that eliminates the linear variables $\hat{\underline{d}}_1$ from the final p (actually $q \geq p$, see below) equations and makes the preceding equations triangular in $\hat{\underline{d}}_1$. The result is $n-q$ equations that are linear in the $n-p$ variables $\hat{\underline{d}}_1$,

$$\tilde{\underline{F}}_1 + \tilde{J}_1\hat{\underline{d}}_1 + \tilde{J}_2\hat{\underline{d}}_2 + \tfrac{1}{2}\tilde{A}_1\{\hat{S}_2^T\hat{\underline{d}}_2\}^2 = \underline{0}\,, \tag{2.7a}$$

$\tilde{J}_1$ being upper triangular, plus the system of q quadratic equations in the p unknowns $\hat{\underline{d}}_2$

$$\tilde{\underline{F}}_2 + \tilde{J}_3\hat{\underline{d}}_2 + \tfrac{1}{2}\tilde{A}_2\{\hat{S}_2^T\hat{\underline{d}}_2\}^2 = \underline{0}\,, \tag{2.7b}$$

where $\{\underline{v}\}^2$, $\underline{v} \in \mathbb{R}^p$, is the vector whose components are $\{(\underline{v}[i])^2;\ i = 1,2,\cdots,p\}$. Here $q \geq p$, with $q = p$ as long as J is nonsingular, or J is singular but J augmented by the p rows $\{\underline{s}_k^T\}$ has full column rank. In practice this means that q generally equals p unless rank $(\underline{F}'(\underline{x}_c)) < n-p$. The root or minimizer of $\underline{M}_T$ is then found from (2.7) by calculating the $\hat{\underline{d}}_2$ which is the root or minimizer of the quadratic system of equations (2.7b), substituting this $\hat{\underline{d}}_2$ into (2.7a) and calculating $\hat{\underline{d}}_1$ (which is not defined uniquely by (2.7a) if $q > p$) by solving a triangular system of linear equations, and multiplying $\hat{\underline{d}}$ by Q to obtain $\underline{d}$.

The cost of solving the tensor model by this process is the standard $2n^3/3$ cost of a QR factorization, plus an additional $n^2p \leq n^{2.5}$ cost for the orthogonal transformation of the variable space, plus the cost of solving the $p \times p$ system of quadratics. The latter is limited to $O(p)$ iterations which each cost $p^3/6$ multiplications and additions, so it is an insignificant

$O(p^4) \leq O(n^2)$ cost. The case $p=1$ is the most frequent in our computational experience, and in this case the quadratic equation is solved or minimized analytically. Thus solving the derivative tensor model costs essentially the same as finding the root of the standard linear model (1.3) by the QR factorization. It is possible to adapt the tensor solution algorithm to use the PLU factorization, or a sparse factorization, instead.

On singular problems with rank $(\underline{F}'(\underline{x}_*)) \geq n-p$, the solution of the tensor model by the above process is usually well posed. The convergence analysis for singular systems of nonlinear equations shows that near $\underline{x}_*$, we can expect the past steps $\{\underline{s}_k\}$ to be in directions near the null vectors of $\underline{F}'(\underline{x}_*)$. In this case, the quadratic term of the tensor model supplies information in the directions where the linear model is lacking. Thus the linear system (2.7a) is well conditioned, and the ill-conditioning of the standard linear model is moved into the linear term of the quadratic equations (2.7b), which still are well posed due to the quadratic term.

In addition, if $\underline{F}'(\underline{x}_c)$ happens to be singular or ill-conditioned at any iteration on any problem, and has p or fewer small or zero singular values, then the solution of the tensor model usually will be well posed for similar reasons.

2.3 *Computational Results with the Derivative Tensor Model*

A computer implementation of a derivative tensor method that is based upon the ideas summarized in Sections 2.1 and 2.2 has been extensively tested. A high level description of the method we have implemented is given in Algorithm 2.1.

ALGORITHM 2.1. An Iteration of the Derivative Tensor Method, given $\underline{x}_c$ and $\underline{F}(\underline{x}_c)$:

1. Calculate $\underline{F}'(\underline{x}_c)$ and decide whether to stop. If not:
2. Select the past points to use in the tensor model from among the $\sqrt{n}$ most recent past points.

3. Calculate the second order term of the tensor model, T_c, so that the tensor model interpolates $\underline{F}(\underline{x})$ at all points selected in step 2.
4. Find the root of the tensor model, or its minimizer (in the l_2 norm) if it has no real root.
5. Select $\underline{x}_+ = \underline{x}_c + \lambda_c \underline{d}_c$, where $\underline{d}_c$ either is the step calculated in step 4 or the Newton step, using a line search to choose λ_c.
6. Set $\underline{x}_c \leftarrow \underline{x}_+$ and $\underline{F}(\underline{x}_c) \leftarrow \underline{F}(\underline{x}_+)$; then go to step 1.

Details of our implementation are given in Frank (1984) and Schnabel and Frank (1984). Note that the Newton step can be calculated easily from (2.7), and occasionally it is used as the search direction in the tensor method. In particular, the Newton step is used in step 5 if Algorithm 2.1 finds that a root $\underline{d}_T$ of the tensor model is not a descent direction for $\|\underline{F}(\underline{x})\|_2$ (a very rare occurrence in practice but not precluded in theory) and the point $\underline{x}_c + \underline{d}_T$ is unacceptable; or if Algorithm 2.1 finds a minimizer of the tensor model at which the l_2 norm of the tensor model is too close to its value at $\underline{x}_c$; or if Algorithm 2.1 fails to find a root or minimizer of the tensor model in $8p$ iterations.

We compared our tensor method to an algorithm that is identical except that the second order term T_c is always zero. That is, the comparison algorithm is a finite difference Newton's method with a line search, except that, when $J_c = \underline{F}'(\underline{x}_c)$ is singular or sufficiently ill-conditioned, the Newton step $-J_c^{-1}\underline{F}(\underline{x}_c)$ is modified to the approximation to the pseudo-inverse step $-(J_c^T J_c + \varepsilon I)^{-1} J_c^T \underline{F}(\underline{x}_c)$ with ε small (see Dennis and Schnabel, 1983).

The Newton and tensor methods were compared on sets of nonsingular and singular test problems. The results are summarized in Table 1. The nonsingular test problems are a standard set in this field, given in Moré, Garbow and Hillstrom (1981); their dimensions range from $n = 2$ to 30. The singular problems

Table 1. Summary of the Results of the Derivative Tensor Method

Problem Set	Number of Problems	Average Ratio, Tensor Method / Standard Method: Iterations	Jacobian evaluations	Function evaluations	Tensor Better	Standard Better	Tie
Problems with $\underline{F}'(\underline{x}_*)$ Nonsingular							
All problems	25	0.811	0.813	0.828	18	1	6
Harder problems*	11	0.662	0.668	0.691	11	0	0
Singular Test Set with Rank $(\underline{F}'(\underline{x}_*)) = n-1$							
All problems	17	0.576	0.609	0.603	15	0	2
Harder problems*	9	0.392	0.429	0.434	9	0	0
Singular Test Set with Rank $(\underline{F}'(\underline{x}_*)) = n-2$							
All problems	13	0.631	0.664	0.729	11	2	0
Harder problems*	7	0.499	0.535	0.542	7	0	0

*Problems where the slower method required at least 10 iterations are simple modifications of these problems constructed to have the same solution $\underline{x}_*$ with rank $(\underline{F}'(\underline{x}_*)) = n-1$ and $n-2$, respectively. The procedure for generating these singular problems is described in Schnabel and Frank (1984).

A significant feature of the test results is that the tensor method is hardly ever less efficient than the standard method, and is almost always more efficient. In fact, on problems requiring ten or more iterations of the standard method, the tensor method is always superior. The gains in efficiency on the nonsingular problems are an average of about 18% if all test problems, including some very easy problems where no gains are likely, are considered, and the average improvement on the harder problems is about 32%. The gains in efficiency on the singular problems average about 40% and 30% in the rank $n-1$ and $n-2$ cases, respectively, while on the harder problems only these figures increase to about 57% and 46% respectively. In addition, the tensor method solved a significantly greater range of problems with rank $(\underline{F}'(\underline{x}_*)) = n-2$ than the standard method; this is not shown in Table 1 which reflects only problems solved by both methods.

The improvements of the tensor method on the problems with rank $(\underline{F}'(\underline{x}_*)) = n-1$ are partially explained by the faster local convergence of the tensor method, which is discussed in Section 2.4 below. In fact, our stopping tolerances were relatively loose; at tighter stopping tolerances the advantages of the tensor method on singular problems are greater. On nonsingular problems, the improvements of the tensor method apparently come from using a model that better interpolates $\underline{F}(\underline{x})$; to our knowledge, the local convergence rate is no better than for Newton's method.

These computational results indicate that the derivative tensor method is consistently as reliable as currently used methods for solving systems of nonlinear equations. In addition, on problems where function evaluation is the dominant cost, it is consistently as efficient and often considerably more efficient, especially on problems with a small rank deficiency in $\underline{F}'(\underline{x}_*)$. The additional cost to an iteration of the tensor method from extra arithmetic operations and computer storage is small. Indeed, in our tests, even on problems with $n = 30$, the number of past points interpolated by the tensor model generally was 1 or 2, so that the additional arithmetic and storage costs are very small. For these reasons, we believe that the derivative tensor method should be considered as a promising alternative to standard methods for general purpose software for solving systems of nonlinear equations.

2.4 *Convergence Analysis for the Derivative Tensor Method*

Frank (1984) has analyzed extensively the local convergence of the derivative tensor method. The most important result is that, when rank $(\underline{F}'(\underline{x}_*)) = n - 1$, the method described in the previous sections is shown to be locally 3-step convergent with q-order 7/6, meaning that if $\|\underline{x}_0 - \underline{x}_*\|$ is sufficiently small, the sequence of iterates $\{\underline{x}_k\}$ converges to $\underline{x}_*$ and, for some $c > 0$, obeys

$$\|\underline{x}_{k+3} - \underline{x}_*\| \leq c\, \|\underline{x}_k - \underline{x}_*\|^{7/6}$$

for all $k > 0$. This rate of convergence is significantly faster than Newton's method which is linearly convergent with constant approaching 1/2 under the same assumptions. For simplicity of analysis, Frank's result is proven for a method that interpolates only the most recent past iterate $(p = 1)$; however it is not expected that interpolating additional past points would impair its performance.

The reasoning behind the three step convergence result is interesting because it helps explain how the tensor method works on singular problems. From an arbitrary starting point close to $\underline{x}_*$, it is shown that the first step provides at least linear convergence, giving an iterate whose error (its difference with $\underline{x}_*$)

is nearly in the direction of the null vector of $\underline{F}'(\underline{x}_*)$. The next step also provides at least linear convergence and also results in an error nearly in the null vector direction. Thus, after two steps, the current iterate and the previous iterate are both close to being along the null vector direction from $\underline{x}_*$, so that the step $\underline{s}_1$ (the difference of these iterates) used in constructing the tensor term is essentially in this direction. Thus the quadratic term of the tensor model provides information in precisely the direction where the linear model is lacking. This causes the third step to be a fast one, in fact giving an order 1.5 improvement which would lead to three step q-order 1.5. (The smaller 7/6 rate comes from allowing for the possibility that the first or second steps are, by luck, too good.) After the third step, the convergence argument does not suggest that the error of the new iterate is close to the null space of $\underline{F}'(\underline{x}_*)$, so the analysis indicates that the three step process repeats. In practice, however, the errors appear to remain close to the null space so that one step, q-superlinear convergence is observed.

When the rank of $\underline{F}'(\underline{x}_*)$ is less than $n-1$, the derivative tensor method probably is not faster than linearly convergent in theory because the model described in Section 3.1 does not approximate enough of $\underline{F}''(\underline{x})$. However the test results of the previous section indicate that fast convergence still is obtained in the case rank $(\underline{F}'(\underline{x}_*)) = n-2$. It would be possible to approximate the necessary portions of $\underline{F}''(\underline{x})$ using previous values of the Jacobian rather than the function; this has not been pursued.

When $\underline{F}'(\underline{x}_*)$ is nonsingular, Frank (1984) shows that the tensor method retains the q-quadratic convergence of Newton's method. This simply means that, close to the solution, the quadratic term of the tensor model has a small effect and does not hurt the convergence. When $n=1$ it can be shown that the derivative tensor method has q-order 2.41, but the tensor method does not interpolate enough information for this result to extend to multi-dimensional problems.

3. SECANT TENSOR METHODS

When analytic Jacobians are unavailable and function evaluation is sufficiently expensive, it is not cost effective to calculate a finite difference Jacobian approximation at each iteration of a method for solving systems of nonlinear equations. Instead, secant methods are used that only occasionally form finite difference Jacobians and otherwise they are based just on the function values at the iterates.

The standard secant method for nonlinear equations, Broyden's method, uses a linear model of $\underline{F}(\underline{x})$ around $\underline{x}_c$ that interpolates only $\underline{F}(\underline{x}_c)$ and $\underline{F}(\underline{x}_{-1})$, where $\underline{x}_{-1}$ is the previous iterate. Thus there are additional previous function values, namely $\underline{F}(\underline{x}_{-2})$, $\underline{F}(\underline{x}_{-3})$,$\cdots$, that a method could also interpolate. In the derivative tensor method discussed in Section 2, the interpolation of these function values was the basis for the tensor term T_c. In the secant case, however, since the first derivative matrix is not known, the value of $\underline{F}(\underline{x})$ at a previous iterate $\underline{x}_{-k}$ is only sufficient to determine a linear model in the direction $\underline{x}_c - \underline{x}_{-k}$. Roughly speaking, only if there are two previous iterates in the same direction from $\underline{x}_c$ is there sufficient information to determine part of a quadratic model. This suggests that it is more difficult to form a quadratic model in the secant method case. Recall, however, that for singular problems the iterates often converge nearly along a single direction, so a quadratic model still may be possible.

Thus, before considering how one might base a tensor secant model upon the interpolation of multiple function values, it is relevant to discuss how a linear secant model can interpolate multiple function values. We address this question briefly, we then indicate when and how we form a secant tensor model, and we discuss some computational results of the secant tensor method.

3.1 *Linear Models with Multiple Secant Equations*

Suppose $\underline{x}_c$ is the current iterate and $\underline{x}_{-1}, \cdots, \underline{x}_{-p}$ are a set of p not necessarily consecutive past iterates chosen as in Section 2.1. That is, $\underline{x}_{-1}$ is the most recent past iterate, and the set of directions $\{\underline{s}_k\} = \{\underline{x}_{-k} - \underline{x}_c ; k = 1, 2, \cdots, p\}$ are linearly independent. Then it is possible to choose the Jacobian approximation $J_c \in \mathbb{R}^{n \times n}$ so that the secant model (1.2) interpolates $\underline{F}(\underline{x}_{-k})$, $k = 1, 2, \cdots, p$. This requires

$$\underline{F}(\underline{x}_{-k}) = \underline{F}(\underline{x}_c) + J_c \underline{s}_k , \quad k = 1, 2, \cdots, p. \tag{3.1}$$

Let S and Y be the matrices whose columns are $\{\underline{s}_k ; k = 1, 2, \cdots, p\}$ and $\{\underline{F}(\underline{x}_{-k}) - \underline{F}(\underline{x}_c) ; k = 1, 2, \cdots, p\}$ respectively. Then Schnabel (1983) shows that the closest matrix J_c to the previous Jacobian J_{-1} that causes (3.1) to be satisfied is

$$J_c = J_{-1} + (Y - J_{-1} S)(S^T S)^{-1} S^T. \tag{3.2}$$

Broyden's method is simply the special case of (3.2) with $p = 1$. The update (3.2) appears to be a rank p change to J_{-1}, but if the linear model at the previous update interpolated all the relevant previous function values then, due to the new function value $\underline{F}(\underline{x}_c)$, (3.2) is a rank one update.

Multiple secant updates along these lines have been proposed by Barnes (1965), Gay and Schnabel (1978), and Schnabel (1983). Frank (1984) implemented a version where the past iterates to interpolate are chosen by the fairly restrictive criteria described in Section 2.1, essentially very strong linear independence and only $\sqrt{n}$ iterates considered. Schnabel (1983) showed that this method is q-superlinearly convergent under the same conditions as Broyden's method. Frank found that on the nonsingular problems from the Moré, Garbow and Hillstrom (1981) test set the multiple secant method was better on 10 problems, worse on 2, and about the same on 21, with an average improvement of 10%. On the singular problems described in Section 3.3, the multiple secant method was only marginally better than Broyden's method.

These results indicate that a properly implemented multiple secant method, using a linear model, is consistently at least as efficient as Broyden's method. Therefore, our secant tensor method, which also is based on interpolating multiple function values, builds upon this linear multiple secant model.

3.2 *Forming the Secant Tensor Model*

To determine a second order model of $\underline{F}(\underline{x})$ using function values only, it is necessary to have more than $p+1$ function values that are nearly in some p dimensional subspace of the variable space. To illustrate the approach taken in forming our secant tensor model, suppose that there are two past iterates, $\underline{x}_{-1}$ and $\underline{x}_{-2}$, and that the steps $\underline{s}_i = \underline{x}_{-i} - \underline{x}_c$, $i = 1,2$, are nearly linearly dependent. That is, $\underline{s}_2 = \alpha\underline{s}_1 + \underline{z}$ where $\underline{z}^T\underline{s}_1 = 0$ and $\|\underline{z}\| \,/\, \|\underline{s}_2\|$ is small. The tensor secant model (1.5) interpolates $\underline{F}(\underline{x})$ at $\underline{x}_{-1}$ and $\underline{x}_{-2}$ if

$$\underline{F}(\underline{x}_{-k}) = \underline{F}(\underline{x}_c) + J_c\underline{s}_k + \tfrac{1}{2}\, T_c\underline{s}_k\underline{s}_k\,, \quad k = 1,2. \tag{3.3}$$

In the situation where $\underline{s}_1$ and $\underline{s}_2$ are nearly collinear, we interpret (3.3) as giving two pieces of information in the direction $\underline{s}_1$ and no new information in the orthogonal direction $\underline{z}$. Therefore we impose the conditions

$$J_c\underline{z} = J_{-1}\underline{z}\,, \quad T_c\underline{s}_1\underline{z} = T_c\underline{z}\underline{z} = \underline{0}\,, \tag{3.4}$$

for we know that our minimum norm methods for choosing J_c and T_c will cause these conditions to hold if the secant equations are in the direction $\underline{s}_1$ only. Combining (3.3) and (3.4) and using $\underline{s}_2 = \alpha\underline{s}_1 + \underline{z}$ gives

$$\underline{F}(\underline{x}_{-1}) = \underline{F}(\underline{x}_c) + J_c\underline{s}_1 + \tfrac{1}{2}\, T_c\underline{s}_1\underline{s}_1 \tag{3.5a}$$

$$\underline{F}(\underline{x}_{-2}) = \underline{F}(\underline{x}_c) + \alpha J_c\underline{s}_1 + J_{-1}\underline{z} + (\alpha^2/2)\, T_c\underline{s}_1\underline{s}_1\,. \tag{3.5b}$$

Equations (3.5) are two linear equations in the two unknown vectors $J_c\underline{s}_1$ and $T_c\underline{s}_1\underline{s}_1$, which are easily solved to yield

$$J_c \underline{s}_1 = \underline{y} = (\alpha^2 \underline{u} - \underline{v}) / (\alpha^2 - \alpha) \tag{3.6a}$$

$$T_c \underline{s}_1 \underline{s}_1 = \underline{t} = 2(\alpha \underline{u} - \underline{v}) / (\alpha - \alpha^2) \tag{3.6b}$$

where $\underline{u} = \underline{F}(\underline{x}_{-1}) - \underline{F}(\underline{x}_c)$ and $\underline{v} = \underline{F}(\underline{x}_{-2}) - \underline{F}(\underline{x}_c) - J_{-1}\underline{z}$.

Equations (3.6) are the secant equations for J_c and T_c, respectively. Given these conditions, we form J_c as in the linear model multiple secant method and T_c as in the derivative tensor method. That is, J_c is given by (3.2) with $Y = \underline{y}$ and $S = \underline{s}_1$, while $T_c = \underline{a}\underline{s}_1\underline{s}_1$ with $\underline{a} = \underline{t}/(\underline{s}_1^T \underline{s}_1)^2$. Thus the tensor model becomes

$$\underline{M}_T(\underline{x}_c + \underline{d}) = \underline{F}(\underline{x}_c) + J_c \underline{d} + \tfrac{1}{2}\underline{a}(\underline{w}^T \underline{d})^2 \tag{3.7}$$

with $\underline{w} = \underline{s}_1$.

The remaining issue is the criterion for choosing $\underline{s}_2$ to be "nearly linearly dependent" on $\underline{s}_1$. The residual $\underline{z}$ cannot be allowed to be too large, or the inaccuracy in $J_c \underline{z}$ may cause the tensor T_c to be entirely inaccurate. Frank (1984) shows that it is necessary that $\|\underline{z}\| \leqslant O(\|\underline{s}_2\|^2)$ for the resultant T_c to be reliable. Furthermore, he indicates that one can expect consecutive iterates to satisfy this condition near the solution of a singular problem. There is no reason, however, to expect this condition to be satisfied for nonsingular problems. Thus it is likely that the quadratic term in the secant tensor method will be used near the roots of singular problems only.

Frank (1984) has generalized the above procedures to use more past iterates. First he chooses p strongly linearly independent directions from past iterates to $\underline{x}_c$ by the same process used in the derivative secant method and the linear model multiple secant method. This set of directions is the generalization of the direction $\underline{s}_1$ in the above example. Then he chooses q directions from additional past iterates to $\underline{x}_c$ that are nearly dependent on the subspace spanned by the first set, in the sense described above. This set is the generalization of the direction

$\underline{s}_2$ in the above example. In our computational tests the second set contained 0 or 1 directions over 99% of the time, so, having discussed the 0-direction case in Section 3.1, we now consider the 1-direction case. It gives a set of equations similar to (3.5), which are easily reduced to the conditions $J_c S = Y$, $T_c \underline{w}\underline{w} = \underline{t}$, for known S, $Y \in \mathbb{R}^{n \times p}$, $\underline{t} \in \mathbb{R}^n$ and some $\underline{w}$ in the span of the columns of S. J_c then is chosen by (3.2) while $T_c = \underline{a}\underline{w}\underline{w}$ with $\underline{a} = \underline{t}/(\underline{w}^T\underline{w})^2$. Thus the secant tensor model again is given by (3.7).

3.3 *Solving the Secant Tensor Model*

Algebraically, the secant tensor model (3.7) is just the special case of the derivative tensor model (2.5) with $p=1$. Thus its root or minimizer is found by the procedure of Section 2.2. Since the tensor term has rank one, the solution process results in finding the root or minimizer of one quadratic equation in one unknown followed by the solution of a system of $n-1$ linear equations in $n-1$ unknowns. Therefore no iterative procedure is required and the cost is essentially the same as for finding the root of a linear model.

It is possible to perform each iteration of Broyden's method in $O(n^2)$ operations by updating a QR factorization of J_c as proposed originally by Gill and Murray (1972). These efficiencies can be extended to the secant tensor method so that it does not cost appreciably more than the $O(n^2)$ implementation of Broyden's method, but this was not done in our implementation.

In the case where the quadratic term of the secant tensor model has rank greater than one (which is very rare in practice), Frank (1984) solves the secant tensor model by a minor generalization of the techniques of Section 2.2.

3.4 *Computational Results with the Secant Tensor Method*

A secant tensor method has been implemented and tested on the same nonsingular and singular problems that were used for the derivative tensor method tests described in Section 2.3. The basic iteration is summarized in Algorithm 3.1 below.

ALGORITHM 3.1. An Iteration of the Secant Tensor Method, given $\underline{x}_c$ and $\underline{F}(\underline{x}_c)$:

1. Decide whether to stop. If not:
2. Select the two sets of past points to use in the tensor model from among the $\sqrt{n}$ most recent past points.
3. Calculate the first and second order terms of the tensor model, J_c and T_c, so that the tensor model interpolates $\underline{F}(\underline{x})$ at all the points selected in step 2.
4. Find the root of the tensor model, or its minimizer (in the l_2 norm) if it has no real root.
5. Select $\underline{x}_+$ by a trust region method that chooses $\underline{x}_+ - \underline{x}_c$ to be a linear combination of the steepest descent direction, and the step calculated in step 4 or the root of the linear part of the model.
6. Set $\underline{x}_c \leftarrow \underline{x}_+$ and $\underline{F}(\underline{x}_c) \leftarrow \underline{F}(\underline{x}_+)$; then go to step 1.

In addition to the strategy shown in Algorithm 3.1, the Jacobian is calculated by finite differences at the initial iteration and whenever the secant algorithm calculates two unsuccessful trial steps in a row. This latter practice is taken from Moré's MINPACK algorithm (Moré, Garbow and Hillstrom, 1980), as is the use of a trust region strategy at step 5.

The secant tensor method described above was compared to a Broyden's method version and to a linear model multiple secant method version of the same code. These were derived by setting $T_c = 0$ in the secant model, and by allowing one or multiple secant equations for the Jacobian approximation (i.e., $p = 1$ or $p > 1$ in (3.1) and (3.2)). The remainder of the code was unchanged.

On the nonsingular test problems from Moré, Garbow and Hillstrom (1981), the secant tensor method was better than Broyden's method on 9 problems, worse on 5, and about the same on 18, with an average improvement in function evaluations of 9%.

These results are marginally worse than the results for the linear model multiple secant method given in Section 3.1. Thus adding multiple interpolation conditions to the linear model seems to help a bit on nonsingular problems, but the tensor term seems to give no additional help.

On the test problems with rank $\underline{F}'(\underline{x}_*) = n-1$, the secant tensor method was better than Broyden's method on 18 problems, worse on 1, and tied on 4, with an average improvement of 25%. On the test problems with rank $\underline{F}'(\underline{x}_*) = n-2$, the secant tensor method was better than Broyden's method on 18 problems, worse on 1, and tied on 1, with an average improvement of 33%. In both of these cases, the linear model multiple secant method was not appreciably better than Broyden's method. Thus the addition of a tensor term seems to help considerably on problems with a low rank singularity at the solution.

In over 99% of the iterations on each test set, the rank of the tensor term was 0 or 1. Implementing a secant tensor method with a rank one tensor requires little additional storage and very few additional arithmetic operations in comparison to Broyden's method. Thus it appears that the gains mentioned above can be obtained at little additional cost to the computer, and a reasonably small increase in the complexity of the code. Therefore, such a secant tensor code might be a useful general purpose alternative to a Broyden's method code in a setting where singular or ill-conditioned problems are solved regularly.

4. EXTENSIONS OF TENSOR METHODS TO OPTIMIZATION PROBLEMS

The nonlinear least squares problem

$$\min_{\underline{x} \in \mathbb{R}^n} \|\underline{F}(\underline{x})\|_2, \quad \underline{F} : \mathbb{R}^n \to \mathbb{R}^m \tag{4.1}$$

can be viewed as an overdetermined version of the nonlinear equations problem (1.1). For nonlinear least squares, the Jacobian matrix is usually computed analytically or approximated by finite differences. Thus it is natural to consider extending the derivative tensor method summarized in Section 2 to solving (4.1).

The derivative tensor method extends to overdetermined systems of equations with very little change. The formation of the tensor model is as before except that the model has m quadratic components rather than n. Assuming no rank deficiency in the Jacobian, the solution procedure outlined in Section 2.2 now reduces the quadratic equations to the solution, in a least squares sense, of $m-n+p$ quadratic equations in p unknowns, followed by the solution of $n-p$ linear equations in $n-p$ unknowns. (If $p=0$ this is just the QR algorithm for solving the linear model in the least squares sense.) The cost in arithmetic operations and computer storage again is hardly more than for a standard method for nonlinear least squares.

A tensor method for nonlinear least squares along these lines is currently being implemented at the University of Colorado. This approach is most closely related to other nonlinear equations based approaches to nonlinear least squares. Of these the most computationally efficient appears to be Moré's trust region Levenberg–Marquardt algorithm (Moré, 1978), implemented in MINPACK (Moré, Garbow and Hillstrom, 1980), and it will be interesting to see how the tensor method compares to it on full rank and rank deficient problems. An alternate approach to nonlinear least squares, related more closely to viewing the problem as a special case of unconstrained optimization, is embodied in the NL2SOL algorithm of Dennis, Gay and Welsch (1981). We also intend to compare the tensor method for nonlinear least squares to NL2SOL.

The general unconstrained optimization problem is

$$\min_{\underline{x} \in \mathbb{R}^n} f(\underline{x}) : \mathbb{R}^n \to \mathbb{R} . \tag{4.2}$$

The necessary condition for a solution of (4.2), $\nabla f(\underline{x}) = \underline{0}$, is a system of n nonlinear equations in n unknowns. The standard local method for (4.2), also called Newton's method, is simply

(1.4) applied to this system of equations.

Thus it is tempting to expect that the tensor method for nonlinear equations can be used for unconstrained optimization calculations simply by applying the methods of Sections 2 and 3 to the system of equations $\underline{\nabla} f(\underline{x}) = \underline{0}$. This approach has several major flaws. One is that all the derivatives of the unconstrained optimization problem are symmetric, so that their approximations should be too, but the tensors T_c derived in Sections 2 and 3 are not 3-way symmetric. More importantly, if an unconstrained optimization problem has a singular Hessian matrix at a local minimizer, then the projection of the third derivative tensor $\nabla^3 f(\underline{x}_*)$ in the null space direction $\underline{v}$ of the Hessian (i.e. $\nabla^3 f(\underline{x}_*)\underline{v}\underline{v}\underline{v}$) must also be 0. Thus approximating the third derivative for unconstrained optimization, the analog of approximating the second derivative as is done in our tensor methods for nonlinear equations, is not expected to help solve singular unconstrained optimization problems. It appears that approximations to both the third and fourth derivative matrices would be required to improve the solution of singular optimization problems.

An approach to tensor methods for unconstrained optimization that makes small rank approximations to the third and fourth derivatives is currently under way at the University of Colorado. The effect of approximating both third and fourth derivative matrices is that, at each iteration, one must solve a small system of cubic equations in addition to the remaining linear equations. The storage and arithmetic overheads remain reasonable. Very preliminary computational results using this approach to solve singular optimization problems are encouraging.

REFERENCES

Barnes, J.G.P. (1965), "An algorithm for solving nonlinear equations based on the secant method", *Comput. J.*, *8*, pp.66-72.

Broyden, C.G., Dennis Jr., J.E. and Moré, J.J. (1973), "On the local and superlinear convergence of quasi-Newton methods", *J. Inst. Math. Appl.*, *12*, pp.223-245.

Decker, D.W., Keller, H.B. and Kelley, C.T. (1983), "Convergence rates for Newton's method at singular points", *SIAM J. Numer. Anal.*, *20*, pp.296-314.

Decker, D.W. and Kelley, C.T. (1980a), "Newton's method at singular points I", *SIAM J. Numer. Anal.*, *17*, pp.66-70.

Decker, D.W. and Kelley, C.T. (1980b), "Newton's method at singular points II", *SIAM J. Numer. Anal.*, *17*, pp.465-471.

Decker, D.W. and Kelley, C.T. (1982), "Convergence acceleration for Newton's method at singular points", *SIAM J. Numer. Anal.*, *19*, pp.219-229.

Decker, D.W. and Kelley, C.T. (1985), "Broyden's method for a class of problems having singular Jacobian at the root", *SIAM J. Numer. Anal.*, *22*, pp.566-574.

Dennis Jr., J.E., Gay, D.M. and Welsch, R.E. (1981), "An adaptive nonlinear least-squares algorithm", *ACM Trans. Math. Software*, *7*, pp.348-368.

Dennis Jr., J.E. and Schnabel, R.B. (1979), "Least change secant updates for quasi-Newton methods", *SIAM Rev.*, *21*, pp.443-459.

Dennis Jr., J.E. and Schnabel, R.B. (1983), *Numerical Methods for Unconstrained Optimization and Nonlinear Equations*, Prentice-Hall (Englewood Cliffs, New Jersey).

Frank, P.D. (1984), *Tensor Methods for Solving Systems of Nonlinear Equations*, Ph.D. Thesis, Department of Computer Science, University of Colorado at Boulder.

Gay, D.M. and Schnabel, R.B. (1978), "Solving systems of nonlinear equations by Broyden's method with projected updates", in *Nonlinear Programming*, *3*, eds. Mangasarian, O.L., Meyer, R.R. and Robinson, S.M., Academic Press (New York), pp.245-281.

Gill, P.E. and Murray, W. (1972), "Quasi-Newton methods for unconstrained optimization", *J. Inst. Math. Appl.*, *9*, pp.91-108.

Griewank, A.O. (1980a), *Analysis and Modification of Newton's Method at Singularities*, Ph.D. Thesis, Australian National University.

Griewank, A.O. (1980b), "Starlike domains of convergence for Newton's method at singularities", *Numer. Math.*, *35*, pp.95-111.

Griewank, A.O. (1985), "On solving nonlinear equations with simple singularities or nearly singular solutions", *SIAM Rev.*, *27*, pp.537-563.

Griewank, A.O. and Osborne, M.R. (1981), "Newton's method for singular problems when the dimension of the null space is >1", *SIAM J. Numer. Anal.*, *18*, pp.145-149.

Griewank, A.O. and Osborne, M.R. (1983), "Analysis of Newton's method at irregular singularities", pp.747-773.

Keller, H.B. (1970), "Newton's method under mild differentiability conditions", *J. Comput. System Sci.*, 4, pp. 15-28.

Kelley, C.T. (1985), "A Shamanskii-like acceleration scheme for nonlinear equations at singular roots", preprint, Department of Mathematics, North Carolina State University.

Kelley, C.T. and Suresh, R. (1983), "A new acceleration method for Newton's method at singular points", *SIAM J. Numer. Anal.*, *20*, pp.1001-1009.

Moré, J.J. (1978), "The Levenberg—Marquardt algorithm: implementation and theory", in *Numerical Analysis, Dundee 1977, Lecture Notes in Mathematic 630*, ed. Watson, G.A., Springer-Verlag (Berlin), pp.105-116.

Moré, J.J., Garbow, B.S. and Hillstrom, K.E. (1980), "User guide for MINPACK-1", Report ANL-80-74, Argonne National Laboratory, Argonne, Illinois.

Moré, J.J., Garbow, B.S. and Hillstrom, K.E. (1981), "Testing unconstrained optimization software", *ACM Trans. Math. Software*, *7*, pp.17-41.

Rall, L.B. (1966), "Convergence of the Newton process to multiple solutions", *Numer. Math.*, *9*, pp.23-37.

Reddien, G.W. (1978), "On Newton's method for singular problems", *SIAM J. Numer. Anal.*, *15*, pp.993-996.

Reddien, G.W. (1979), "Newton's method and high order singularities", *Comput. Math. Appl.*, *5*, pp.79-86.

Schnabel, R.B. (1983), "Quasi-Newton methods using multiple secant equations", Technical Report CU-CS-247-83, Department of Computer Science, University of Colorado at Boulder.

Schnabel, R.B. and Frank, P. (1984), "Tensor methods for nonlinear equations", *SIAM J. Numer. Anal.*, *21*, pp.815-843.

Robert B. Schnabel
Department of Computer Science
University of Colorado
Boulder
Colorado 80309
U.S.A.

Paul D. Frank
Boeing Computer Services
Tukwila
Washington 98042
U.S.A.

10 Numerical Methods for Bifurcation Problems

A. D. JEPSON* and A. SPENCE†

ABSTRACT

Bifurcation theory is the study of nonlinear equations with multiple solutions. A nonlinear equation is said to exhibit bifurcation if the number of solutions changes as a parameter varies. Examples arise in a wide range of applications, for example, instabilities in fluid flow and exothermic reactions. This paper contains a summary of recent work on the application of singularity theory to the design of systematic numerical methods for the solution of bifurcation problems with many parameters.

1. INTRODUCTION

The steady states of many physical systems can be modelled by nonlinear multiparameter problems of the form

$$\underline{f}(\underline{x},\lambda,\underline{\alpha}) = \underline{0}, \quad \underline{f}: X \times \mathbb{R} \times \mathbb{R}^p \to Y, \tag{1.1}$$

where $\underline{f}$ is a smooth function and X and Y are Banach spaces. Such equations arise, for example, in nonlinear elasticity, exothermic reactions and laser physics. One celebrated example in fluid mechanics is the Taylor problem, which is the study of the steady axisymmetric flow of an incompressible viscous fluid between concentric circular cylinders (Benjamin and Mullin, 1981).

* Research supported by NSERC Canada

† Research supported by SERC United Kingdom

In numerical studies (see Cliffe, 1983, and Cliffe and Spence, 1986), a finite element method is used to discretize the steady Navier-Stokes equations governing the flow and a system like (1.1) is derived, where $\underline{x} \in \mathbb{R}^n$ represents the velocity and pressure of the flow, λ is the Reynolds number and there is one control parameter, the aspect ratio of the domain. In this case $X = Y = \mathbb{R}^n$ with n taking values in the range 1000-5000. A completely different type of equation, where $X = Y = \mathbb{R}$ but $\underline{\alpha} \in \mathbb{R}^4$, arises in a simple model of an exothermic reaction in a continuous flow stirred tank reactor (CSTR) which is discussed in Section 3. Other examples, involving cases when $X \neq Y$, may be found in the books of Keller and Antman (1969), Rabinowitz (1977), Poston and Stewart (1978) and Golubitsky and Schaeffer (1985). The reason for distinguishing one parameter in (1.1) is that in many experiments only one parameter can be easily changed independently, and this parameter is varied "quasistatically" while the control parameters are held fixed. This distinction simplifies the analysis but still allows reinterpretation of the results if a different choice for λ is made. Finally, in this introductory paragraph we remark that problems like (1.1) often arise in steady-state studies, and that typically the equilibria of a system need to be found before its dynamical behaviour can be properly understood.

The main feature of problems like (1.1) is the existence of multiple solutions $\underline{x}$ for given values of $(\lambda, \underline{\alpha})$; it is this multiplicity which is so important physically and makes the problems so interesting mathematically. Roughly speaking, a *bifurcation* is a change in the number of solutions as a parameter varies. We shall see in Section 2 that a necessary condition for a bifurcation at $(\underline{x}_0, \lambda_0, \underline{\alpha}_0)$ is that the (Fréchet) derivative $\underline{f}_{\underline{x}}^0 := \underline{f}_{\underline{x}}(\underline{x}_0, \lambda_0, \underline{\alpha}_0) : X \rightarrow Y$ must be singular. We call such points *singular points*, though the term *bifurcation point* is also used. If $\dim \mathrm{Null}\,(\underline{f}_{\underline{x}}^0) = 1$ then $(\underline{x}_0, \lambda_0, \underline{\alpha}_0)$ is a *simple* singular point. We shall only consider simple singular points here though multiple singular points can also be treated (see references given in Section 5). If $\underline{f}_{\underline{x}}^0$ is nonsingular then $(\underline{x}_0, \lambda_0, \underline{\alpha}_0)$ is said to be a *regular* point. It is usual to present solutions of such problems in the form of a *bifurcation diagram*, i.e. a graph $\{(\ell(\underline{x}), \lambda);\ \ell : X \to \mathbb{R}\}$ for any given $\underline{\alpha}$, where ℓ is some functional of the state $\underline{x}$. Obviously if $X = \mathbb{R}$ then we take $\ell(x) = x$. Examples of several bifurcation diagrams are given in Figure 1. Each of the small sketches shows a possible form of the dependence of the solution(s) of (1.1) on λ, and the surfaces of the figure separate the different kinds of form that can occur. The problem of calculating the singular points and a bifurcation diagram of (1.1) is called a *bifurcation problem*.

As can be seen from the reference section in this paper only recently has there been a concerted attempt at a systematic study of the numerical analysis of bifurcation problems, though several bifurcation problems have been treated successfully in the past, see for example, Anselone and Moore (1966), and Bauer, Reiss and Keller (1970). Some early theoretical work is that of Keller and Langford (1972), Keener and Keller (1973), and Weiss (1975). More recently work has concentrated mainly on the derivation of systems for the calculation of singular points of various types, and there has also been some analysis of the effect of discretization errors. A good reference text for the "State-of-the-Art" in numerical methods for bifurcation problems before 1984 is the conference proceedings edited by Küpper, Mittelmann and Weber (1984). Other books which deal specifically with numerical methods are Mittelmann and Weber (1980), Kubíček and Marek (1983), and Rheinbolt (1986).

It is not the aim of this article to summarize the various algorithms, approaches or theories derived for bifurcation problems over the past ten years. Such a survey could be extracted from Küpper, Mittelmann and Weber (1984). Rather we shall outline a way of looking at nonlinear multi-parameter problems which holds great promise for the future development of efficient, systematic numerical methods. The approach is based on ideas from singularity theory which have proved extremely successful in explaining complicated bifurcation phenomena in many interesting physical problems (see Golubitsky and Schaeffer, 1985). Singularity theory provides a local analysis of solution behaviour near a singular point, and the essential idea of the approach described here is to ally this local information with continuation algorithms which can compute solution paths of nonlinear problems for wide ranges of parameter values. The work described in this paper is a combination of ideas from Jepson and Spence (1985a,b, 1986b). Related ideas may be found in the work of Beyn (1984), and Beyn and Bohl (1984), and it is worth noting that singularity theory was also used to analyse discretization errors by Fujii and Yamaguti (1980), and Brezzi and Fujii (1982).

To motivate the approach to be described, consider first the type of information one might try to obtain from system (1.1). Possible questions are:

(i) For what values of $\underline{\alpha}$ are there "interesting" or "singular" phenomena ?

(ii) For a given $\underline{\alpha}$ can all the solutions of (1.1) be found, including disconnected branches ?

We propose that the numerical strategy outlined here allows reasonably confident answers to these questions. To illustrate the sort of information the approach is able to provide consider the following example (see Golubitsky and Keyfitz, 1980). Let

$$f(x,\lambda,\underline{\alpha}) := x^3+\lambda^2+\alpha_1+\alpha_2 x+\alpha_3\lambda x = 0 \tag{1.2}$$

where $x \in \mathbb{R}$ (i.e. $X=\mathbb{R}$, $p=3$ in (1.1)). In Figure 1 the parameter space for $\alpha_3>0$ is divided into five open regions $\{R_1, R_2, \cdots, R_5\}$ by two surfaces, or varieties, B and H. These surfaces are given by

$$B = \{\underline{\alpha} \in \mathbb{R}^3 : \{(x,\lambda): f = f_\lambda = f_x = 0\} \neq \emptyset\} \tag{1.3a}$$

$$H = \{\underline{\alpha} \in \mathbb{R}^3 : \{(x,\lambda): f = f_x = f_{xx} = 0\} \neq \emptyset\} \tag{1.3b}$$

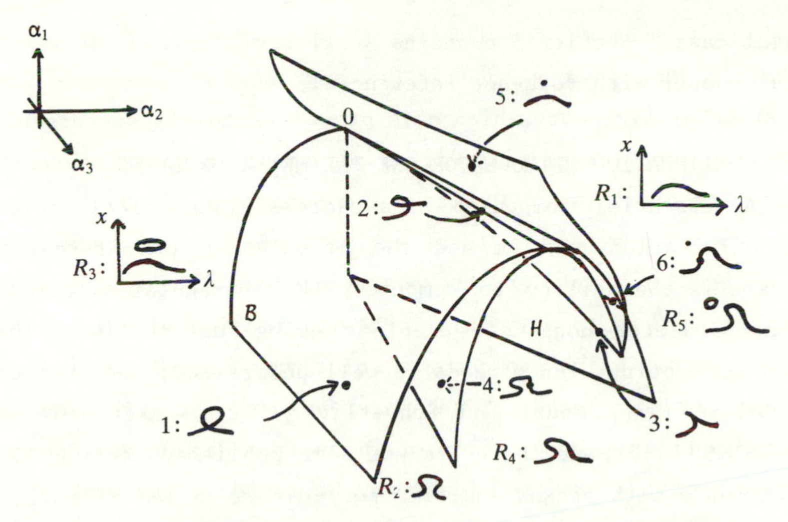

FIG. 1 Parameter space plot for f given by (1.2)

and are even about $\alpha_3 = 0$. The qualitative shape of the bifurcation diagrams for (1.2) is the same for all values of $\underline{\alpha}$ in any one region. These shapes are sketched in Figure 1. For example, for $\underline{\alpha} \in R_1$ there are no singular points; for $\underline{\alpha} \in R_3$ there are two disconnected solution curves, one forming a closed loop (isola); for $\underline{\alpha} \in R_4$ there is an S-shaped curve. The varieties B and H are themselves divided into open regions by the paths passing through $\underline{0}$ and the points 2, 3 and 6. Again bifurcation diagrams for (1.2) are qualitatively similar (see Jepson and Spence, 1985b) in each open region of B and H and on each of the paths, after $\underline{0}$ has been removed.

The presentation of the solutions of (1.2) using a parameter space plot is a good way of providing qualitative information. If quantitative information can also be obtained then we can expect to answer questions (i) and (ii) positively.

The plan of the paper is as follows. Section 2 contains a discussion of continuation methods. Section 3 describes the basis of a numerical approach to solving (1.1) when $X = Y = \mathbb{R}$, which is given in detail in Jepson and Spence (1985b). Section 4 contains a discussion of the extension of these results to the general case. Section 5 contains a list of topics not covered in this paper with relevant references.

2. CONTINUATION METHODS FOR ONE-PARAMETER PROBLEMS

An essential component of a successful numerical approach to a bifurcation problem and the calculation of bifurcation diagrams is the ability to compute paths of regular solutions of one-parameter nonlinear problems using continuation. The theory of continuation methods is well understood, see, for example, Keller (1977), Menzel and Schwetlick (1978), Wacker (1978), and Rheinboldt (1980), and there are now some published codes — AUTO (which deals with steady and time periodic solutions, Doedel, 1981) and PITCON (Rheinboldt and Burkardt, 1983; Rheinboldt, 1986). In this section we merely go over the main ideas.

Consider the problem

$$\underline{g}(\underline{x}, \lambda) = \underline{0}, \quad \underline{g} : \mathbb{R}^n \times \mathbb{R} \to \mathbb{R}^n, \tag{2.1}$$

where $\underline{g}$ is sufficiently smooth and where, for convenience, we take $X = Y = \mathbb{R}^n$ (though the theory readily extends to the Banach space case). The key theorem on which every continuation method depends is the Implicit Function Theorem (see, for example, Golubitsky and Schaeffer, 1985):

> "Suppose that $\underline{g}(\underline{x}_0, \lambda_0) = \underline{0}$ and that $\underline{g}_{\underline{x}}^0 := \underline{g}_{\underline{x}}(\underline{x}_0, \lambda_0)$ is nonsingular. Then there exist neighbourhoods U of $\underline{x}_0$ in $\mathbb{R}^n$ and V of λ_0 in $\mathbb{R}$ such that, for every λ in V, (2.1) has a unique solution $\underline{x} = \underline{x}(\lambda) \in U$. Moreover, if $\underline{f}$ is in C^s so is $\underline{x}(\lambda)$. Finally for λ in V, $\underline{x}(\lambda) - \underline{x}(\lambda_0) = o(\lambda - \lambda_0)$."

Thus, if $(\underline{x}_0, \lambda_0)$ is a regular point of (2.1), then locally the solution $\underline{x}$ is a function of λ and this provides some justification for presenting the solutions of (2.1) as plots of $\underline{x}$ against λ. One immediate corollary is that at a bifurcation we must have that $\underline{g}_{\underline{x}}^0$ is singular.

The main steps in a simple continuation method are readily

described. Assume $(\underline{x}_0, \lambda_0)$ is a known regular solution of $\underline{g}(\underline{x}, \lambda) = \underline{0}$ and that a solution $\underline{x}_1$, say, is required for $\lambda_1 = \lambda_0 + \Delta\lambda$ with $\Delta\lambda$ given. Differentiation of (2.1) with respect to λ gives

$$\underline{g}_{\underline{x}}\underline{x}_\lambda + \underline{g}_\lambda = \underline{0} \tag{2.2}$$

where $\underline{x}_\lambda := d\underline{x}/d\lambda$. A "predicted" value for $\underline{x}_1$, say $\underline{x}_1^{(0)}$, is given by

$$\underline{x}_1^{(0)} := \underline{x}_0 + \Delta\lambda\underline{x}_\lambda^0$$

where $\underline{x}_\lambda^0$ is found from (2.2), i.e. $\underline{x}_\lambda^0 = -(\underline{g}_{\underline{x}}^0)^{-1}\underline{g}_\lambda^0$. (Here we have used the notation $\underline{g}_\lambda^0 := \underline{g}_\lambda(\underline{x}_0, \lambda_0)$.) The solution $\underline{x}_1$ is now found from $\underline{g}(\underline{x}, \lambda_1) = \underline{0}$ using a nonlinear equation "solver", say Newton's method, with starting value $\underline{x}_1^{(0)}$. This approach is commonly referred to as a *predictor-solver* scheme and the above method is called the Euler-Newton method. Clearly there are important numerical questions, which we do not discuss here, relating to the choice of $\Delta\lambda$, the choice of predictor, and the efficient implementation of the solver step. Nonetheless such an approach should allow the computation of a connected path of regular points of (2.1).

This procedure will break down near a singular point (theoretically the widths of the neighbourhoods U and V shrink to zero, and numerically $\Delta\lambda$ needs to be reduced successively in order for the solver to converge). However, the addition of a second equation usually overcomes this difficulty. Consider the pair

$$\underline{G}(\underline{y}, s) := \begin{cases} \underline{g}(\underline{x}, \lambda) &= \underline{0} \\ N(\underline{x}, \lambda, s) &= 0 \end{cases}, \quad \underline{y} = (\underline{x}, \lambda), \tag{2.3}$$

where $N : \mathbb{R}^n \times \mathbb{R} \times \mathbb{R} \to \mathbb{R}$ is a normalizing equation and s is a user-introduced parameter, which is often related to arc-length. Can we apply the Implicit Function Theorem to $\underline{G}(\underline{y}, s) = \underline{0}$ and use s to parameterize the solution $\underline{y} = (\underline{x}, \lambda)$? Clearly the answer is yes, provided the Jacobian

$$\underline{G}_{\underline{y}} = \begin{bmatrix} \underline{g}_{\underline{x}} & \underline{g}_{\lambda} \\ N_{\underline{x}}^{T} & N_{\lambda} \end{bmatrix} \tag{2.4}$$

is nonsingular. Standard theory for bordered matrices indicates the possibility that $\underline{G}_{\underline{y}}$ will be nonsingular even if $\underline{g}_{\underline{x}}$ is singular. However, for this to be so, we need to restrict attention to the case $\dim \mathrm{Null}\,(\underline{g}_{\underline{x}}^{0}) \leqslant 1$, and so at a singular point $(\underline{x}_0, \lambda_0)$ we make the assumptions that there exist $\underline{\phi}_0$ and $\underline{\psi}_0$ such that

$$\mathrm{Null}\ (\underline{g}_{\underline{x}}^{0}) = \mathrm{span}\,\{\underline{\phi}_0\}\,, \qquad \underline{\phi}_0 \in \mathbb{R}^n \setminus \{\underline{0}\}\,, \tag{2.5a}$$

$$\mathrm{Range}\ (\underline{g}_{\underline{x}}^{0}) = \{\underline{y} \in \mathbb{R}^n : \underline{\psi}_0^T \underline{y} = 0\}\,, \qquad \underline{\psi}_0 \in \mathbb{R}^n \setminus \{\underline{0}\}\,. \tag{2.5b}$$

The following theorem on the nonsingularity of $\underline{G}_{\underline{y}}$ is easily proved (Keller, 1977).

THEOREM 2.6. Let $\underline{G}$ be defined by (2.3) and assume that (2.5) holds if $(\underline{x}_0, \lambda_0)$ is a singular point of $\underline{g} = \underline{0}$.

(i) If $(\underline{x}_0, \lambda_0)$ is a regular point of $\underline{g} = \underline{0}$ then $\underline{G}_{\underline{y}}$ is nonsingular iff

$$N_{\lambda}^{0} - N_{\underline{x}}^{0T}\,(\underline{g}_{\underline{x}}^{0})^{-1}\,\underline{g}_{\lambda}^{0} \neq 0\,. \tag{2.7}$$

(ii) If $(\underline{x}_0, \lambda_0)$ is a singular point of $\underline{g} = \underline{0}$ then $\underline{G}_{\underline{y}}$ is nonsingular iff

$$\underline{\psi}_0^T \underline{g}_{\lambda}^{0} \neq 0\,, \tag{2.8}$$

and

$$N_{\underline{x}}^{0T}\,\underline{\phi}_0 \neq 0\,. \tag{2.9}$$

If (2.8) holds the conditions (2.7) and (2.9) provide no difficulty since one can always choose N in (2.3) such that they hold. If (2.5) and (2.8) hold then we call $(\underline{x}_0, \lambda_0)$ a *simple turning point* (or *limit* or *fold point*) of $\underline{g}(\underline{x}, \lambda) = \underline{0}$. Thus Theorem 2.6 implies that regular points and simple turning points of $\underline{g}(\underline{x}, \lambda) = \underline{0}$ are regular points of $\underline{G}(\underline{y}, s) = \underline{0}$. Hence, provided an initial value $(\underline{x}_0, \lambda_0)$ is known, a predictor-solver scheme will compute connected components of a bifurcation diagram containing regular and simple turning points.

In most cases the addition of the extra equation in (2.3) involves little extra numerical work and most continuation algorithms use a system like (2.3), even for computing paths of regular points. It is important to note that often $\underline{g}_{\underline{x}}$ has some special structure which it is advantageous to preserve, say when $\underline{g}$ arises from a discretization of a partial differential equation, and care should be taken not to lose this structure when solving systems involving $\underline{G}_{\underline{y}}$. Techniques have been devised for such cases, and, in addition to the references already quoted, we refer the reader to Chan (1984), and Mackens and Jarausch (1984). Another point to note is that some techniques work, either explicitly or implicitly, with $\underline{g}_{\underline{x}}$ in spite of its singularity at a turning point. This has advantages since it is important to recognize the presence of a singular point and also approximations to $\underline{\phi}_0$ and $\underline{\psi}_0$ are then readily found.

In the remainder of this paper we shall assume that we have at our disposal a continuation algorithm which can compute connected paths of regular solutions of $\underline{g}(\underline{x},\lambda)=\underline{0}$, and which detects turning points. The accurate calculation of simple turning points is discussed in the next two sections.

3. SCALAR PROBLEMS

In this section we consider the numerical calculation of singular points of (1.1) when $X = Y = \mathbb{R}$ i.e.

$$f(x,\lambda,\underline{\alpha}) = 0, \quad f : \mathbb{R}\times\mathbb{R}\times\mathbb{R}^p \to \mathbb{R}, \tag{3.1}$$

which we call the *scalar* case. Such problems arise naturally, for example in the model of the CSTR, or are derived from some reduction process applied to more general problems (see Section 4). The main aims of the numerical approach to solving (3.1) are (a) to completely decompose the control parameter space $\mathbb{R}^p$ of (3.1) into regions for which the bifurcation diagram is qualitatively similar, and (b) to provide initial points on each connected component of the bifurcation diagram at any given $\underline{\alpha} \in \mathbb{R}^p$.

The material in this section is based on that in Jepson and Spence (1985b) which uses singularity theory (see Golubitsky and Schaeffer, 1985) to construct equations and inequalities characterizing various types of singular points of the scalar equation. These defining conditions have several numerically important properties which are gathered together in the concept of *well-formulated defining conditions* (Definition 3.10), and which, allied with continuation algorithms (Section 2), provide an extremely powerful tool in the numerical solution of nonlinear problems like (3.1).

We spend some time on the simplest case because several important ideas are introduced. For the moment drop the dependence of f on $\underline{\alpha}$ and consider the calculation of a singular point of

$$f(x,\lambda) = 0, \qquad f : \mathbb{R} \times \mathbb{R} \to \mathbb{R}. \tag{3.2}$$

A natural system to consider is

$$\underline{F}(\underline{y}) := (f(x,\lambda), f_x(x,\lambda)) = \underline{0}, \quad \underline{y} = (x,\lambda) \in \mathbb{R}^2. \tag{3.3}$$

By definition a solution $\underline{y}_0$ is a regular point if $\underline{F}_{\underline{y}}^0$ is non-singular, which, in view of $f_x^0 = 0$, is equivalent to

$$f_\lambda^0 \neq 0, \quad \text{and} \quad f_{xx}^0 \neq 0. \tag{3.4a,b}$$

If (3.4a,b) are satisfied then we call $\underline{y}_0 = (x_0, \lambda_0)$ a *quadratic* turning point, the reason for which is clear when the graph of $f(x,\lambda) := x^2 - \lambda = 0$ is drawn. We call (3.3) an *extended* system and (3.4a,b) are called *side-constraints*. Together (3.3) and (3.4) are the *defining conditions* for a quadratic turning point.

Quadratic turning points have several nice properties. One is that Newton's method applied to (3.3) converges quadratically under (3.4). Another is that a sensitivity analysis easily shows that they are structurally *stable* under small perturbations. Consider a perturbation of $f(x,\lambda)$, which we write as

$$\hat{f}(x,\lambda,\varepsilon) := f(x,\lambda) + \varepsilon p(x,\lambda). \tag{3.5}$$

Then if we define

$$\hat{\underline{F}}(\underline{y},\varepsilon) := (f + \varepsilon p\ ,\ f_x + \varepsilon p_x)$$

we have, with an obvious notation, $\hat{\underline{F}}_{\underline{y}} = \underline{F}_{\underline{y}} + \varepsilon \underline{p}_{\underline{y}}$. By hypothesis $\hat{\underline{F}}_{\underline{y}}(\underline{y}_0, 0)$ is nonsingular and so the Implicit Function Theorem ensures the existence of a unique solution, $\underline{y}(\varepsilon)$, to $\hat{\underline{F}}(\underline{y},\varepsilon) = \underline{0}$ for ε small enough. Hence $\hat{f}(x,\lambda,\varepsilon) = 0$ has a unique quadratic turning point $(x(\varepsilon)\ ,\ \lambda(\varepsilon))$ satisfying the stability result

$$x(\varepsilon) - x_0 = O(\varepsilon), \quad \lambda(\varepsilon) - \lambda_0 = O(\varepsilon). \tag{3.6}$$

This type of argument is common in the stability analysis of elastic structures where the $\varepsilon p(x,\lambda)$ term represents a "lumping together" of all the imperfections in any model. Alternatively, one might consider $\varepsilon = h^r$ where h is a stepsize and $\hat{f}$ is a "discretization" of f. The results given by (3.6) provide convergence results (see Brezzi and Fujii, 1982).

Now let us change our perspective slightly and let ε in $\hat{f}(x,\lambda,\varepsilon) = 0$ represent a specific control parameter. Obviously the above results are unchanged, but provided the side constraints

$$f_\lambda(x(\varepsilon)\ ,\ \lambda(\varepsilon), \varepsilon) \neq 0 \tag{3.7a}$$

$$f_{xx}(x(\varepsilon)\ ,\ \lambda(\varepsilon), \varepsilon) \neq 0 \tag{3.7b}$$

still hold, there is no need for ε to remain small, and repeated applications of the Implicit Function Theorem allow ε to take a possibly wide range of values. Effectively under (3.7) ε parameterizes a curve of quadratic turning points. If we return to the notation in (3.1) then a curve of quadratic turning points $(x(\alpha)\ ,\ \lambda(\alpha))$ satisfying

$$\underline{F}(\underline{y},\alpha) := \begin{cases} f(x,\lambda,\alpha) = 0 \\ f_x(x,\lambda,\alpha) = 0, \end{cases} \qquad x,\lambda,\alpha \in \mathbb{R},$$

can be computed by continuation in α (or more probably in an arc-length parameter), providing the side-constraints corresponding to (3.7) hold with α replacing ε. One nice feature is that since the side-constraints (3.4) appear in the Jacobian $\underline{F}_{\underline{y}}$ they can be monitored during the continuation and checked for any sign changes, which clearly indicate the presence of points where a side-constraint fails.

Obviously one needs a strategy to follow when a zero in a side-constraint is detected, and a complete systematic procedure for the problem (3.1) is described in Jepson and Spence (1985b) which has its basis in the singularity theory of Golubitsky and Schaeffer (1985).

In Golubitsky and Schaeffer (1985) the possible types of behaviour of solutions of (3.2) near a singular point are classified according to a certain notion of equivalence called *contact equivalence*. (Roughly speaking, two problems are contact equivalent if the bifurcation diagram of one can be transformed to the bifurcation diagram of the other without changing the type and number of singular points at any stage.) The equivalence classes obtained are referred to simply as "singularities" and with each singularity is associated a number called the *codimension*. If this number is finite then the equivalence class contains a *polynomial* canonical form. (We have already hinted at this in the case of quadratic turning points since $f(x,\lambda) := x^2 - \lambda$ satisfies (3.3) and (3.4) at $(0,0)$; conversely, any f satisfying (3.3) and (3.4) at (x_0,λ_0) is contact equivalent to $x^2 - \lambda$.)

In Jepson and Spence (1985b) the singularities of codimension less than 4 are arranged in a hierarchy (Figure 2) where q is the codimension. The (q,j)-singularity is the singularity at $(x,\lambda) = (0,0)$ of the polynomial that is displayed in the

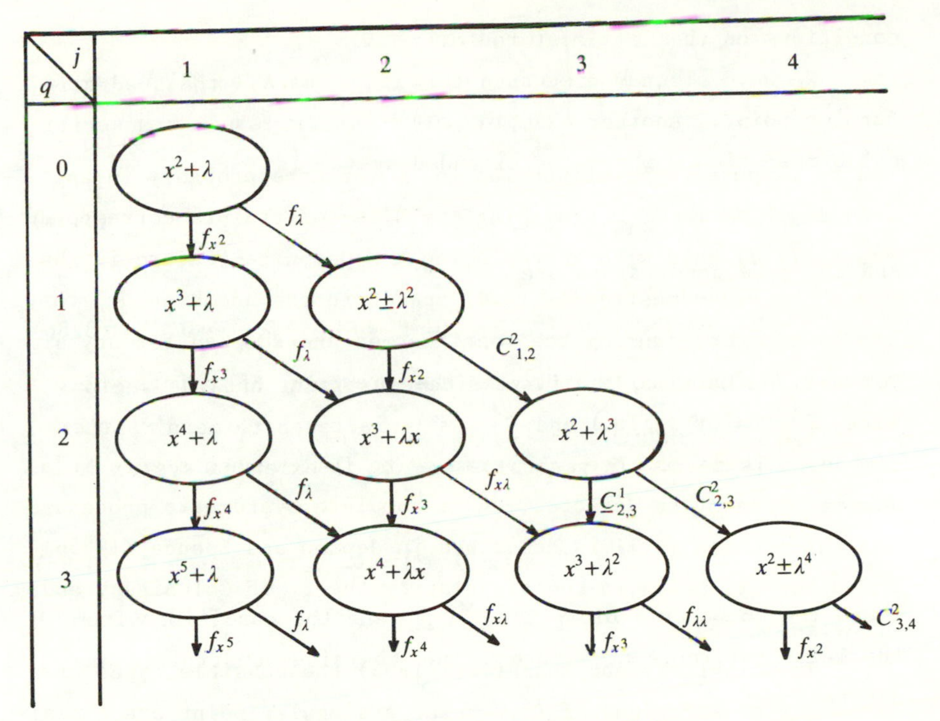

FIG. 2 Hierarchy of singularities

Here $C^2_{1,2} = f^2_{x\lambda} - f_{xx}f_{\lambda\lambda}$, $\underline{C}^1_{2,3} = (f_{xx}, f_{x\lambda})$ **and** $C^2_{2,3}$ **and** $C^2_{3,4}$ **are given in Jepson and Spence (1985b), where they are referred to as** D_3 **and** D_4 **respectively**

(q, j)-node of the figure. The structure in the singularity hierarchy reflects the fact that the singularities are partially ordered, and it is convenient to use standard terminology regarding this ordering. The main point of the hierarchy is that it provides defining conditions for singular points in the following way. For (3.1) to have the (q, j)-singularity at $(x_0, \lambda_0, \underline{\alpha}_0)$ not only f and f_x but also the labels on all the links of the routes from the top left hand corner of the figure to the (q, j)-node must vanish at $(x_0, \lambda_0, \underline{\alpha}_0)$, and the labels on the links leaving the node must not vanish. In this case, if there is more than one route, then the conditions on each route are equivalent, but there are good numerical reasons for preferring the conditions on the left-most route.

We have already discussed the (0,1) case — the quadratic turning point. Another example is the (1,2) case — transcritical bifurcation — where the extended system is

$$\underline{H}_{1,2} := (f, f_x, f_\lambda)^T = \underline{0} \tag{3.8a}$$

and the side constraints are

$$C^1_{1,2} := f_{xx} \neq 0, \quad C^2_{1,2} := f^2_{x\lambda} - f_{xx} f_{\lambda\lambda} \neq 0. \tag{3.8b}$$

For the (2,3) case, the first side-constraint has the vector form $\underline{C}^1_{2,3} = (f_{xx}, f_{x\lambda})$ and $\underline{C}^1_{2,3} \neq \underline{0}$ is taken to mean *neither* component is zero. In general, the (q, j)-extended system is denoted by

$$\underline{H}_{q,j}(x, \lambda, \underline{\alpha}) = \underline{0} \tag{3.9a}$$

where the components of $\underline{H}$ are f, f_x and the conditions from the left-most route of Figure 2, and the side-constraints are denoted by

$$C^k_{q,j}(x, \lambda, \underline{\alpha}) \neq 0, \quad k = 1, 2. \tag{3.9b}$$

Together (3.9a) and (3.9b) are the defining conditions for the (q, j)-singularity.

One can now use these defining conditions to compute paths of (q, j)-singular points, to detect higher codimensional singularities on these paths, and to compute paths of lower codimensional singularities passing through a given (q, j)-singular point. This process of moving up and down the hierarchy allows one to find the singular points of highest codimension and to decompose control parameter space into regions of qualitatively similar behaviour. Thus, for example, if f is given by (1.2) one can compute Figure 1.

The mathematical properties of the defining conditions which help produce a useful, computationally convenient tool for general nonlinear equations are summarized as (i) they are *specific* for (q, j)-singularities, (ii) the extended system is usually *regular* at (q,j)-singularities, and (iii) the side-constraints are *transversal* (i.e. generically, at least one changes sign at a singularity of higher codimension). Another useful property is that the derivative(s) in $C^k_{q,j}$ occur in the Jacobian of $\underline{H}_{q,j}$. Before making these remarks precise we need to clarify the term "usually" in (ii). On physical grounds we must expect to be able to compute only observable or structurally stable phenomena. For problem (3.2) the side-constraints (3.4) guarantee stability, but for multiparameter problems the situation is more complicated and an extra assumption is required. The stability of a singularity $(x_0, \lambda_0, \underline{\alpha}_0)$ of (3.1) under arbitrary perturbation (cf. (3.5)) is now expressed as "$f(x_0,\lambda_0,\underline{\alpha}_0)$ is *universally unfolded* by $\underline{\alpha}$" or "$f(x_0,\lambda_0,\underline{\alpha})$ is a *universal unfolding* of $f(x_0,\lambda_0,\underline{\alpha}_0)$", (see Golubitsky and Schaeffer, 1985). Essentially this means that the parameters $\underline{\alpha}$ enter into f in such a way as to provide all possible perturbations of $f(x_0,\lambda_0,\underline{\alpha}_0) = 0$. Hence the term "usually" in (ii) above includes all universally unfolded (q,j)-singularities.

The above mathematical properties are gathered together in the concept of *well-formulated defining conditions*:

DEFINITION 3.10. Let

$$\underline{H}_{q,j}(x,\lambda,\underline{\alpha}) = \underline{0}, \quad \underline{H}: \mathbb{R}\times\mathbb{R}\times\mathbb{R}^p \to \mathbb{R}\times\mathbb{R}^{q+1}, \tag{3.11a}$$

and

$$C^k_{q,j}(x,\lambda,\underline{\alpha}) \neq 0, \quad k = 1,2, \tag{3.11b}$$

be smooth functions. Then (3.11a,b) are said to be *well-formulated defining conditions* for (3.1) to have a (q,j)-singularity at $(x_0,\lambda_0,\underline{\alpha}_0)$ if: (i) (*Specificity*). (3.11a,b) hold if and only if $(x_0,\lambda_0,\underline{\alpha}_0)$ is a (q,j)-singularity. (ii) (*Regularity*). If $(x_0,\lambda_0,\underline{\alpha}_0)$ is a (q,j)-singularity of (3.1) that is universally unfolded by $\underline{\alpha}\in\mathbb{R}^p$ and if $q \leqslant p$ then the matrix

$$\partial\underline{H}^0_{q,j}/\partial(x,\lambda,\underline{\alpha}): \mathbb{R}\times\mathbb{R}\times\mathbb{R}^q \to \mathbb{R}\times\mathbb{R}^{q+1}$$

has rank $q+2$. (iii) (*Transversality of side-constraints*).

Assume that $q+1 \leqslant \min(3,p)$ and that $(x(s), \lambda(s), \underline{\alpha}(s))$ is a path of (q,j)-singular points passing through a $(q+1, j')$-singular point which is universally unfolded by $\alpha_1, \alpha_2, \cdots, \alpha_{q+1}$ where $j' = j$ or $j+1$. Then for $k=1$ or 2, $C^k_{q,j}$ changes sign at the $(q+1, j')$-singular point. (The vector valued side-constraint $\underline{C}^1_{2,3}$ is transversal if at least one component is transversal.)

The fact that the defining conditions obtained from the singularity hierarchy satisfy Definition 3.10 is proved in Jepson and Spence (1985b).

The power of a numerical technique based on well-formulated conditions is illustrated by the numerical results for the CSTR problem given in that paper. The nonlinear problem has the form

$$f(x,\lambda,\underline{\alpha}) := \alpha_3 - (1+\lambda)x + \lambda\alpha_1/(1+\lambda\alpha_2 A(x, \alpha_4)) = 0$$

where

$$A(x,\alpha_4) := e^{-\alpha_4 x/(1+x)}$$

and $\underline{\alpha} \in \mathbb{R}^4$. The parameter space is decomposed *quantitatively* for large ranges of parameter values, and pictures like Figure 1 can be computed which are in a certain sense maps of parameter space.

A hierarchy for symmetric problems satisfying $f(-x, \lambda, \underline{\alpha}) = -f(x,\lambda,\underline{\alpha})$ is given in Jepson, Spence and Cliffe (1986). It is used in the numerical solution of the Taylor problem.

Finally, we remark that all this theory and analysis would not be very useful if it were restricted to scalar problems. The fact that all the ideas, results and numerical strategies extend to more general problems is the main result of the next section.

4. VECTOR PROBLEMS

In this section we consider (1.1) when X and Y are Banach spaces, which we refer to as the *vector* problem. In many cases $X = Y = \mathbb{R}^n$ with $\underline{f}$ arising as a discretization of an ordinary or partial differential equation, and much recent work has been concerned with algorithms for the calculation of singular points in such finite dimensional cases (see Küpper *et al*, 1984).

Rather than summarising that work here we consider how the ideas introduced in Section 3 might be extended to the vector case. Because of its usefulness to computation, the main aim of this section is to show how the concept of well-formulated defining conditions (Definition 3.10) may be extended to vector problems. As a consequence of this extension the techniques and procedures developed for scalar problems can be applied to vector problems and the same type of information about the solutions may be obtained (see, for example, Beyn, 1984; Jepson, Spence and Cliffe, 1986; Jepson and Spence, 1986b).

First we make some assumptions on $\underline{f}$. For the problem

$$\underline{f}(\underline{x},\lambda,\underline{\alpha}) = \underline{0}, \quad \underline{f} : X \times \mathbb{R} \times \mathbb{R}^p \to Y, \tag{4.1}$$

we assume that at any singular point $(\underline{x}_0, \lambda_0, \underline{\alpha}_0)$,

$$\underline{f}_{\underline{x}}^0 \text{ is a Fredholm operator of index zero,} \tag{4.2a}$$

with

$$\text{Null}\,(\underline{f}_{\underline{x}}^0) = \text{span}\{\underline{\phi}_0\}, \qquad \underline{\phi}_0 \in X \setminus \{\underline{0}\}, \tag{4.2b}$$

$$\text{Range}\,(\underline{f}_{\underline{x}}^0) = \{\underline{y} \in Y : \underline{\psi}_0^* \underline{y} = 0\}, \quad \underline{\psi}_0^* \in Y^* \setminus \{\underline{0}\}, \tag{4.2c}$$

where Y^* is the dual space of Y. (If $X = Y = \mathbb{R}^n$ then $\underline{\psi}_0^* \underline{y}$ is usually written $\underline{\psi}_0^T \underline{y}$ with $\underline{\psi}_0^T$ being the left null vector.)

Under these assumptions it is well-known in nonlinear analysis that one can reduce (4.1) to an equivalent scalar problem using the Liapunov-Schmidt reduction. We present a generalization of this reduction here. Assume that closed subspaces $X_1, X_2 \subset X$ and $Y_1, Y_2 \subset Y$ are known, satisfying

$$X = X_1 \oplus X_2, \quad \dim X_2 = 1, \tag{4.3a}$$

$$Y = Y_1 \oplus Y_2, \quad \dim Y_2 = 1, \tag{4.3b}$$

and that associated projections P and Q are given satisfying

$$P : X \to X, \quad \text{Range}\,(P) = X_2, \quad \text{Null}(P) = X_1 \tag{4.4a}$$

$$Q : Y \to Y, \quad \text{Range}\,(Q) = Y_2, \quad \text{Null}(Q) = Y_1. \tag{4.4b}$$

The main assumption is that P and Q are chosen so that

$$(I-Q)\,\underline{f}_{\underline{x}}^{0}(I-P) : X_1 \to Y_1 \text{ is nonsingular.} \qquad (4.5)$$

(Note that for $X = Y = \mathbb{R}^n$ these assumptions will present no difficulty since information will be available from the continuation method to enable suitable choices for P and Q to be made.) Problem (4.1) can now be rewritten as

$$(I-Q)\,\underline{f}(\underline{u}+\underline{v}\,,\lambda\,,\underline{\alpha}) = \underline{0} \in Y_1\,, \quad \underline{u} \in X_1,\ \underline{v} \in X_2, \qquad (4.6a)$$

$$Q\underline{f}(\underline{u}+\underline{v}\,,\lambda\,,\underline{\alpha}) = \underline{0} \in Y_2 \qquad (4.6b)$$

and, because of (4.5), we can apply the Implicit Function Theorem to (4.6a) to show the existence of a unique solution $\underline{u}=\underline{u}(\underline{v}\,,\lambda\,,\underline{\alpha})$ for $(\underline{u}+\underline{v}\,,\lambda\,,\underline{\alpha})$ near $(\underline{x}_0\,,\lambda_0\,,\underline{\alpha}_0)$. Substitution into (4.6b) gives

$$Q\underline{f}(\underline{u}(\underline{v}\,,\lambda\,,\underline{\alpha})+\underline{v}\,,\lambda\,,\underline{\alpha}) = \underline{0}\,. \qquad (4.7)$$

If we now introduce basis vectors $\underline{x}_2$ and $\underline{y}_2$ for X_2 and Y_2 respectively, along with mappings $V_2 : \mathbb{R} \to X_2$ and $W_2 : Y_2 \to \mathbb{R}$ defined by

$$V_2\varepsilon = \varepsilon\underline{x}_2\,, \quad W_2(\eta\underline{y}_2) = \eta\,, \qquad (4.8)$$

then (4.7) can be rewritten as

$$\bar{h}(\varepsilon\,,\lambda\,,\underline{\alpha}) := W_2 Q\underline{f}(\underline{u}(V_2\varepsilon\,,\lambda\,,\underline{\alpha})+V_2\varepsilon\,,\lambda\,,\underline{\alpha}) = 0\,. \qquad (4.9)$$

The reduction process is now finished. The zeros of $\bar{h}$ are in one-to-one correspondence with the zeros of $\underline{f}$ and this equivalence allows the local solution behaviour of (4.1) to be studied via the scalar equation (4.9). This leads us to make the following definition:

DEFINITION 4.10. Suppose $(\underline{x}_0\,,\lambda_0\,,\underline{\alpha}_0)$ is a solution of (4.1) such that (4.2) – (4.5) hold. Then $(\underline{x}_0\,,\lambda_0\,,\underline{\alpha}_0)$ is a $(q\,,j)$-singular point of (4.1) if $(\varepsilon_0\,,\lambda_0\,,\underline{\alpha}_0) := (V_2^{-1}P\underline{x}_0\,,\lambda_0\,,\underline{\alpha}_0)$ is a $(q\,,j)$-singular point of (4.9). Furthermore the parameters $\underline{\alpha}$ are a universal unfolding for the singularity at $(\underline{x}_0\,,\lambda_0\,,\underline{\alpha}_0)$ if the singularity of $\bar{h}$ at $(\varepsilon_0\,,\lambda_0\,,\underline{\alpha}_0)$ is universally unfolded by $\underline{\alpha}$.

A point to note is that the reduction process works for any P and Q provided (4.5) holds and so clearly $\bar{h}$ is not unique. However an important result (Jepson and Spence, 1986a) is that the type and unfolding behaviour of the singular point of $\bar{h}$, corresponding to $(\underline{x}_0, \lambda_0, \underline{\alpha}_0)$ of $\underline{f}$, is independent of P and Q.

A natural idea is to exploit the equivalence of (4.9) and (4.1) to set up defining conditions for (q, j)-singular points of (4.1) using the theory in Section 3. Computationally this would probably be very inefficient because of the need to solve a nonlinear equation for $\underline{u}$ for every evaluation of $\bar{h}$. (For $(\varepsilon, \lambda, \underline{\alpha})$, $\underline{v} = V_2\varepsilon$, find $\underline{u}$ from (4.6a), then evaluate $\bar{h}$.) The following modification provides a more efficient method. If (4.6a) is embedded in a family of problems

$$(I-Q)\underline{f}(\underline{u}+\underline{v}, \lambda, \underline{\alpha}) = \underline{c} \in Y_1 \qquad (4.11)$$

for $\underline{c}$ near $\underline{0}$, then the reduction procedure is as before with $\underline{u} = \underline{u}(\underline{v}, \lambda, \underline{\alpha}, \underline{c})$ and the reduced equation

$$h(\varepsilon, \lambda, \underline{\alpha}, \underline{c}) = 0 \qquad (4.12)$$

is obtained. Clearly $h(\varepsilon, \lambda, \underline{\alpha}, \underline{0}) = \bar{h}(\varepsilon, \lambda, \underline{\alpha})$. An iteration to find a zero of h can now be based on the *pair* (4.11) and (4.12). (For $(\underline{x}, \lambda, \underline{\alpha})$, $\varepsilon = V_2^{-1}P\underline{x}$, $h(\varepsilon, \lambda, \underline{\alpha}, \underline{c}) = W_2 Q\underline{f}(\underline{x}, \lambda, \underline{\alpha})$ with $\underline{c} = (I-Q)\underline{f}(\underline{x}, \lambda, \underline{\alpha})$.) Thus it is more efficient to set up defining conditions for singularities of (4.1) by including (4.11) explicitly along with the defining conditions for the scalar problem (4.12). Hence defining conditions for a (q, j)-singular point of (4.1) are

$$\underline{F}_{q,j} := \begin{pmatrix} \underline{c} \\ \\ \underline{H}_{q,j} \end{pmatrix} \equiv \begin{pmatrix} (I-Q)\underline{f}(\underline{x}, \lambda, \underline{\alpha}) \\ \\ \underline{H}_{q,j}(\varepsilon, \lambda, \underline{\alpha}, \underline{c}) \end{pmatrix} = \underline{0}, \qquad (4.13a)$$

$$C^k_{q,j} \neq 0, \quad k = 1, 2, \qquad (4.13b)$$

where $\underline{H}_{q,j}$ and $C^k_{q,j}$ are obtained from the scalar hierarchy as described in Section 3. The fact that (4.13) inherits the well-formulated properties from the scalar defining conditions is the content of the following theorem (Jepson and Spence, 1986b).

THEOREM 4.14. If (4.2) and (4.5) are satisfied then the following three properties hold:

(i) $(\underline{x}_0, \lambda_0, \underline{\alpha}_0)$ satisfies (4.13) if and only if $(\underline{x}_0, \lambda_0, \underline{\alpha}_0)$ is a (q,j)-singularity of (4.1).

(ii) If $q \leqslant p$ and if this singularity is universally unfolded by $\alpha_1, \alpha_2, \cdots, \alpha_q$ then $\partial \underline{F}^0_{q,j} / \partial(\underline{x}, \lambda, \underline{\alpha})$ is nonsingular.

(iii) Assume that $q+1 \leqslant \min(3,p)$ and that $(\underline{x}(s), \lambda(s), \underline{\alpha}(s))$ is a path of (q,j)-singular points passing through a $(q+1, j')$-singular point which is universally unfolded by $\alpha_1, \alpha_2, \cdots, \alpha_{q+1}$ where $j' = j$ or $j+1$. Then for $k=1$ or 2, $C^k_{q,j}$ changes sign at the $(q+1, j')$-singular point.

This is an important and useful result in that it implies that the strategies and techniques derived for scalar equations can be carried over to the vector case. Of course there are technical changes because of the presence of (4.11) but these do not effect any of the theoretical results. (Naturally, these "technical changes" will prove important when considering the cost and efficiency of the approach in any practical example.)

The final point to discuss in this section is the use of (4.13) to actually calculate singular points when $X = Y = \mathbb{R}^n$. It happens, when the details are worked out, that the reduction process can be interpreted in different ways and several familiar methods are recovered. There is not enough space here to describe the evaluation of $\underline{F}_{q,j}$ or $\partial \underline{F}_{q,j} / \partial(\underline{x}, \lambda, \underline{\alpha})$ so we simply state some results and indicate where proofs are to be found. The case of a quadratic turning point $(q,j) = (0,1)$ is discussed in Jepson and Spence (1984). For suitable P and Q one recovers both the system

$$\left.\begin{aligned} \underline{f}(\underline{x},\lambda) &= \underline{0} \\ \underline{f}_{\underline{x}}(\underline{x},\lambda)\underline{\phi} &= \underline{0} \\ \ell(\underline{\phi})-1 &= 0 \end{aligned}\right\} \tag{4.15}$$

(Seydel, 1979; Moore and Spence, 1980), where $\ell(\underline{\phi}) = 1$ is a scaling condition, and the system

$$\left.\begin{aligned} \underline{f}(\underline{x},\lambda) &= \underline{0} \\ \underline{\psi}^T \underline{f}_{\underline{x}}(\underline{x},\lambda)\underline{\phi} &= 0 \end{aligned}\right\} \tag{4.16}$$

where $\underline{\psi}$ and $\underline{\phi}$ are defined (approximately) to be singular vectors of $\underline{f}_{\underline{x}}$ (cf. Abbott, 1978; Pönish and Schwetlick, 1981; Griewank and Reddien, 1984; Spence and Jepson, 1984). For both systems the two side-constraints are

$$\underline{\psi}_0^T \underline{f}_\lambda^0 \neq 0 \tag{4.17a}$$

and

$$\underline{\psi}_0^T \underline{f}_{\underline{x}\underline{x}}^0 \underline{\phi}_0 \underline{\phi}_0 \neq 0, \tag{4.17b}$$

where $\underline{\psi}_0$ and $\underline{\phi}_0$ are as in (4.2b,c) (cf. (3.4a,b)). Theorems on the regularity of solutions of (4.15) or (4.16) subject to (4.17a,b) and (4.2) follow quickly from Theorem 4.14(ii).

The case of a transcritical bifurcation point is discussed in Jepson and Spence (1986b) where, using (4.11) and the defining conditions (3.8a,b) applied to (4.12), the methods in Spence and Jepson (1984) and Weber (1982) are recovered. Much of the effort in previous papers has involved proving the regularity of the extended system at the appropriate singular points, which can be quite tedious for singular points of high codimension. The use of part (iii) of Theorem 4.14 reduces this effort considerably.

If one considers the derivation of some of the known methods it is not surprising that many are equivalent in the sense that they can be derived from the same reduction process, though they may not be equivalent with regard to work done on a computer! This unification, however, does provide quick analysis and justification of some of the past methods which had involved some fairly lengthy calculations, see, for example, Spence and Werner (1982) and Jepson and Spence (1985a).

The techniques described in these two papers were used to produce numerical solutions for the Taylor problem, where $X = Y = \mathbb{R}^n$ and $\alpha \in \mathbb{R}$, see Cliffe (1983) and Cliffe and Spence (1986). The presentation of numerical results in terms of plots of parameter space are given for $\underline{\alpha} \in \mathbb{R}^2$ in Jepson, Spence and Cliffe (1986), though a different hierarchy is used because of the presence of a symmetry. Other work on the vector problem using singularity theory is described in Beyn (1984) and Beyn and Bohl (1984). There are also some interesting examples of vector problems in Rheinboldt (1986).

5. OTHER TOPICS

There are many other aspects of the numerical solution of bifurcation problems which we have not had space to discuss in this paper. Therefore we provide a few references that introduce some other important topics.

Analysis of discretization errors: Fujii and Yamaguti (1980); Beyn (1980); Brezzi, Rappaz and Raviart (1981); Moore and Spence (1981); Descloux (1984); Moore, Spence and Werner (1986).

Hopf bifurcation: Doedel (1981); Griewank and Reddien (1983); Brezzi, Ushiki and Fujii (1984); Descloux (1984); Jepson and Keller (1984); Roose and Hlavacek (1985).

Multiple bifurcation: Menzel (1984); Jepson and Decker (1986).

Problems with symmetry: Abbott (1978); Brezzi, Rappaz and Raviart (1981); Werner and Spence (1984); Cliffe and Spence (1986); Jepson, Spence and Cliffe (1986).

Techniques for large sparse systems: Chan (1984); Mackens and Jarausch (1984); Mittelmann and Weber (1985); Weber (1985); Bolstad and Keller (1986); Rheinboldt (1986).

REFERENCES

Abbott, J.P. (1978), "An efficient algorithm for the determination of certain bifurcation points", *J. Comput. Appl. Math.*, *4*, pp.19-27.

Anselone, P.M. and Moore, R.H. (1966), "An extension of the Newton-Kantorovich method for solving nonlinear equations with an application to elasticity", *J. Math. Anal. Appl.*, *13*, pp.476-501.

Bauer, L., Reiss, E.L. and Keller, H.B. (1970), "Axisymmetric buckling of hollow spheres and hemispheres", *Comm. Pure Appl. Math.*, *23*, pp.529-568.

Benjamin, T.B. and Mullin, T. (1981), "Anomalous modes in the Taylor experiment", *Proc. Roy. Soc. London Ser. A*, *377*, pp.221-249.

Beyn, W.J. (1980), "On discretizations of bifurcation problems", in Mittelmann *et al* (1980), pp.46-73.

Beyn, W.J. (1984), "Defining equations for singular solutions and numerical applications", in Küpper *et al* (1984), pp.42-56.

Beyn, W.J. and Bohl, E. (1984), "Organizing centers for discrete reaction diffusion models", in Küpper *et al* (1984), pp.57-67.

Bolstad, J.H. and Keller, H.B. (1986), "A multigrid continuation method for elliptic problems with folds", *SIAM J. Sci. Statist. Comput.*, *7*, pp.1081-1104.

Brezzi, F. and Fujii, H. (1982), "Numerical imperfections and perturbations in the approximation of nonlinear problems", in *The Mathematics of Finite Elements and Applications IV*, ed. Whiteman, J.R., Academic Press (London), pp.431-452.

Brezzi, F., Rappaz, J. and Raviart, P.A. (1981), "Finite dimensional approximation of nonlinear problems, Part II: Limit Points", *Numer. Math.*, *37*, pp.1-28; Part III: Simple Bifurcation Points", *Numer. Math.*, *38*, pp.1-30.

Brezzi, F., Ushiki, S. and Fujii, H. (1984), "'Real' and 'ghost' bifurcation dynamics in difference schemes for ODEs", in Küpper *et al* (1984), pp.79-104.

Chan, T. (1984), "Techniques for large sparse systems arising from continuation methods", in Küpper *et al* (1984), pp.116-128.

Cliffe, K.A. (1983), "Numerical calculations of two-cell and single-cell Taylor flows", *J. Fluid Mech.*, *135*, pp.219-233.

Cliffe, K.A. and Spence, A. (1986), "Numerical calculations of bifurcations in the finite Taylor Problem", in *Numerical Methods for Fluid Dynamics*, eds. Morton, K.W. and Baines, M., Clarendon Press (Oxford), pp.177-197.

Descloux, J. (1984), "On Hopf and subharmonic bifurcations", in Küpper *et al* (1984), pp.145-161.

Doedel, E.J. (1981), "AUTO: a program for the automatic bifurcation analysis of autonomous systems", *Congr. Numer.*, *30*, pp.265-284.

Fujii, H. and Yamaguti, M. (1980), "Structure of singularities and its numerical realization in nonlinear elasticity", *J. Math. Kyoto Univ.*, *20*, pp.489-590.

Golubitsky, M. and Keyfitz, B. (1980), "A qualitative study of the steady-state solutions for a continuous flow stirred tank chemical reactor", *SIAM J. Math. Anal.*, *11*, pp.316-339.

Golubitsky, M. and Schaeffer, D. (1985), *Singularities and Groups in Bifurcation Theory*, Springer (New York).

Griewank, A. and Reddien, G. (1983), "The calculation of Hopf points by a direct method", *IMA J. Numer. Anal.*, *3*, pp.295-303.

Griewank, A. and Reddien, G.W. (1984), "Characterization and computation of generalized turning points", *SIAM J. Numer. Anal.*, *21*, pp.176-185.

Jepson, A.D. and Decker, D.W. (1986), "Convergence cones near bifurcation", *SIAM J. Numer. Anal.*, *23*, pp.959-975.

Jepson, A.D. and Keller, H.B. (1984), "Steady state and periodic solution paths: their bifurcations and computations", in Küpper *et al* (1984), pp.219-246.

Jepson, A.D. and Spence, A. (1984), "Singular points and their computation", in Küpper *et al* (1984), pp.195-209.

Jepson, A.D. and Spence, A. (1985a), "Folds in solutions of two parameter systems and their calculation. Part I", *SIAM J. Numer. Anal.*, *22*, pp.347-368.

Jepson, A.D. and Spence, A. (1985b), "The numerical solution of nonlinear equations having several parameters I: Scalar equations", *SIAM J. Numer. Anal.*, *22*, pp.736-759.

Jepson, A.D. and Spence, A. (1986a). "On a reduction process for nonlinear equations", *SIAM J. Math. Anal.*, to be published.

Jepson, A.D. and Spence, A. (1986b), "The numerical solution of nonlinear equations having several parameters II: Vector equations", in preparation.

Jepson, A.D., Spence, A. and Cliffe, K.A. (1986), "The numerical solution of nonlinear equations having several parameters III: Problems with symmetry", in preparation.

Keener, J.P. and Keller, H.B. (1973), "Perturbed bifurcation theory", *Arch. Rational Mech. Anal.*, *50*, pp.159-175.

Keller, H.B. (1977), "Numerical solution of bifurcation and nonlinear eigenvalue problems", in Rabinowitz (1977), pp.359-384.

Keller, H.B. and Langford, W.F. (1972), "Iterations, perturbations and multiplicities for nonlinear bifurcation problems", *Arch. Rational Mech. Anal.*, *48*, pp.83-108.

Keller, J.B. and Antman, S. (eds.) (1969), *Bifurcation Theory and Nonlinear Eigenvalue Problems*, Benjamin (New York).

Kubíček, M. and Marek, M. (1983), *Computational Methods in Bifurcation Theory and Dissipative Structures*, Springer (New York).

Küpper, T., Mittelmann, H.D. and Weber, H. (eds.) (1984), *Numerical Methods for Bifurcation Problems*, Birkhäuser (Basel).

Mackens, W. and Jarausch, H. (1984), "Numerical treatment of bifurcation branches by adaptive condensation", in Küpper *et al* (1984), pp.296-309.

Menzel, R. (1984), "Numerical determination of multiple bifurcation points", in Küpper *et al* (1984), pp.310-318.

Menzel, R. and Schwetlick, H. (1978), "Zur Lösung parameterabhängiger nichtlinear Gleichungen mit singulären Jacobi-Matrizen", *Numer. Math.*, *30*, pp.65-79.

Mittelmann, H.D. and Weber, H. (eds.) (1980), *Bifurcation Problems and their Numerical Solution*, Birkhäuser (Basel).

Mittelmann, H.D. and Weber, H. (1985), "Multi-grid solution of bifurcation problems", *SIAM J. Sci. Statist. Comput.*, *6*, pp.49-60.

Moore, G. and Spence, A. (1980), "The calculation of turning points of nonlinear equations", *SIAM J. Numer. Anal.*, *17*, pp.567-576.

Moore, G. and Spence, A. (1981), "The convergence of operator approximations at turning points", *IMA J. Numer. Anal.*, *1*, pp.23-38.

Moore, G., Spence, A. and Werner, B. (1986), "Operator approximation and symmetry-breaking bifurcation", *IMA J. Numer. Anal.*, *6*, pp.331-336.

Pönisch, G. and Schwetlick, H. (1981), "Computing turning points of curves implicitly defined by nonlinear equations depending on a parameter", *Computing*, *26*, pp.107-121.

Poston, T. and Stewart, I. (1978), *Catastrophe Theory and its Applications*, Pitman (London).

Rabinowitz, P.H. (ed.) (1977), *Applications of Bifurcation Theory*, Academic Press (New York).

Rheinboldt, W.C. (1980), "Solution fields of nonlinear equations and continuation methods", *SIAM J. Numer. Anal.*, *17*, pp.221-237.

Rheinboldt, W.C. (1986), *Numerical Analysis of Parameterized Nonlinear Equations*, Wiley (New York).

Rheinboldt, W.C. and Burkardt, J.V. (1983), "A locally parameterized continuation process", *ACM Trans. Math. Software*, *9*, pp.215-235, 236-241.

Roose, D. and Hlavacek, V. (1985), "A direct method for the computation of Hopf bifurcation points", *SIAM J. Appl. Math.*, *45*, pp.879-894.

Seydel, R. (1979), "Numerical computation of branch points in nonlinear equations", *Numer. Math.*, *33*, pp.339-352.

Spence, A. and Jepson, A.D. (1984), "The numerical calculation of cusps, bifurcation points and isola formation points in two parameter problems", in Küpper *et al* (1984), pp.502-514.

Spence, A. and Werner, B. (1982), "Non-simple turning points and cusps", *IMA J. Numer. Anal.*, *2*, pp.413-427.

Wacker, H. (ed.) (1978), *Continuation Methods*, Academic Press (New York).

Weber, H. (1982), "Zur Verzweigung bei einfachen Eigenwerten", *Manuscripta Math.*, *38*, pp.77-86.

Weber, H. (1985), "Multigrid bifurcation iteration", *SIAM. J. Numer. Anal.*, *22*, pp.262-279.

Weiss, R. (1975), "Bifurcation in difference approximations to two-point boundary value problems", *Math. Comp.*, *29*, pp.746-760.

Werner, B. and Spence, A. (1984), "The computation of symmetry-breaking bifurcation points, "*SIAM J. Numer. Anal.*, *21*, pp.388-399.

A.D. Jepson
Department of Computer Science
University of Toronto
Toronto M5S 1A7
Canada

A. Spence
School of Mathematics
University of Bath
Claverton Down
Bath BA2 7AY
England

11 On the Iterative Solution of Differential and Integral Equations Using Secant Updating Techniques

ANDREAS GRIEWANK

ABSTRACT

We discuss the application of quasi-Newton methods based on secant updating to certain mildly nonlinear operator equations. On two point boundary value problems and weakly singular integral equations of the second kind mesh independent superlinear convergence is observed. While the evaluation of derivatives is always avoided, savings in linear algebra computations are only achieved by the standard low rank updates.

1. INTRODUCTION

The discretization of differential and integral equations leads to systems of linear or nonlinear algebraic equations in a large and possibly varying number of variables. Typically the Jacobians, i.e. derivative matrices, of these systems have some special structure, for example a certain sparsity pattern. Sparse Jacobians can often be evaluated, stored and factorized at greatly reduced costs so that Newton's method can be implemented quite efficiently. When the underlying problem is linear this amounts to a direct solution of the resulting vector equation.

In cases where the Jacobian is not easily evaluated or hard to factorize Newton's method may no longer be the optimal choice. To avoid the factorization, iterative methods for the solution of linear systems have received much attention, especially since the advent of vector and parallel computers. While suitably preconditioned conjugate gradient methods work very well on many positive definite systems (see Johnson, Michelli and Paul, 1983, for example), the situation seems much less clear for non-symmetric problems. The observation of Winther (1977) that Broyden's method can be employed as an efficient iterative equation solver was not accepted, as it was thought to require more storage than the GCR method (Elman, 1982).

The conjugate gradient method is widely understood in its dual role as a linear equations solver and a descent method for unconstrained optimization. In contrast, the method of Broyden (1965) is almost exclusively thought of as a quasi-Newton method for solving a nonlinear system whose Jacobian matrix is not available. Analysis and testing of Broyden's method and the related "Variable Metric Schemes" for unconstrained optimization have largely focused on small to medium sized problems with dense Jacobians or Hessians. While the original motivation for these classical secant methods was to avoid the explicit evaluation of derivative matrices, they were also found to reduce the required linear algebra operations per step, in comparison to Newton's method.

Naturally the replacement of exact Jacobians or Hessians by secant approximations destroys the quadratic convergence of Newton's method. However, Broyden, Dennis and Moré (1973) and Dennis and Moré (1974) showed in the early seventies that the classical secant methods exhibit locally Q-superlinear convergence, which means that the ratio between successive residual norms tends to zero. For the particular case of the BFGS method on a uniformly convex function, Powell (1976) proved convergence from arbitrary starting points and initial Hessian approximations. Otherwise the global convergence properties of most quasi-Newton methods are theoretically uncertain but in practice entirely satisfactory, provided certain safeguards such as line searches or trust regions are employed (Dennis and Schnabel, 1983).

On a dense problem in a moderate number of variables, a standard quasi-Newton code will rarely take more than twice the number of steps required by a comparable Newton implementation. While the resulting overall computing time depends strongly on the cost of evaluating the problem functions, most users appreciate the convenience of not having to provide derivative matrices. For these reasons quasi-Newton methods are now the mainstay of software libraries for nonlinear optimization calculations. However, most standard routines are not very suitable for the solution of large structured problems.

The performance of the conjugate gradient method on quadratic functionals in Hilbert space has been analyzed by Daniel (1967), Hayes (1954), Horwitz and Sarachik (1968), Werner (1974), Winther (1980) and others. By Dixon's (1972) theorem these results apply also to variable metric methods with line searches because they generate theoretically the same iterates as conjugate gradients. However, for this very reason at least in the quadratic case, there is no incentive for using a method that requires much more storage and many more arithmetic operations. For the nonquadratic case in very many variables, several researchers (e.g. Buckley and LeNir, 1983) have developed "limited memory" variable metric methods as a compromise between conjugate gradients and fully blown variable metric methods. Examples of Stoer (1979) and Powell (1984) demonstrated that the convergence rate of variable metric methods may only be linear on Hilbert space problems. For structural finite element calculations Strang (1980) has suggested a Shamanskii like procedure (Brent, 1973), where the exact Hessian is re-evaluated every five steps or so, and within each period the Hessian is appended by rank two updates according to the BFGS formula (Dennis and Moré, 1983). See also the survey by Rheinboldt and Riks (1984).

One obvious deficiency of the classical low rank updates is that they immediately destroy any sparsity that might be present in the Jacobian or Hessian. This means not only that the storage requirement is unnecessarily high but also that *a priori* available derivative information is ignored. On the other hand the imposition of sparsity on the approximating Jacobian or Hessian usually implies that the updates are of full rank, so that the computation of the next step becomes as expensive as for Newton's method. The partitioned updating technique developed by Griewank and Toint (1982a) overcomes some of these difficulties. It has been implemented in the Harwell routine VE∅8 and apparently solves a variety of problems satisfactorily.

One salient aspect of VE∅8 is that on discretizations of partial differential equations, e.g. the minimal surface problem, the speed of convergence appears to be the same for all sufficiently fine discretizations. For Newton's method this "mesh independence principle" has recently been established by Allgower *et al* (1986). They actually show that the individual Newton iterates are close approximations to the corresponding Newton iterates on the underlying infinite dimensional problem.

In this paper we will concentrate on quasi-Newton methods for the solution of operator equations that are in general not self-adjoint. Obviously practical algorithms can only be applied to discretized equations and even then we have to discount the effects of round-off. However, only properties that extend at least conceptually to the underlying infinite dimensional problem can be expected to be mesh independent. Without discussing any specific discretization we will assume that they leave problem parameters (like Lipschitz constants and induced operator norms) largely unaffected.

To illustrate our analysis we consider the following two sample problems. Firstly there is the weakly singular integral equation

$$x(t) + \alpha \int_0^1 \frac{\phi(t,\tau)\,x(t)}{|t-\tau|^{1-\alpha}}\, d\tau = b(t), \quad 0 \leq t \leq 1, \tag{1.1}$$

where $\alpha \in (0,1)$ and the functions ϕ and b are sufficiently smooth. Since the integral operator is linear, analytic derivatives are clearly available. However, we find in Section 3 that the resulting dense linear systems can be solved quite economically by Broyden's method. Moreover the speed of convergence can be directly related to the exponent α, which determines the rate at which the singular values of the integral operator decline towards zero. Broyden's method was first applied to integral equations by Kelley and Sachs (1985a).

Secondly we consider a system of first order differential equations

$$\dot{\underline{x}}(t) + \underline{\phi}(t, \underline{x}(t)) = \underline{b}(t), \quad 0 \leq t \leq 1, \tag{1.2}$$

with $\underline{\phi} : [0,1] \times \mathbb{R}^m \to \mathbb{R}^m$, subject to linear homogeneous boundary conditions

$$A_0 \underline{x}(0) + A_1 \underline{x}(1) = \underline{0} \tag{1.3}$$

with $A_1, A_2 \in \mathbb{R}^{m \times m}$. Here our aim is to avoid the computation of the Jacobian $\underline{\phi}_{\underline{x}} = \nabla_{\underline{x}} \underline{\phi} \in \mathbb{R}^{m \times m}$ whose analytic evaluation or approximation by differencing is required by all standard codes. If the boundary conditions are nonlinear the corresponding $m \times 2m$ Jacobian can also be approximated by Broyden updating without affecting the convergence result described in Section 4. The idea of solving ordinary differential equations by pointwise Broyden updates is due to Hart and Soul (1973). They showed numerically that the sparse Broyden method due to Schubert (1970) is not competitive in this application. For computational results on boundary value problems and stiff initial value problems see Deuflhard (1978) and Brown *et al* (1985) respectively. Kelley and Sachs (1985b, 1986) have extended pointwise updates to elliptic partial differential equations and to control problems.

The paper is organized as follows. In Section 2 we develop a general framework for the application of secant methods to operator equations. Section 3 analyses the classical rank one update in a Hilbert space setting in a way that is suitable to integral operators. In Section 4 we consider pointwise Broyden updates with damping which may, for example, be used in the solution of the boundary value problem (1.2). The paper concludes with a brief summary and discussion.

2. LOCAL CONVERGENCE OF QUASI-NEWTON METHODS FOR OPERATOR EQUATIONS

The two sample problems and many other operator equations (Brezzi, Rappaz and Raviart, 1980, 1981a,b) can be written in the form

$$\underline{F}(\underline{x}) \equiv L\underline{x} + \underline{f}(\underline{x}) = \underline{b} \, . \tag{2.1}$$

Here the leading linear operator L and its possibly nonlinear perturbation $\underline{f}(\underline{x})$ map some Banach space X into another one Y. We assume that $\underline{f}$ has a Fréchet derivative $\underline{f}'(\underline{x})$ such that, at a particular root of interest $\underline{x}_* \in \underline{F}^{-1}(\underline{b})$ for two constants σ and ω, we have

$$\sigma^{-1} \equiv \|(L+\underline{f}'(\underline{x}_*))^{-1}\| < \infty \tag{2.2}$$

and

$$\|\underline{f}'(\underline{x}) - \underline{f}'(\underline{x}_*)\| \leq \omega \|\underline{x} - \underline{x}_*\| \tag{2.3}$$

for all $\underline{x}$ in some neighbourhood of $\underline{x}_*$. Here $\|\cdot\|$ denotes the norms on X and Y as well as the corresponding induced norms on the Banach spaces $\mathcal{L}[X,Y]$ and $\mathcal{L}[Y,X]$ of bounded linear operators between X and Y. The right hand side of (2.3) could be replaced by $\omega(\ln \|\underline{x}-\underline{x}_*\|)^{-2}$ but for simplicity we use the much stronger condition of Lipschitz continuity. As is the case for the boundary value problem, L may be an unbounded differential operator and often $\{L^{-1}\underline{f}(\underline{x}) : \underline{x} \in X\}$ will be compact.

Now let us suppose that $\underline{f}'(\underline{x})$ is either hard to evaluate or the resulting linear operator $L + \underline{f}'(\underline{x})$ is not easily inverted. Instead of applying Newton's method exactly, one might then prefer to replace $\underline{f}'(\underline{x}_k)$ at the current iterate $\underline{x}_k \in X$ by a suitable approximation $B_k \in \mathcal{L}[X,Y]$. Provided $L + B_k$ has a bounded inverse, the next quasi-Newton iterate is

$$\underline{x}_{k+1} \equiv (L+B_k)^{-1}(B_k\underline{x}_k - \underline{f}(\underline{x}_k) + \underline{b}) , \qquad (2.4)$$

which satisfies the equivalent relation

$$L\underline{x}_{k+1} + \underline{f}(\underline{x}_k) + B_k(\underline{x}_{k+1} - \underline{x}_k) = \underline{b} \quad . \qquad (2.5)$$

Almost all iterative methods for solving the operator equation (2.1) can be written in this form.

In order to guarantee local convergence of the above iteration in the nonlinear case, one has to ensure that the discrepancies

$$D_k \equiv B_k - \underline{f}'(\underline{x}_*) \in \mathcal{L}[X,Y] \qquad (2.6)$$

are sufficiently small. With σ as defined in (2.2) we have by the Banach Perturbation Lemma (Ortega and Rheinboldt, 1970)

$$\|(L+B_k)^{-1}\| \leq 1/(\sigma - \|D_k\|) , \qquad (2.7)$$

so that all steps are well defined if $\|D_k\| \leq \sigma/5$. Moreover, subtracting $\underline{x}_*$ from both sides of (2.4), we derive by standard arguments

$$q_k \equiv \rho_{k+1}/\rho_k \leq 5(\|D_k\| + \omega\rho_k)/(4\sigma)$$

where here and throughout the paper

$$\rho_k \equiv \|\underline{x}_k - \underline{x}_*\| \qquad \text{for all} \quad k \geq 0 . \qquad (2.8)$$

Hence, provided that inequality (2.3) is satisfied in the neighbourhood $\{\underline{x} : \|\underline{x} - \underline{x}_*\| \leq \rho_0\}$, the conditions

$$\rho_0 \leq \frac{1}{20}\,\sigma/\omega \quad \text{and} \quad \|D_k\| \leq \frac{1}{5}\sigma , \quad k \geq 0, \qquad (2.9)$$

ensure Q-linear convergence as on every iteration we have $q_k \leq 5/16 < 1$.

Obviously $\|D_k\| \leq \sigma/5$ may hold for constant $B_k = B_0$. For example the chord method defined by $B_k = \underline{f}'(\underline{x}_0)$ with $\rho_0 \leq \sigma(20\omega)^{-1}$ satisfies (2.9) and therefore converges Q-linearly (Ortega and Rheinboldt, 1970). The key idea of secant methods is to keep updating the B_k so that they form better and better approximations to $\underline{f}'(\underline{x}_*)$ without requiring the evaluation of $\underline{f}'(\underline{x}_k)$ and the refactorization of $L+\underline{f}'(\underline{x}_k)$ at each step. In the finite dimensional case such methods achieve Q-superlinear convergence, i.e. $q_k \to 0$, whenever they converge at least linearly to a regular root. In the infinite dimensional case this need not be true as we will see in Section 3. Firstly, we will discuss some basic properties of secant updates with bounded deterioration and deduce a corresponding local convergence result of the form given by Dennis and Moré (1977).

In many applications it is known *a priori* that the derivative $\underline{f}'(\underline{x})$ belongs to some closed subspace $C \subset \mathcal{L}[X,Y]$; for example C could consist of all matrices with a certain symmetry or sparsity pattern. Naturally the approximations B_k should then also be restricted to C. After a nontrivial step $\underline{s}_k \equiv \underline{x}_{k+1} - \underline{x}_k$ from $\underline{x}_k$ to $\underline{x}_{k+1}$ we look for a new approximation $B_{k+1} \in C$ that satisfies the so-called "secant condition"

$$B_{k+1}\underline{s}_k = \underline{y}_k \equiv \underline{f}(\underline{x}_{k+1}) - \underline{f}(\underline{x}_k) \,. \tag{2.10}$$

When $\underline{f}(\underline{x})$ is linear this equation holds for $B_{k+1} = \underline{f}'(\underline{x}_*)$ and in all cases it is satisfied when B_k is the operator

$$J_k \equiv \int_0^1 \underline{f}'(\underline{x}_k + \tau\underline{s}_k)\, d\tau \,. \tag{2.11}$$

Further it follows from (2.3) that

$$\|J_k - \underline{f}'(\underline{x}_*)\| \leq \omega\rho_k \tag{2.12}$$

provided $q_k \leq 1$ as we may assume.

In general the closed affine variety

$$[\underline{y}_k/\underline{s}_k] \equiv \{B \in C : B\underline{s}_k = \underline{y}_k\} \tag{2.13}$$

will contain infinitely many operators in addition to J_k. Also J_k is usually unknown, and we have no other information about $\underline{f}'(\underline{x}_*)$ as storing previous secant pairs $\{\underline{s}_j, \underline{y}_j\}$ for $j < k$ is too costly. Now, since B_k is presumably already a reasonable approximation to $\underline{f}'(\underline{x}_*)$ and J_k could be any element of $[\underline{y}_k/\underline{s}_k]$, we would like to ensure that

$$\|B_{k+1} - B\| \leq \|B_k - B\| \quad \text{for all} \quad B \in [\underline{y}_k/\underline{s}_k]. \tag{2.14}$$

It turns out that this "nondeterioration" condition determines B_{k+1} uniquely in most settings of interest.

As an immediate consequence of (2.14) we obtain from (2.12) by the triangle inequality the so-called "bounded deterioration" property of Dennis and Moré (1977)

$$\|D_k\| \leq \|D_{k-1}\| + 2\omega\rho_{k-1} \leq \|D_0\| + 2\sum_{j=0}^{k-1} \omega\rho_j .$$

Now, as long as $q_j \leq 1/3$ for all $j < k$, the sum on the right is bounded by $3\omega\rho_0$ so that $\|D_k\| \leq \|D_0\| + 3\omega\rho_0$. Conversely, if this upper bound is smaller than $\sigma/5$, it follows that $q_k \leq 1/3$. In this way one obtains by induction the following local convergence result which is essentially due to Dennis (1971).

PROPOSITION 2. If $x_0 \in X$ and B_0 are chosen such that

$$\omega\rho_0 \leq \sigma/20 \quad \text{and} \quad \|D_0\| \leq \sigma/20 , \tag{2.15}$$

then any iteration of the form (2.4) with the B_k's satisfying (2.14) converges Q-linearly with $q_k \leq 1/3$ for all $k \geq 0$.

Obviously the bound on q_k could be improved or even expressed as a function of $\|D_0\|$ and ρ_0. Usually the proof of linear convergence is only the preparatory stage, before the actual asymptotic rate of convergence is determined. As already mentioned, achieving Q-superlinear convergence, i.e. $q_k \to 0$, has been widely viewed as the main benefit of secant updating. However, if a particular secant method is applied to a family of discretizations of the same underlying problem, it may be important to know how its performance depends on the mesh size.

To motivate our concern, we consider the application of Newton's method to discretizations of an operator equation with a singular Jacobian at the root of interest. Since the roots of the discretizations will in general not be exactly singular, in each case the theoretical rate of convergence of Newton's method is quadratic. However, in practice the convergence rate will look linear (Griewank, 1985a) until the problem has been solved up to the discretization error. As we will see in Section 3 a similar effect can occur for quasi-Newton methods, even if the underlying problem is nonsingular.

In order to preclude the possibility that the onset of superlinear convergence is significantly delayed by grid refinements, one could look for bounds on the q_k that are independent of the mesh. It seems that such bounds cannot be obtained, but it is possible to bound the errors $\rho_k \equiv \|\underline{x}_k - \underline{x}_*\|$ directly.

To this end we note firstly that by (2.1), (2.5) and (2.10)

$$\underline{r}_k \equiv \underline{y}_k - B_k \underline{s}_k = \underline{F}(\underline{x}_{k+1}) - \underline{b} \,. \tag{2.16}$$

Moreover under the assumption of Proposition 2 we have

$$\|\underline{s}_k\| \leqslant \rho_k + \rho_{k+1} \leqslant 4\rho_k/3$$

and

$$\rho_{k+1} \leqslant \|\underline{F}(\underline{x}_{k+1}) - \underline{b}\| \, 20/(19\sigma) \,,$$

so that (2.16) gives

$$q_k \leqslant (1.5/\sigma) \, \|\underline{r}_k\| \, / \, \|\underline{s}_k\| \,. \tag{2.17}$$

This inequality confirms that Q-superlinear convergence is implied by the so-called Dennis and Moré (1974) condition $\|\underline{r}_k\| / \; \underline{s}_k\| \to 0$. Unfortunately the mere fact that the $\|\underline{r}_k\| / \|\underline{s}_k\|$ tend to zero does not indicate when (i.e. for which k) these ratios become small. However, in certain situations, including our examples, one can obtain the stronger result that there exist two constants $p \geqslant 2$ and $c_p \geqslant 0$ such that

$$\frac{1}{k}\sum_{j=0}^{k-1} \|\underline{r}_j\|/\|\underline{s}_j\| \leq \tfrac{2}{3}\sigma\left(c_p/k\right)^{1/p} \tag{2.18}$$

for all $k>0$. Thus the averages of the ratios $\|\underline{r}_k\|/\|\underline{s}_k\|$ decline at a certain rate, and it follows from (2.17) by the arithmetic-geometric mean inequality that

$$\frac{\rho_k}{\rho_0} = \prod_{j=0}^{k-1} q_j \leq \left(c_p/k\right)^{k/p} \tag{2.19}$$

Provided that c_p does not depend on the discretization width, we have now obtained a uniform bound on ρ_k. Since it implies R-superlinear convergence, i.e. $\rho_k^{1/k} \to 0$, we will refer to the property (2.19) as "special R-superlinear convergence". Asymptotically we would like p to be as small as possible but apparently it cannot be less than 2.

When $\underline{f}(\underline{x})$ is linear, i.e. $\omega=0$, the initial condition $\omega\rho_0 \leq \sigma/20$ holds for arbitrary ρ_0 so the convergence result becomes global with respect to $\underline{x}$. Moreover, the condition $\|D_0\| \leq \sigma/20$ can also be dropped, provided some provision is made for the unlikely event that some $L+B_k$ may fail to have a bounded inverse. In the case of integral equations of the second kind we have $L=I$, and $\underline{f}(\underline{x})$ as well as all B_k are compact. Then any singular $L+B_k$ must have a proper null vector which we may use as the next step $\underline{s}_k$. Thus even without a line search one obtains global and superlinear convergence for the extension of Broyden's method to Hilbert spaces. In the nonlinear case $\underline{x}_0$ must be restricted to a level set on which $\underline{F}(\underline{x})$ is invertible but the condition on $\|D_0\|$ can be removed by a suitable line search (Griewank, 1986).

3. THE RANK ONE BROYDEN METHOD IN HILBERT SPACE

Suppose that the domain X and range Y are both real separable Hilbert spaces, and that the Frechet derivative $\underline{f}'(\underline{x})$ has no particular structure, so $C=\mathcal{L}[X,Y]$. Then it follows from Proposition 3.1 in Griewank (1985c) that the nondeterioration condition (2.14) uniquely characterizes the celebrated Broyden formula

$$B_{k+1} = B_k + \underline{r}_k \underline{s}_k^T / \underline{s}_k^T \underline{s}_k \, , \tag{3.1}$$

$\underline{r}_k$ being defined in (2.16). Here $\underline{s}^T$ denotes the linear functional associated with any $\underline{s} \in X$ by the Riesz representation theorem.

In the finite dimensional setting the Broyden formula is usually derived as the least change update with respect to the so-called Frobenius matrix norm. This inner product norm also plays a central role in the classical proof of superlinear convergence (Broyden, Dennis and Moré, 1973). However, that analysis cannot be generalized to proper Hilbert spaces because the Frobenius norm, being the ℓ^2 norm of the singular value sequence, is infinite for most operators, e.g. the identity.

To illustrate the behaviour of Broyden's method in the infinite dimensional case, let us consider the linear example

$$\underline{F}(\underline{x}) \equiv (I + B_*)\, \underline{x} = \underline{0} \, . \tag{3.2}$$

Here $X = Y = l^2$ and the singly infinite matrix B_* vanishes except for its subdiagonal elements, $\{\lambda_i : i = 0, 1, 2, \cdots\}$ say, where

$$1 > \lambda_0 \geqslant \lambda_1 \geqslant \cdots \geqslant \lambda_j \geqslant \cdots \geqslant \lambda_* = \lim \lambda_j \geqslant 0 \, . \tag{3.3}$$

With $L = I$, $\underline{f}'(\underline{x}) = B_*$ and $\underline{b} = \underline{0}$ this problem fits into the framework described in Section 2, as $\|B_*\| = \lambda_0 < 1$ and consequently $\|(I + B_*)^{-1}\| \leqslant 1/(1 - \lambda_0) < \infty$. Starting with $B_0 = 0$ and $\underline{x}_0$ the first column of $(I + B_*)^{-1}$ one finds (Griewank, 1985c) that for all k

$$\|\underline{F}(\underline{x}_{k+1})\| = \|\underline{r}_k\| = \lambda_k \|\underline{s}_k\| = \lambda_k \|\underline{F}(\underline{x}_k)\| \, . \tag{3.4}$$

Hence we see that the rate of convergence is directly determined by the singular values of the initial discrepancy $D_0 = -B_*$. In particular we have Q-superlinear convergence if and only if D_0 is compact, i.e. $\lambda_* = 0$. Moreover since $\|\underline{r}_k\| / \|\underline{s}_k\| = \lambda_k$ exactly, one obtains special R-superlinear convergence as defined in (2.19) provided the singular values λ_k are p-summable for some $p \geqslant 2$.

It is shown by Griewank (1985c) that these last observations remain true for general Hilbert space operators if $\underline{f}(\underline{x})$ is nonlinear. The crucial constant c_p occurring in (2.19) may be chosen as

$$c_p = \frac{1}{40} + \left(\frac{3}{2\sigma}\right)^p \sum_{k=0}^{\infty} \lambda_k^p \tag{3.5}$$

where the λ_k's are still the singular values of $D_0 = B_0 - \underline{f}'(\underline{x}_*)$.

Typically discretization reduces the λ_k's and sets all but a finite number of them to zero. Hence for all $p \geqslant 2$ the constant c_p is finite on any grid but it would grow unbounded through mesh refinement if the singular values of the underlying operator were not p-summable. In the worst case, where D_0 is not even compact, superlinear convergence occurs only as an artefact of the discretization. However, Broyden's method still succeeds in asymptotically eliminating all $\lambda_k > \lambda_*$ as it is shown by Griewank (1985c) that

$$\overline{\lim_k}\, q_k \leqslant \overline{\lim_k}\, [1.5\, \|\underline{r}_k\| \,/\, (\sigma \|\underline{s}_k\|)] \leqslant 1.5\, \lambda_*/\sigma\,.$$

In the case of the linear example (3.2) this relation follows from (2.17) and (3.4).

A class of problems where the exponent p is known explicitly is the weakly singular integral equation of the form

$$x + fx = b \tag{3.6}$$

where

$$x \in X \equiv \mathcal{L}^2[0,1]\,, \quad b \in Y \equiv X\,, \tag{3.7}$$

and for some $\alpha \in (0,1)$

$$(fx)(t) \equiv \alpha \int_0^1 \frac{\phi(t,\tau)\, x(\tau)}{|t-\tau|^{1-\alpha}}\, d\tau\,, \quad 0 \leqslant t \leqslant 1\,, \tag{3.8}$$

with

$$\phi(t,\tau) \in C^2([0,1]\times[0,1]).$$

For fixed ϕ and arbitrary α the linear operator f is compact and its norm is $O(\alpha^0)$ due to the factor α in (3.8). Moreover, the singular values of $f'(x_*)$ are p-summable when p satisfies $p > 1/\alpha$ and $p \geq 2$ (Johnson *et al*, 1979). Hence, if we begin with the choice $B_0 = 0$, the initial discrepancy $D_0 = -f'(x_*)$ also has p-summable singular values.

This choice of B_0 is computationally advantageous, because the resulting B_k are of rank k, so each linear system in $I+B_k$ can be solved cheaply. In contrast, a discretization of the full operator $I+f$ may yield a nonsymmetric dense linear system whose direct solution would be quite costly. Applying Broyden's method can be regarded as an iterative technique for solving this dense system. Since the R-superlinear convergence estimate (2.19) holds, we expect the number of steps, required by Broyden's method to reduce the error ρ_k below a certain tolerance, to be nearly constant for all sufficiently fine discretizations. At least with uniform spacing on $[0,1]$ one must choose a large number of grid points, since the solution $x_* = (I+f)^{-1}b$ is typically nonsmooth at the boundary even when b is analytic (Chandler, 1979; Brunner, 1987).

Expecting the number n of variables to be considerably larger than the number of iterations, we employ the following implementation of Broyden's method. The approximate Jacobians are stored in the factorized form $B_k = R_k S_k^T$, where the two $n\times k$ matrices R_k and S_k are given by

$$R_k \equiv [\underline{r}_j/\|\underline{s}_j\|]_{j=0}^{k-1} \quad \text{and} \quad S_k \equiv [\underline{s}_j/\|\underline{s}_j\|]_{j=0}^{k-1}.$$

Then the Sherman-Morrison-Woodbury formula implies

$$(I+B_k)^{-1} = I - R_k E_k^{-1} S_k^T \tag{3.9}$$

where

$$E_k \equiv I + S_k^T R_k \in \mathbb{R}^{k\times k}. \tag{3.10}$$

Hence a linear system of the form $(I+B_k)\underline{s} = -\underline{F}(\underline{x}_k)$ can be solved in $2nk+k^2$ arithmetic operations provided an LU factorization of E_k is available. Except when some B_k and consequently E_k have rank less than k, the LU factorizations of the E_k can be computed recursively in $\frac{1}{3}k^3+O(k^2)$ operations. Thus the total cost of computing the first k Broyden iterates is essentially $k^2n+\frac{2}{3}k^3$ arithmetic operations compared with $\frac{3}{2}k^2n+O(k^3)$ for the generalized conjugate residual method (Eisenstat *et al*, 1983) which requires the same amount of storage. Here we have not included the cost of applying the kernel f to the first k iterates which amounts to some kn^2 arithmetic operations. This effort completely dominates the work for computing the Broyden iterates, when k is much smaller than n as we expect.

The efficacy of our procedure was verified on the integral equation (3.6) with $\phi(t,\tau)=t+\tau^2-1$ and $b(t)=\cos(t\pi/4)$. The operator was discretized by the so-called product midpoint rule (Brunner, 1987) on a uniform grid of 3, 9, 27, 81, 243 and 729 intervals. All calculations were started from $\underline{x}_0=\underline{0}$ where the residual norm is of magnitude one. Figure 1 displays the number of iterations required to reduce the residual $\|\underline{F}(\underline{x}_k)-\underline{b}\|$ below 10^{-15} for $\alpha=.5$, .4, .3, .2 and .1, except that in the last case the algorithm had to be stopped at the hundredth iteration. In all other cases the number of steps became constant once the discretization was sufficiently fine. In the second worst case $\alpha=.2$ the method took only 52 steps even when there were 729 variables.

The ability to solve such a large dense nonsymmetric linear system is important, because for a uniform mesh the discretization error declines slower than $1/n$ even for higher order schemes like the product Simpson rule (Chandler, 1980). While graded meshes yield more accuracy with fewer grid points (Brunner, 1987), uniform meshes have the advantage that f can be stored more or less like a Toeplitz matrix with $\phi(t,\tau)$ being re-evaluated at n^2 arguments for each matrix vector product.

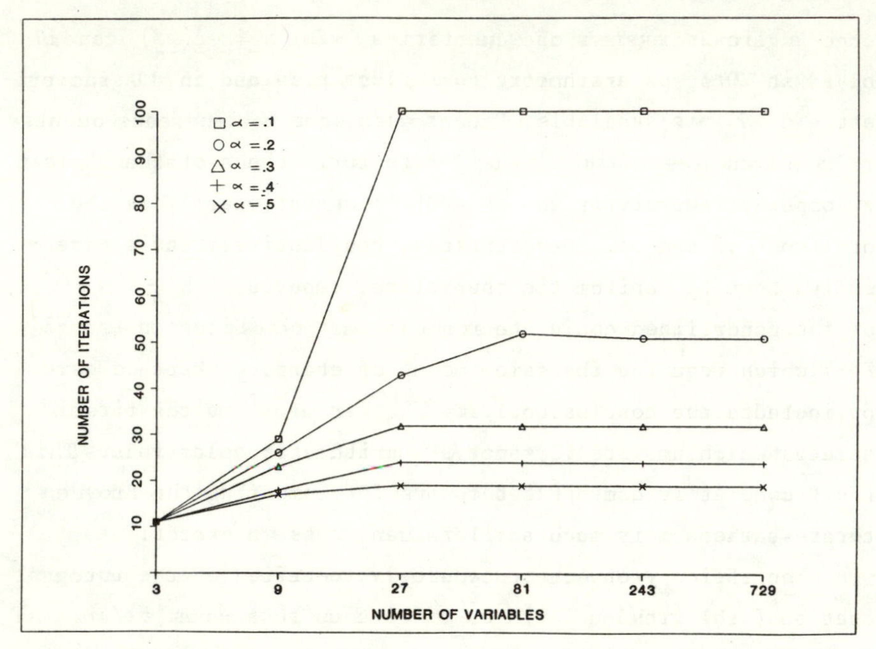

FIG. 1 Number of iterations for the integral equation (3.6)

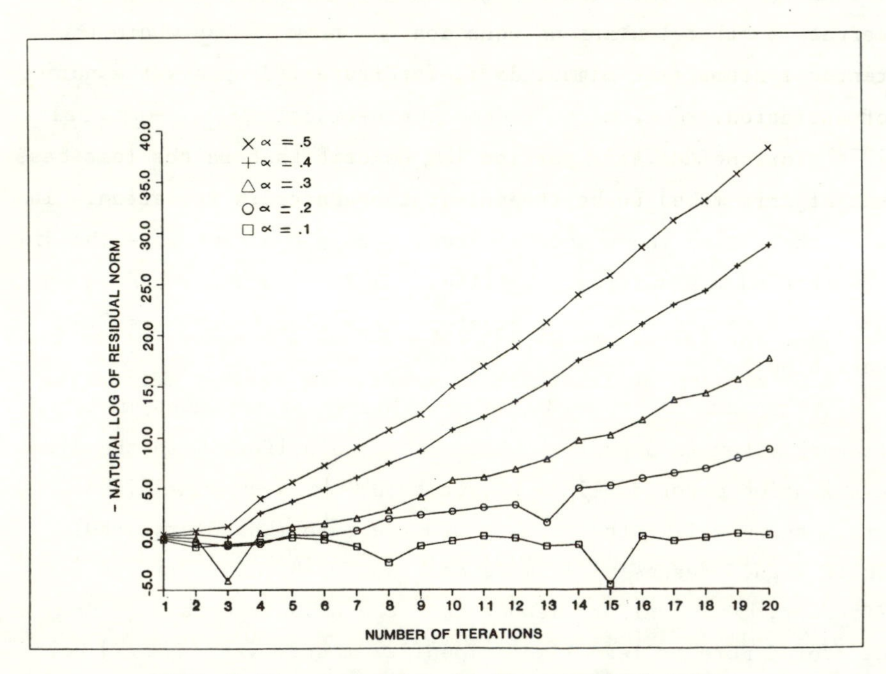

FIG. 2 The reduction in the residual norm

Figure 2 displays the quantities $-\ln(\|\underline{F}(\underline{x}_k)-\underline{b}\|)$ over the first 20 steps again for $\alpha=.1, .2, .3, .4$ and $.5$. Assuming that $\|\underline{F}_k-\underline{b}\|\approx\rho_k$ and $c_p\approx 1$ for $p=1/\alpha$, one would expect because of (2.19) that $-\ln(\|\underline{F}(\underline{x}_k)-\underline{b}\|)\approx\alpha k \ln k$. The actual numbers are somewhat larger for $\alpha=.5$ and $\alpha=.4$ but clearly smaller for $\alpha=.3, .2$ and $.1$. Nevertheless, qualitatively the numerical results seem to confirm the theoretical results.

Considering a more general linear system of the form $(L+f)\,x=b$ when the inversion of L is cheap, we reach the following tentative conclusion. If $L^{-1}f$ is nonsymmetric but has rapidly declining (ideally square summable) singular values, then Broyden's method applied to $(I+L^{-1}f)\,x=L^{-1}b$ is likely to achieve satisfactory accuracy in a reasonable number of steps. Even when the λ_k converge only slowly to zero, it may be conjectured that the Broyden iterates x_k are up to a constant optimal in the Krylov spaces spanned by the vectors $(L^{-1}f)^j L^{-1}b$ for $0\leqslant j<k$. This question as well as the relation of Broyden's method to other iterative equation solvers for nonsymmetric linear systems (see Elman, 1982, for instance) has yet to be investigated.

To conclude this section let us briefly consider equations of the form (3.6) where the integral operator is nonlinear, i.e.

$$f(x)\,(t) = \int_0^1 \phi(t,\tau,x(\tau))\,d\tau, \qquad 0\leqslant t\leqslant 1.$$

Provided the kernel $\phi\colon [0,1]^2\times\mathbb{R}\to\mathbb{R}$ has a bounded derivative ϕ_x with respect to its third argument, the Fréchet derivative f' takes the values

$$(f'(x)v)\,(t) = \int_0^1 \phi_x(t,\tau,x(\tau))\,v(\tau)\,d\tau, \qquad v\in X. \tag{3.11}$$

Hence the only algebraic derivative required for the implementation of Newton's method is the scalar function ϕ_x. Should it be difficult or laborious to calculate, ϕ_x could be replaced by the secant approximation

$$B_k(t,\tau) \equiv \frac{\phi(t,\tau,x_k(\tau)) - \phi(t,\tau,x_{k-1}(\tau))}{x_k(\tau) - x_{k-1}(\tau)} . \qquad (3.12)$$

Just as in the univariate secant method, the resulting quasi-Newton step s_k depends only on the current and the previous iterate x_k and x_{k-1}, so one would expect the same order of convergence, namely $(1+\sqrt{5})/2 \approx 1.62$. However the computation of $s_k = x_{k+1} - x_k$ by solving (2.4) would be as expensive as for Newton's method. Therefore the rank one update (3.1) with $r_k = y_k - B_k s_k$ and

$$y_k = \left\{ \int_0^1 \phi(t,\tau,x_{k+1}(\tau)) - \phi(t,\tau,x_k(\tau))\,d\tau\,,\quad 0 \leq t \leq 1 \right\} \in Y$$

may still be preferable. This method was used by Kelley and Sachs (1985a) to solve Hammerstein integral equations. Again superlinear convergence occurs if and only if $D_0 = -f'(x_*)$ is compact.

This choice between the low rank update (3.1) and the high rank update (3.12) seems typical for most large scale problems. The first option economizes on linear algebra but ignores a lot of structure. The second incorporates this information but, being of high rank, it requires as many matrix operations per step as Newton's method. Naturally we would expect much better derivative approximations and hence fewer iterations from the second choice.

4. POINTWISE BROYDEN UPDATE FOR SUBSTITUTION OPERATORS

The two point boundary value problem (1.2) shows that in some calculations one may face the task of approximating the derivative of a substitution operator

$$\underline{f}(\underline{x})(t) = \underline{\phi}(t,\underline{x}(t))\,, \quad 0 \leq t \leq 1\,, \qquad (4.1)$$

with the domain and range

$$X \equiv C([0,1]:\mathbb{R}^m) \quad \text{and} \quad Y_0 \equiv \mathcal{L}^2([0,1]:\mathbb{R}^m)\,. \qquad (4.2)$$

Provided $\underline{\phi}:[0,1]\times \mathbb{R}^m \to \mathbb{R}^m$ has a jointly Lipschitz continuous Fréchet derivative $\underline{\phi}_{\underline{x}} = \nabla_{\underline{x}}\underline{\phi}$ with respect to $\underline{x}$, the Fréchet derivative $\underline{f}'(\underline{x})$ applied to an element $\underline{v}\in X$ is given by

$$(\underline{f}'(\underline{x})\underline{v})(t) = \underline{\phi}_{\underline{x}}(t,\underline{x}(t))\,\underline{v}(t), \quad 0\leq t\leq 1. \tag{4.3}$$

Hence we may identify $\underline{f}'(\underline{x})$ with the matrix valued function $\{\underline{\phi}_{\underline{x}}(t,\underline{x}(t)) : 0\leq t\leq 1\}$, which is an element of

$$C = C([0,1]: \mathbb{R}^{m\times m}). \tag{4.4}$$

Naturally one would restrict the approximating Jacobians B_k to the same subspace $C\subset \mathcal{L}[X,Y]$, on which the induced operator norm is given by

$$\|C\| = \sup_{\underline{x}\in X} \frac{\|C\underline{x}\|_2}{\|\underline{x}\|_\infty} = \left[\int_0^1 \|C(t)\|_2^2\,dt\right]^{\frac{1}{2}}, \quad C\in C, \tag{4.5}$$

where $\|C(t)\|_2$ is the Euclidean norm of $C(t)\in \mathbb{R}^{m\times m}$.

After a step from $\underline{x}_k(\cdot)$ to $\underline{x}_{k+1}(\cdot) = \underline{x}_k(\cdot)+\underline{s}_k(\cdot)$, the secant condition (2.10) imposes the "pointwise" identity

$$B_{k+1}(t)\,\underline{s}_k(t) = \underline{y}_k(t) \equiv \underline{\phi}(t,\underline{x}_{k+1}(t)) - \underline{\phi}(t,\underline{x}_k(t)) \tag{4.6}$$

for all $t\in[0,1]$. It can be shown that the only update satisfying this relation and the nondeterioration condition (2.14) is given by the pointwise Broyden formula

$$B_{k+1}(t) = B_k(t) + \underline{r}_k(t)\,\underline{s}_k(t)^T/\underline{s}_k(t)^T\,\underline{s}_k(t), \quad 0\leq t\leq 1, \tag{4.7}$$

where $\underline{r}_k(t) = \underline{y}_k(t) - B_k(t)\,\underline{s}_k(t)$. We see that, due to the domain of t, this update is no longer of low rank, so the calculation of the next step $\underline{s}_{k+1}(t)$ is as expensive as in the case of Newton's method.

Strictly speaking, the update (4.7) is not defined when the vector function $\underline{s}_k(t)$ vanishes for any particular value $\hat{t}$ of t. Theoretically this provides no difficulty, because with the convention $B_{k+1}(\hat{t}) \equiv B_k(\hat{t})$ at all such $\hat{t}$, the new matrix valued function $B_{k+1}(t)$ is still bounded, though possibly discontinuous. However, one must expect numerical instabilities whenever the denominator $\|\underline{s}_k(t)\|_2^2$ is small. To avoid such difficulties one can replace $\{\|\underline{s}_k(t)\|_2^2 : 0 \leq t \leq 1\}$ by a positive continuous function $\{\mu_k(t) : 0 \leq t \leq 1\}$ such that

$$\|\underline{s}_k(t)\|_2^2 \leq \mu_k(t) \leq \|\underline{s}_k\|_\infty^2 = \max_{0 \leq t \leq 1} \|\underline{s}_k(t)\|_2^2 . \tag{4.8}$$

The resulting new approximation $B_{k+1}(t)$ still has the nondeterioration property (2.14), but it satisfies the secant condition (4.6) only where $\mu_k(t) = \|\underline{s}_k(t)\|_2^2$. At all other values of t the correction is damped, which certainly makes sense where $\|\underline{s}_k(t)\|_2$ is very small. In test calculations the maximal damping $\mu_k(t) \equiv \|\underline{s}_k\|_\infty^2$ was found to slow the convergence considerably even though superlinear convergence occurs eventually. A suitable compromise damping has yet to be developed. The following results apply for any $\mu_k(t)$ within the bracket (4.8).

With $Y \equiv Y_0 \times \mathbb{R}^m$ and

$$L\underline{x} \equiv \left[\frac{d}{dt}\underline{x},\ A_0\underline{x}(0) + A_1\underline{x}(1)\right] \in Y , \quad \underline{x} \in X , \tag{4.9}$$

the boundary value problem (1.2),(1.3) fits into the framework developed in Section 2. The regularity assumptions (2.2) and (2.3) at a particular solution function $\underline{x}_*(t) \in X$ are quite usual (Keller, 1976). Again one could relax the continuity conditions on $\underline{f}'$ or equivalently $\underline{\phi}_{\underline{x}}$ considerably. The quasi-Newton iteration (2.5) requires the solution of the linear differential equations

$$\dot{\underline{x}}_{k+1}(t) + B_k(t)\,\underline{x}_{k+1}(t) = \underline{b}(t) - \underline{\phi}(t, \underline{x}_k(t)) + B_k(t)\,\underline{x}_k(t) \tag{4.10}$$

subject to $A_0\underline{x}(0) + A_1\underline{x}(1) = \underline{0}$. According to Proposition 2, we obtain local and Q-superlinear convergence in the supremum norm $\|\underline{x}\|_\infty = \max\{\|\underline{x}(t)\|_2 : 0 \leq t \leq 1\}$. While this is perhaps not very surprising, it is shown by Griewank (1985b) that it follows that

$$\sum_{k=0}^{\infty} \|\underline{r}_k\|_2^2 / \|\underline{s}_k\|_\infty^2 \leq 4.5\,\rho_0^2\,\omega^2 + 2\|D_0\|^2 . \tag{4.11}$$

This desirable property is crucially dependent on the choice of the infinity and two norms on X and Y_0 respectively. By Hölder's inequality (4.11) and (2.15) imply (2.18) with $c_p = .01$ and $p = 2$. Hence the initial conditions

$$\omega \|\underline{x}_0(t) - \underline{x}_*(t)\| \leq \sigma/20 \quad \text{for all} \quad t \in [0\,,\,1] \tag{4.12}$$

and

$$\int_0^1 \|B_0(t) - \underline{\phi}_{\underline{x}}(t, \underline{x}_*(t))\|_2^2 \, dt \leq (\sigma/20)^2 \tag{4.13}$$

ensure that the iteration (4.10) converges with

$$\|\underline{x}_k(t) - \underline{x}_*(t)\| \leq (1/\sqrt{3k})^k \rho_0\,, \quad t \in [0\,,1]. \tag{4.14}$$

Since (4.11) implies the Dennis and Moré condition

$$\|\underline{r}_k\|_2 / \|\underline{s}_k\|_\infty \to 0\,,$$

we have also Q-superlinear convergence.

The remarkable property of the special R-superlinear convergence estimate (4.14) is again its mesh independence. The Lipschitz constant ω and the bound $1/\sigma$ on the norm of $(L + \underline{f}'(\underline{x}_*))^{-1}$ will vary only slightly under consistent and stable discretizations. Similarly the initial conditions in the corresponding discrete norms will not become more stringent as the mesh size gets smaller.

Test calculations in Hart and Soul (1973), Kelly and Sachs (1985b), and Griewank (1985b) show that on two-point boundary value problems, the pointwise Broyden update yields rapid convergence, rarely taking more than twice the number of steps required by Newton's method. None of the test problems had a very complicated right hand side and so far no general purpose code is available.

Since the unit interval $[0, 1]$ can be replaced by any compact domain Ω in Euclidean space, pointwise Broyden updates are also applicable to systems of partial differential equations. However, for proper superlinear convergence to occur, the leading derivative term must apparently be truly linear. On quasi-linear PDE's or the worse case of algebraic differential equations one can only expect "weak" superlinear convergence in that $\|\underline{r}_{k+1}\|_{\bar{q}} / \|\underline{r}_k\|_q \to 0$ for $\bar{q} > q \geq 1$. The reason for this apparent deficiency is that the pointwise updates guarantee only $\|\underline{r}_k\|_2 / \|\underline{s}_k\|_\infty \to 0$ but not $\|\underline{r}_k\|_q / \|\underline{s}_k\|_q \to 0$ for some common $q \geq 1$. Therefore we need a smoothing operator L^{-1} to map $\mathcal{L}^2(\Omega)$ back into $C(\Omega)$. However these questions are still under investigation, and the results of numerical experiments are far from conclusive. The situation is similar in the case of pointwise variable metric updates for elliptic problems. Griewank and Toint (1982b) establish local and superlinear convergence for particular discretizations, but so far no mesh independent result has been published.

5. SUMMARY AND DISCUSSION

There is theoretical and numerical evidence that quasi-Newton methods based on secant updating techniques can efficiently solve discretizations of certain differential and integral equations. Typically one has to choose between low rank updates that economize on linear algebra or pointwise updates that economize on storage and promise better derivative approximations. Superlinear convergence occurs whenever the approximated operator is a compact perturbation of a leading linear term. If the singular values of the perturbation are p-summable for some $p \geq 2$, then Broyden's method and similar variable metric schemes for unconstrained optimization achieve a special mesh independent rate of superlinear convergence. The properties of secant methods for quasi-linear differential equations and differential-algebraic systems are still under investigation. So far the Harwell routine VE08 written by Toint appears to be the only production code based on pointwise updating techniques. As a next step the development of a derivative-free two-point boundary value code would seem to be a worthwhile and realistic project.

REFERENCES

Allgower, E.L., Böhmer, K., Potra, F.A. and Rheinboldt, W.C. (1986), "A mesh-independence principle for operator equations and their discretizations", *SIAM J. Numer. Anal.*, *23*, pp.160-169.

de Boor, C., de Hoog, F. and Keller, H.B. (1983), "The stability of one-step schemes for first-order two-point boundary value problems", *SIAM J. Numer. Anal.*, *20*, pp.1139-1146.

Brent, R.P. (1973), "Some efficient algorithms for solving systems of nonlinear equations", *SIAM J. Numer. Anal.*, *10*, pp.327-344.

Brezzi, F., Rappaz, J. and Raviart, P.A. (1980, 1981a,b), "Finite dimensional approximation of nonlinear problems, Part I. Branches of nonsingular solutions", *Numer. Math.*, *36*, pp.1-25; "Part II. Limit points", *Numer. Math.*, *37*, pp.1-28; "Part III. Simple bifurcation points", *Numer. Math.*, *38*, pp.1-30.

Brown, P.N., Hindmarsh, A.C. and Walker, H.F. (1985), "Experiments with quasi-Newton methods in solving stiff ODE systems", *SIAM J. Sci. Statist. Comput.*, *6*, pp.297-313.

Broyden, C.G. (1965), "A class of methods for solving nonlinear simultaneous equations", *Math. Comp.*, *19*, pp.577-593.

Broyden, C.G., Dennis, J.E. and Moré, J.J. (1973), "On the local and superlinear convergence of quasi-Newton methods", *J. Inst. Math. Appl.*, *12*, pp.223-245.

Brunner, H. (1987), "Collocation methods for one-dimensional Fredholm and Volterra integral equations", in *The State of the Art in Numerical Analysis* (this volume), pp.563-600.

Buckley, A. and LeNir, A. (1983), "QN-like variable storage conjugate gradients", *Math. Programming*, *27*, pp.155-175.

Chandler, G.A. (1979), *Superconvergence of Numerical Methods for Second Kind Integral Equations*, Ph.D. Thesis, Australian National University, Canberra.

Chandler, G.A. (1980), "Superconvergence for second kind integral equations", in *The Application and Numerical Solution of Integral Equations*, eds. Anderssen, R.S., de Hoog, F.R. and Lukas, M.A., Sijthoff and Noordhoff (Alphen aan den Rijn), pp.103-117.

Daniel, J.W. (1967), "The conjugate gradient method for linear and nonlinear operator equations", *SIAM J. Numer. Anal.*, *4*, pp.10-26.

Dennis, J.E., Jr. (1971), "Toward a unified convergence theory for Newton-like methods", in *Nonlinear Functional Analysis and Applications*, ed. Rall, L.B., Academic Press (New York), pp.425-472.

Dennis, J.E. and Moré, J.J. (1974), "A characterization of superlinear convergence and its application to quasi-Newton methods", *Math. Comp.*, *28*, pp.549-560.

Dennis, J.E. and Moré, J.J. (1977), "Quasi-Newton methods, motivation and theory", *SIAM Rev.*, *19*, pp.46-89.

Dennis, J.E. and Schnabel, R.B. (1983), *Numerical Methods for Unconstrained Optimization and Nonlinear Equations*, Prentice-Hall (Englewood Cliffs, N.J.).

Deuflhard, P. (1978), "Nonlinear equation solvers in boundary value problem codes", in *Codes for Boundary Value Problems in Ordinary Differential Equations, Lecture Notes in Computer Science*, *76*, eds. Childs, B., Scott, M., Daniel, J.W., Denman, E. and Nelson, P., Springer-Verlag (Berlin), pp.40-66.

Dixon, L.C.W. (1972), "Variable metric algorithms: necessary and sufficient conditions for identical behaviour on nonquadratic functions", *J. Optim. Theory Appl.*, *10*, pp.34-40.

Eisenstat, S.C., Elman, H.C. and Schultz, M.H. (1983), "Variational iterative methods for nonsymmetric systems of linear equations", *SIAM J. Numer. Anal.*, *20*, pp.345-357.

Elman, H.C. (1982), *Iterative Methods for Large, Sparse Nonsymmetric Systems of Linear Equations*, Ph.D. Thesis, Department of Computer Science, Yale University, New Haven.

Griewank, A. (1985a), "On solving nonlinear equations with simple singularities or nearly singular solutions", *SIAM Rev.*, *27*, pp.537-563.

Griewank, A. (1985b), "The solution of boundary value problems by Broyden-based secant methods", Technical Report CMA-R22-85, Centre for Mathematical Analysis, Australian National University (to appear in *Proceedings of the Computational Theory and Applications Conference*, *University of Melbourne*, eds. Noye, J. and May, R., North-Holland).

Griewank, A. (1985c), "The local convergence of Broyden-like methods on Lipschitzian problems in Hilbert spaces", Technical Report CMA-R45-85, Centre for Mathematical Analysis, Australian National University (to appear in *SIAM J. Numer. Anal.*).

Griewank, A. (1986), "The 'global' convergence of Broyden-like methods with a suitable line search", *J. Austral. Math. Soc. Ser. B*, *28*, pp.75-92.

Griewank, A. and Toint, Ph.L. (1982a), "Partitioned variable metric updates for large structured optimization problems", *Numer. Math.*, *39*, pp.119-137.

Griewank, A. and Toint, Ph.L. (1982), "Local convergence analysis for partitioned quasi-Newton updates", *Numer. Math.*, *39*, pp.429-448.

Hart, W.E. and Soul, S.O.W. (1973), "Quasi-Newton methods for discretized nonlinear boundary problems", *J. Inst. Math. Appl.* *11*, pp.351-359.

Hayes, R.M. (1954), "Iterative methods of solving linear problems on Hilbert space", in *Contributions to the Solution of Linear Systems and the Determination of Eigenvalues*, *NBS Appl. Math. Series*, *39*, ed. Taussky, O., pp.71-103.

Horwitz, L.B. and Sarachik, P.E. (1968), "Davidon's method in Hilbert space", *SIAM J. Appl. Math.*, *16*, pp.676-695.

Johnson, W.B., König, H., Maurey, B. and Retherford, J.R. (1979), "Eigenvalues of p-summing and ℓ_p-type operators in Banach spaces", *J. Funct. Anal.*, *32*, pp.353-380.

Johnson, O.G., Micchelli, C.A. and Paul, G. (1983), "Polynomial preconditioners for conjugate gradient calculations", *SIAM J. Numer. Anal.*, *20*, pp.362-376.

Keller, H.B. (1976), *Numerical Solution of Two Point Boundary Value Problems: Regional Conference Series in Applied Mathematics*, *24*, SIAM Publications (Philadelphia).

Kelley, C.T. and Sachs, E.W. (1985a), "Broyden's method for approximate solution of nonlinear integral equations", *J. Integral Equations*, *9*, pp.25-43.

Kelley, C.T. and Sachs, E.W. (1985b), "A quasi-Newton method for elliptic boundary value problems" (to appear in *SIAM J. Numer. Anal.*)

Kelley, C.T. and Sachs, E.W. (1986), "A pointwise quasi-Newton method for unconstrained optimal control problems" (submitted to *SIAM J. Control Optim.*).

Moré, J.J. and Trangenstein, J.A. (1976), "On the global convergence of Broyden's method", *Math. Comp.*, *30*, pp.523-540.

Ortega. J.M. and Rheinboldt, W.C. (1970), *Iterative Solution of Nonlinear Equations in Several Variables*, Academic Press (New York).

Powell, M.J.D. (1976), "Some global convergence properties of a variable metric algorithm for minimization without exact line searches", in *Nonlinear Programming*, *SIAM-AMS Proceedings*, *Vol. IX*, eds. Cottle, R.W. and Lemke, C.E., SIAM Publications (Philadelphia), pp.53-72.

Powell, M.J.D. (1984), "On the rate of convergence of variable metric algorithms for unconstrained optimization", in *Proceedings of the International Congress of Mathematicians*, *Warsaw 1983*, *Vol. 2*, eds. Ciesielski, Z. and Olech, C., Polish Scientific Publishers (Warsaw), pp.1525-1539.

Rheinboldt, W.C. and Riks, E. (1984), "Solution techniques for nonlinear finite element equations", in *State-of-the-Art Surveys on Finite Element Technology*, eds. Noor, A.K. and Pilkey, W.D., ASME Publications (New York), pp.183-223.

Schubert, L.K. (1970), "Modification of a quasi-Newton method for nonlinear equations with a sparse Jacobian", *Math. Comp.*, *24*, pp.27-30.

Stoer, J. (1979), "Two examples on the convergence of certain rank-2 minimization methods for quadratic functionals in Hilbert space", *Linear Algebra Appl.*, *28*, pp.217-222.

Strang, G. (1980), "The quasi-Newton method in finite element calculations", in *Computational Methods in Nonlinear Mechanics*, ed. Oden, J.T., North-Holland (Amsterdam), pp.451-456.

Werner, J. (1974), "Über die Konvergenz des Davidon-Fletcher-Powell-Verfahrens für streng konvexe Minimierungsaufgaben im Hilbertraum", *Computing*, *12*, pp.167-176.

Winther, R. (1977), *A Numerical Galerkin Method for a Parabolic Problem*, Ph.D. Dissertation, Cornell University, Ithaca.

Winther, R. (1980), "Some superlinear convergence results for the conjugate gradient method", *SIAM J. Numer. Anal.*, *17*, pp.14-17.

Andreas Griewank
Mathematics Department
Southern Methodist University
Dallas
Texas 75275
U.S.A.

12 Methods for Nonlinear Constraints in Optimization Calculations

M. J. D. POWELL

ABSTRACT

We review the main techniques that are used for nonlinear equality constraints in sequential quadratic programming optimization algorithms, when the objective and constraint functions are smooth and their first derivatives can be calculated. Many of them are closely related to the Newton iteration for satisfying the first order conditions for optimality, so the Lagrangian function and linear approximations to the constraints occur frequently. There is also the important problem of forcing convergence efficiently from poor initial estimates of solutions, and here *line searches*, *trust regions*, L_1 *penalty functions* and *differential penalty functions* are employed. Moreover, to achieve fast convergence, suitable approximations to second derivatives of the Lagrangian function are required, which depend on estimates of Lagrange multipliers. Recent research on these questions is described and discussed.

1. INTRODUCTION

The number of publications on algorithms for constrained optimization calculations is growing rapidly. Therefore this paper is intended to provide a path for the nonspecialist through this work to some recent research. We aim for new techniques and ideas that seem to be highly useful for satisfying nonlinear constraints. The path is so narrow that it does not include inequality constraints, because they are not needed in order to describe the given recent work. Therefore the reader who requires a balanced view of the subject should also look elsewhere; in particular the books by Bertsekas (1982a), Fletcher (1981), and Gill, Murray and Wright (1981), and the proceedings of a NATO workshop (Powell, 1982) are recommended. Some of the techniques that we consider have not yet been extended to inequality constraints.

We study general algorithms that adjust the components of the vector $\underline{x} \in \mathbb{R}^n$ to minimize a given objective function $\{F(\underline{x}): \underline{x} \in \mathbb{R}^n\}$, subject to the constraints

$$c_i(\underline{x}) = 0, \qquad i = 1,2,\cdots,m. \tag{1.1}$$

All functions are real, and, for every $\underline{x}$, we assume that their values and first derivatives can be calculated, and that the constraint gradients $\{\underline{\nabla} c_i(\underline{x});\ i = 1,2,\cdots,m\}$ are linearly independent, so we have $m \leqslant n$.

All the algorithms that are mentioned are iterative, and each iteration tries to improve an estimate of the solution, $\underline{x}_k$ say, where k is the iteration number. A trial step $\underline{d}_k$ is derived from a model of the main calculation, which usually depends on an $n \times n$ symmetric matrix B_k, that provides suitable second derivative information and that in some methods is forced to be positive definite. For example, in the unconstrained case, $\underline{d}_k$ may be obtained by minimizing the quadratic approximation

$$Q_k(\underline{d}) = F(\underline{x}_k) + \underline{d}^T \underline{\nabla} F(\underline{x}_k) + \tfrac{1}{2} \underline{d}^T B_k \underline{d}, \quad \underline{d} \in \mathbb{R}^n, \tag{1.2}$$

to $F(\underline{x}_k + \underline{d})$. In order to achieve convergence from poor starting approximations, the change to $\underline{x}_k$ depends on a merit function $\{\Phi(\underline{x}): \underline{x} \in \mathbb{R}^n\}$ say, that ideally takes its least value at the required solution. Therefore most algorithms provide $\underline{x}_{k+1} = \underline{x}_k + \underline{d}_k$ as the new vector of variables only if $\Phi(\underline{x}_k + \underline{d}_k) < \Phi(\underline{x}_k)$. In any case $\underline{x}_{k+1}$ has the form

$$\underline{x}_{k+1} = \underline{x}_k + \alpha_k \underline{d}_k , \qquad (1.3)$$

where the step-length α_k is positive in a *line search method*, but may be zero in a *trust region method*. Finally a typical iteration calculates B_{k+1} from B_k and from differences between available first derivatives.

If $\underline{d}_k$ is derived by minimizing an approximation to the merit function, then we call the algorithm a *penalty function method*. Two of these methods are considered in Section 2, including Fletcher's (1981, 1985) very successful procedure for constrained optimization that is based on the L_1 *penalty function*. We note in passing the importance of the Lagrangian function

$$L(\underline{x},\underline{\lambda}) = F(\underline{x}) - \sum_{i=1}^{m} \lambda_i c_i(\underline{x}), \quad \underline{x} \in \mathbb{R}^n, \quad \underline{\lambda} \in \mathbb{R}^m , \qquad (1.4)$$

and in particular that it is suitable to regard B_k as an approximation to a second derivative matrix $\nabla^2_{xx} L(\underline{x},\underline{\lambda})$ of the Lagrangian function.

The calculation of the trial step $\underline{d}_k$ that minimizes the quadratic function (1.2) subject to the constraints

$$c_i(\underline{x}_k) + \underline{d}_k^T \underline{\nabla} c_i(\underline{x}_k) = 0 , \qquad i = 1,2,\cdots,m , \qquad (1.5)$$

is studied in Section 3. This trial step is used by many optimization algorithms. Our main purpose is to identify the contribution from B_k to $\underline{d}_k$ when the conditions (1.5) are imposed. Our conclusions are an extension of the remark that B_k is irrelevant when $m = n$, because in this case, remembering the assumption that constraint gradients are linearly independent, the system (1.5) defines $\underline{d}_k$.

These conclusions are important to Section 4, because there we review recent papers on *reduced second derivative matrices*. The main idea is to replace B_k by an $(n-m) \times (n-m)$ symmetric matrix

$$M_k \approx Z_k^T \nabla_{xx}^2 L(\underline{x}_k, \underline{\lambda}) Z_k, \qquad (1.6)$$

where Z_k is chosen to have $(n-m)$ linearly independent columns that are orthogonal to the constraint gradients $\{\underline{\nabla} c_i(\underline{x}_k);\ i=1,2,\cdots,m\}$. This technique is particularly valuable when $(n-m) \ll n$, and when many of the constraints are linear.

Section 5 considers the merit functions that are used to force convergence from poor starting approximations. We note the *Maratos effect*, which is that some highly suitable changes to the variables can cause both the objective function $F(\underline{x})$ and the constraint violations $\{|c_i(\underline{x})|;\ i=1,2,\cdots,m\}$ to increase. Therefore it can be inefficient to employ a merit function of the form

$$\Phi(\underline{x}) = F(\underline{x}) + W(\underline{c}(\underline{x})), \quad \underline{x} \in \mathbb{R}^n, \qquad (1.7)$$

where $W(\underline{c})$ is non-negative for all $\underline{c} \in \mathbb{R}^m$ and satisfies $W(\underline{0}) = 0$. Further, if we also require W to be continuously differentiable, then $\underline{\nabla}\Phi(\underline{x}) = \underline{\nabla}F(\underline{x})$ whenever $\underline{c}(\underline{x}) = \underline{0}$, so the required solution, $\underline{x}^*$ say, is not a minimum of expression (1.7) in the usual case when $\underline{\nabla}F(\underline{x}^*) \neq \underline{0}$. In Section 5 we recall from Fletcher (1973) and from Di Pillo and Grippo (1979) that these difficulties can be avoided by allowing each value of $\Phi(\underline{x})$ to depend on first derivatives of $F(\underline{x})$ and $\{c_i(\underline{x});\ i=1,2,\cdots,m\}$.

Section 6 concerns the choice of α_k in equation (1.3), which is guided by the merit function $\{\Phi(\underline{x}): \underline{x} \in \mathbb{R}^n\}$. We compare *line search methods* with *trust region methods*. When line searches are employed the merit function must satisfy $\Phi(\underline{x}_k + \alpha \underline{d}_k) < \Phi(\underline{x}_k)$ for sufficiently small positive α, which may require the adjustment of some parameters of Φ; then α_k is chosen to give the reduction $\Phi(\underline{x}_{k+1}) < \Phi(\underline{x}_k)$ in the merit function. In a trust region method the calculation of $\underline{d}_k$ includes a bound of the form $\|\underline{d}_k\| \leq \Delta_k$, and now we require the merit function to provide $\Phi(\underline{x}_k + \underline{d}_k) < \Phi(\underline{x}_k)$ if Δ_k is sufficiently small;

therefore some iterations may set $\underline{x}_{k+1} = \underline{x}_k$ and $\Delta_{k+1} < \Delta_k$. We note that trust regions have some advantages over line searches, but that it is sometimes necessary to relax the conditions (1.5). Of course in both cases an algorithm terminates if $\underline{d}_k = \underline{0}$, which should happen only if $\underline{x}_k$ satisfies the first order conditions

$$\begin{aligned} &\underline{\nabla} F(\underline{x}_k) \in \text{span}\,\{\underline{\nabla} c_i(\underline{x}_k); \quad i = 1,2,\cdots,m\} \\ &c_i(\underline{x}_k) = 0\,, \quad i = 1,2,\cdots,m \end{aligned} \tag{1.8}$$

for a solution of the given constrained optimization calculation.

Finally, Section 7 offers a personal view of the current state of research on general algorithms for nonlinear constraints. It is suggested that too little attention is given to some of the needs of computer users.

2. PENALTY FUNCTIONS AND LAGRANGE MULTIPLIERS

In the late 1960's, when algorithms for unconstrained optimization were in much better shape than methods that included nonlinear constraints explicitly, the least squares penalty function

$$\Phi_2(\underline{x}) = F(\underline{x}) + r \sum_{i=1}^{m} [c_i(\underline{x})]^2\,, \qquad \underline{x} \in \mathbb{R}^n\,, \tag{2.1}$$

was often used to convert the equality constrained optimization problem to a sequence of unconstrained calculations. Specifically, a monotonic increasing and divergent sequence of values of the positive parameter r was chosen, and, for each r, the vector of variables that minimizes expression (2.1), $\underline{x}(r)$ say, was calculated, because, if $\|\underline{x}(r)\|$ remains finite as $r \to \infty$, then all limit points of $\{\underline{x}(r)\}$ are solutions of the constrained problem. Such properties are given by Fiacco and McCormick (1968), but they do not take advantage of the fact that first derivatives of the functions $F(\underline{x})$ and $\{c_i(\underline{x});\ i = 1,2,\cdots,m\}$ are available separately.

This fact is useful when minimizing expression (2.1) for

very large r, because the second derivative matrix

$$\nabla^2 \Phi_2(\underline{x}) = \nabla^2 F(\underline{x}) + 2r \sum_{i=1}^{m} [\underline{\nabla} c_i(\underline{x})\, \underline{\nabla} c_i(\underline{x})^T + c_i(\underline{x})\, \nabla^2 c_i(\underline{x})] \quad (2.2)$$

is usually highly ill-conditioned. However, as pointed out by Broyden and Attia (1984), the matrix

$$\nabla^2 F(\underline{x}) + 2r \sum_{i=1}^{m} c_i(\underline{x})\, \nabla^2 c_i(\underline{x}) \quad (2.3)$$

need not have any large eigenvalues, the reason being that, if $\|\underline{x}(r)\|$ is bounded, then the factors $\{2rc_i(\underline{x}(r))\,;\ i=1,2,\cdots,m\}$ are also bounded, because the constraint gradients are linearly independent and the definition of $\underline{x}(r)$ implies the identity

$$\underline{\nabla}\Phi_2(\underline{x}(r)) = \underline{\nabla} F(\underline{x}(r)) + 2r \sum_{i=1}^{m} c_i(\underline{x}(r))\, \underline{\nabla} c_i(\underline{x}(r)) = \underline{0}\,. \quad (2.4)$$

Therefore, when calculating the least value of $\{\Phi_2(\underline{x}) : \underline{x} \in \mathbb{R}^n\}$ for large r, it is suitable to let the trial step from $\underline{x}_k$ be the value of $\underline{d}$ that minimizes the quadratic function

$$F(\underline{x}_k) + \underline{d}^T \underline{\nabla} F(\underline{x}_k) + \tfrac{1}{2}\underline{d}^T B_k \underline{d} + r \sum_{i=1}^{m} [c_i(\underline{x}_k) + \underline{d}^T \underline{\nabla} c_i(\underline{x}_k)]^2\,, \quad \underline{d} \in \mathbb{R}^n\,, \quad (2.5)$$

where B_k is an approximation to the matrix (2.3). We see that this expression includes the first and second derivatives of the Taylor series expansion of $\{\Phi_2(\underline{x}_k + \underline{d}) : \underline{d} \in \mathbb{R}^n\}$ about $\underline{d} = \underline{0}$, in a way that restricts ill-conditioning to linear approximations to the constraints. Thus one advantage is that, for infinite r, the trial step can be derived by minimizing expression (1.2) subject to the constraints (1.5). Further, Broyden and Attia (1984) and Gould (1986) show that arbitrarily large values of r need not cause ill-conditioning when the trial step is calculated, provided that B_k remains finite.

This method of determining $\underline{d}_k$ agrees with the prototype algorithm that is outlined in the third paragraph of Section 1.

Then in the fourth paragraph of Section 1 it is suggested that B_k contains suitable information for the main calculation if it is an approximation to the second derivative matrix of the Lagrangian function (1.4). Therefore now we relate the terms $\{-2rc_i(\underline{x});\ i=1,2,\cdots,m\}$ in the matrix (2.3) to the Lagrange multipliers $\lambda_1^*,\cdots,\lambda_m^*$ that are defined by the equation

$$\underline{\nabla}F(\underline{x}^*) = \sum_{i=1}^{m} \lambda_i^* \underline{\nabla} c_i(\underline{x}^*) . \tag{2.6}$$

We deduce from equation (2.4) that the terms $\{-2rc_i(\underline{x}(r));\ i=1,2,\cdots,m\}$ converge to these multipliers as $\underline{x}(r)\rightarrow\underline{x}^*$. Thus it is suitable to let B_k be an estimate of $\nabla^2_{xx}L(\underline{x},\underline{\lambda})$.

We note a close relation between the least squares penalty function (2.1) and the very successful REQP algorithm for constrained optimization that was developed by Bartholomew-Biggs (1982). It is that the search direction of a typical iteration of REQP is the $\underline{d}$ that minimizes the quadratic function (2.5) (Powell, 1983).

Because the least squares penalty function method for nonlinear constraints requires a sequence of unconstrained calculations, many algorithms employ the L_1 penalty function

$$\Phi_1(\underline{x}) = F(\underline{x}) + r \sum_{i=1}^{m} |c_i(\underline{x})| , \quad \underline{x}\in\mathbb{R}^n , \tag{2.7}$$

instead. Its derivative discontinuities at the constraint boundaries provide the very useful property that, if r is *any* positive parameter that satisfies the conditions

$$r > |\lambda_i^*| , \quad i=1,2,\cdots,m , \tag{2.8}$$

then a minimum of Φ_1 occurs at the required solution $\underline{x}^*$, which can be proved by means of the inequality

$$\Phi_1(\underline{x}) \geqslant \Phi_1(P\underline{x}) \geqslant \Phi_1(\underline{x}^*) , \tag{2.9}$$

where $\underline{x}$ is any point in a sufficiently small neighbourhood of $\underline{x}^*$, and where $P\underline{x}$ is the nearest feasible point to $\underline{x}$ with respect to

Euclidean distance. Therefore the basic technique of Fletcher's L_1 algorithm for constrained optimization (see Fletcher, 1985, for instance) is to minimize Φ_1 for a sufficiently large value of r, except that inequality constraints are usually included too, which is easy to do in this case. By analogy with our use of expression (2.5), it is now suitable to obtain a trial step from $\underline{x}_k$ by minimizing the function

$$F(\underline{x}_k) + \underline{d}^T \underline{\nabla} F(\underline{x}_k) + \tfrac{1}{2}\underline{d}^T B_k \underline{d} + r \sum_{i=1}^{m} |c_i(\underline{x}_k) + \underline{d}^T \underline{\nabla} c_i(\underline{x}_k)|, \quad \underline{d} \in \mathbb{R}^n, \tag{2.10}$$

which is a quadratic programming calculation. A comparison of expressions (2.5) and (2.10) when r is very large suggests correctly that it is still appropriate to let B_k be an approximation to the second derivative matrix of the Lagrangian function.

In order that the functions (2.5) and (2.10) are bounded below, it is usual to choose B_k to satisfy $\underline{z}^T B_k \underline{z} > 0$ for every nonzero vector $\underline{z}$ that is orthogonal to the constraint gradients $\{\underline{\nabla} c_i(\underline{x}_k);\ i = 1,2,\cdots,m\}$. Assuming that this condition holds, there is a unique value of $\underline{d}$ that minimizes the quadratic (1.2) subject to the constraints (1.5). From now on we call this vector $\underline{d}_k$, because it is the trial step of most algorithms for constrained optimization calculations. In addition to expression (1.5), $\underline{d}_k$ must satisfy the equation

$$\underline{\nabla} Q_k(\underline{d}_k) = \underline{\nabla} F(\underline{x}_k) + B_k \underline{d}_k = \sum_{i=1}^{m} \lambda_i \underline{\nabla} c_i(\underline{x}_k), \tag{2.11}$$

where $\{\lambda_i;\ i = 1,2,\cdots,m\}$ are the Lagrange multipliers of the quadratic programming problem that defines $\underline{d}_k$.

These multipliers are relevant to the remark that $\underline{d}_k$ is often the trial step of an iteration of Fletcher's L_1 algorithm. Specifically, recalling that inequality (2.8) implies that $\underline{x}^*$ is a minimum of the function (2.7), we replace the functions $\{F(\underline{x}):\ \underline{x} \in \mathbb{R}^n\}$ and $\{c_i(\underline{x}):\ \underline{x} \in \mathbb{R}^n\}$ by $\{Q_k(\underline{d}):\ \underline{d} \in \mathbb{R}^n\}$ and $\{c_i(\underline{x}_k) + \underline{d}^T \underline{\nabla} c_i(\underline{x}_k):\ \underline{d} \in \mathbb{R}^n\}$ respectively. It follows that $\underline{d}_k$

minimizes the function (2.10) if r is above the threshold $\|\underline{\lambda}\|_\infty = \max\{|\lambda_i|;\ i = 1,2,\cdots,m\}$.

If this threshold is very high, then usually the constraint gradients $\{\underline{\nabla} c_i(\underline{x}_k);\ i = 1,2,\cdots,m\}$ are nearly linearly dependent and there is a strong tendency for the step $\underline{d}_k$ to be far too long for the validity of the model that was used in its calculation. In this case, therefore, it is often more suitable to obtain the trial step by minimizing expression (2.5) or (2.10) for a moderate value of r, in order that the step does not have to satisfy the linear approximations (1.5) to the original constraints. This is only one of several advantages of penalty functions, but we find later that some useful extensions have been proposed to the quadratic programming definition of $\underline{d}_k$. One should also keep in mind that the efficiency of a penalty function method can be damaged seriously by a poor choice of r.

3. THE CALCULATION OF $\underline{d}_k$

We have already defined $\underline{d}_k$ to be the vector that minimizes the quadratic function

$$Q_k(\underline{d}) = F(\underline{x}_k) + \underline{d}^T \underline{\nabla} F(\underline{x}_k) + \tfrac{1}{2}\underline{d}^T B_k \underline{d}, \qquad \underline{d} \in \mathbb{R}^n, \tag{3.1}$$

subject to the linear constraints

$$c_i(\underline{x}_k) + \underline{d}^T \underline{\nabla} c_i(\underline{x}_k) = 0, \qquad i = 1,2,\cdots,m. \tag{3.2}$$

In order to identify the dependence of $\underline{d}_k$ on B_k, which is important to the study of reduced second derivative approximations in Section 4, we consider the *null space method* for calculating $\underline{d}_k$ (see Gill, Murray and Wright, 1981, for instance).

This method requires the QR factorization

$$A_k = \underbrace{\left[\, Y_k \;\vdots\; Z_k \,\right]}_{m \quad\ n-m} \underbrace{\begin{bmatrix} R_k \\ \cdots \\ 0 \end{bmatrix}}_{m} \begin{matrix} \updownarrow m \\ \updownarrow n-m \end{matrix} \tag{3.3}$$

where A_k is the $n \times m$ matrix whose columns are the constraint gradients $\{\underline{\nabla} c_i(\underline{x}_k);\ i=1,2,\cdots,m\}$, where the columns of Y_k are an orthonormal basis of the column space of A_k, where the columns of Z_k are an orthonormal basis of the null space of A_k^T, and where R_k is nonsingular and upper triangular. We express $\underline{d}_k$ in the form

$$\underline{d}_k = Y_k \underline{\eta}_k + Z_k \underline{\zeta}_k, \tag{3.4}$$

and address the problem of calculating $\underline{\eta}_k \in \mathbb{R}^m$ and $\underline{\zeta}_k \in \mathbb{R}^{n-m}$. Writing condition (3.2) as $\underline{c}(\underline{x}_k) + A_k^T \underline{d}_k = \underline{0}$, it follows from expressions (3.3) and (3.4) that we have the identity

$$R_k^T \underline{\eta}_k = -\underline{c}(\underline{x}_k), \tag{3.5}$$

so $\underline{\eta}_k$ can be found by forward substitution. We see also that the constraints (3.2) are independent of $\underline{\zeta}_k$. Therefore $\underline{\zeta}_k$ is the vector that minimizes the quadratic function

$$\begin{aligned}\hat{Q}_k(\underline{\zeta}) = {} & [F(\underline{x}_k) + \underline{\eta}_k^T Y_k^T \underline{\nabla} F(\underline{x}_k) + \tfrac{1}{2}\underline{\eta}_k^T Y_k^T B_k Y_k \underline{\eta}_k] \\ & + \underline{\zeta}^T [Z_k^T \underline{\nabla} F(\underline{x}_k) + Z_k^T B_k Y_k \underline{\eta}_k] + \tfrac{1}{2}\underline{\zeta}^T Z_k^T B_k Z_k \underline{\zeta}, \qquad \underline{\zeta} \in \mathbb{R}^{n-m}.\end{aligned} \tag{3.6}$$

Hence $\underline{\zeta}_k$ is bounded and uniquely defined if and only if the $(n-m) \times (n-m)$ matrix $M_k = Z_k^T B_k Z_k$ is positive definite, which justifies the condition on B_k that is given at the beginning of the paragraph that includes equation (2.11).

Expression (3.6) shows the entire contribution from B_k to the calculation of the trial step. Therefore $Z_k^T B_k Z_k$ is relevant whenever $m<n$, but $\underline{d}_k$ is independent of $Y_k^T B_k Y_k$. Further, if $\underline{c}(\underline{x}_k) = \underline{0}$, which implies $\underline{\eta}_k = \underline{0}$, then $\underline{d}_k$ is also independent of $Z_k^T B_k Y_k$. We make use of these remarks in the next section.

The *null space method* is particularly efficient when $(n-m) \ll n$. A useful alternative when $m \ll n$ is the *range space method*, which calculates $\underline{d}_k$ from the $(m+n) \times (m+n)$ system of linear equations

$$\left[\begin{array}{c:c} B_k & -A_k \\ \hdashline A_k^T & 0 \end{array}\right] \left[\begin{array}{c} \underline{d}_k \\ \cdots \\ \underline{\lambda} \end{array}\right] = \left[\begin{array}{c} -\underline{\nabla}F(\underline{x}_k) \\ \\ -\underline{c}(\underline{x}_k) \end{array}\right], \tag{3.7}$$

that are equivalent to the relations (2.11) and (3.2). The form (3.7) is instructive because of its close relation to the Newton iteration for satisfying the first order conditions that hold at the solution of our constrained optimization problem. Specifically, these conditions are the nonlinear system

$$\left.\begin{array}{l} \underline{\nabla}F(\underline{x}) = \sum_{i=1}^{m} \lambda_i \underline{\nabla}c_i(\underline{x}) \\ \\ \underline{c}(\underline{x}) = \underline{0} \end{array}\right\} \tag{3.8}$$

in the unknowns $\underline{x} \in \mathbb{R}^n$ and $\underline{\lambda} \in \mathbb{R}^m$, and, if the Newton step is from $\{\underline{x}_k, \underline{\lambda}_k\}$ to $\{\underline{x}_k + \underline{\delta}_k, \underline{\lambda}_k + \underline{\eta}_k\}$, then $\underline{\delta}_k$ and $\underline{\eta}_k$ are defined by the equations

$$\left[\begin{array}{c:c} B_k & -A_k \\ \hdashline A_k^T & 0 \end{array}\right] \left[\begin{array}{c} \underline{\delta}_k \\ \cdots\cdots \\ \underline{\lambda}_k + \underline{\eta}_k \end{array}\right] = \left[\begin{array}{c} -\underline{\nabla}F(\underline{x}_k) \\ \\ -\underline{c}(\underline{x}_k) \end{array}\right], \tag{3.9}$$

where B_k is the matrix

$$B_k = \nabla^2 F(\underline{x}_k) - \sum_{i=1}^{m} (\underline{\lambda}_k)_i \nabla^2 c_i(\underline{x}_k) \tag{3.10}$$

(see Powell, 1983, for instance). It follows that, if the matrix of expression (3.1) is a sufficiently good approximation to the second derivative matrix of the Lagrangian function, then usually the iteration $\underline{x}_{k+1} = \underline{x}_k + \underline{d}_k$ has a superlinear rate of convergence.

However, the dependence of many algorithms for constrained optimization on Lagrange multiplier estimates does cause difficulties. Therefore the view that is taken by Goodman (1985) of the first order conditions (3.8) may lead to substantial improvements to current algorithms. He writes these conditions in the form

$$\left.\begin{array}{r} Z(\underline{x})^T \underline{\nabla} F(\underline{x}) = \underline{0} \\ \underline{c}(\underline{x}) = \underline{0} \end{array}\right\}, \tag{3.11}$$

where, as in expression (3.3), $Z(\underline{x})$ is an $n \times (n-m)$ matrix with orthonormal columns that are orthogonal to the constraint gradients $\{\underline{\nabla} c_i(\underline{x});\ i=1,2,\cdots,m\}$. We see that $\underline{\lambda} \in \mathbb{R}^m$ has been eliminated, leaving a system of n nonlinear equations in the unknowns $\underline{x} \in \mathbb{R}^n$.

Goodman (1985) considers using the Newton iteration to solve the system (3.11), and gives careful attention to the difficulty that $Z(\underline{x})$ is not uniquely defined, because it can be replaced by $\Omega(\underline{x}) Z(\underline{x})$, where $\Omega(\underline{x})$ is any $(n-m) \times (n-m)$ orthogonal matrix. The actual Newton step depends on this freedom, not through the choice of $Z(\underline{x}_k)$, but through the freedom at $\underline{x}_k$ in the first derivatives of the elements of $Z(\underline{x})$ with respect to the variables, the total number of degrees of freedom being about $\frac{1}{2}(n-m)(n-m-1)$. Remembering that $Z(\underline{x})^T Z(\underline{x}) = I$, he shows that these degrees of freedom can be chosen to satisfy the equations

$$\dot{Z}_l(\underline{x}_k)^T Z(\underline{x}_k) = Z(\underline{x}_k)^T \dot{Z}_l(\underline{x}_k) = 0, \quad l = 1,2,\cdots,n, \tag{3.12}$$

where the elements of the $n \times (n-m)$ matrix $\dot{Z}_l(\underline{x})$ are defined to be the first derivatives with respect to x_l of the elements of $Z(\underline{x})$. Further, the conditions (3.12) take up all the freedom in the Newton step from $\underline{x}_k$.

Goodman (1985) relates this Newton step, $\hat{\underline{\delta}}_k$ say, to the vector $\underline{\delta}_k$ that is defined by the system (3.9). He shows that, if $\underline{\lambda}_k$ in the matrix (3.10) is the value of $\underline{\lambda}$ that minimizes the expression

$$\|\underline{\nabla} F(\underline{x}_k) - A_k \underline{\lambda}\|_2, \quad \underline{\lambda} \in \mathbb{R}^m, \tag{3.13}$$

then $\hat{\underline{\delta}}_k = \underline{\delta}_k$. Thus the numerical solution of the system (3.11) includes implicit estimates of Lagrange multipliers.

4. REDUCED SECOND DERIVATIVE MATRICES

In most of my own research on sequential quadratic programming methods for nonlinearly constrained optimization, I have required the $n \times n$ matrix B_k of the quadratic form (3.1) to be positive definite, which helps the calculation of the trial step (particularly when some of the constraints are inequalities), and which usually allows the sequence $\{\underline{x}_k: k = 1,2,3,\cdots\}$ to converge superlinearly (see Powell, 1983, for instance). However, we have deduced from expression (3.6) that, if $\underline{c}(\underline{x}_k) = \underline{0}$, then $\underline{d}_k$ can be calculated using the $(n-m) \times (n-m)$ *reduced second derivative matrix*

$$M_k = Z_k^T B_k Z_k \tag{4.1}$$

instead of B_k. Several recent papers, including Coleman and Conn (1984) and Nocedal and Overton (1985), have addressed this idea, which is particularly attractive when n is large, A_k is sparse, $B_k \approx \nabla^2 L$ is full, and $(n-m) \ll n$, because then it can provide a very great reduction in the number of nonzero matrix elements that occur. Further, the idea is highly successful when all the constraints are linear (see Gill, Murray and Wright, 1981, for instance), but nonlinearity of the constraints introduces at least two severe difficulties, namely that it is often very inefficient to satisfy $\underline{c}(\underline{x}_k) = \underline{0}$ to high accuracy, and the matrix Z_k cannot be independent of k.

We have noted already that, if $\underline{c}(\underline{x}_k) \neq \underline{0}$, then the submatrix $Z_k^T B_k Y_k$ of expression (3.6) is important to $\underline{d}_k$, but, following usual practice, we assume throughout this section that only M_k is available and we set $Z_k^T B_k Y_k = 0$. Thus, if $Z_k^T \underline{\nabla} F(\underline{x}_k) = \underline{0}$, we have $\underline{\zeta}_k = \underline{0}$ in expression (3.4), which forces $\underline{d}_k$ to be in the column space of A_k. For example, if we were minimizing the function

$$F(\underline{x}) = x_1 - 2x_1 x_2 + x_2^2, \qquad \underline{x} \in \mathbb{R}^2, \tag{4.2}$$

subject to $x_1 = 0$, and if the components of $\underline{x}_k$ were (θ, θ) for some $\theta \in \mathbb{R}$, then the second component of $\underline{d}_k$ would be zero. Hence, remembering equation (1.3), the new vector of variables satisfies $\|\underline{x}_{k+1} - \underline{x}^*\| \geq 2^{-\frac{1}{2}} \|\underline{x}_k - \underline{x}^*\|$, and the rate of convergence is no better than linear. However, due to condition (3.2) and the linearity of the constraint, the choice $\underline{x}_{k+1} = \underline{x}_k + \underline{d}_k$ would yield $\underline{c}(\underline{x}_{k+1}) = \underline{0}$, so the matrix $M_{k+1} = Z_{k+1}^T B_{k+1} Z_{k+1}$ can provide all the information that is needed to calculate a good value of $\underline{x}_{k+2}$.

This simple example indicates typical behaviour for nonlinear constraints, which is that, when $Z_k^T B_k Y_k = 0$ and $\|\underline{x}_k - \underline{x}^*\| \ll 1$, then the iteration provides $\|\underline{x}_{k+1} - \underline{x}^*\| = O(\|\underline{c}(\underline{x}_k)\|)$ and $\|\underline{c}(\underline{x}_{k+1})\| \ll \|\underline{c}(\underline{x}_k)\|$ or $\|\underline{x}_{k+1} - \underline{x}^*\| \ll \|\underline{x}_k - \underline{x}^*\|$. Thus the choice $Z_k^T B_k Y_k = 0$ can allow pairs of iterations to converge Q-superlinearly (Powell, 1978b), which is usually fast enough in practice. Further, there seems to be no need to reduce the constraint violations $\{\|\underline{c}(\underline{x}_k)\|;\ k = 1, 2, 3, \cdots\}$ that occur automatically, but many iterations of the *reduced gradient method* (see Lasdon, 1982, for instance) include further changes to the variables to ensure that constraint violations are sufficiently small.

An alternative to assuming that $Z_k^T B_k Y_k = 0$, when the matrix (4.1) gives all the available second derivative information, is preferred by Coleman and Conn (1982, 1984). They recommend the trial step $\hat{\underline{d}}_k = \underline{h}_k + \underline{v}_k$, where the *horizontal step* $\underline{h}_k$ minimizes expression (3.1) subject to $A_k^T \underline{d} = \underline{0}$, and where the *vertical step* $\underline{v}_k$ is the shortest vector in $\mathbb{R}^n$ that satisfies $\underline{c}(\underline{x}_k + \underline{h}_k) + A_k^T \underline{v}_k = \underline{0}$. Thus these vectors have the values

$$\left.\begin{aligned} \underline{h}_k &= -Z_k M_k^{-1} Z_k^T \underline{\nabla} F(\underline{x}_k) \\ \underline{v}_k &= -A_k (A_k^T A_k)^{-1} \underline{c}(\underline{x}_k + \underline{h}_k) \end{aligned}\right\} . \tag{4.3}$$

The use of $\underline{c}(\underline{x}_k + \underline{h}_k)$ in expression (4.3) instead of $\underline{c}(\underline{x}_k)$ in equation (3.5) gives the algorithm of Coleman and Conn a

Q-superlinear convergence property (Byrd, 1984). Further, there is a tendency for constraint violations to be smaller than before, which reduces the importance of the term $Z_k^T B_k Y_k$ that is suppressed.

The remainder of this section considers techniques for generating M_{k+1} from M_k and from the changes in first derivatives that occur during the calculation of $\underline{x}_{k+1}$ from $\underline{x}_k$. We seek suitable extensions of the usual procedures for updating $n \times n$ second derivative approximations (see Fletcher, 1981, for instance). Here two important features are that the difference $(B_{k+1} - B_k)$ is of low rank, and the freedom in B_{k+1} that remains after including information from changes in first derivatives is usually chosen to make $\|B_{k+1} - B_k\|$ small.

It is sensible to keep $\|B_{k+1} - B_k\|$ small when second derivatives are continuous, but there is a similar case for keeping $\|M_{k+1} - M_k\|$ small only if Z_k has some continuity properties too. It would be ideal if there were a continuous function $Z(\underline{x})$ from $\mathbb{R}^n$ to the set of $n \times (n-m)$ matrices with orthogonal columns, such that $A(\underline{x})^T Z(\underline{x}) = 0$ for all $\underline{x}$ and $Z_k = Z(\underline{x}_k)$ for all k. However, Byrd and Schnabel (1986) point out that such continuous functions need not exist, except in the special cases $m = n-1$, $m = 1$ and $n = 4$, and $m = 1$ and $n = 8$. For example, discontinuities are found in Householder's method for calculating QR factorizations due to a choice of sign in each elementary rotation matrix. Therefore Coleman and Sorensen (1984) recommend a bias towards previous decisions in the discrete decisions of the QR factorization, and Gill, Murray, Saunders, Stewart and Wright (1985) propose a technique that keeps Z_k reasonably close to Z_{k-1}. Thus Z_k depends not only on $\underline{x}_k$ but also on the factorizations of previous iterations. Nevertheless we take the view that we can write $Z_k = Z(\underline{x}_k)$, where $\{Z(\underline{x}) : \underline{x} \in \mathbb{R}^n\}$ is differentiable. Perhaps this view can be clarified by the kind of argument that is given in the paragraph that includes equation (3.12).

The usual way of including in the $n \times n$ matrices $\{B_k : k = 1,2,3,\cdots\}$ some relevant properties of $\nabla^2_{xx} L(\underline{x},\underline{\lambda})$ is to choose B_{k+1} to satisfy the equation

$$\begin{aligned} B_{k+1}\underline{\delta}_k = & \left[\underline{\nabla}F(\underline{x}_k + \underline{\delta}_k) - \sum_{i=1}^{m} \lambda_i \underline{\nabla} c_i(\underline{x}_k + \underline{\delta}_k)\right] \\ & - \left[\underline{\nabla}F(\underline{x}_k) - \sum_{i=1}^{m} \lambda_i \underline{\nabla} c_i(\underline{x}_k)\right], \end{aligned} \tag{4.4}$$

where $\underline{\delta}_k = \underline{x}_{k+1} - \underline{x}_k$ if this vector is nonzero, and where $\underline{\lambda} \in \mathbb{R}^m$ is an estimate of the vector of Lagrange multipliers, for example the vector that minimizes expression (3.13) or the multipliers of equation (2.11). Therefore, if one chooses $\underline{\delta}_k = Z_{k+1}\underline{\zeta}$ for some $\underline{\zeta} \in \mathbb{R}^{n-m}$, it would be suitable to base the calculation of $M_{k+1} = Z^T_{k+1} B_{k+1} Z_{k+1}$ on the condition

$$\begin{aligned} M_{k+1}\underline{\zeta} = & Z^T_{k+1}\left[\underline{\nabla}F(\underline{x}_k + \underline{\delta}_k) - \sum_{i=1}^{m} \lambda_i \underline{\nabla} c_i(\underline{x}_k + \underline{\delta}_k)\right] \\ & - Z^T_{k+1}\left[\underline{\nabla}F(\underline{x}_k) - \sum_{i=1}^{m} \lambda_i \underline{\nabla} c_i(\underline{x}_k)\right]. \end{aligned} \tag{4.5}$$

Further, by allowing changes to this equation that are of magnitude $o(\|\underline{\zeta}\|)$, which retains many theoretical convergence properties, one can deduce some other equations that are suitable for guiding the calculation of M_{k+1} (Nocedal and Overton, 1985). In particular, because of the assumption

$$\|Z_{k+1} - Z_k\| = O(\|\underline{x}_{k+1} - \underline{x}_k\|), \tag{4.6}$$

and because the square bracket terms of expression (4.5) tend to zero as $\underline{x}_k \to \underline{x}^*$, we may employ the equation

$$M_{k+1}\underline{\zeta}_k = Z^T_{k+1}\underline{\nabla}F(\underline{x}_{k+1}) - Z^T_k \underline{\nabla}F(\underline{x}_k) \tag{4.7}$$

when we have the relation

$$\underline{x}_{k+1} - \underline{x}_k = Z_k \underline{\zeta}_k + o(\|\underline{x}_{k+1} - \underline{x}_k\|). \tag{4.8}$$

It is remarkable that, due to the identities

$$Z_k^T A_k = Z_{k+1}^T A_{k+1} = 0 ,$$

no Lagrange multiplier estimates occur in equation (4.7). It follows that the information on constraint curvature, which can be crucial to fast convergence, reaches M_{k+1} through the difference between Z_k and Z_{k+1}, especially when $\{\underline{\nabla}F(\underline{x}):\ \underline{x}\in\mathbb{R}^n\}$ is constant. There seems to be a connection between expression (4.7) and the estimation of the Jacobian matrix of the first part of the nonlinear system (3.11).

Condition (4.8) is of practical value only if the term $o(\|\underline{x}_{k+1}-\underline{x}_k\|)$ is specified as a tolerance that can be tested numerically. This need is considered by Nocedal and Overton (1985), who recommend setting $M_{k+1}=M_k$ if the ratio

$$\|A_k^T(\underline{x}_{k+1}-\underline{x}_k)\| \,/\, \|\underline{x}_{k+1}-\underline{x}_k\|$$

is unacceptably large. Alternatively, at the expense of doubling the number of first derivatives that are calculated on an iteration, Coleman and Conn (1984) prefer to replace equation (4.7) by the condition

$$M_{k+1}\underline{\zeta}_k = Z_k^T\left[\underline{\nabla}F(\underline{x}_k+\underline{h}_k) - \sum_{i=1}^{m} \lambda_i \underline{\nabla}c_i(\underline{x}_k+\underline{h}_k) - \underline{\nabla}F(\underline{x}_k)\right], \tag{4.9}$$

where $\underline{h}_k$ is the horizontal step of expression (4.3), so $\underline{\zeta}_k$ has the value $-M_k^{-1}Z_k^T\underline{\nabla}F(\underline{x}_k)$, and where $\underline{\lambda}\in\mathbb{R}^m$ minimizes expression (3.13).

An observation of Byrd and Schnabel (1986) avoids the doubtful assumption that $Z_k=Z(\underline{x}_k)$, where $\{Z(\underline{x}):\ \underline{x}\in\mathbb{R}^n\}$ is differentiable. It is that, if Z_k and $M_k=Z_k^TB_kZ_k$ are known, then the $n\times n$ matrix $Z_kZ_k^TB_kZ_kZ_k^T$ is available too, and it is independent of the freedom in Z_k, because $P_k=Z_kZ_k^T$ is the unique orthogonal projection operator from $\mathbb{R}^n$ to the null space of A_k^T. Therefore they address the question of generating M_{k+1} from equation (4.5) and the matrix $P_kB_kP_k$. Because $Z_{k+1}^TP_kB_kP_kZ_{k+1}$

seems to be the most suitable available approximation to $Z_{k+1}^T B_k Z_{k+1}$, they suggest the calculation of M_{k+1} by applying an updating formula to $Z_{k+1}^T Z_k M_k Z_k^T Z_{k+1}$, which removes the dependence on the freedom in Z_k. Further, the product $Z_k^T Z_{k+1}$ is generated automatically, if Z_{k+1} is chosen in the way that is proposed by Gill, Murray, Saunders, Stewart and Wright (1985).

5. MERIT FUNCTIONS

The purpose of a merit function $\{\Phi(\underline{x}): \underline{x} \in \mathbb{R}^n\}$ is to provide an automatic way of deciding whether to accept a trial change $\underline{d}_k$ to an estimate $\underline{x}_k$ of the solution of a constrained optimization calculation. We recall from Section 1 that it is usual to set $\underline{x}_{k+1} = \underline{x}_k + \underline{d}_k$ only if the inequality

$$\Phi(\underline{x}_k + \underline{d}_k) < \Phi(\underline{x}_k) \tag{5.1}$$

is satisfied, but sometimes this test is replaced by the stronger condition

$$\Phi(\underline{x}_k + \underline{d}_k) \leqslant \Phi(\underline{x}_k) - \theta \rho_k \tag{5.2}$$

for some constant $\theta \in (0, \frac{1}{2})$, where ρ_k is a prediction of the reduction $[\Phi(\underline{x}_k) - \Phi(\underline{x}_k + \underline{d}_k)]$ in the merit function. Thus many algorithms converge successfully even when the starting point $\underline{x}_1$ is far from the required solution $\underline{x}^*$. It has been shown by Han (1977) that the L_1 penalty function (2.7) is suitable for forcing global convergence, and it is used as a merit function by many optimization algorithms, including the procedures of Byrd, Schnabel and Shultz (1985), Coleman and Conn (1982) and Powell (1978a), for example.

We see that $\{\Phi_1(\underline{x}): \underline{x} \in \mathbb{R}^n\}$ is a special case of the function

$$\Phi(\underline{x}) = F(\underline{x}) + W(\underline{c}(\underline{x})), \quad \underline{x} \in \mathbb{R}^n, \tag{5.3}$$

mentioned in Section 1, where $\{W(\underline{c}): \underline{c} \in \mathbb{R}^m\}$ is non-negative and satisfies $W(\underline{0}) = 0$. Such functions can take their least value at $\underline{x}^*$, but unfortunately the *Maratos effect* may occur (Maratos, 1978), which means that $W(\underline{x}_k + \underline{d}_k) > W(\underline{x}_k)$ is possible even when

$$\|\underline{x}_k + \underline{d}_k - \underline{x}^*\| \ll \|\underline{x}_k - \underline{x}^*\|.$$

For example, if $n=2$, $F(\underline{x}) = 2(x_1^2 + x_2^2 - 1) - x_1$, $m=1$ and the constraint is $x_1^2 + x_2^2 - 1 = 0$, then the values

$$\underline{x}_k = \begin{bmatrix} \cos\theta \\ \sin\theta \end{bmatrix}, \quad \underline{d}_k = \begin{bmatrix} \sin^2\theta \\ -\cos\theta\sin\theta \end{bmatrix} \tag{5.4}$$

yield $F(\underline{x}_k + \underline{d}_k) > F(\underline{x}_k)$, $c(\underline{x}_k) = 0$ and $c(\underline{x}_k + \underline{d}_k) > 0$, so the trial step increases all merit functions of the form (5.3), even though it gives the quadratic rate of convergence

$$\|\underline{x}_k + \underline{d}_k - \underline{x}^*\| \approx \tfrac{1}{2}\|\underline{x}_k - \underline{x}^*\|^2. \tag{5.5}$$

Three techniques have been proposed for avoiding the inefficiencies of the Maratos effect. Condition (5.1) can be relaxed on some iterations (Chamberlain, Lemaréchal, Pedersen and Powell, 1982), or $\underline{d}_k$ can be replaced by $\hat{\underline{d}}_k$ which is calculated using $\underline{c}(\underline{x}_k + \underline{d}_k)$ to satisfy $\|\underline{c}(\underline{x}_k + \hat{\underline{d}}_k)\| = o(\|\hat{\underline{d}}_k\|^2)$ (Fletcher, 1982; Mayne and Polak, 1982), or one can give up the form (5.3) of the merit function. Some other useful merit functions are mentioned below, and, unlike expression (2.7), they are continuously differentiable.

They are derived from the first and second order sufficiency conditions that are usually satisfied at $\underline{x}^*$. In particular these conditions imply that the function

$$\Phi^*(\underline{x}) = F(\underline{x}) - \sum_{i=1}^{m} \lambda_i^* c_i(\underline{x}) + r \sum_{i=1}^{m} [c_i(\underline{x})]^2, \quad \underline{x} \in \mathbb{R}^n, \tag{5.6}$$

has a local minimum at $\underline{x}^*$ for sufficiently large r where $\underline{\lambda}^*$ is still defined by equation (2.6) (see Fletcher, 1981, for instance), but $\underline{\lambda}^*$ is unknown during the minimization calculation. Therefore the *differentiable exact penalty function*

$$\Phi_F(\underline{x}) = F(\underline{x}) - \sum_{i=1}^{m} \lambda_i(\underline{x})\, c_i(\underline{x}) + r \sum_{i=1}^{m} [c_i(\underline{x})]^2, \quad \underline{x} \in \mathbb{R}^n, \tag{5.7}$$

is proposed by Fletcher (1973), where, as in expression (3.13), $\underline{\lambda}(\underline{x})$ is chosen to minimize the function

$$\|\underline{\nabla}F(\underline{x}) - \sum_{i=1}^{m} \lambda_i \underline{\nabla} c_i(\underline{x})\|_2 , \qquad \underline{\lambda} \in \mathbb{R}^m , \tag{5.8}$$

so it has the value

$$\underline{\lambda}(\underline{x}) = [A(\underline{x})^T A(\underline{x})]^{-1} A(\underline{x})^T \underline{\nabla} F(\underline{x}) , \qquad \underline{x} \in \mathbb{R}^n . \tag{5.9}$$

The merit function (5.7) avoids the Maratos effect because, assuming some differentiability, the second order sufficiency conditions give the relation

$$\Phi_F(\underline{x}) - \Phi_F(\underline{x}^*) \sim \|\underline{x} - \underline{x}^*\|^2 , \qquad \underline{x} \in N^* , \tag{5.10}$$

for sufficiently large r, where $N^* \subset \mathbb{R}^n$ is a neighbourhood of $\underline{x}^*$.

The algorithms of Boggs and Tolle (1984) and of Powell and Yuan (1986a) are line search procedures that combine the trial step of Section 3 with the merit function (5.7), except that Boggs and Tolle (1984) replace the term $r\|\underline{c}(\underline{x})\|^2$ of $\Phi_F(\underline{x})$ by $r\underline{c}(\underline{x})^T [A(\underline{x})^T A(\underline{x})]^{-1} \underline{c}(\underline{x})$. The numerical results and theory of these papers suggest that the algorithms are successful at avoiding the inefficiencies that are caused by the nondifferentiability of expression (2.7). Powell and Yuan (1986a) give particular attention to the automatic choice of a suitable value of r without calculating any second derivatives of the functions F and $\{c_i;\ i = 1,2,\cdots,m\}$, which requires some care because $\{\underline{\lambda}(\underline{x}):\ \underline{x} \in \mathbb{R}^n\}$ depends on first derivatives of these functions.

It is onerous, however, to compute the vector (5.9) whenever a value of Φ_F is required. Therefore, in order to guide the choice of the step-length along $\underline{d}_k$ from $\underline{x}_k$, Schittkowski (1983) and Gill, Murray, Saunders and Wright (1986) consider a line search merit function that, when all constraints are equalities, takes the form

$$\phi_k(\alpha) = F(\underline{x}_k + \alpha \underline{d}_k) - [\underline{\lambda}_k + \alpha(\underline{\mu}_k - \underline{\lambda}_k)]^T \underline{c}(\underline{x}_k + \alpha \underline{d}_k)$$
$$+ r\|\underline{c}(\underline{x}_k + \alpha \underline{d}_k)\|^2, \quad \alpha \in \mathbb{R}, \tag{5.11}$$

so it differs from $\Phi_F(\underline{x}_k + \alpha \underline{d}_k)$ only in the Lagrange multiplier estimates. Here $\underline{\mu}_k$ is the value of $\underline{\lambda}$ that is defined by the system (3.7), $\underline{\lambda}_1 = \underline{\mu}_1$, and, for $k \geq 2$, $\underline{\lambda}_k = \underline{\lambda}_{k-1} + \alpha_{k-1}(\underline{\mu}_{k-1} - \underline{\lambda}_{k-1})$, where α_{k-1} is the step length of the formula $\underline{x}_k = \underline{x}_{k-1} + \alpha_{k-1}\underline{d}_{k-1}$. This technique can reduce substantially the number of times that the gradients $\underline{\nabla}F(\underline{x})$ and $\{\underline{\nabla}c_i(\underline{x});\ i = 1,2,\cdots,m\}$ are calculated, but Powell (1985) finds among several numerical experiments a few cases where the test $\phi_k(1) < \phi_k(0)$ is satisfied but $\|\underline{x}_k + \underline{d}_k - \underline{x}^*\| / \|\underline{x}_k - \underline{x}^*\|$ is large.

Another interesting merit function that depends on the gradients $\underline{\nabla}F(\underline{x})$ and $\{\underline{\nabla}c_i(\underline{x});\ i = 1,2,\cdots,m\}$ is proposed by Di Pillo and Grippo (1979). It has the form

$$\Phi_{DG}(\underline{x}, \underline{\lambda}) = F(\underline{x}) - \sum_{i=1}^{m} \lambda_i c_i(\underline{x}) + r \sum_{i=1}^{m} [c_i(\underline{x})]^2$$
$$+ \|M(\underline{x})[\underline{\nabla}F(\underline{x}) - A(\underline{x})\underline{\lambda}]\|_2^2, \quad \underline{x} \in \mathbb{R}^n, \quad \underline{\lambda} \in \mathbb{R}^m, \tag{5.12}$$

where, for each $\underline{x}$, $M(\underline{x})$ is a full rank matrix whose dimensions are usually $n \times n$ or $m \times n$. This merit function can be used to control the adjustment of both $\underline{x}$ and $\underline{\lambda}$ because, for sufficiently large r, it has a minimum when $\underline{x} = \underline{x}^*$ and $\underline{\lambda} = \underline{\lambda}^*$. A disadvantage of Φ_{DG}, however, if $M(\underline{x})$ is set to a large multiple of the $n \times n$ unit matrix, is that it can also have local minima at all values of $\underline{x}$ and $\underline{\lambda}$ that satisfy the first order conditions (3.8), which include maxima of the constrained optimization calculation. Therefore it is better to let $M(\underline{x})$ be an $m \times n$ matrix, in order that in nondegenerate cases the last two terms of expression (5.12) influence only $2m$ of the $(m+n)$ degrees of freedom in $\underline{x}$ and $\underline{\lambda}$. Thus it is possible that one can move continuously from a local maximum of the given constrained problem to the required solution $(\underline{x}^*, \underline{\lambda}^*)$ in a way that reduces expression (5.12) monotonically.

This valuable property is achieved often. Indeed, if $M(\underline{x})A(\underline{x})$ is an $m \times m$ nonsingular matrix for all $\underline{x}$, then, under some nondegeneracy and continuity assumptions, it can be shown that for sufficiently large r, all local minima of $\{\Phi_{DG}(\underline{x},\underline{\lambda}):\underline{x} \in \mathbb{R}^n,\ \underline{\lambda} \in \mathbb{R}^m\}$ are local minima of the optimization calculation (see Bertsekas, 1982a, for instance). In particular we consider the choice

$$M(\underline{x}) = [A(\underline{x})^T A(\underline{x})]^{-1} A(\underline{x})^T, \quad \underline{x} \in \mathbb{R}^n, \tag{5.13}$$

which is mentioned in Bertsekas (1982b). Because expression (5.12) is a strictly convex quadratic function of $\underline{\lambda}$ for each $\underline{x}$, and because $M(\underline{x})A(\underline{x}) = I$, at each local minimum of Φ_{DG} the vector $\underline{\lambda}$ has the value

$$\hat{\underline{\lambda}}(\underline{x}) = M(\underline{x})\underline{\nabla}F(\underline{x}) + \tfrac{1}{2}\underline{c}(\underline{x}) = \underline{\lambda}(\underline{x}) + \tfrac{1}{2}\underline{c}(\underline{x}), \tag{5.14}$$

where $\underline{\lambda}(\underline{x})$ is still expression (5.9). It follows that $(\underline{x},\underline{\lambda})$ is a local minimum of Φ_{DG} if and only if $\underline{\lambda} = \hat{\underline{\lambda}}(\underline{x})$ and $\underline{x}$ is a local minimum of the function

$$\begin{aligned} \Phi_{DG}(\underline{x},\hat{\underline{\lambda}}(\underline{x})) &= F(\underline{x}) - \hat{\underline{\lambda}}(\underline{x})^T\underline{c}(\underline{x}) + r\|\underline{c}(\underline{x})\|^2 + \|\tfrac{1}{2}\underline{c}(\underline{x})\|^2 \\ &= \Phi_F(\underline{x}) - \tfrac{1}{4}\sum_{i=1}^{m}[c_i(\underline{x})]^2, \quad \underline{x} \in \mathbb{R}^n. \end{aligned} \tag{5.15}$$

Thus the local minima of Φ_{DG} and Φ_F are identical if r is replaced by $(r-\frac{1}{4})$ in expression (5.7). Therefore it is suitable to regard the merit function of Di Pillo and Grippo (1979) as a generalization of Fletcher's (1973) function that removes the restriction (5.9) on the value of $\underline{\lambda}$. This remark is discussed in Section 7.

6. LINE SEARCHES AND TRUST REGIONS

As mentioned already, a typical iteration of a line search algorithm for optimization subject to nonlinear constraints calculates a trial step $\underline{d}_k$ from $\underline{x}_k$ (see Section 3), and then the approximation to $\underline{x}^*$ is updated by the formula

$$\underline{x}_{k+1} = \underline{x}_k + \alpha_k \underline{d}_k, \qquad (6.1)$$

where α_k is a positive step-length that is chosen to give a sufficient reduction in a merit function (see Section 5). Although this kind of procedure is employed by several successful computer programs, we begin this section by noting some disadvantages of line searches, in order to indicate that a different approach, namely *trust regions*, deserves careful consideration. Details of line searches, such as ensuring that $\underline{d}_k$ is a descent direction of the merit function and adjusting α_k, can be found elsewhere, for example see Gill, Murray and Wright (1981).

Some of these disadvantages are shown in an example of Powell (1970), in which two nonlinear equations in two unknowns cannot be solved by Newton's iteration with line searches. Here each iteration searches along the Newton direction for the point that minimizes the sum of squares of residuals of the equations, $\{\Phi(\underline{x}): \underline{x} \in \mathbb{R}^2\}$ say. Convergence occurs to a point where the Jacobian matrix is singular and the equations do not hold, but $\|\underline{\nabla}\Phi(\underline{x})\|$ stays bounded away from zero. Such convergence occurs from a wide range of starting points. What happens is that the lengths of the search directions diverge, these directions become orthogonal to $\underline{\nabla}\Phi(\underline{x})$, and the step-lengths tend to zero. This example can be regarded as a constrained optimization calculation in which the function $\{F(\underline{x}): \underline{x} \in \mathbb{R}^n\}$ is identically zero and $m=n$. Thus, due to the constraints (3.2), the search direction of the example is the one that is calculated in Section 3; further, the merit function of the example is proportional to expression (5.7).

The convergence to an inadequate vector of variables in this example is a consequence of some features of the algorithm that probably would not be copied by one who carried out the calculation by hand. In particular, one would not choose very long search directions, partly because one would be concerned with the validity of the linear approximations to the constraints, and more strongly because, having noticed that many consecutive iterations require $\alpha_k \ll 1$, one would deduce that the Newton iteration gives trial steps that are far too big. Further, the value of $\underline{\nabla}\Phi(\underline{x})$ would indicate the possibility of reductions in the merit function on most iterations that are much larger than the reductions that occur. Thus failure to solve the equations automatically can point the way to more efficient algorithms.

These considerations led to the use of *trust regions* in penalty function algorithms, where, as in Section 2, the trial step, $\underline{\hat{d}}_k$ say, is intended to reduce the merit function $\{\Phi(\underline{x}): \underline{x} \in \mathbb{R}^n\}$. The key feature of a trust region method is a bound

$$\|\underline{d}\| \leqslant \Delta_k \tag{6.2}$$

on the trial step, where Δ_k is adjusted automatically from iteration to iteration, the aim being to make the value $\underline{x}_{k+1} = \underline{x}_k + \underline{\hat{d}}_k$ acceptable without Δ_k being unnecessarily small (see Fletcher, 1981, for instance). In a penalty function trust region method one just includes the constraint (6.2) in the otherwise unconstrained minimization calculation that gives the search direction, for example the minimization of expression (2.5) or (2.10). Then the choice of Δ_{k+1} usually depends on the ratio

$$[\Phi(\underline{x}_k) - \Phi(\underline{x}_k + \underline{\hat{d}}_k)] \,/\, [P(\underline{0}) - P(\underline{\hat{d}}_k)], \tag{6.3}$$

where $\{P(\underline{d}): \underline{d} \in \mathbb{R}^n\}$ is the function that is minimized to determine $\underline{\hat{d}}_k$.

Some other techniques for computing trial steps in trust region algorithms have been proposed recently, that are similar to the calculation of $\underline{d}_k$ in Section 3, except that the linear constraints (3.2) are modified if necessary so that they do not conflict with inequality (6.2). For example, Vardi (1985) defines $\underline{d}_k$ by minimizing expression (3.1) subject to the bound

(6.2) and the conditions

$$\theta_k c_i(\underline{x}_k) + \underline{d}^T \underline{\nabla} c_i(\underline{x}_k) = 0, \quad i = 1, 2, \cdots, m, \tag{6.4}$$

where θ_k is a real number from the interval

$$(0, 1] \cap (0, \Delta_k / \|A_k (A_k^T A_k)^{-1} \underline{c}(\underline{x}_k)\|]. \tag{6.5}$$

The value

$$\underline{d} = -\theta_k A_k (A_k^T A_k)^{-1} \underline{c}(\underline{x}_k) \tag{6.6}$$

shows that the constraints (6.2) and (6.4) are consistent. In this case, as described in Byrd, Schnabel and Shultz (1985), one can compute the trial step by the following extension of the null space method of Section 3. Letting the new $\underline{d}_k$ have the form (3.4), the vector $\underline{\eta}_k$ is defined by equation (3.5), after replacing the right hand side of this equation by $-\theta \underline{c}(\underline{x}_k)$. Then $\underline{\zeta}_k$ minimizes expression (3.6) subject to the bound $\|Y_k \underline{\eta}_k + Z_k \underline{\zeta}_k\| \leq \Delta_k$, which reduces to the inequality

$$\|\underline{\zeta}_k\| \leq [\Delta_k^2 - \|\underline{\eta}_k\|^2]^{\frac{1}{2}} \tag{6.7}$$

when the 2-norm is employed. Thus finding $\underline{\zeta}_k$ is analogous to the calculation of the trial step of a trust region algorithm for unconstrained optimization (see Moré, 1983, for instance).

Celis, Dennis and Tapia (1985) and Powell and Yuan (1986b), however, prefer to have more freedom in $\underline{d}$ than is allowed by the equations (6.4). Therefore they replace these equations by the inequality

$$\|\underline{c}(\underline{x}_k) + A_k^T \underline{d}\|_2 \leq \psi_k, \tag{6.8}$$

where $\psi_k = \|\underline{c}(\underline{x}_k) + A_k^T \bar{\underline{d}}_k\|_2 < \|\underline{c}(\underline{x}_k)\|_2$ for some known vector $\bar{\underline{d}}_k$ that satisfies $\|\bar{\underline{d}}_k\| \leq \Delta_k$. The two papers suggest different choices of $\bar{\underline{d}}_k$. The trial step is now derived from the minimization of the quadratic function (3.1) subject to the constraints (6.2) and (6.8), but this calculation has to be iterative, and it is not practicable to complete it. Therefore, as in the unconstrained case (Moré, 1982), one requires an efficient technique for adjusting $\underline{d}$, and a test that indicates when the current $\underline{d}$ is an adequate trial step. Further, there are difficulties from local minima when B_k is not positive definite. These questions are a subject of current research.

Several ideas for calculating trial steps have been mentioned, and each one is of some use, because the trial step should depend on the choice of the merit function $\{\Phi(\underline{x}): \underline{x} \in \mathbb{R}^n\}$. The point of this remark is that, if further progress from $\underline{x}_k$ is required, then, for sufficiently small Δ_k, the trial step must satisfy the condition $\Phi(\underline{x}_k + \underline{d}_k) < \Phi(\underline{x}_k)$, an increase in the penalty parameter r of the merit function being allowed if necessary. Therefore, if for example the merit function is Φ_F (see equation (5.7)), then a move from $\underline{x}_k$ along the trial step should provide an initial decrease in the penalty term $\|\underline{c}(\underline{x})\|^2$ of Φ_F. This condition is provided by the constraint (6.8) on $\underline{d}$, but it may not hold when $\underline{d}$ minimizes expression (2.10) subject to $\|\underline{d}\| \leq \Delta_k$, even if Δ_k is very small. From this point of view condition (6.4) is particularly versatile, because it ensures that all the nonzero constraint residuals $\{|c_i(\underline{x})|;\ i = 1, 2, \cdots, m\}$ decrease initially if one moves from $\underline{x}_k$ along $\underline{d}$.

7. DISCUSSION

The remarks of the last two paragraphs of Section 6 show that some trust region algorithms become complicated when the trial step is not calculated by the minimization of an approximation to the merit function. Therefore we ask whether there is a need for so much sophistication. A ready answer to such questions is that general optimization calculations with nonlinear constraints are so important that all promising ideas deserve to be investigated, but we consider further the advantages of the given techniques. In addition to the motivation for trust regions in Section 6, we note that no positivity conditions on B_k are needed when trust regions are employed, which can be particularly valuable when B_k is known to be sparse. Moreover, a full trust region step when Δ_k is very small usually moves more efficiently to constraint boundaries than a step along a search direction that also changes the variables by about Δ_k by choosing the step-length to be much less than one. It seems, therefore, that it is worthwhile to develop some algorithms that achieve global convergence by the use of trust regions.

Is it also worthwhile to employ differentiable merit functions? The avoidance of the Maratos effect due to the relation (5.10) has been mentioned already, and the numerical results of Powell and Yuan (1986a) indicate that differentiable merit functions are sometimes much more efficient than nondifferentiable ones. A more fundamental reason for preferring Φ_F, however, comes from the point of view that minimizing $\{F(\underline{x}):\ \underline{x} \in \mathbb{R}^n\}$ subject to the constraints (1.1) should be the same as minimizing $\{F(\underline{x}) + \Sigma \mu_i c_i(\underline{x}):\ \underline{x} \in \mathbb{R}^n\}$ subject to these constraints for any values of the parameters $\{\mu_i;\ i = 1, 2, \cdots, m\}$. An obvious way of achieving this equivalence is by subtracting multiples of the constraints from the objective function so that the modified objective function is a close approximation to the Lagrangian function. Thus we may obtain the first two terms on the right hand side of expression (5.7), and the last term is just a convenient way of providing convexity near $\underline{x}^*$.

If we are now persuaded to develop an algorithm whose merit function is Φ_F, why do we not calculate a trial step by minimizing an approximation to Φ_F? One reason is that the dependence of $\Phi_F(\underline{x})$ on $\underline{\nabla} F(\underline{x})$ and $\{\underline{\nabla} c_i(\underline{x});\ \ i = 1, 2, \cdots, m\}$ makes it difficult to form an adequate approximation, but let us suppose that all the derivatives that we need are available. Then the approximation to Φ_F would be smooth, so an $n \times n$ second derivative matrix would probably be required when minimizing the approximation. Thus we would fail to take advantage of the very precise guidance on the trial step that is given by linear approximations to the constraints, and we would be unable to include the reduced second derivative techniques that are the subject of Section 4. Therefore there seems to be a good case for separating the calculation of the trial step from the merit function, but it should be added that the purpose of these three paragraphs is to defend a procedure that does not enjoy the strong advantage of simplicity.

The penalty function methods, however, are much more satisfactory at present when the constraint gradients $\{\underline{\nabla} c_i(\underline{x}_k);$ $i = 1,2,\cdots,m\}$ are nearly or exactly linearly dependent. Several difficulties arise in this case. One is that the calculation of the trial step is ill-conditioned if it includes the constraints (1.5) or (6.4), but this point may be of minor importance because in practice inequality conditions occur more often than equations, and the minimization of expression (2.10) or the form (6.8) of the constraints usually avoids this ill-conditioning. A related difficulty is that it becomes very hazardous to employ the matrix Z_k as suggested in Sections 3 and 4 because, in addition to the freedom in Z_k that has been mentioned already, the null space of A_k^T is poorly defined. In other words the space that is spanned by the columns of Z_k is highly sensitive to rounding errors and to changes in the variables. In practice one might ignore this difficulty, or one might remove the ill-conditioning by dropping some of the constraints.

A more serious problem when the constraint gradients are nearly linearly dependent is that some Lagrange multiplier estimates, including the values (5.9), tend to be unbounded. The L_1 penalty function is particularly suitable for avoiding this difficulty, because, if one expresses the minimization of the function (2.10) as a quadratic programming calculation, and if one takes the Lagrange multipliers of this calculation as the Lagrange parameter estimates, then the moduli of the estimates are at most r (Fletcher, 1981). The estimates may be required in order to update B_k. However, when Φ_F is preferred to Φ_1, then the Lagrange multiplier estimates are more important because they also occur in the definition of the merit function. In this case it becomes attractive to use the Di Pillo and Grippo (1979) merit function (5.12) instead, in order to have the freedom to keep $\underline{\lambda}$ bounded. For example one might choose $\underline{\lambda}$ to minimize expression (5.8) subject to $||\underline{\lambda}|| \leqslant \Lambda$, where Λ is a parameter of the calculation.

These remarks suggest that it would be useful to develop methods for updating B_k that do not require Lagrange multiplier estimates. The study of the equations (3.11) by Goodman (1985) and the updating formula (4.7) may provide some suitable procedures, but it is unfortunate (and perhaps inevitable) that these approaches degrade when the difficulties of the previous two paragraphs occur.

Any user of optimization algorithms who has read thus far may have become irritated and impatient by the attention that has been given to academic questions that are less important in practice than some of the present needs of computer users. The given topics describe recent research, and, if this review were extended towards related fields that include many publications, then we would be obliged to consider more theory, such as analyses of local convergence that often depend on conditions that are seldom achieved in practice. There are, however, some good recent papers on practicalities; for example, see Gill, Murray, Saunders and Wright (1985).

Two practical questions that deserve much more attention are avoiding the calculation of first derivatives, and reducing the amount of computation of the routine parts of algorithms for nonlinear constraints. Substantial progress towards the first of these needs may require some new ideas, but, although techniques like "warm starts" in sequential quadratic programming procedures (see Gill, Murray, Saunders and Wright, 1985, for instance) have been known to be useful for several years, little attention has been given by others to efficient implementations of optimization algorithms. Here we have in mind research on actual computer programs that are available to users generally, rather than studies of abstract conceptions of efficiency that are likely to be useful, but that initially yield only theoretical results.

It has been shown that a complete state-of-the-art algorithm for nonlinear constraints includes several different components, and that many individual components are major subjects of research. Therefore much knowledge, experience and effort are required to provide good computer programs for general constrained calculations. Although many researchers publish numerical results to demonstrate the advantages of their ideas, most work of this kind is too specialized to be presented as a major contribution to a working program. It follows from these trends that there are now relatively few researchers who are developing the kind of software that makes advances in optimization algorithms of real value to computer users.

REFERENCES

Bartholomew-Biggs, M.C. (1982), "Recursive quadratic programming methods for nonlinear constraints", in *Nonlinear Optimization 1981*, ed. Powell, M.J.D., Academic Press (London), pp.213-221.

Bertsekas, D.P. (1982a), *Constrained Optimization and Lagrange Multiplier Methods*, Academic Press (New York).

Bertsekas, D.P. (1982b), "Augmented Lagrangian and differentiable exact penalty methods", in *Nonlinear Optimization 1981*, ed. Powell, M.J.D., Academic Press (London), pp.223-234.

Boggs, P.T. and Tolle, J.W. (1984), "A family of descent functions for constrained optimization", *SIAM J. Numer. Anal.*, *21*, pp.1146-1161.

Broyden, C.G. and Attia, N.F. (1984), "A smooth sequential penalty function method for solving nonlinear programming problems", in *System Modeling and Optimization*, *Lecture Notes in Control and Information Sciences*, *59*, eds. Balakrishnan, A.V. and Thomas, M., Springer-Verlag (Berlin), pp.237-245.

Byrd, R.H. (1984), "On the convergence of constrained optimization methods with accurate Hessian information on a subspace", Report CU-CS-270-84, University of Colorado, Boulder.

Byrd, R.H. and Schnabel, R.B. (1986), "Continuity of the null space basis and constrained optimization", *Math. Programming*, *35*, pp.32-41.

Byrd, R.H., Schnabel, R.B. and Shultz, G.A. (1985), "A trust region algorithm for nonlinearly constrained optimization", Report CU-CS-313-85, University of Colorado, Boulder.

Celis, M.R., Dennis, J.E. and Tapia, R.A. (1985), "A trust region strategy for nonlinear equality constrained optimization", in *Numerical Optimization 1984*, eds. Boggs, P.T., Byrd, R.H. and Schnabel, R.B., SIAM Publications (Philadelphia), pp.71-82.

Chamberlain, R.M., Lemaréchal, C., Pedersen, H.C. and Powell, M.J.D. (1982), "The watchdog technique for forcing convergence in algorithms for constrained optimization", *Math. Programming Stud.*, *16*, pp.1-17.

Coleman, T.F. and Conn, A.R. (1982), "Nonlinear programming via an exact penalty function: global analysis", *Math. Programming*, *24*, pp.137-161.

Coleman, T.F. and Conn, A.R. (1984), "On the local convergence of a quasi-Newton method for the nonlinear programming problem", *SIAM J. Numer. Anal.*, *21*, pp.755-769.

Coleman, T.F. and Sorensen, D.C. (1984), "A note on the computation of an orthonormal basis for the null space of a matrix", *Math. Programming*, *29*, pp.234-242.

Di Pillo, G. and Grippo, L. (1979), "A new class of augmented Lagrangians in nonlinear programming", *SIAM J. Control Optim.*, *17*, pp.618-628.

Fiacco, A.V. and McCormick, G.P. (1968), *Nonlinear Programming: Sequential Unconstrained Minimization Techniques*, John Wiley and Sons (New York).

Fletcher, R. (1973), "An exact penalty function for nonlinear programming with inequalities", *Math. Programming*, *5*, pp.129-150.

Fletcher, R. (1981), *Practical Methods of Optimization, Vol. 2: Constrained Optimization*, John Wiley and Sons (Chichester).

Fletcher, R. (1982), "Second order corrections for non-differentiable optimization", in *Numerical Analysis*, *Lecture Notes in Mathematics*, *912*, ed. Watson, G.A., Springer-Verlag (Berlin), pp.85-114.

Fletcher, R. (1985), "An L_1 penalty method for nonlinear constraints", in *Numerical Optimization 1984*, eds. Boggs, P.T., Byrd, R.H. and Schnabel, R.B., SIAM Publications (Philadelphia), pp.26-40.

Gill, P.E., Murray, W., Saunders, M.A., Stewart, G.W. and Wright, M.H. (1985), "Properties of a representation of a basis for the null space", *Math. Programming*, *33*, pp.172-186.

Gill, P.E., Murray, W., Saunders, M.A. and Wright, M.H. (1985), "Software and its relationship to methods", in *Numerical Optimization 1984*, eds. Boggs, P.T., Byrd, R.H. and Schnabel, R.B., SIAM Publications (Philadelphia), pp.139-159.

Gill, P.E., Murray, W., Saunders, M.A. and Wright, M.H. (1986), "Some theoretical properties of an augmented Lagrangian merit function", Report SOL 86-6, Stanford University.

Gill, P.E., Murray, W. and Wright, M.H. (1981), *Practical Optimization*, Academic Press (London).

Goodman, J. (1985), "Newton's method for constrained optimization", *Math. Programming*, *33*, pp.162-171.

Gould, N.I.M. (1986), "On the accurate determination of search directions for simple differentiable penalty functions", *IMA J. Numer. Anal.*, *6*, pp.357-372.

Han, S.P. (1977), "A globally convergent method for nonlinear programming", *J. Optim. Theory Appl.*, *22*, pp.297-309.

Lasdon, L.S. (1982), "Reduced gradient methods", in *Nonlinear Optimization 1981*, ed. Powell, M.J.D., Academic Press (London), pp.243-250.

Maratos, N. (1978), "Exact penalty functions for finite dimensional and control optimization problems", Ph.D. thesis, University of London.

Mayne, D.Q. and Polak, E. (1982), "A superlinearly convergent algorithm for constrained optimization problems", *Math. Programming Stud.*, *16*, pp.45-61.

Moré, J.J. (1982), "Recent developments in algorithms and software for trust region methods", in *Mathematical Programming: The State of the Art*, eds. Bachem, A., Grötschel, M. and Korte, B., Springer-Verlag (Berlin), pp.258-287.

Nocedal, J. and Overton, M.L. (1985), "Projected Hessian updating algorithms for nonlinearly constrained optimization", *SIAM J. Numer. Anal.*, *22*, pp.821-850.

Powell, M.J.D. (1970), "A hybrid method for nonlinear equations", in *Numerical Methods for Nonlinear Algebraic Equations*, ed. Rabinowitz, P., Gordon and Breach (London), pp.87-114.

Powell, M.J.D. (1978a), "A fast algorithm for nonlinearly constrained optimization calculations", in *Numerical Analysis Dundee, 1977, Lecture Notes in Mathematics, 630*, ed. Watson, G.A., Springer-Verlag (Berlin), pp.144-157.

Powell, M.J.D. (1978b), "The convergence of variable metric methods for nonlinearly constrained optimization calculations", in *Nonlinear Programming 3*, eds. Mangasarian, O.L., Meyer, R.R. and Robinson, S.M., Academic Press (New York), pp.27-63.

Powell, M.J.D. (ed.) (1982), *Nonlinear Optimization 1981*, Academic Press (London).

Powell, M.J.D. (1983), "Variable metric methods for constrained optimization", in *Mathematical Programming: The State of the Art*, eds. Bachem, A., Grötschel, M. and Korte, B., Springer-Verlag (Berlin), pp.288-311.

Powell, M.J.D. (1985), "The performance of two subroutines for constrained optimization on some difficult test problems", in *Numerical Optimization 1984*, eds. Boggs, P.T., Byrd, R.H. and Schnabel, R.B., SIAM Publications (Philadelphia), pp.160-177.

Powell, M.J.D. and Yuan, Y. (1986a), "A recursive quadratic programming algorithm that uses differentiable exact penalty functions", *Math. Programming*, *35*, pp.265-278.

Powell, M.J.D. and Yuan, Y. (1986b), "A trust region algorithm for equality constrained optimization", Report DAMTP 1986/NA2, University of Cambridge.

Schittkowski, K. (1983), "On the convergence of a sequential quadratic programming method with an augmented Lagrangian line search function", *Math. Operationsforschung u. Statistik. Ser. Optimization*, *14*, pp.197-216.

Vardi, A. (1985), "A trust region algorithm for equality constrained minimization: convergence properties and implementation", *SIAM J. Numer. Anal.*, *22*, pp.575-591.

M.J.D. Powell
Department of Applied Mathematics
and Theoretical Physics
University of Cambridge
Silver Street
Cambridge CB3 9EW
England

13 The Influence of Vector and Parallel Processors on Numerical Analysis

IAIN S. DUFF

ABSTRACT

It is now ten years since the first CRAY-1 was delivered to Los Alamos National Laboratory. Since then, supercomputers with vector processing capability have become widespread and important in the solution of problems in many areas of science and engineering involving large-scale computing. Their influence on numerical analysis has been less dramatic but we indicate the extent of that influence.

In the last year or so, advanced "superminis" that exhibit various more general forms of parallelism have been developed and marketed. We identify these and give some general principles which algorithm designers are using to take advantage of these parallel architectures. We argue that parallel processors are having a much stronger influence on numerical analysis than vector processors and illustrate our claims with examples from several areas of numerical analysis including linear algebra, optimization, and the solution of partial differential equations.

1. INTRODUCTION

It is interesting to reflect that, if a paper of this title had been included in the proceedings of the previous State-of-the-Art conference (Jacobs, 1977), its content would have been considerably different from the following text. It is not that parallelism was not studied. There were many early papers on the subject (for example, Miranker and Liniger, 1967, and Traub, 1973) and much analysis principally on abstract machines like the paracomputer (Schwartz, 1980) where unlimited parallelism could be allowed without concern for communication or synchronization costs. Nor was it that parallel machines did not exist. Burroughs delivered the ILLIAC IV to Illinois in 1972 (Bouknight *et al.*, 1972) and there was much activity in developing numerical algorithms for that machine (for example, Sameh, 1977). However, the predominant trend in research ten years ago was that of complexity analysis, and results of the form "the solution of $A\underline{x}=\underline{b}$ can be performed in $O(\log_2^2 n)$ time on a machine with $2n^{3.31}/\log_2^2 n$ processors" (Csanky, 1976) would have been considered at great length in a survey paper written at that time.

However, the situation in 1986 is radically different. The emphasis has changed to the construction of algorithms for specific architectures, the design of algorithmic principles for a range of parallel or vector machines, the establishment of a parallel programming methodology, and aids to portability such as language developments, language extensions, and the development of schedulers. A major reason for such a change of emphasis is the availability of machines: both high-powered vector supercomputers and, more recently, several general-purpose superminis with parallel architectures. We discuss such machines in Section 2, distinguishing between vector and parallel architectures but avoiding any detailed description of architectures. We refer the reader to Dongarra and Duff (1985) for a consumer's guide to commercially available hardware. The survey paper by Ortega and Voigt (1985) contains a summary of developments in hardware and algorithms just prior to the arrival in the marketplace of these machines.

It is also worth mentioning that several projects at universities and research laboratories are now stimulating much research and are promising to make real advances in the design of parallel architecture machines. Indeed, some of the current commercial machines have been developed directly from such projects. Examples of projects include the CEDAR project at Illinois, the NYU ULTRA project and the allied IBM RP3 effort, the CMU Warp, LCAP at IBM Kingston, the Connection project at MIT, and the Caltech hypercube.

The influence of vector computers is presented in Section 3 and a discussion of some basic concepts in parallel computing in Section 4. Then, in Section 5, we define several techniques used in algorithms which exploit parallelism; in each case we give concrete examples of such algorithms, with one or two explored in detail. In the following three sections, we briefly consider the influence of both vector and parallel processors in the areas of linear algebra, partial differential equations, and optimization, before concluding with some overall remarks on current and future trends in Section 9.

2. VECTOR AND PARALLEL PROCESSORS

The most general categorization of parallel architectures is due to Flynn (1966). Two of his main categories are SIMD (Single Instruction Multiple Data) and MIMD (Multiple Instruction Multiple Data) machines. Broadly speaking, these subdivisions correspond to vector computers and more general parallel computers. We do not, however, discuss any categorization in detail because we do not feel that it is particularly germane to our present concern. Indeed, it is not our intention to give a rigorous classification or to list all machines which are available (see, for example, Dongarra and Duff, 1985). We do, however, make a distinction between vector and parallel machines. Although we subdivide the latter between shared and local memory architectures, we make limited use of this distinction in the later sections.

We define a vector processor as one which can process operations on vectors with great efficiency but which does not exhibit a more flexible form of parallelism. For example, when we discuss vector machines *per se* we assume that it is not possible for them to execute quite different programs concurrently. Examples of vector machines are those of Cray Research (CRAY-1,

CRAY X-MP, and CRAY-2), Control Data (CYBER 205), Fujitsu (FACOM VP 50, 100, 200, and 400 ··· also marketed by Amdahl and Siemens), Hitachi (S-810 and 820), and NEC (SX-1, SX-2). We concentrate primarily on the influence of these high-powered and expensive supercomputers in Section 3 although we note that there are several other cheaper but less powerful vector machines now available, for example the IBM 3090 VF machine, the Alliant, several FPS machines, and the so-called Crayettes from Convex and Scientific Computer Systems.

Some of these vector machines have models with more than one processor with access to a shared memory (CRAY X-MP, CRAY-2, and Alliant). Other machines with a parallel architecture whose processors share a common memory include the ELXSI 6400, FLEX/32, Sequent Balance 21000, and the ENCORE Multimax. Denelcor, whose HEP computer played a significant role in the development of algorithms and methodology for shared memory machines (for example, Kowalik, 1985), has now been liquidated, a testimony to the current competitiveness in this area of scientific computing. Although we have lumped the shared memory machines together in this way, we should stress that many architectural differences exist which can have an important effect on the design and performance of algorithms. For example, access to common memory may be via a global bus or a more costly but faster switch, which itself may have a range of configurations. Furthermore, each processor may have a cache, and cache management may vary greatly from machine to machine (for example, Montry and Benner, 1985). However, in the spirit of the rest of this paper we regard such differences as affecting only the implementation details specific for the machine itself and so do not consider them further.

In local memory parallel architectures, each processor has its own memory and communicates with other processors through message passing. The local memory machines can be further subdivided according to the connections between processors.

Popular configurations include a linear array, a ring (for example, the CDC Cyberplus), a two-dimensional lattice with toroidal topology (the Goodyear MPP and the ICL DAP), or a more complex connectivity as in the BBN Butterfly. A cube-connected cycle configuration has been analysed, and algorithms have been designed for it (Preparata and Vuillemin, 1981), but we know of no commercially available machine with this architecture. Perhaps the currently most popular local memory topology is that of a hypercube, distinguished by the fact that each processor in a hypercube with 2^n processors is connected to n processors which can be identified by flipping a single bit of a binary string of length n which labels the processor. As well as the logarithmic growth in connection complexity, the hypercube is also easily scaled to reasonably many processors (>1000) and has the attractive property that a message can be passed from one processor to any other by communication paths whose length is logarithmic in the number of processors. An attractive feature is the parallel communication feature whereby messages can be sent from all the processors to all other processors concurrently in $2\log_2 n$ steps. Examples of computers exhibiting this topology are the Ametek System 14, INTEL iPSC, NCUBE, the Thinking Machines Connection Machine, and the FPS T-Series machines.

Two comments are worth making when comparing vector supercomputers with parallel machines. The first is that, although the latter are usually much slower than the former, they are far cheaper, typically costing between \$500K and \$1.5M as opposed to the more than \$10M price tag on most supercomputers. Their manufacturers claim a better price-performance characteristic compared with the supercomputers. A second difference is that many of the superminis are built from widely available chips and indeed owe their genesis to research projects of universities or noncommercial laboratories. Thus most of the companies are small, new, and presently financed by venture capital rather than sales or a major company. The viability of these companies will therefore depend on establishing a niche in the marketplace. This again justifies our determination to stand back from discussing detailed implementations, since by the time of publication any architecture being so addressed could well be obsolete. Our main reason for the foregoing catalogue of available computers is to emphasize the prevalence of vector and parallel architectures. As we said in the introduction, this availability is one of the most important factors in the influence of vector and parallel computers on numerical analysis.

3. THE INFLUENCE OF VECTOR PROCESSORS

Without doubt the advent of vector machines has had a profound influence on computation in science and engineering. Indeed, one can argue that the field of scientific computation owes its birth to the existence of such machines. Certainly, since the first CRAY was delivered to Los Alamos in 1976, increasingly complex calculations have been undertaken using vector supercomputers whose computational rates now approach one Gigaflop (a thousand Megaflops or one thousand million floating point operations per second). In spite of this impact, we would argue that the infuence on numerical analysis due to such machines is relatively small. Certainly there are few numerical analysis research papers in which the use of such architectures is the primary concern. The reason for the lack of influence lies both in the very nature of such supercomputers and in the lack of in-house access by most numerical analysts.

By definition, a supercomputer is a powerful computing engine which means that, even in nonvector mode, its capacity for computation exceeds that of other computers. Thus, without paying much attention to optimizing or designing algorithms to exploit the architecture, the use of a supercomputer should give considerable computational gains. Much of the influence of vector supercomputers in computational science has been due to this fact alone. Indeed vectorization is sometimes considered as merely an added bonus. In this context, it is not surprising that the influence on numerical analysis has not been dramatic. In particular applications, however, significant attempts have been made to exploit vectorization, and the computational science literature abounds with papers on this subject. Note that our definition is not machine specific, and indeed machines presently considered supercomputers will not be so considered in a few years time.

Since many vector machines require the use of fairly long vectors (well over 1000 in some instances) to exploit their hardware fully, much effort has been spent in reorganizing the data so that inner-loop operations are performed on long vectors (for example, see Rizzi, 1985). This is usually a top-down procedure and often results in a significant redesign of the entire method. In this sense, the influence of vector machines on computational science is very great, but few numerical analysts would regard this as numerical analysis, although some would argue that such considerations should be of more concern. The other major way in which vector processing has influenced computational science has been in the choice of solution algorithms employed, for example the choice of method for the solution of linear equations may be very dependent on the architecture being used.

However, the concept of a vector is a rather simple one in a mathematical sense and so has not caught the interest of most numerical analysts. When dealing with vectors, the natural tendency has been to think in terms of a bottom-up approach, and a primary concern has been to develop techniques to assist in the portability of codes while maintaining efficiency over a range of architectures. Much use has been made of the BLAS (Lawson *et al.*, 1979) and most manufacturers have implemented efficient versions for their machines, usually using assembler level coding.

As an example of the use of the BLAS, we consider the product of a matrix A with a vector $\underline{c}$. One can view this product as a linear combination of the columns of A, so that

$$A\underline{c} = \sum c_j A_{\cdot j} \tag{3.1}$$

where $A_{\cdot j}$ is the jth column of A, or one can obtain the ith component of the product as a scalar product between $\underline{c}$ and the ith row of A. That is

$$(A\underline{c})_i = A^T_{i\cdot}\underline{c} \tag{3.2}$$

where $A^T_{i\cdot}$ is the ith row of A. In the former case (3.1), use can be made of the SAXPY routine from the BLAS, which adds a multiple of one vector to another. In case (3.2), the scalar product routine, SDOT, is appropriate. On some vector machines, the use of SAXPY is much to be preferred. For example, on the CYBER 205, the SAXPY routine is intrinsically more efficient than the SDOT code (Louter-Nool, 1985) and accesses the columns, which are stored contiguously, rather than the rows which are not. On the other hand, a hardware peculiarity in the CRAY-1 allows the SDOT routine to perform significantly faster than SAXPY on that machine (Duff and Reid, 1982). Thus, although efficient kernels can be used to effect matrix-vector multiplication on vector machines, the choice of kernel can be very machine dependent.

This fact, coupled with a desire for portability and efficiency over a wide range of parallel and vector machines, has led many people to use larger computational modules when designing code for vector machines. An obvious candidate is matrix-vector multiplication itself. Dongarra *et al.* (1984) have formalised the higher-level modules in a proposal for an extended set of BLAS, sometimes called $O(n^2)$ BLAS since the arithmetic involved is $O(n^2)$ for vectors of length n rather than the $O(n)$ of the original BLAS. The proposal includes BLAS for matrix-vector multiplication (where the matrix can be general, symmetric, triangular, or banded), rank-one and rank-two updates to matrices, and the solution to triangular systems. It is possible to formulate many of the routines from LINPACK and EISPACK using these kernels (for example, Dongarra *et al.*, 1986a), and it is hoped that the establishment of an agreed set of extended BLAS will persuade manufacturers to provide efficient code for them in a similar way to the current situation for the original BLAS. The omens are good since already many manufacturers supply efficient code for matrix-vector multiplication.

It is worth reflecting that much debate has gone into the definition of what should be included in a set of extended BLAS. For example, Gustavson (private communication, 1985) has found that, in order to obtain high performance in the solution of equations on the IBM 3090 VF, it is necessary to code the kernel as a matrix-matrix product, while the frontal code, MA32, at Harwell (Duff, 1983) uses a double Gaussian elimination step (rank-two change to an unsymmetric matrix) to obtain nearly maximum performance on both a CRAY-1 and a CRAY-2 (Dave and Duff, 1986). It is inevitable that any proposal will not cover all needs or present usage, but the general feeling and that of the authors of the proposal is that the important thing is to establish a standard which manufacturers will recognize.

A form of parallelism more general than simple vectorization consists of the independent execution of different instantiations of the same computational sequence. It is possible to exploit this very common form of general parallelism on vector computers by vectorizing across the independent streams. We can illustrate this with the ADI method as used in the solution of elliptic or parabolic equations. At each half-step of the ADI method it is required to solve a set of k tridiagonal systems, each of order n,

$$T_i \underline{x}_i = \underline{b}_i , \qquad i = 1, 2, \cdots, k, \quad (3.3)$$

which are independent of each other. Now, the inherently recursive nature of the solution of a tridiagonal system has made it somewhat of a *cause célèbre* in the world of vectorization and, although algorithms exist which are superior to straightforward Gaussian elimination (see Section 6), highly vectorized codes do not exist. However, if one chooses as vectors the corresponding components from each system, for example $[\underline{x}_1]_1[\underline{x}_2]_1[\underline{x}_3]_1\cdots[\underline{x}_k]_1$, then the systems (3.3) can be solved as a single block tridiagonal system with n blocks of order k and vectors of length k.

It is difficult to interpret more general forms of parallelism in this way but, as we shall see in Section 5, many of these forms are of a similar kind to that described above.

4. BASIC ISSUES IN PARALLELISM

Since it is recognized that much of the intended audience of this paper may have little prior experience of concepts and issues in parallelism, some of the basic terms are discussed in this section.

Returning to vector computers for a moment, a common approximation for the time to perform a calculation on a vector of length n is

$$t_n = (n + n_{1/2})\, r_\infty^{-1} \text{ microseconds}, \tag{4.1}$$

where r_∞ is the maximum asymptotic rate of computation, and $n_{1/2}$ is the vector length at which half the asymptotic rate is attained (Hockney and Jesshope, 1981). We use the notation $n_{1/2}$ since it is now standard in the literature. We would like to stress, however, that the quantity $n_{1/2}$ is independent of n. The formula (4.1) gives rise to a performance curve of the form shown in Figure 4.1 whose shape gives quantitative justification to our desire for long vectors in Section 3. Typical values for r_∞ and $n_{1/2}$ on a 2-pipe CYBER 205 utilizing linked triads in 64-bit arithmetic are 200 and 200 respectively. The size of $n_{1/2}$ is important when designing and comparing algorithms on vector machines.

Another issue pertinent to both vector and parallel machines is the effect of vectorization (or parallelism) on the

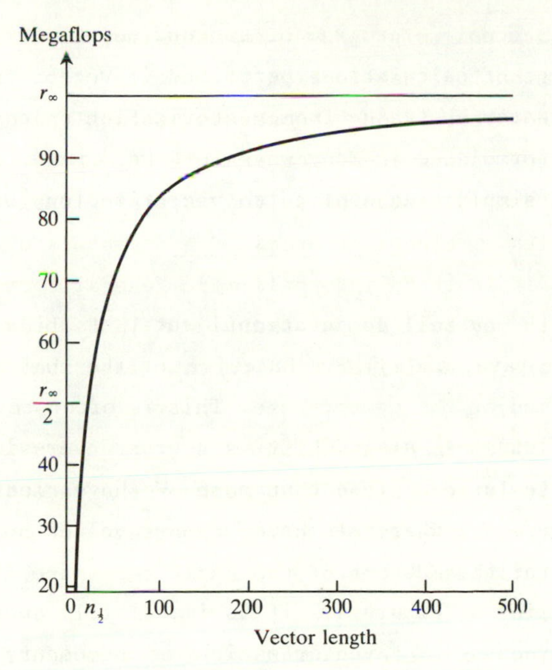

FIG. 4.1 Typical performance curve for vector machines

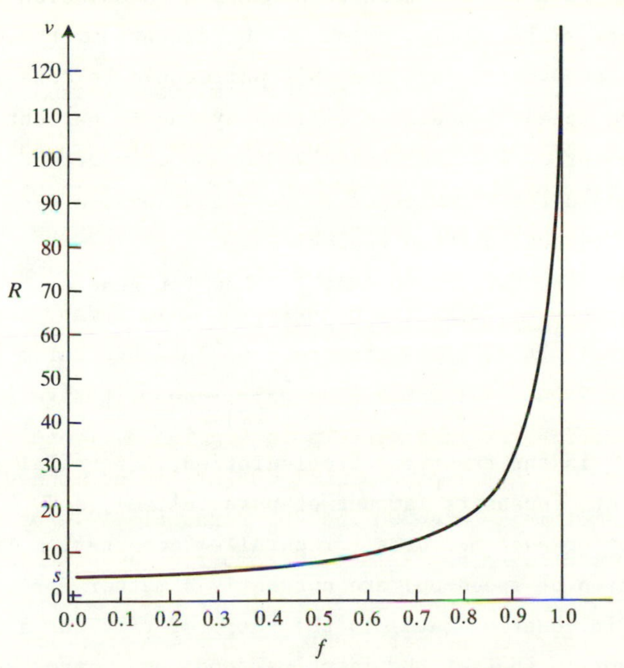

FIG. 4.2 Graph of Amdahl's Law for machine with $v = 130$ and $s = 4$

run time of a complete program or subroutine. Clearly the higher the percentage of calculations performed in vector or parallel mode, the greater the gains from vectorization or parallelism. This can be formulated in many ways and we choose one of the earliest and simplest as applied to vectorization, viz.

$$R = \frac{1}{fv^{-1} + (1-f)\,s^{-1}}, \tag{4.2}$$

where R is the overall computational rate, v the vector rate, s the scalar rate, and f the fraction of the number of operations performed at the vector rate. This is often termed Amdahl's Law (Amdahl, 1967) and, although it is a gross over-simplification, it is adequate for our present purpose. We have plotted equation (4.2) in Figure 4.2 where we have chosen values of v and s appropriate for the CRAY-1.

On looking at Figure 4.2 it is immediately apparent that a very high percentage of vectorization is necessary if overall computational rates comparable to the vector rate are to be attained. For example, even if 90 percent of the calculation were at the vector rate and this part could be done in zero time, then the speed-up would be limited by the 10 percent scalar part to a factor of 10. Fortunately, for some important calculations, for example the solution of linear equations, a very high percentage of the calculation vectorizes.

The counterpart to Amdahl's Law for general parallelism is Ware's Law (Ware, 1972) given by

$$S = \frac{1}{fp^{-1} + (1-f)}, \tag{4.3}$$

where f is the fraction of calculation in parallel mode, p the number of processors (amount of parallelism), and S the speed-up. Performance measures for parallel computation (including a definition of speed-up) are currently a matter for some debate. The S in equation (4.3) is the ratio of time for a code on one processor to that of the identical code on p processors. Some people prefer to define speed-up as the ratio that compares the best parallel algorithm with the best sequential one, which may of course be different. Indeed some people define a similar measure using the increase in problem size, for the same time of solution, as the number of processors increases. Of course equation (4.3) is quite inadequate in the parallel case (see, for example, Buzbee, 1984) since communication and synchronization delays are absent, but we use it to illustrate the risk of making over-optimistic extrapolation. For example, a speed-up of 1.8 on 2 processors may seem very encouraging but, if Ware's Law holds, the corresponding speed-up on 32 processors would be 7.2 and on 1024 processors would be 8.9, a far from exciting prospect.

The major issue in both vectorization and parallelism is that of access to data. This comes in many different guises. The phenomenon of bank conflict on vector machines is well recognized. Memory is divided into banks (sometimes as few as 4 or 8) and the same bank cannot be accessed on consecutive instructions, a delay of typically 4 clock cycles being imposed. Contiguous vectors are allocated cyclically to the banks so no conflict arises, but for sparse vectors (accessed through an index vector) or poor dimensioning, the effects of bank conflict can be severe. Bank conflict can also occur when different tasks on different processors require access to the same bank (for example, see Butel, 1985, and Meurant, 1985). On the Cyber 205, a penalty is paid for noncontiguous access so the access to data is of great importance. Some quite bizarre skew storage schemes for matrices have been proposed (for example, Jalby *et al.*, 1984, and te Riele, 1985), but we feel these are aberrations caused by poorly designed architectures and do not consider them further. On many shared memory machines a memory hierarchy exists, for example a small local memory, large registers, or a cache. Efficient data access in such an environment is clearly very important (for example, Dave and Duff, 1986, and Jalby and Meier, 1986).

An area where data access is more complicated and quite critical is that of fast Fourier transforms (FFTs). Indeed, general-purpose machines based on communication topologies specifically geared to FFTs have been designed, an example being the BBN butterfly. Special purpose boxes for FFTs and signal processing applications have existed for some time.

On a local memory machine, the distribution of and access to data is naturally one of the primary concerns since poor placement can greatly increase the amount of data traffic between processors (the routing delays) and cause severe degradation in performance. A simple example of this is seen in the Choleski factorization of a matrix on the hypercube where the distribution of the matrix to the nodes can have a significant effect on subsequent efficiency of solution. George *et al.* (1986a) and Geist and Heath (1985) have studied this effect and show that it is better to assign a column at a time in a wrap around fashion than to assign blocks of matrix columns to hypercube nodes. George *et al.* also show distribution by submatrices to be inferior.

For other algorithms (for example QR factorization) a more efficient technique may be to store by block rows (Chamberlain and Powell, 1986).

The ADI method on a hypercube (for example, Saad and Schultz, 1985c, and Johnsson *et al.*, 1985) is a case where data are accessed differently in different parts of the algorithm.

A parallel implementation of an algorithm for solving a problem of size n is said to be "consistent" if the number of elementary mathematical operations required by this implementation as a function of n is the same order of magnitude as required by the usual implementation on a serial computer (Lambiotte and Voigt, 1975). Sameh and Kuck (Kuck, 1978) use the term redundancy to describe the same phenomenon. A parallel algorithm requiring $O(n \log_2 n)$ operations would not be consistent if the best serial algorithm required $O(n)$ operations. The importance of this on a vector machine is that each arithmetic operation takes a definite amount of time, albeit much less in vector than in scalar mode. Thus, even if the $O(n \log_2 n)$ algorithm operates in vector mode while the $O(n)$ one is in scalar, at a certain problem size the asymptotic performance dominates and the parallel algorithm is poorer. A major difference between vector and parallel architectures is that the consistency argument does not hold in the same way for general parallelism since some operations are arguably free of cost. An example of this phenomenon is illustrated by Stone's method for solving tridiagonal systems in Section 5.4.

In a typical parallel computation, several calculations proceed simultaneously but from time to time must communicate unless the computation is "embarrassingly" parallel (Karp, 1986), which means that it consists mainly of independent calculations with no synchronization or communication. Some Monte Carlo algorithms fall into this category. Communication often takes the form of passing updated data from one processor to the other, either as an explicit message on local memory machines, or by a flag (event posting), or by explicit synchronization (using locks to protect critical sections) to control access to shared data on shared memory architectures. An important measure of parallelism is the amount of calculation performed without interruption and is often termed the granularity (Kung, 1980). Roughly speaking, the granularity is the amount of computational work that can be performed by a processor between synchronization points. It is large if there is much work and small otherwise. Large granularity and low synchronization overheads are ideal goals, although the penalties for not attaining them vary significantly on different architectures. When considered as a single unit, the extended BLAS discussed in Section 3 are an example of large granularity when compared with the original BLAS. Clearly a significant penalty can be paid in synchronization since one calculation may be held at a synchronization point until another has finished. This has led to the development of asynchronous algorithms, which we discuss in Section 5.5.

As an example of the influence of granularity, we consider

the solution of sparse systems using multifrontal schemes. By utilizing sparsity alone, an elimination tree can be constructed and used to drive a parallel algorithm where the only synchronization required is between son and father nodes in the tree. If the nodes are regarded as atomic, then the level of parallelism reduces to one at the root and usually increases only slowly as we progress away from the root. If, however, we recognize that parallelism can be exploited within the calculations at each node (corresponding to one or a few steps of Gaussian elimination on a full submatrix), much greater parallelism can be achieved. In Table 4.1, we give some results of Duff and Johnsson (1986) illustrating this effect. Of course, the increase in parallelism comes at the cost of smaller granularity, and the most efficient balance between these opposing effects will depend on the computer architecture.

Table 4.1 Illustration of effect of change in granularity. Values given are, in each case, the maximum speed-up.

	30 × 30 grid	10 × 100 grid
No parallelism within nodes	3.7	7.5
Parallelism within nodes	30	47

There are many other issues in parallelism, some of particular importance on local memory machines. Load balancing describes the distribution of a computation over several processors where the goal is to keep all processors equally busy. Although load balancing is of concern on shared memory machines, the currently far greater number of processors in local memory architectures creates added interest in load balancing although the full implications are currently a matter for some debate (see, for example, Iqbal *et al.*, 1986). Particular applications where load balancing is of concern are multisectioning (Section 5.2) and global optimization (Section 8).

The perimeter effect relates to the ratio of the amount of computation within a processor to the amount of data that must be transferred between the processor and other processors. In some cases, for example when solving grid-based problems by subdividing the grid between processors, the amount of data transferred (the "boundary") grows sublinearly with the computation. The implication is that the overheads of a message passing environment become less significant as problem size (and granularity) increases. Currently, this effect is rarely encountered because of the sizes of problems being run on message passing architectures. Indeed there is some debate concerning this phenomenon, because of the efficiency of passing long messages, and because the implications of local memory size can prevent an arbitrary increase in problem size. McBryan and Van de Velde (1986) have, however, observed the perimeter effect when solving large systems of hyperbolic partial differential equations on the Caltech hypercube, and Lichnewsky (1982) has observed the phenomenon when solving linear systems arising in finite-element calculations.

Our aim in this section has been to introduce some of the basic issues in parallel processing of concern to the numerical analyst, and we will feel free to use the terms without further comment in later sections. We have not, however, been completely exhaustive and recommend the book by Hockney and Jesshope (1981) for further background reading. The books by Kronsjö (1985) and Schendel (1984) discuss the design of parallel algorithms in some detail, although they are more theoretical than our present approach.

5. TECHNIQUES FOR THE DEVELOPMENT OF PARALLEL ALGORITHMS

In this section, we classify several commonly used techniques for exploiting parallelism. In doing so we are extending earlier work of Voigt (1985), van Leeuwen (1985), and te Riele (1985). Like them, we preface our categorization with the caveat that any particular algorithm may exhibit traits of more than one category. Our main intent in following this line is to give numerical analysts a feeling for the issues which influence parallel algorithm development. By giving concrete examples of each issue, we illustrate both the particular technique and its influence on numerical analysis. Because of the detail with which we explore these techniques, we have chosen to identify each in a separate subsection. In all cases we have kept references to specific parallel architectures to a minimum so that the generality of each paradigm is emphasized.

5.1 *Vectorization*

We discussed vectorization in Section 3. It is clearly a very powerful tool and, since some computers combine aspects of both vector and parallel machines (for example, CRAY X-MP, CRAY-2, and ALLIANT), the advent of more general forms of parallelism will not remove the need for vectorization. Indeed all today's supercomputers use vectorization as their principal means of achieving high-speed computation. Additionally, as we illustrated with the example of ADI in Section 3, it is often possible to reformulate parallelism as vectorization in a rather straightforward manner. A technique which uses the same philosophy as the hardware of vector processors is pipelining. We discuss a pipelined technique for QR factorizations in Section 6.

5.2 *Divide and Conquer*

Without doubt, divide and conquer is the most widespread technique used in the development of parallel algorithms. The idea is the first that would occur to anyone contemplating parallel algorithm design. One first partitions the problem into several subproblems and then solves the subproblems separately. Unless the subproblems are independent, data must be communicated between them. For the success of a divide and conquer technique, it is necessary that the gains in solving the subproblems in parallel are not outweighed by the work required to construct the solution to the whole problem from that of the subproblems.

There are many areas in numerical analysis where divide and conquer methods are employed. In linear algebra, they are used in the solution of banded and tridiagonal systems (Wang, 1981) and in obtaining eigenvalues of symmetric tridiagonal matrices (Cuppen, 1981). Bisection or more general subdivision techniques include using Sturm sequences to isolate eigenvalues (Barlow *et al.*, 1983), root-finding techniques (Kronsjö, 1985), and line-search methods (Schnabel, 1984). Problems defined over two or three space dimensions often are amenable to a subdivision process where the only information that needs to be communicated is data on the boundaries of the subdivisions. In the solution of partial differential equations, techniques of this kind are termed domain decomposition (see for example, Glowinski *et al.*, 1982, and Chan and Resasco, 1985). In structures problems, the term substructuring is used (Noor *et al.*, 1978), and in the solution of linear systems we use the term dissection (George, 1973). The original motivation for these techniques was not to design algorithms for parallelism but rather to solve very large problems by splitting them into tractable subproblems. As in "sectioning methods", which can be viewed as a one-dimensional form of substructuring, it is the task of the numerical analyst to control and combine the solutions of the subproblems in order to solve the main calculation. The added bonus of easy parallelism has spurred much work in this area, particularly in domain decomposition for the solution of partial differential equations.

Similar decomposition techniques are used in other areas, including approximation theory (see Cox, 1987, for example), global optimization (Schnabel, 1984), and combinatorial optimization (Kindervater and Lenstra, 1985; Roucairol, 1986), in addition to many nonnumerical and semi-numerical areas, such as sorting (Tseng and Lee, 1984) and convex-hull problems (Evans and Mai, 1985), which are outside the scope of this present discussion.

We will now indicate the use of divide and conquer in

more detail by looking further at algorithms in two of the above areas, namely tridiagonal eigensystems and parallel multisectioning.

The divide-and-conquer algorithm for symmetric tridiagonal eigensystems was originally proposed by Cuppen (1981) and has been developed for shared memory multiprocessors by Dongarra and Sorensen (1986) and for local memory machines by Ipsen and Jessup (1986). The idea is simply to observe that any $n \times n$ tridiagonal symmetric matrix, T, can be partitioned as

$$T = \begin{pmatrix} T_1 & \beta \underline{e}_k \underline{e}_1^T \\ \beta \underline{e}_1 \underline{e}_k^T & T_2 \end{pmatrix} = \begin{pmatrix} \hat{T}_1 & 0 \\ 0 & \hat{T}_2 \end{pmatrix} + \beta \begin{pmatrix} \underline{e}_k \\ \underline{e}_1 \end{pmatrix} (\underline{e}_k^T \ \underline{e}_1^T)$$

where T_1, T_2, $\hat{T}_1$, and $\hat{T}_2$ are tridiagonal matrices, β is the $(k, k+1)$th entry of T, and $\underline{e}_1$ and $\underline{e}_k$ are the first and kth columns of the $(n-k) \times (n-k)$ and $k \times k$ identity matrices respectively. If the eigendecompositions of $\hat{T}_1$ and $\hat{T}_2$ are given by

$$\hat{T}_1 = Q_1 D_1 Q_1^T \quad \text{and} \quad \hat{T}_2 = Q_2 D_2 Q_2^T$$

then T can be written as

$$T = \begin{pmatrix} Q_1 & 0 \\ 0 & Q_2 \end{pmatrix} \left\{ \begin{pmatrix} D_1 & 0 \\ 0 & D_2 \end{pmatrix} + \beta \begin{pmatrix} \underline{q}_1 \\ \underline{q}_2 \end{pmatrix} (\underline{q}_1^T \ \underline{q}_2^T) \right\} \begin{pmatrix} Q_1^T & 0 \\ 0 & Q_2^T \end{pmatrix},$$

where $\underline{q}_1$ is the last row of Q_1 and $\underline{q}_2$ is the first row of Q_2. Thus, the eigendecomposition of T has been reduced to the eigenproblem for a rank-one change to a diagonal matrix. This has been considered by Bunch *et al.* (1978), among others. Naturally, to obtain more parallelism the algorithm should be applied recursively by subdividing $\hat{T}_1$ and $\hat{T}_2$, and so on. The results of Dongarra and Sorensen (1986) show good speed-up on both the Denelcor HEP and the Alliant FX/8. An interesting feature of this work is that the divide-and-conquer algorithm can save work by deflation in the rank-one update problems to the extent that, even on a sequential machine, it can outperform the QR implementation in the TQL1 routine of EISPACK. For a general symmetric matrix, one must first use orthogonal similarity transformations to reduce it to symmetric tridiagonal form. Parallel methods for this calculation are discussed by Dongarra and Sorensen (1986) and Moler (private communication, 1986).

Multisectioning methods require a good adaptive load-balancing strategy since it is not known in advance where the work will be. When using the method to obtain the eigenvalues of a symmetric matrix using Sturm sequences, each processor can work independently on a different subinterval and the main problem lies in ensuring that a similar amount of work is required in each subinterval. Ipsen (private communication, 1986) does this by first running all processors on one large interval that is known to contain all the eigenvalues (obtained, for example, from Gerschgorin bounds) in order to obtain an approximate eigenvalue distribution that is used to determine the apportionment of subintervals to processors during the main phase. Bernstein and Goldstein (1986) have also developed a parallel Sturm sequence algorithm.

5.3 *Reordering*

The second main technique used in designing parallel algorithms is that of reordering the problem or data to enhance the underlying parallelism. Perhaps the area in which this technique has been most used is in the solution of linear equations from finite-difference discretizations of partial differential equations. For example, when using the SOR method to solve for a five-point discretization of the Laplacian on an $m \times n$ grid, the natural pagewise ordering of the points forbids parallelism since a value at a point is computed from its immediate predecessor in the same row. This is easily overcome by reordering the points by diagonals so that all points on each diagonal can be computed in parallel. When using successive line over-relaxation (SLOR), parallelism is most easily obtained by updating first the odd rows and then the even ones (called a zebra ordering). A red-black, or chequerboard, ordering is very beneficial for vectorization since vector lengths can be increased from n to $mn/2$. Much work has been done to evaluate the effect of different orderings on the convergence of SOR (see, for example, Adams and Jordan, 1984, O'Leary, 1984, and Adams, 1985) which seems remarkably resilient to such assaults. The use of SOR or incomplete LU factorization (ILU) preconditionings in conjunction with conjugate gradients or similar methods can also be similarly implemented in parallel by the use of multicolour orderings on the original grid (Adams, 1983). Lichnewsky (1984) combines dissection with multicolouring to develop preconditioners appropriate for multiple-processor vector machines. Erhel *et al.* (1985) investigated the effect of reorderings on preconditioned conjugate gradients in the context of vectorization.

In the one-dimensional case, if a red-black ordering is performed on a tridiagonal system being solved by a direct method, cyclic reduction results. This reordering can also be obtained from a nested-dissection technique. Thus we see one instance of the relationship between divide and conquer and reordering, illustrating the point we made in Section 5.1 concerning the overlap between categories for parallel algorithms.

Although nested dissection and minimum-degree orderings are very similar in behaviour for the amount of arithmetic and storage when used as a pivoting strategy for the solution of grid-based problems using sparse Gaussian elimination, they give fairly different levels of parallelism when used to construct an elimination tree for the parallel implementation of multifrontal methods (Duff, 1986). We illustrate this point with some results from Duff and Johnsson (1986) in Table 5.3.1.

Another reordering of a computation, particularly suitable for local memory machines, is the wavefront version of Choleski's LL^T factorization (see, for example, Kung *et al.*, 1981, and O'Leary and Stewart, 1985). We indicate the order of computation of the entries of L^T in Figure 5.3.1, where all calculations of the same number are performed in parallel and data are transferred

Table 5.3.1 Comparison of two orderings for generating an elimination tree for multifrontal solution. The problem is generated by a 5-point discretization of a 10×100 grid.

Ordering	Minimum degree	Nested dissection
Number of levels in tree	52	15
Number of pivots on longest path	232	61
Maximum speed-up	9	47

from a calculation at stage i to stage $i+1$ only. An important thing about the wavefront method is that, as soon as the operations with the first pivot have been performed on entry (2,2) (at step 3), this entry can then immediately be used as the second pivot and calculations using the second pivot row and column can sweep down the matrix immediately following the calculations using the first pivot. Subsequent pivots can follow in like fashion so that the whole decomposition is effected in $3n-2$ parallel steps. A similar strategy can be adopted in solving time-dependent problems where different waves can simultaneously be computing at different time steps. This windowing technique is discussed by Saltz and Naik (1985).

```
1 2 3 4 5 6 7 •
  3 4 5 6 7 •
    5 6 7 •
      7 •
        •
          •
```

FIG. 5.3.1 Order of computation in the wavefront method

The use of ADI techniques on hypercubes poses a rather interesting ordering problem since the data must be transposed between half sweeps in order to maintain parallelism in both directions. On some machines (for example the CYBER 205) it is best not to transpose the data, but rather to solve the systems in the second direction as a very large concatenated single tridiagonal system. However, Saad and Schultz (1985c) have found it sometimes preferable to do the transposition on the hypercube and keep the parallelism over the tridiagonal systems (see Section 3) in both directions. McBryan and Van de Velde (1986) discuss in some detail algorithms for matrix transposition on hypercubes.

In the QR method, reordering to obtain parallelism is obtained by using the suggestion of pairwise pivoting of Gentleman (1976). Clearly this freedom allows several Givens rotations to proceed at once, the overall efficiency being determined by the ordering chosen. Sameh and Kuck (1978) and Modi and Clarke (1984) have described possible orderings, and the latter scheme can be shown to require only about $\log_2 m + (n-1)\log_2(\log_2 m)$ steps on a system with m rows and n columns, when $m \gg n \gg 1$. Pairwise elimination can also be applied to Gaussian elimination in a stable fashion (Sameh, 1985; Sorensen, 1985).

5.4 *Recursive Doubling*

Recursive doubling techniques were originally suggested for vectorization but, as we shall shortly indicate, they give rise to inconsistent methods and so are asymptotically inferior. They have, however, been suggested as being more viable on parallel architectures where extra work may not be penalized. As an example of recursive doubling suppose that we wish to compute

$$\prod_{i=1}^{k} r_i, \qquad k = 1, 2, \cdots, n,$$

for given r_i, $i = 1, 2, \ldots, n$. A straightforward computation yields the n products in $n-1$ multiplications. If, however, we place $r_1, r_2, \cdots, r_n$ in a vector, $\underline{r}_0$ say, and denote by $\text{shift}_k(\underline{r})$ the operator that shifts the vector $\underline{r}$ down by 2^k positions and places 1's in the first 2^k positions, then the sequence of operations

$$\underline{r}_{k+1} = \text{shift}_k(\underline{r}_k) * \underline{r}_k \qquad k = 0, 1, \cdots,$$

will produce a vector $\underline{r}_k$, after $\lceil \log_2 n \rceil$ steps, that holds the desired products. That is, the successive $\underline{r}_k$, $k = 0, 1, \cdots$, are given by

$$\begin{array}{lllcl}
r_1 & r_1 & r_1 & & r_1 \\
r_2 & r_1 r_2 & r_1 r_2 & & r_1 r_2 \\
r_3 & r_2 r_3 & r_1 r_2 r_3 & & r_1 r_2 r_3 \\
r_4 & r_3 r_4 & r_1 r_2 r_3 r_4 & & \cdot \\
r_5 & r_4 r_5 & r_2 r_3 r_4 r_5 & & \cdot \\
r_6 & \cdot & \cdot & \cdots\cdots\cdots & \cdot \\
r_7 & \cdot & \cdot & & \cdot \\
\cdot & \cdot & \cdot & & \cdot \\
\cdot & \cdot & \cdot & & \cdot \\
\cdot & \cdot & \cdot & & \cdot \\
r_n & r_{n-1} r_n & r_{n-3} r_{n-2} r_{n-1} r_n & & r_1 r_2 \cdots r_n .
\end{array}$$

Unfortunately, the complexity of each vector multiply is $O(n)$, giving $O(n \log_2 n)$ multiplications in total as opposed to $O(n)$ for the straightforward algorithm. Van Leeuwen (1985) gives a list of functions amenable to recursive doubling.

An algorithm for solving tridiagonal systems based on recursive doubling was proposed by Stone (1973). If one defines

$$q_0 = 1, \; q_1 = a_1, \; q_i = a_i q_{i-1} - c_i b_{i-1} q_{i-2}, \qquad i = 2, \cdots, n,$$

then the LU factorization of the matrix

$$\begin{array}{cccccc}
a_1 & b_1 & & & & \\
c_2 & a_2 & b_2 & & & \\
 & c_3 & a_3 & \cdot & & \\
 & & \cdot & \cdot & \cdot & \\
 & & & \cdot & \cdot & \\
 & & & & \cdot &
\end{array}$$

includes the entries

$$u_i = \frac{q_i}{q_{i-1}},$$

where the u_i are the diagonal entries of U. We see this by comparing the expression for generating the q_i above with the normal factorization equation

$$u_i = a_i - \frac{c_i b_{i-1}}{u_{i-1}} .$$

Stone observed that the recurrence for q_i can be written as

$$Q_i \equiv \begin{pmatrix} q_i \\ q_{i-1} \end{pmatrix} = \begin{pmatrix} a_i & -c_i b_{i-1} \\ 1 & 0 \end{pmatrix} \begin{pmatrix} q_{i-1} \\ q_{i-2} \end{pmatrix} \equiv G_i Q_{i-1} = \prod_{j=2}^{i} G_j Q_1$$

from which the use of recursive doubling immediately follows. Because of its instability and inconsistency this method fell from favour but is now being revived, together with an investigation of methods for stabilization, in a more parallel environment. Doubling algorithms for Toeplitz systems are discussed by Brent *et al.* (1980) and Morf (1980).

5.5 *Asynchronous Techniques*

As discussed in Section 4, the need for synchronizing parallel processes can be a major problem in achieving effective levels of parallelism. Therefore techniques have been developed where processes can execute asynchronously without having to suspend computation while awaiting the results from another process.

It is fairly easy to devise an asynchronous version of almost any iterative process. For example, Kronsjö (1985) discusses an asynchronous version of the Newton-Raphson iteration. However, it is usually more difficult to analyse the convergence or stability of such a technique.

An algorithm of this type, whose initial development predated the current interest in parallelism, is chaotic relaxation (Chazan and Miranker, 1969; Miellou, 1975; Baudet, 1978) where, for example, new values of the variables on a five-point discretized mesh are calculated from the four neighbouring values without regard to when these values were calculated. Experimentation has been done on parallel architectures (for example, at Los Alamos, Hiromoto, private communication, 1985) that confirms the earlier analysis of Chazan and Miranker (1969) and indicates the feasibility of such techniques.

We discuss in more detail a suggested asynchronous algorithm for checking the convergence of an iterative process (Saltz *et al.*, 1986). Here we assume that a region has been divided into subdomains, each assigned to a processor. Iterations continue on each subdomain with neighbouring subdomains communicating with each other, as mentioned in Section 5.2. There is therefore a local synchronization between neighbouring subdomains but this is not very costly if there is reasonable load balancing. Checking for convergence does, however, present more of a problem, and it is this part of the algorithm for which an asynchronous method is used. Each processor can conduct a convergence check on data in its subdomain leading to either a nonconvergent result or a tentatively converged result (tentative because later data coming in from the boundaries may cause future nonconvergence). Each processor keeps track of its number of iterations and a "header" is defined as the iteration count at the beginning of a tentatively converged sequence. A separate processor is designated as a "host". The algorithm then proceeds as follows, with action (2) being performed by the host and actions (1) and (3) by the other processors, all in an asynchronous fashion.

(1) As soon as a subdomain (assigned to processor k, say) reaches tentative convergence, processor k sends its header, i_k, to the host.

(2) Host waits until all processors have reported and then calculates

$$i_{\max} = \max\{i_k\}\ .$$

It compares $i_{\max}$ with its previously calculated value (if any) and, if equal, the host broadcasts a stop to all processors and stops itself. Otherwise, the host broadcasts the new value to all processors and repeats action (2).

(3) When the processor receives prompt from host, it sends

back its current header if tentatively convergent, or the next one if it is not.

Saltz *et al.* (1986) have programmed this method on a hypercube and find that the number of unnecessary calculations because of processors continuing iterations after convergence is very low (3–5%).

5.6 *Explicit Methods*

Clearly explicit methods are very suitable for vectorization or parallelism. Thus, if

$$x_i^{n+1} = f_i(\underline{x}^n, \underline{x}^{n-1}, \cdots), \qquad i = 1, 2, \cdots,$$

then each new x_i can be calculated in a parallel or vector operation.

An example of an explicit method for solving one-dimensional time-dependent partial differential equations is Euler's method, but its stability is only assured by restricting the time step drastically. The Crank-Nicolson method is unconditionally stable but requires the solution of a tridiagonal system at each step. Evans and Abdullah (1983) suggest combining two steps of a semi-explicit method of Saul'yev (1964) to get an unconditionally stable method which is only implicit because a set of 2×2 systems must be solved at each stage. These systems are trivially invertible to yield an explicit method, termed the group explicit method by Evans and Abdullah. Indeed any method that is implicit because of the solution of small blocks may also exploit parallelism by solving its implicit systems in parallel. Gelenbe *et al.* (1982) experiment with both implicit and explicit methods for the one-dimensional heat equation using two processors and develop a probabilistic model to assist in their analysis.

A simple example of converting an implicit algorithm to an explicit one is the use of a Neumann series expansion

$$(I-A)^{-1} = I + A + A^2 + \cdots$$

to approximate the inverse of a matrix. Thus, the solution of a set of equations is sometimes possible by matrix-vector multiplication only — an explicit process. An example is given by van der Vorst (1982), who uses a truncated Neumann series approximation to $(I-L)^{-1}$, L a lower triangular matrix, as a preconditioner for conjugate gradients.

5.7 *"New" Algorithms*

It is often claimed that no new algorithms have been developed because of the advent of parallel processors. To some extent this is correct, although it is largely because such machines have only just recently become readily available, and much of the current effort is in trying to implement known techniques and modifications of them on the new architectures. Certainly many methods with obvious implications for parallelism have been proposed over the last few years. We have seen some examples in earlier sections and would maintain that much of the present interest in block techniques, including block preconditioning methods (Concus *et al.*, 1985), has been generated because of their application to parallel and vector machines. Indeed claims have been made for new algorithms which are just block factorizations. For example, the WZ algorithm of Evans and Hatzopoulos (1979) is identical to block 2×2 pivoting and can also be used as a splitting for SOR like methods or as a preconditioner (Evans and Sojoodi-Haghighi, 1982).

New interest in old algorithms has been sparked by the advent of parallel architectures. For example, Gauss-Jordan elimination avoids the need for the recursive back-substitution step and is preferable to Gaussian elimination on some architectures in spite of its larger number of arithmetic operations (Bowgen *et al.*, 1984). Indeed, the use of explicit inverses and a resurrection of Cramer's rule (Swarztrauber, 1979) have also been suggested on parallel machines.

One novel idea is the use of multi-splittings (O'Leary and

White, 1983), where several iterations with different splittings of the same matrix are executed in parallel. Some analysis of this idea has been done, but no results have been reported of practical experience on parallel architectures.

Recently there has been development of a class of methods for solving partial differential equations called cellular automata or micro-random-walk methods. The technique is based on monitoring the motion of particles in cells and the principal calculations involved are nonnumerical in the sense that most of the work involves manipulation of bit vectors. Because of their possible importance in the solution of numerical problems, it is clearly appropriate to mention them here. Much work in this area has been done by Wolfram on the Connection machine, and cellular automata have also been studied by Frisch *et al.* (1986a, 1986b), *inter alios*. Indeed it may well be that the existence of parallel architectures gives rise to the further development of this whole class of methods.

6. INFLUENCE ON LINEAR ALGEBRA

In the next three sections, we survey the influence of vector and parallel architectures on three major areas in numerical analysis. As with many new developments in the field, the area in which most effect has been felt is linear algebra. This is both because of the very well-defined nature of many linear algebra algorithms and their central role in other numerical areas. Indeed, some people have argued that the advent of vector and parallel machines has revitalised numerical linear algebra in a similar way to the development of sparse matrix techniques in the 1970s. We discuss the influence on linear algebra in this section by indicating some of the areas in which parallel or vector algorithms have been developed. From bibliographies on vector and parallel computing, for example that of the Bochum Computing Centre (Bernutat-Buchmann *et al.*, 1983), it is evident that the other areas most influenced are the solution of partial differential equations and optimization. Indeed, out of around 1000 numerical references in the Bochum bibliography, 192 are on linear algebra, 116 on partial differential equations, and 57 on optimization (te Riele, 1985). We discuss the influence on partial differential equations in Section 7 and on optimization in Section 8. In our conclusions in Section 9, we touch briefly on other areas of numerical analysis.

Since we are discussing the influence on partial differential equations in the next section, we will leave a discussion of linear algebra algorithms particularly geared to discretizations of partial differential equations until then. Here we do not explicitly consider the source of the linear algebra problem. Earlier reviews in this area have been given by Gentleman (1978), Heller (1978), Sameh (1983), and Ipsen and Saad (1985).

Tridiagonal systems play a critical role in numerical analysis so it is not surprising that they have received much attention. They have particularly been of interest in the study of parallelism because of the inherently recursive and sequential nature of such systems. We have already discussed several algorithms for tridiagonal systems which can exploit vectorization or parallelism, including Stone's method (Stone, 1973) and cyclic or odd-even reduction (Buneman, 1969). Another variant arises when we effectively perform an odd-even reduction on all equations instead of on alternate ones. This is called odd-even elimination, and one stage of this algorithm on the matrix

$$\begin{bmatrix} a_1 & b_1 & & & & \\ c_2 & a_2 & b_2 & & & \\ & c_3 & a_3 & b_3 & & \\ & & \cdot & \cdot & \cdot & \\ & & & \cdot & \cdot & \\ & & & & & \cdot \\ & & & & & \cdot \quad \cdot \\ & & & & & \cdot \quad \cdot \end{bmatrix}$$

eliminates the entries adjacent to the diagonal in the way that produces the form

$$\begin{bmatrix}
a_1^1 & 0 & b_1^1 & & & & & & \\
0 & a_2^1 & 0 & b_2^1 & & & & & \\
c_3^1 & 0 & a_3^1 & 0 & b_3^1 & & & & \\
 & c_4^1 & 0 & a_4^1 & 0 & & & & \\
 & & \cdot & 0 & \cdot & & & & \\
 & & & \cdot & 0 & & 0 & \cdot & \\
 & & & & \cdot & & \cdot & 0 & \cdot \\
 & & & & & & 0 & \cdot & 0 \\
 & & & & & & \cdot & 0 & \cdot
\end{bmatrix}.$$

Next we eliminate similarly the new nonzero off-diagonal entries, and after $\log_2 n$ such reduction steps, the resulting system is diagonal. Like Stone's method, odd-even elimination, termed PARACR by Hockney and Jesshope (1981), is not consistent and is unstable, but further work on this algorithm continues. A divide-and-conquer strategy applied to tridiagonal matrices gives a partitioning scheme where most of the work can be performed on separate tridiagonal systems, the communication between them being kept low. An example of a partitioning scheme is that of Wang (1981), discussed in detail by Duff, Erisman and Reid (1986).

Partitioning schemes can be extended to more general banded systems (Johnsson, 1985; Dongarra and Johnsson, 1986) and are suitable for implementation on both shared and local memory machines. Another possibility is to consider the banded system as a block tridiagonal system and use block cyclic reduction (Hockney and Jesshope, 1981) in an analogous fashion to the point case discussed earlier. Wavefront methods akin to those discussed in Section 5.3 have also been proposed and developed for banded matrices, particularly those whose band is large (Saad and Schultz, 1985b).

For full systems of equations, the use of the extended BLAS can have a marked effect on vector machines (Dongarra and

Eisenstat, 1984), although we do not know of good applications of this approach to local memory parallel architectures. On message passing architectures, several implementations of Choleski's method (Geist and Heath, 1985) and Gaussian elimination (Geist, 1985) have been proposed. These divide the columns in an *a priori* way across the processors (the cyclic reuse of processors is recommended on the hypercube by, for example, Geist and Heath, 1985) and perform a column oriented reduction, passing information from a reduced column to its successors. On systolic or data-flow architectures, wavefront algorithms as discussed in Section 5.3 have been proposed (Kung *et al.*, 1981; O'Leary and Stewart, 1985). There is, of course, abundant small granularity parallelism in Gaussian elimination, and one way of exploiting this is to perform pairwise pivoting, where, for example, at the first minor step rows $1,3,5,\cdots$ are used to produce zeros in the first column of rows $2,4,\cdots$ respectively. Pairwise pivoting has been discussed for ring architectures by Sameh (1985) and its stability analysed by Sorensen (1985). Block elimination methods (for example, the WZ algorithm of Evans and Abdullah, 1983) have also been proposed and studied on local memory machines like the ICL DAP and the loosely-coupled Loughborough NEPTUNE (4 fully connected VAX 750s). Gauss-Jordan elimination has been proposed for the DAP, and a hybrid algorithm that uses block Gaussian elimination with Gauss-Jordan on the individual blocks has been developed (Bowgen *et al.*, 1984). The often discredited techniques of Cramer's rule (Swarztrauber, 1979) and explicit inverses have also been considered.

The use of a QR decomposition for solving full matrix equations, or more commonly for use in the least-squares solution of overdetermined systems, has received much attention. Pairwise ordering of the kind mentioned for Gaussian elimination in the previous paragraph can be used, and indeed many orderings have been proposed to exploit the independence of orthogonal rotations between different pairs of rows. Ordering schemes have been suggested by several people including Sameh and Kuck (1978), and Modi and Clarke (1984), and experiments have been performed on a range of parallel architectures. In pipelined QR, the matrix R is developed by reducing each row in turn using Givens rotations and is stored in a linear array (pipe) divided into segments. Each row of the matrix in turn is passed through the pipe with the only synchronization being that a row cannot enter a segment until it has been cleared by the previous row in the sequence (Dongarra *et al.*, 1986b). Another technique which has been programmed for the hypercube by Chamberlain and Powell (1986) is to partition the matrix into block rows and perform most of the reduction within each block in an asynchronous fashion, communicating information between the blocks when required. Dongarra *et al.* (1986b) discuss and compare three methods for QR decomposition, namely the use of the extended BLAS in LINPACK routines, the windowed Householder method which is similar in philosophy to the wavefront methods discussed in Section 5.3, and the pipelined Givens method. Parlett and Schreiber (1986) describe a block QR method suitable for implementation on systolic arrays.

For sparse systems, variants of the distributed Choleski algorithm (George *et al.*, 1986b) and the pipelined Givens method (Heath and Sorensen, 1986) can be developed as trivial extensions of the algorithms for full systems. Multifrontal techniques are also proving popular, either with automatic subdivision yielding small granularity (Duff, 1986; Liu, 1985), or using larger granularity by generating a simple subdivision of an underlying region followed by use of the frontal method on each region (Benner, 1986; Geist, private communication, 1985; Berger *et al.*, 1985). In general, partitioning methods can be used to split a matrix into subproblems (for example, Duff *et al.*, 1986) that can then be handled in parallel. A good splitting or tearing is one which produces several approximately equal subproblems with only a few variables in the tear set. Several algorithms have been proposed but we know of no extensive comparisons or implementations on parallel architectures. Also there are unresolved stability problems when this approach is used on general systems. There are several approaches available for sparse systems arising from discretized partial differential equations but we defer discussion of these until the next section.

We discussed a divide-and-conquer technique for the symmetric tridiagonal eigenproblem in Section 5.2. A Sturm sequence approach using multisectioning has also been proposed (Barlow *et al.*, 1983). Jacobi methods have made a comeback because of the ease with which $[n/2]$ sets of transformations can be performed in parallel (for example, Sameh, 1971, Modi and Pryce, 1985, and Karp and Greenstadt, 1986). Methods which only use the matrix in forming matrix-vector products, such as simultaneous iteration or Lanczos methods, can exploit any parallelism both in the matrix-vector product and in the solution of any small auxiliary eigensystem. Although such methods are particularly useful for large sparse systems (see, for example, Parlett, 1980, and Cullum and Willoughby, 1985), they have recently been recommended for solving large dense eigenproblems when only a few eigenpairs are required (Grimes *et al.*, 1985).

One research area in which many algorithms and much theory have been developed is the implementation of numerical linear algebra algorithms on systolic architectures (see, for example, Robert and Tchuente, 1982, and Bentley and Kung, 1983). Much work has been done on algorithms for eigenvalues and the singular value decomposition, in addition to linear equations (see, for example, Heller and Ipsen, 1983, Kung, 1984, Brent and Luk, 1985, and Brent *et al.*, 1985). We have not, however, discussed this work here, since we are unaware of any commercially available systolic architecture computer and a main theme of our presentation is the influence of readily available machines. The wavefront method mentioned in Section 5.3 is, however, suitable for systolic architectures.

7. INFLUENCE ON PARTIAL DIFFERENTIAL EQUATIONS

An obvious way of exploiting parallelism in the solution of partial differential equations is to develop methods for the parallel solution of the resulting discretized systems. Most iterative methods, for example conjugate gradients, require one or more matrix-vector multiplications, some scalar products, and some vector additions at each iteration. Although it is not trivial (see, for example, the discussion of a parallel version of conjugate gradients by Meurant, 1985, or Saad and Schultz, 1985a), it is usually possible to obtain high levels of vectorization or parallelism. For fast convergence, some form of preconditioning is normally required and it is usually this part of the algorith where parallel implementation is difficult. We discussed the use of a truncated Neumann series by van der Vorst (1982, 1986) in Section 5.6, and more recently Meurant (1985) and Seager (1986) have discussed more general parallel preconditionings.

We considered block and point SOR techniques in Section 5.3. Here parallelism can be obtained through multicolourings of the underlying grid (Adams and Jordan, 1986; Adams, 1985; Schreiber and Tang, 1982). Other splittings for exploitation of parallelism are possible, for example the QIF method of Evans and Sojoodi-Haghighi (1982) or the multisplitting techniques of O'Leary and White (1983). Chaotic relaxation (Chazan and Miranker, 1969) has also been tried on parallel architectures (Hiromoto, private communication, 1985). Spectral methods can make ready use of any efficient implementation of FFTs, and efficient use of both vector and parallel architectures is possible (Ortega and Voigt, 1985). Indeed the central role of the FFT algorithm is clear from the fact that 111 references in the 1983 Bochum bibliography were to methods for implementing vectorized or parallel versions of FFTs (Swarztrauber, 1986). FACR techniques (Hockney and Jesshope, 1981) also benefit from the parallel implementation of FFTs.

Multigrid techniques have also been examined with a view

to designing and implementing algorithms on parallel architectures. Many approaches have been suggested including those of Gannon and van Rosendale (1982) and Greenbaum (1985).

ADI methods have a quite natural form of parallelism when used in the solution of either time-dependent problems or elliptic problems. A technique for time-dependent problems called windowing, akin to the wavefront method of Section 5.3, has been proposed for parallel implementation (Saltz and Naik, 1985). Here simultaneous computation is performed on different time steps.

Clearly finite-element calculations admit ready parallelism both in independent computation within each element (for example, Adams and Voigt, 1984) and in the assembly process (Berger *et al.*, 1985). Domain decomposition techniques are becoming increasingly more refined and give immediate parallelism for either finite-difference or finite-element discretizations (see, for example, Glowinski *et al.*, 1982, and Keyes and Gropp, 1985).

8. INFLUENCE ON OPTIMIZATION

A central problem in numerical optimization is often the solution of a set of linear equations. We discussed it in Section 6 and so will here consider other ways in which optimization techniques can exploit parallelism.

An obvious area where divide-and-conquer strategies can be used to good effect is that of global optimization (Schnabel, private communication, 1985). Other allied problems are obtaining feasible points, and nonlinear minimax problems (Schnabel, 1984, and Lootsma, 1984).

Two ways in which many optimization techniques can benefit from parallelism are within the function evaluations themselves (for example, these might involve the solution of partial differential equations), and in the parallel evaluation of difference approximations to gradients, where, for example, the necessary function evaluations could be computed on separate processors, and gains could be achieved if these evaluations were expensive (thus keeping the granularity large). Such parallelism could also be exploited in the sparse case although there many less evaluations would be required.

Recently, methods based on holding and updating a set of conjugate directions (Powell, 1964) have been proposed in a parallel context (Han, 1986) where independent line searches along each direction can be conducted in parallel and can be used effectively to compute a new estimate of the solution. Indeed, a line search itself can be implemented using parallel multisection techniques although there are few instances when sufficient accuracy is required to merit such a thorough search.

A natural candidate for parallel exploitation is that of partial separability (Griewank and Toint, 1984). Obviously, full separability is embarrassingly parallel and the more separable a problem is, the more amenable it is to parallel techniques.

Dixon (1985) discusses the use of optimization techniques in the solution of partial differential equations. He proposes exploiting the parallel nature of finite elements so that the parallelism then lies with the differential equation rather than the optimization technique. It is an example of parallel function evaluation.

Often the minimum values of several functions

$$F(\underline{x}, p)$$

are required for a range of the parameter p. Sometimes individual problems in this parametric family can be solved independently; at other times the dependence may not be total. In both cases, parallelism can be exploited.

Finally, there are many problems in combinatorial optimization which are amenable to parallelism, many of them using branch-and-bound techniques, for example the travelling-salesman problem, and the knapsack problem. A good review of this area is given by Kindervater and Lenstra (1985) so we will not discuss it further here.

9. CONCLUSIONS

We hope that we have illustrated the wide and strong influence that parallel computers are having on numerical analysis. We have chosen some particular algorithmic paradigms and particular areas in numerical analysis but make no claims that these are exhaustive. Indeed parallel algorithms have been proposed for quadrature, the solution of ordinary differential equations, and approximation problems, although their influence has been much less strong than in the areas discussed in Sections 6 to 8.

When polling several colleagues for suggestions on the theme of this paper, more than one suggested that by far the most important influence of parallelism on numerical analysis is that it is difficult to get funding without mentioning it in a proposal. The effect of this is somewhat two-edged. It does indeed mean that there is an increased amount of research on parallel algorithms and some of it is good, but it also means that much inferior work is supported and many poor reports distributed. Such is the danger of the bandwagon.

It seems appropriate, when discussing the current state of the art, to stick one's neck out and make some comments on likely future trends. A major problem with the flurry of new and perhaps poorly-tested ideas is that it soon becomes very difficult to see the wood for the trees. Few comparisons have been conducted between competing methods. For example, Karp and Greenstadt (1986) have suggested that even well implemented versions of parallel Jacobi methods will be inferior to an optimized QR algorithm on almost all architectures, even for near-diagonal matrices. This somewhat contentious claim warrants further investigation. Indeed I suspect that there will be much more work in the future aimed at weeding out the less successful suggestions for parallel algorithms (on any architecture) in a similar way to the demise of recursive doubling on vector machines. Further consolidation should result in the equivalence or near equivalence of suggested approaches, such as the WZ algorithm and 2 × 2 block elimination. Another exciting trend could well arise from the development of methods like that of cellular automata for solving partial differential equations that we mentioned in Section 5.7. Here essentially nonnumerical algorithms can be used to solve numerical problems. Additionally these methods require high degrees of parallelism for efficient implementation.

First, perhaps the real crunch question will be answered in the next ten years: can high levels of architectural parallelism be used in the solution of real problems. We live in hope.

Acknowledgements

I would like to thank the people who responded to a request of mine for information on parallelism in other areas than linear algebra and to thank my colleagues and visitors to Harwell, John Reid, Nick Gould, and Ilse Ipsen for their comments on an early draft. The first draft of this paper was written while the author was visiting INRIA at Rocquencourt near Paris, and I am grateful to Alain Lichnewsky and Francois Thomasset and the CAPRAN project for their support and comments. I am also grateful to Mike Powell, the editor of the State-of-the-Art Proceedings, for his many detailed comments.

REFERENCES

Adams, L.M. (1983), "An M-step preconditioned conjugate gradient method for parallel computation", in *Proceedings International Conference on Parallel Processing, Bellaire, Michigan, August 1983*, IEEE Computer Society Press (New York), pp.36-43.

Adams, L.M. (1986), "Reordering computations for parallel execution", *Comm. Appl. Numer. Methods 2*, pp.263-271.

Adams, L.M. and Jordan, H.F. (1986), "Is SOR color-blind?", *SIAM J. Sci. Statist. Comput. 7*, pp.490-506.

Adams, L.M. and Voigt, R.G. (1984), "A methodology for exploiting parallelism in the finite element process", in *High-speed Computation*, ed. Kowalik, J.S., Springer-Verlag (Berlin), pp.373-392.

Amdahl, G. (1967), "Validity of the single processor approach to achieving large scale computing capabilities", in *AFIPS Conference Proceedings 30*, Thompson Books (Washington DC), pp.483-485.

Barlow, R.H., Evans, D.J. and Shanehchi, J. (1983), "Parallel multisection applied to the eigenvalue problem", *Comput. J. 26*, pp.6-9.

Baudet, G.M. (1978), *The Design and Analysis of Algorithms for Asynchronous Multiprocessors*, Ph.D. Thesis (Report CMU-CS-78-116), Carnegie Mellon University, Pittsburgh.

Benner, R.E. (1986), "Shared memory, cache, and frontwidth considerations in multifrontal algorithm development", Report SAND85-2752, Sandia National Laboratories, Albuquerque.

Bentley, J.L. and Kung, H.T. (1983), "An introduction to systolic algorithms and architectures", *Naval Research Reviews 35 (2)*, pp.3-16.

Berger, P., Dayde, M. and Fraboul, C. (1985), "Experience in parallelizing numerical algorithms for MIMD architectures use of asynchronous methods", *Rech. Aérospat. 5*, pp.325-340.

Bernstein, H.J. and Goldstein, M. (1986), "Parallel implementation of bisection for the calculation of eigenvalues of tridiagonal symmetric matrices", *Computing 37*, pp.85-91.

Bernutat-Buchmann, U., Rudolph, D. and Schloßer, K.-H. (1983), *Parallel Computing I, Eine Bibliographie, Second Edition*, Computing Centre (Bochum).

Bouknight, W., Denenberg, S., McIntyre, D., Randall, J., Sameh, A. and Slotnick, D. (1972), "The ILLIAC IV System", *Proc. IEEE 60*, pp.369-388.

Bowgen, G.S., Hunt, D.J. and Liddell, H.M. (1984), "The solution of n linear equations on a p processor parallel computer", Report 2.30, DAP Support Unit, Queen Mary College, London.

Brent, R.P., Gustavson, F.G. and Yun, D.Y.Y. (1980), "Fast solution of Toeplitz systems of equations and computation of Padé approximants", *J. Algorithms 1*, pp.259-295.

Brent, R.P. and Luk, F. (1985), "The solution of singular-value and symmetric eigenvalue problems on multiprocessor arrays", *SIAM J. Sci. Statist. Comput. 6*, pp.69-84.

Brent, R.P., Luk, F. and Van Loan, C.F. (1985), "Computation of the singular value decomposition using mesh-connected processors", *J. VLSI Comput. Syst. 1*, pp.242-270.

Bunch, J.R., Nielsen, C.P. and Sorensen, D.C. (1978), "Rank-one modification of the symmetric eigenproblem", *Numer. Math. 31*, pp.31-48.

Buneman, O. (1969), "A compact non-iterative Poisson solver", Report 294, Institute for Plasma Research, Stanford University.

Butel, R. (1985), "Conflicts between 2 vector transfers in CRAY-XMP computers", Report 418, INRIA, Rocquencourt, France.

Buzbee, B.L. (1984), "Plasma simulation and fusion calculation", in *High-speed Computation*, ed. Kowalik, J.S., Springer-Verlag (Berlin), pp.417-423.

Chamberlain, R.M. and Powell, M.J.D. (1986), "QR factorization for linear least squares problems on the hypercube", Report CCS 86/10, Chr. Michelsen Institute, Bergen, Norway.

Chan, T.F. and Resasco, D.C. (1985), "A domain-decomposed fast Poisson solver on a rectangle", Report YALEU/DCS/RR-409, Yale University.

Chazan, D. and Miranker, W. (1969), "Chaotic relaxation", *Linear Algebra Appl.* *2*, pp.199-222.

Concus, P., Golub, G.H. and Meurant, G. (1985), "Block preconditioning for the conjugate gradient method", *SIAM J. Sci. Statist. Comput.* *6*, pp.220-252.

Cox, M.G. (1987), "Data approximation by splines in one and two independent variables", in *The State-of-the-Art in Numerical Analysis*, this volume, pp.111-138.

Csanky, L. (1976), "Fast parallel matrix inversion algorithms", *SIAM J. Comput.* *5*, pp.618-623.

Cullum, J.K. and Willoughby, R.A. (1985), *Lanczos Algorithms for Large Symmetric Eigenvalue Computations, Vol. 1. Theory, Vol. 2. Programs*, Birkhäuser (Boston).

Cuppen, J.J.M. (1981), "A divide and conquer method for the symmetric tridiagonal eigenproblem", *Numer. Math.* *36*, pp.177-195.

Dave, A.K. and Duff, I.S. (1986), "Sparse matrix calculations on the CRAY-2", Report CSS 197, AERE Harwell (*Parallel Comput.*, to be published).

Dixon, L.C.W. (1985), "An introduction to parallel computers and parallel computing", Report 166, Numerical Optimisation Centre, The Hatfield Polytechnic, England.

Dongarra, J.J., Du Croz, J., Hammarling, S. and Hanson, R.J. (1984), "A proposal for an extended set of Fortran Basic Linear Algebra Subprograms", Report TM 41, Mathematics and Computer Science Division, Argonne National Laboratory.

Dongarra, J.J. and Duff, I.S. (1985), "Advanced Computer architectures", Report TM 57, Mathematics and Computer Science Division, Argonne National Laboratory.

Dongarra, J.J. and Eisenstat, S.C. (1984), "Squeezing the most out of an algorithm in CRAY Fortran", *ACM Trans. Math. Software* *10*, pp.221-230.

Dongarra, J.J. and Johnsson, S.L. (1986), "Solving banded matrices in prallel", *Parallel Comput.*, to be published.

Dongarra, J.J., Kaufman, L. and Hammarling, S. (1986a), "Squeezing the most out of eigenvalue solvers on high-performance computers", *Linear Algebra Appl.* *77*, pp.113-136.

Dongarra, J.J., Sameh, A.H. and Sorensen, D.C. (1986b), "Implementation of some concurrent algorithms for matrix factorization", *Parallel Comput.* *3*, pp.25-34.

Dongarra, J.J. and Sorensen, D.C. (1986), "A fully parallel algorith, for the symmetric eigenvalue problem", Report TM 62, Mathematics and Computer Science Division, Argonne National Laboratory.

Duff, I.S. (1983), "Enhancements to the MA32 package for solving sparse unsymmetric equations", Report AERE R11009 HMSO, London.

Duff, I.S. (1986), "Parallel implementation of multifrontal schemes", *Parallel Comput.* *3*, pp.193-204.

Duff, I.S., Erisman, A.M. and Reid, J.K. (1986), *Direct Methods for Sparse Matrices*, Oxford University Press (London).

Duff, I.S. and Johnsson, S.L. (1986), "Node orderings and concurrency in sparse problems: an experimental investigation", Harwell Report, to be published.

Duff, I.S. and Reid, J.K. (1982), "Experience of sparse matrix codes on the CRAY-1", *Comput. Phys. Comm.* *26*, pp.293-302.

Erhel, J., Lichnewsky, A. and Thomasset, F. (1985), "Vectorizing finite element methods", Report 383, INRIA, Rocquencourt, France.

Evans, D.J. and Abdullah, A.R.B. (1983), "A new explicit method for the diffusion equation", in *Numerical Methods in Thermal Problems, Volume III*, eds. Lewis, R.W., Johnson, J.A. and Smith, W.R., Pineridge Press (Swansea), pp.330-347.

Evans, D.J. and Hatzopoulos, M. (1979), "A parallel linear system solver", *Internat. J. Comput. Math.* *7*, pp.227-238.

Evans, D.J. and Mai, S.-W. (1985), "Two parallel algorithms for the convex hull problem in a two dimensional space", *Parallel Comput.* *2*, pp.313-326.

Evans, D.J. and Sojoodi-Haghighi, R. (1982), "Parallel iterative methods for solving linear equations", *Internat. J. Comput. Math.* *11*, pp.247-284.

Flynn, M.J. (1966), "Very high-speed computing systems", *Proc. IEEE* *54*, pp.1901-1909.

Frisch, U., Hasslacher, B. and Pomeau, Y. (1986a), "Lattice-gas automata for the Navier Stokes equation", *Phys. Rev. Lett.* *56*, pp.1505-1508.

Frisch, U., d'Humières, D. and Lallemand, P. (1986b), "Lattice gas models for 3D hydrodynamics", *Europhysics Letters* *2*, pp.291-297.

Gannon, D. and van Rosendale, J. (1982), "Highly parallel multigrid solvers for elliptic PDEs: an experimental analysis", ICASE Report, NASA Langley, Virginia.

Geist, G.A. (1985), "Efficient parallel LU factorization with pivoting on a hypercube processor", Report ORNL-6211, Oak Ridge National Laboratory, Tennessee.

Geist, G.A. and Heath, M.T. (1985), "Parallel Cholesky factorization on a hypercube multiprocessor", Report ORNL-6190, Oak Ridge National Laboratory, Tennessee.

Gelenbe, E., Lichnewsky, A. and Staphylopatis, A. (1982), "Experience with the parallel solution of PDEs on a distributed computing system", *IEEE Trans. Comput.* *C-31*, pp.1157-1164.

Gentleman, W.M. (1976), "Row elimination for solving sparse linear systems and least squares problems", in *Numerical Analysis, Dundee 1975, Lecture Notes in Mathematics 506*, ed. Watson, G.A., Springer-Verlag (Berlin), pp.122-133.

Gentleman, W.M. (1978), "Some complexity results for matrix computations on parallel processors", *J. Assoc. Comput. Mach. 25*, pp.112-115.

George, A. (1973), "Nested dissection of a regular finite-element mesh", *SIAM J. Numer. Anal. 10*, pp.345-363.

George, A., Heath, M.T. and Liu, J. (1986a), "Parallel Cholesky factorization on a shared-memory multiprocessor", *Linear Algebra Appl. 77*, pp.165-187.

George, A., Heath, M., Liu, J. and Ng, E. (1986b), "Sparse Cholesky factorization on a local-memory multiprocessor", Report CS-86-01, York University, Ontario.

Glowinski, R., Periaux, J. and Dinh, Q.V. (1982), "Domain decomposition methods for nonlinear problems in fluid dynamics", Report 147, INRIA, Rocquencourt, France.

Greenbaum, A. (1985), "A multigrid method for multiprocessors", Report UCRL-92211, Lawrence Livermore National Laboratory, California.

Griewank, A. and Toint, Ph.L. (1984), "Numerical experiments with partially separable optimization problems", in *Numerical Analysis, Proceedings, Dundee 1983, Lecture Notes in Mathematics 1066*, ed. Griffiths, D.F., Springer-Verlag (Berlin), pp.203-220.

Grimes, R., Krakauer, H., Lewis, J., Simon, H. and Wei, S.H. (1985), "The solution of large dense generalized eigenvalue problems on the CRAY X-MP/24 with SSD", Report ETA-TR-32, Boeing Computer Services, Seattle.

Han, S-P. (1986), "Optimization by updated conjugate subspaces", in *Numerical Analysis, Pitman Research Notes in Mathematics Series 140*, eds. Griffiths, D.F. and Watson, G.A., Longman Scientific and Technical (Burnt Mill, Essex), pp.82-97.

Heath, M.T. and Sorensen, D.C. (1986), "A pipelined Givens method for computing the QR factorization of a sparse matrix", *Linear Algebra Appl. 77*, pp.189-203.

Heller, D. (1978), "A survey of parallel algorithms in numerical linear algebra", *SIAM Rev. 20*, pp.740-777.

Heller, D. and Ipsen, I.C.F. (1983), "Systolic networks for orthogonal decompositions", *SIAM J. Sci. Statist. Comput. 4*, pp.261-269.

Hockney, R.W. and Jesshope, C.R. (1981), *Parallel Computers*, Adam Hilger (Bristol).

Ipsen, I.C.F. and Jessup, E. (1986), "Solving the symmetric tridiagonal eigenvalue problem on the hypercube", Computer Science Report, Yale University, to be published.

Ipsen, I.C.F. and Saad, Y. (1985), "The impact of parallel architectures on the solution of the eigenvalue problem", Report YALEU/DCS/RR-444, Yale University.

Iqbal, M.A., Saltz, J.H. and Bokhari, S.H. (1986), "Performance tradeoffs in static and dynamic load balancing strategies", Report 86-13, ICASE, NASA Langley, Virginia.

Jacobs, D.A.H. (ed.) (1977), *The State of the Art in Numerical Analysis*, Academic Press (London).

Jalby, W., Frailong, J-M. and Lenfant, J. (1984), "Diamond schemes: an organization of parallel memories for efficient array processing", Report 342, INRIA, Rocquencourt, France.

Jalby, W. and Meier, U. (1986), "Optimizing matrix operations on a parallel multiprocessor with a memory hierarchy", Center for Supercomputing Research and Development Report, University of Illinois at Urbana-Champaign, to be published.

Johnsson, S.L. (1985), "Solving narrow banded systems on ensemble architectures", Report YALEU/DCS/RR-418, Yale University.

Johnsson, S.L., Saad, Y. and Schultz, M.H. (1985), "Alternating direction methods on multiprocessors", Report YALEU/DCS/RR-382, Yale University.

Karp, A.H. (1986), "A parallel processing challenge", *IMANA Newsletter 10* (2), pp.25-26.

Karp, A.H. and Greenstadt, J. (1986), "An improved parallel Jacobi method for diagonalizing a symmetric matrix", Report G320-3484, IBM Scientific Center, Palo Alto.

Keyes, D.E. and Gropp, W.D. (1985), "A comparison of domain decomposition techniques for elliptic partial differential equations and their parallel implementation", Report YALEU/DCS/RR-448, Yale University.

Kindervater, G.A.P. and Lenstra, J.K. (1985), "An introduction to parallelism in combinatorial optimization", in *Parallel Computers and Computations*, *CWI Syllabi 9*, eds. van Leeuwen, J. and Lenstra, J.K., Math Centrum (Amsterdam), pp.163-184.

Kowalik, J.S. (ed.) (1985), *Parallel MIMD Computation: HEP Supercomputer and its Applications*, MIT Press (Cambridge, Mass.).

Kronsjö, L. (1985), *Computational Complexity of Sequential and Parallel Algorithms*, John Wiley and Sons (Chichester).

Kuck, D.J. (1978), *The Structure of Computers and Computations*, John Wiley and Sons (New York).

Kung, H.T. (1980), "The structure of parallel algorithms", in *Advances in Computers, Volume 19*, ed. Yovits, M., Academic Press (New York), pp.65-112.

Kung, S.-Y. (1984), "On supercomputing with systolic/wavefront array processors", *Proc. IEEE 72*, pp.867-884.

Kung, S.-Y., Arun, K.S., Bhaskar Rao, D.V. and Hu, Y.H. (1981), "A matrix data flow language/architecture for parallel matrix operations based on computational wavefront concept", in *VLSI Systems and Computations*, eds. Kung, H.T., Sproull, B. and Steele, G., Computer Science Press (Rockville, Maryland), pp.235-244.

Lambiotte, J.J. and Voigt, R.G. (1975), "The solution of tridiagonal linear systems on the CDC STAR-100 computer", *ACM Trans Math. Software 1*, pp.308-329.

Lawson, C.L., Hanson, R.J., Kincaid, D.R. and Krogh, F.T. (1979), "Basic linear algebra subprograms for Fortran usage", *ACM Trans. Math. Software 5*, pp.308-323.

van Leeuwen, J. (1985), "Parallel computers and algorithms", in *Parallel Computers and Computations*, *CWI Syllabi 9*, eds. van Leeuwen, J. and Lenstra, J.K., Math. Centrum (Amsterdam), pp.1-32.

Lichnewsky, A. (1982), "Sur la résolution de systèmes linéaires issus de la méthode des éléments finis par une machine multiprocesseurs", Report 119, INRIA, Rocquencourt, France.

Lichnewsky, A. (1984), "Some vector and parallel implementations for preconditioned conjugate gradient algorithms", in *High-speed Computation*, ed. Kowalik, J.S., Springer-Verlag (Berlin), pp.343-359.

Liu, J.W.H. (1985), "Computational models and task scheduling for parallel sparse Cholesky factorization", Report CS-85-01, York University, Ontario.

Lootsma, F.A. (1984), "Parallel unconstrained optimization methods", Report 84-30, Department of Mathematics and Informatics, Delft University of Technology, The Netherlands.

Louter-Nool, M. (1985), "BLAS on the CYBER 205", Report NM-R8524, Centre for Mathematics and Computer Science, Amsterdam.

McBryan, O.A. and Van de Velde, E.F. (1986), "Hypercube algorithms and implementations", Report DOE/ER/03077-271, Courant Mathematics and Computing Laboratory, New York.

Meurant, G. (1985). "Multitasking the conjugate gradient on the CRAY X-MP/48", Report NA-85-33, Department of Computer Science, Stanford University.

Miellou, J.C. (1975), "Algorithmes de rélaxation chaotique à retards", *Rev. Française Automat. Informat. Recherche Opérationnelle 9*, pp.55-82.

Miranker, W.L. and Liniger, W.M. (1967), "Parallel methods for the numerical integration of ODEs", *Math. Comp. 21*, pp.303-320.

Modi, J.J. and Clarke, M.R.B. (1984), "An alternative Givens ordering", *Numer. Math. 43*, pp.83-90.

Modi, J.J. and Pryce, J.D. (1985), "Efficient implementation of Jacobi's diagonalisation method on the DAP", *Numer. Math. 46*, pp.443-454.

Montry, G.R. and Benner, R.E. (1985), "The effects of cacheing on multitasking efficiency and programming strategy on an ELXSI 6400", Report SAND85-2728, Sandia National Laboratories, Albuquerque.

Morf, M. (1980), "Doubling algorithms for Toeplitz and related equations", in *International Conference on Acoustics, Speech, and Signal Processing, Denver, April 1980*, IEEE Press (New York), pp.954-959.

Noor, A.K., Kamel, H.A. and Fulton, R.E. (1978), "Substructuring techniques — status and projections", *Computers and Structures 8*, pp.621-632.

O'Leary, D.P. (1984), "Ordering schemes for parallel processing of certain mesh problems", *SIAM J. Sci. Statist. Comput. 5*, pp.620-632.

O'Leary, D.P. and Stewart, G.W. (1985), "Data-flow algorithms for parallel matrix computations", *Comm. ACM* *28*, pp.840-853.

O'Leary, D.P. and White, R.E. (1983), "Multi-splittings of matrices and parallel solution of linear systems", Report 1362, Computer Science Center, University of Maryland.

Ortega, J.M. and Voigt, R.G. (1985), "Solution of partial differential equations on vector and parallel computers", *SIAM Rev.* *27*, pp.149-240.

Parlett, B.N. (1980), *The Symmetric Eigenvalue Problem*, Prentice-Hall (Englewood Cliffs, New Jersey).

Parlett, B.N. and Schreiber, R. (1986), "Block reflectors: computation and applications", Technical Report, SAXPY Computer Corporation, Sunnyvale, California (*SIAM J. Numer. Anal.*, to be published).

Powell, M.J.D. (1964), "An efficient method for finding the minimum of a function of several variables without calculating derivatives", *Comput. J.* *7*, pp.155-162.

Preparata, F.P. and Vuillemin, J. (1981), "The cube-connected cycles: a versatile network for parallel computation", *Comm. ACM* *24*, pp.300-309.

te Riele, H.J.J. (1985), "Applications of supercomputers in mathematics", Report NM-N8502, Centre for Mathematics and Computer Science, Amsterdam.

Rizzi, A. (1985), "Vector coding the finite-volume procedure for the CYBER 205", *Parallel Comput.* *2*, pp.295-312.

Robert, Y. and Tchuente, M. (1983), "Calcul en parallele sur des reseaux systoliques", in *Actes Colloque AFCET-GAMNI-ISINA* (*EDF Bull. Direction Études Rech. Sér. C. Math. Inform.*), ed. Bossavit, A., Electricité de France (Grenoble), pp.125-128.

Roucairol, C. (1986), "A study of parallel branch and bound algorithms for the travelling salesman problem", 8th European Conference on Operational Research (September 1986, Lisbon), to be published.

Saad, Y. and Schultz, M.H. (1985a), "Parallel implementations of preconditioned conjugate gradient methods", Report YALEU/DCS/RR-425, Yale University

Saad, Y. and Schultz, M.H. (1985b), "Parallel direct methods for solving banded linear systems", Report YALEU/DCS/RR-387, Yale University.

Saad, Y. and Schultz, M.H. (1985c), "Data communication in hypercubes", Report YALEU/DCS/RR-428, Yale University.

Saltz, J.H. and Naik, V.K. (1985), "Towards developing robust algorithms for solving partial differential equations on MIMD machines", Report 85-39, ICASE, NASA Langley, Virginia.

Saltz, J.H., Naik, V.K. and Nicol, D.M. (1986), "Reduction of the effects of the communication delays in scientific algorithms on message passing MIMD architectures", Report 86-4, ICASE, NASA Langley, Virginia.

Sameh, A.H. (1971), "On Jacobi and Jacobi-like algorithms for a parallel machine", *Math. Comp.* *25*, pp.579-590.

Sameh, A.H. (1977), "Numerical parallel algorithms — a survey", in *High Speed Computer and Algorithm Organization*, eds. Kuck, D.J., Lawrie, D.H. and Sameh, A.H., Academic Press (New York), pp.207-228.

Sameh, A.H. (1983), "An overview of parallel algorithms in numerical linear algebra", in *Actes Colloque AFCET-GAMNI-ISINA (EDF Bull. Direction Études Rech. Ser. C. Math. Inform.)* ed. Bossavit, A., Electricité de France (Grenoble), pp.129-134.

Sameh, A.H. (1985), "On some parallel algorithms on a ring of processors", *Comp. Phys. Comm.* *37*, pp.159-166.

Sameh, A.H. and Kuck, D.J. (1978), "On stable parallel linear system solvers", *J. Assoc. Comput. Mach.* *25*, pp.81-91.

Saul'yev, V.K. (1964), *Integration of Equations of Parabolic Type by the Method of Nets*, Macmillan (New York).

Schendel, U. (1984), *Introduction to Numerical Methods for Parallel Computers*, Ellis Horwood (Chichester).

Schnabel, R.B. (1984), "Parallel computing in optimization", Report CU-CS-282-84, University of Colorado, Boulder.

Schreiber, R. and Tang, W-P. (1982), "Vectorizing the conjugate gradient method", working paper, Department of Computer Science, Stanford University.

Schwartz, J.T. (1980), "Ultracomputers", *ACM Trans. Program. Lang. Syst.* *2*, pp.484-521.

Seager, M.K. (1986), "Parallelizing conjugate gradient for the CRAY X-MP, *Parallel Comput.* *3*, pp.35-47.

Sorensen, D.C. (1985), "Analysis of pairwise pivoting in Gaussian elimination", *IEEE Trans. Comput.* *C-34*, pp.274-278.

Stone, H.S. (1973), "An efficient parallel algorithm for the solution of a tridiagonal linear system of equations", *J. Assoc. Comput. Mach.* *20*, pp.27-38.

Swarztrauber, P.N. (1979), "A parallel algorithm for solving general tridiagonal equations", *Math. Comp.* *33*, pp.185-199.

Swarztrauber, P.N. (1986), "Vector-concurrent FFTs", *Parallel Comput.*, to be published.

Traub, J.F. (ed.) (1973), *Complexity of Sequential and Parallel Numerical Algorithms*, Academic Press (New York).

Tseng, S.S. and Lee, R.C.T. (1984), "A new parallel sorting algorithm based upon min-mid-max operations", *BIT 24*, pp.187-195.

Voigt, R.G. (1985), "Where are the parallel algorithms?", in *AFIPS Conference Proceedings 54*, AFIPS Press (Reston, Virginia), pp.329-334.

van der Vorst, H.A. (1982), "A vectorizable variant of some ICCG methods", *SIAM J. Sci. Statist. Comput.* *3*, pp.350-356.

van der Vorst, H.A. (1986), "The performance of FORTRAN implementations for preconditioned conjugate gradients on vector computers", *Parallel Comput.* *3*, pp.49-58.

Wang, H.H. (1981), "A parallel method for tridiagonal equations", *ACM Trans. Math. Software 7*, pp.170-183.

Ware, W. (1972), "The ultimate computer", *IEEE Spectrum*, March, pp.84-91.

Iain S. Duff
Computer Science and Systems Division
Harwell Laboratory
Didcot
Oxon OX11 0RA
England

14 Developments in Stability Theory for Ordinary Differential Equations

J. D. LAMBERT

ABSTRACT

During the ten years that have elapsed since the last State of the Art conference, there has been a considerable development in nonlinear stability theory of numerical methods for ordinary differential equations. In this paper we first discuss the need for such a development and then give a survey of its main features.

1. INTRODUCTION

Ten years ago, the only stability definitions in common use in connection with the numerical solution of ordinary differential equations were zero-stability, relevant to the case when the steplength tends to zero, and absolute (or relative) stability, relevant to the case of a fixed non-zero steplength. The various definitions relevant to stiffness, such as A-, $A(\alpha)$-, A_0-, L- and stiff-stability could be seen as special cases of absolute stability. Absolute stability is based on an overly restrictive linear test equation, and there was clearly a need to develop a stability theory based on a more realistic nonlinear test equation. In the first part of this paper we shall elaborate on this need and comment on how the theory based on the linear test equation was frequently abused; in the second part, we shall survey the main features of the nonlinear stability theory. Although nonlinear stability theory originated in connection with linear multistep methods (G-stability), the major developments have been in the area of Runge-Kutta methods. In this latter context, the reader's attention is drawn to the excellent treatise by Dekker and Verwer (1984), a book to which the present author acknowledges his indebtedness. Only developments in stability theory relevant to the problem of stiffness are covered in this paper, though there have been some similar developments in connection with explicit methods (Dahlquist and Jeltsch (1979)).

2. A GENERAL FRAMEWORK

By a *problem* we shall mean the initial value problem

$$\dot{\underline{y}} = \underline{f}(t,\underline{y}), \quad \underline{y}(t_0) = \underline{y}_0 , \quad t_0 \leqslant t \leqslant T , \quad \underline{f} : \mathbb{R}\times\mathbb{R}^m \rightarrow \mathbb{R}^m \qquad (2.1)$$

and by a *method* we shall mean a numerical method of the class defined by

$$\begin{aligned} \underline{y}_\mu &= \underline{\eta}_\mu(h) , \quad \mu = 0,1,\cdots,k-1 \\ \sum_{j=0}^{k} \alpha_j \underline{y}_{n+j} &= h\underline{\phi}_f(t_n, \underline{y}_{n+k}, \underline{y}_{n+k-1}, \cdots, \underline{y}_n) \end{aligned} \qquad (2.2)$$

where h is the steplength and k the stepnumber. This class includes linear multistep methods, predictor-corrector methods in $P(EC)^r E$ mode, hybrid methods and Runge-Kutta methods.

The essence of a stability definition for a method is that we impose conditions C_p on the *problem* (2.1) such that the exact solution $\underline{y}(t)$ possesses a certain "stability" property; we then seek to impose conditions on the *method* (2.2) such that the numerical solution $\{\underline{y}_n\}$, yielded by the method applied to the problem satisfying C_p, possesses an analogous "stability" property. We shall find it useful to encapsulate this structure in the following diagram, which we shall call the *syntax diagram* for a general stability definition for (2.2).

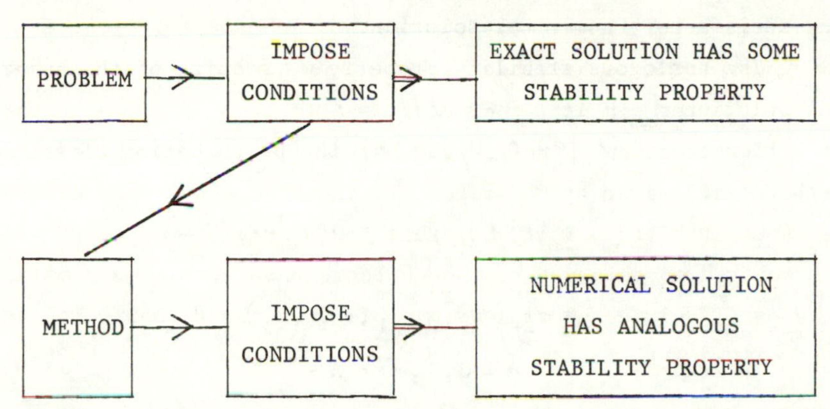

The rightmost box on the lower line normally defines the stability property, and the box to its left the conditions for the method to possess that stability property; but it can also be the case that the middle box on the lower line defines the stability property, in which case the box to its right is interpreted as a consequence of the property. Clearly, the choice between these two interpretations of the syntax diagram is largely a matter of taste, and examples of both can be found in the literature. We now illustrate this point, and the use of the syntax, in the case of zero-stability.

The stability property we require of the exact solution of (2.1) is *total stability*:

Denote by $(\underline{\delta}(t), \underline{\delta})$ the perturbation of the problem (2.1) given by

$$\dot{\underline{z}} = \underline{f}(t, \underline{z}) + \underline{\delta}(t), \quad \underline{z}(t_0) = \underline{y}_0 + \underline{\delta},$$

and let the solution of this perturbed problem be denoted by $\underline{z}(t)$. Consider two such perturbations $(\underline{\delta}(t), \underline{\delta})$, $(\tilde{\underline{\delta}}(t), \tilde{\underline{\delta}})$ and the corresponding solutions $\underline{z}(t)$, $\tilde{\underline{z}}(t)$. Then, if

$$\|\underline{\delta}(t) - \tilde{\underline{\delta}}(t)\| \leq \varepsilon, \quad \|\underline{\delta} - \tilde{\underline{\delta}}\| \leq \varepsilon \Rightarrow \|\underline{z}(t) - \tilde{\underline{z}}(t)\| \leq S\varepsilon$$

where S and ε are positive constants, then the problem (2.1) is totally stable. This requirement is not very demanding, and indeed is implied by the classical Lipschitz condition. It is clear that without total stability, we have no hope of obtaining any satisfactory numerical solution.

The analogous stability property we require of the numerical solution $\{\underline{y}_n\}$ is *zero-stability*:

Denote by $\{\underline{\delta}_n | n = 0, 1, \ldots, N\}$ the perturbation of the method (2.2) given by

$$\underline{z}_\mu = \underline{\eta}_\mu(h) + \underline{\delta}_\mu, \quad \mu = 0, 1, \cdots, k-1$$

$$\sum_{j=0}^{k} \alpha_j \underline{z}_{n+j} = h\underline{\phi}_f(t_n, \underline{z}_{n+k}, \cdots, \underline{z}_k) + \underline{\delta}_{n+k}, \quad n = 0, 1, \cdots, N-k$$

and let the solution of this perturbed difference equation be denoted by $\{\underline{z}_n\}$. Consider two such perturbations $\{\underline{\delta}_n\}$ and $\{\underline{\tilde{\delta}}_n\}$ and the corresponding solutions $\{\underline{z}_n\}$ and $\{\underline{\tilde{z}}_n\}$. Then if

$$\|\underline{\delta}_n - \underline{\tilde{\delta}}_n\| \leqslant \varepsilon^* \Rightarrow \|\underline{z}_n - \underline{\tilde{z}}_n\| \leqslant S^* \varepsilon^*, \quad n = 0, 1, \cdots, N$$

where S^* and ε^* are positive constants, then the method (2.2) is zero-stable; this is implied by the familiar root condition, leading to the following syntax for the definition of zero-stability

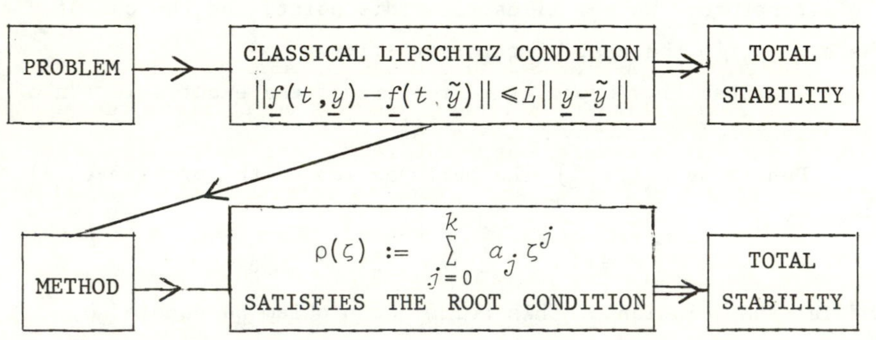

Note that some authors would regard the middle box on the lower line as *defining* zero-stability, in which case the rightmost box would be interpreted as a consequence.

The concept of *absolute stability* is linked with the notion of a linear constant coefficient test equation (see, for example, Lambert (1973)). Its syntax is given as follows:

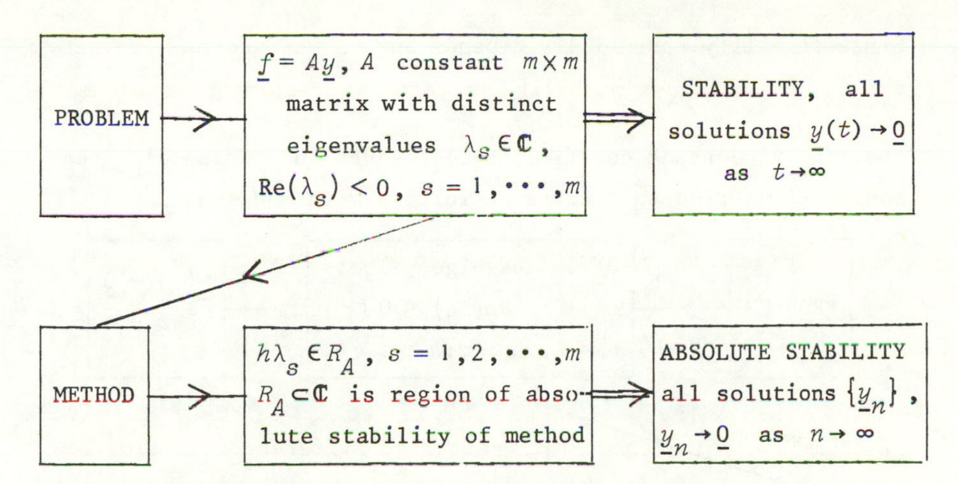

Absolute stability spawns several other definitions (relevant to stiffness) depending on the size of the region R_A. Thus, for example, we have :

A-stability : $R_A \supset \{h\lambda \mid \mathrm{Re}(h\lambda) < 0\}$

$A(\alpha)$-stability : $R_A \supset \{h\lambda \mid -\alpha < \pi - \arg(h\lambda) < \alpha\}$

A_0-stability : $R_A \supset \{h\lambda \mid \mathrm{Im}(h\lambda) = 0,\ \mathrm{Re}(h\lambda) < 0\}$.

3. LIMITATIONS AND ABUSES OF ABSOLUTE STABILITY

The restriction to a linear constant coefficient test equation $\dot{\underline{y}} = A\underline{y}$ is a severe one, and it is not surprising that attempts were — and still are — made to interpret absolute stability in the context of a more general equation. A typical argument applied to the general problem (2.1) goes as follows:

In some neighbourhood of the exact solution $\underline{y}(t)$ of (2.1), solutions of $\dot{\underline{y}} = \underline{f}(t, \underline{y})$ are well-represented by those of the variational equation

$$\dot{\underline{y}} = \underline{f}(t, \underline{y}(t)) + [\underline{y} - \underline{y}(t)]\, \partial \underline{f}/\partial \underline{y}(t, \underline{y}(t)) .$$

Assume that the Jacobian $\partial \underline{f}/\partial \underline{y}$ can be "frozen", that is locally regarded as a constant matrix A, yielding an equation of the form

$$\dot{\underline{y}} = A\underline{y} + \underline{\phi} .$$

Since stability essentially depends on A, we ignore $\underline{\phi}$ to get

$$\dot{\underline{y}} = A\underline{y} ,$$

the linear constant coefficient test equation. This false reasoning gives rise to a false definition with the syntax

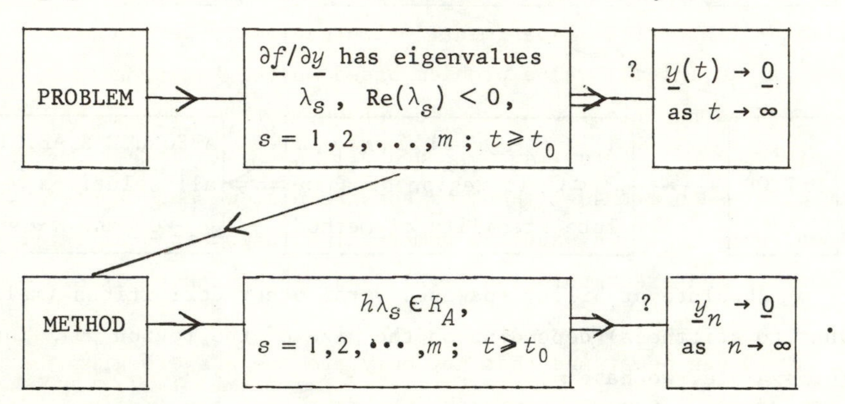

The literature contains several examples illustrating the invalidity of the queried implication in the top line. We give here a new example, which will be used for other purposes later in this paper.

EXAMPLE 1

$$\dot{\underline{y}} = A(t)\underline{y} , \quad t \geq 0 , \quad \underline{y} \in \mathbb{R}^2$$

$$A(t) = \begin{bmatrix} -1/[4(1+t)] & 1/(1+t)^2 \\ -1/4 & -1/[4(1+t)] \end{bmatrix} . \qquad (3.1)$$

The eigenvalues of $A(t)$ are $\lambda_{1,2} = (-1/4 \pm i/2)/(1+t)$, so that $\mathrm{Re}(\lambda_{1,2}) < 0$ for all $t \geq 0$, and the "frozen" Jacobian argument would indicate that for all solutions $\underline{y}(t)$, $\|\underline{y}(t)\| \to 0$ as $t \to \infty$; it would also suggest that the solutions would be oscillatory. However, the general solution is

$$\underline{y}(t) = A_1 \underline{Y}_1 + A_2 \underline{Y}_2$$

where

$$\begin{aligned} \underline{Y}_1 &= [(1+t)^{-3/4} , -\tfrac{1}{2}(1+t)^{1/4}]^T \\ \underline{Y}_2 &= [(1+t)^{-3/4} \ln(1+t), (1+t)^{1/4}[1 - \tfrac{1}{2}\ln(1+t)]]^T . \end{aligned} \qquad (3.2)$$

Thus, in general, $\|\underline{y}(t)\| \to \infty$ as $t \to \infty$. Moreover, the solution is not oscillatory, and the predictions of the "frozen" Jacobian argument are incorrect on both counts.

The fallacy of the queried implication on the lower line of the syntax for the false definition can readily be established by numerical experiment. Alternatively, we can argue that if an initial value problem based on the equation (3.1) is integrated numerically by any convergent method for which R_A contains $h\lambda_{1,2}$ for all $h < h_0$ and all $t \geqslant t_0$ (and there certainly exist such methods) then, by convergence, we can choose h such that $h\lambda_{1,2} \in R_A$ and the numerical solution is arbitrarily close to the exact solution (3.2), and hence cannot satisfy $\underline{y}_n \to \underline{0}$ as $n \to \infty$.

It can be argued that not only are the answers given by the "frozen" Jacobian approach incorrect, but the implied questions are also ill-founded. The approach attempts to link decaying solutions of the general problem (2.1) with negativity of the real parts of the eigenvalues of the Jacobian, and likewise oscillatory solutions with the complexity of the eigenvalues. By "decaying solutions" one can only mean $\underline{y}(t) \to \underline{0}$ as $t \to \infty$, for *all* solutions; whilst it is easy enough to demand this of the test equation $\dot{\underline{y}} = A\underline{y}$, it is much too global a property ever to expect of the general *nonlinear* equation $\dot{\underline{y}} = \underline{f}(t, \underline{y})$. A link between complex eigenvalues of the Jacobian and oscillatory solutions is similarly not to be expected. When the Jacobian varies significantly with t, the usual interpretation that is put on the "frozen" Jacobian argument is that the Jacobian should be regarded as piecewise constant, and that is an assumption which is *local* in t; we cannot expect that to imply the *global* (in t) property of periodicity of the exact solutions. It is not, in this author's opinion, going too far to regard the "frozen" Jacobian argument as an example of that old adage, "Ask a stupid question and you'll get a stupid answer"!

Despite the above criticism, it has to be admitted that the concept of absolute stability, applied to general problems, works well in practice most of the time, and instances where it leads to misleading results are rare. Nevertheless, there can be no doubt that there existed, ten years ago, a strong motivation for seeking an improved stability theory which did not depend on the linear constant coefficient test equation.

4. THE ONE-SIDED LIPSCHITZ CONDITION AND THE LOGARITHMIC NORM

What conditions can we impose on the *general* problem (2.1) in order to control the qualitative behaviour of the exact solutions in some meaningful way? The classical Lipschitz condition is not useful in this context, essentially because it cannot distinguish between equations such as $\dot{y} = -100y$ and $\dot{y} = 100y$. What turns out to be a much more useful concept is the *one-sided Lipschitz condition*, defined as follows:

Let $\langle \cdot, \cdot \rangle$ be an inner product on $\mathbb{R}^m$ and let $\|\cdot\|$ be the corresponding inner-product norm. The condition is that

$$\langle \underline{f}(t,\underline{y}) - \underline{f}(t,\underline{\tilde{y}}) \ , \underline{y} - \underline{\tilde{y}} \rangle \leqslant v(t) \, \|\underline{y} - \underline{\tilde{y}}\|^2 \qquad (4.1)$$

for all $\underline{y}$, $\underline{y} \in M$, a convex region containing the exact solution of (2.1). $v(t)$ is the *one-sided Lipschitz constant* of $\underline{f}$ with respect to $\underline{y}$.

A simple argument yields a useful consequence of (4.1): Let $\underline{y}(t)$, $\underline{\tilde{y}}(t)$ be two solutions of (2.1) contained in M, satisfying initial conditions $\underline{y}(t_0) = \underline{y}_0$, $\underline{\tilde{y}}(t_0) = \underline{\tilde{y}}_0$, and define

$$\phi(t) := \|\underline{y}(t) - \underline{\tilde{y}}(t)\|^2 \ .$$

Then

$$\dot{\phi}(t) = 2\langle \underline{\dot{y}} - \underline{\dot{\tilde{y}}}, \underline{y} - \underline{\tilde{y}} \rangle = 2\langle \underline{f}(t,\underline{y}) - \underline{f}(t,\underline{\tilde{y}}), \ \underline{y} - \underline{\tilde{y}} \rangle \leqslant 2v(t)\,\phi(t) \ ,$$

a differential inequality which can be integrated by means of the integrating factor

$$\eta(t) = \exp(-2 \int_0^t v(\tau)\, d\tau)$$

to give

$$d/dt\,[\phi(t)\,\eta(t)] \leqslant 0\,,$$

and it follows that $\phi(t)\,\eta(t)$ is monotonic decreasing for all $t \in [t_0\,, T]$. It then follows that

$$\|\underline{y}(t_2) - \tilde{\underline{y}}(t_2)\| \leqslant \exp\left(\int_{t_1}^{t_2} v(\tau)\,d\tau\right) \|\underline{y}(t_1) - \tilde{\underline{y}}(t_1)\| \qquad (4.2)$$

for all $t_0 \leqslant t_1 \leqslant t_2 \leqslant T$.

NOTES

(i) f satisfies one-sided Lipschitz condition

$$\underset{\not\Leftarrow}{\Longrightarrow}\; f \text{ satisfies classical Lipschitz condition,}$$

(ii) $v(t)$ can be negative. If $v(t) \leqslant 0$ for all $t \in [t_0, T]$, then by (4.2)

$$\|\underline{y}(t_2) - \tilde{\underline{y}}(t_2)\| \leqslant \|\underline{y}(t_1) - \tilde{\underline{y}}(t_1)\| \qquad (4.3)$$

for all $t_0 \leqslant t_1 \leqslant t_2 \leqslant T$; that is, the solutions behave *contractively*, and the problem (and the function $\underline{f}$) are said to be *dissipative*. In the increasingly popular terminology of dynamical systems this means that, as time evolves, the solution curves are coming together in the phase-portrait of the equation (Hirsch and Smale, 1974).

(iii) The conclusion (4.3) holds only for $\underline{y}$, $\tilde{\underline{y}} \in M$ (cf. remarks at the end of the previous section). However, if $\underline{f}$ is linear in $\underline{y}$, $\underline{f} = A(t)\underline{y}$, then we may take $M = \mathbb{R}^m$, and (4.3) holds globally.

The question is now how to find a one-sided Lipschitz constant $v(t)$. Once again the appropriate tool is to hand in the shape of the *logarithmic norm* (Dahlquist (1959)), $\mu[A]$ of a matrix A defined by

$$\mu[A] := \lim_{\delta \to 0} \{[\|I + \delta A\| - 1]/\delta\}\,.$$

NOTES

(i) The above limit exists for all matrices, for all subordinate norms.

(ii) $\mu[A]$ is *not* a norm; it can be negative.

(iii) $\mu[A]$ is norm-dependent:

$$\|\cdot\| = L_1\text{-norm} \;:\; \mu_1 = \max_j \, [a_{jj} + \textstyle\sum_{i\neq j} a_{ij}]$$

$$\|\cdot\| = L_2\text{-norm} \;:\; \mu_2 = \text{max eigenvalue of } \tfrac{1}{2}(A + A^T)$$

$$\|\cdot\| = L_\infty\text{-norm} \;:\; \mu_\infty = \max_i \, [a_{ii} - \textstyle\sum_{j\neq i} a_{ij}]$$

(iv) The smallest possible one-sided Lipschitz constant of $\underline{f} = A\underline{y}$ is $\mu[A]$. (Note that A can be a function of t.)

Thus if $\mu[A(t)] \leqslant 0$, $t \in [t_0, T]$, (in the norm corresponding to the inner product defining the one-sided Lipschitz constant) we have contractivity of the solutions of $\dot{\underline{y}} = A(t)\underline{y}$ in that norm. The result generalizes to the general nonlinear case and to any norm (see, for example, Dekker and Verwer (1984)):

Let $\|\cdot\|$ be any norm, let $\underline{y}(t)$, $\tilde{\underline{y}}(t) \in M$ be any two solutions of $\dot{\underline{y}} = \underline{f}(t, \underline{y})$ satisfying initial conditions $\underline{y}(t_0) = \underline{y}_0$, $\tilde{\underline{y}}(t_0) = \tilde{\underline{y}}_0$, and let $v(t)$ be a piecewise continuous function such that for all $t \in [t_0, T]$

$$\mu[\partial\underline{f}/\partial\underline{y}] \leqslant v(t)$$

for all $\underline{y} \in M$; then

$$\|\underline{y}(t_2) - \tilde{\underline{y}}(t_2)\| \leqslant \exp\left(\int_{t_1}^{t_2} v(\tau)\, d\tau\right) \|\underline{y}(t_1) - \tilde{\underline{y}}(t_1)\| \tag{4.4}$$

for all $\underline{y}, \tilde{\underline{y}} \in M$, $t_0 \leqslant t_1 \leqslant t_2 \leqslant T$.

Thus, if $\mu[\partial\underline{f}/\partial\underline{y}] < 0$, we are assured of contractivity of the exact solutions, for the *general nonlinear problem*.

We illustrate the preceding analysis with reference to Example 1, using the L_2-norm throughout. The eigenvalues $\sigma_{1,2}$ of $\frac{1}{2}(A + A^T)$, where A is given by (3.1) turn out to be

$$\sigma_1 = \;\; 1/[2(1+t)^2] - 1/[4(1+t)] - 1/8$$

$$\sigma_2 = -1/[2(1+t)^2] - 1/[4(1+t)] + 1/8$$

and their graphs as functions of t for $t \geqslant 0$ are displayed in Figure 1.

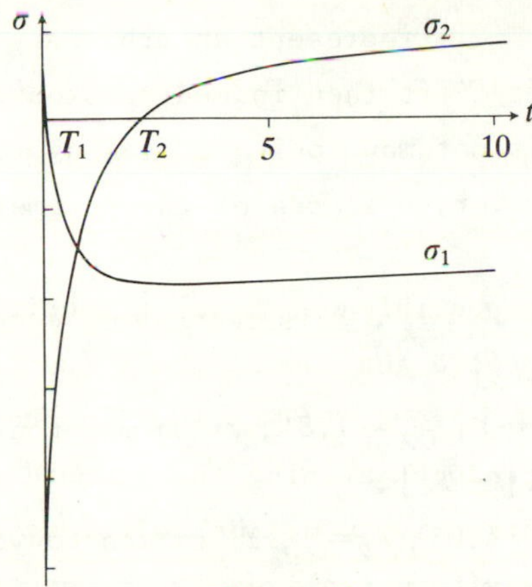

Figure 1

Since $\mu_2[A] = \max[\sigma_1, \sigma_2]$ it is clear from Figure 1 that

$$\begin{aligned} \mu_2[A] &\geqslant 0 \quad \text{if} \quad t \in [0, T_1] \quad \text{or} \quad t \in [T_2, \infty) \\ \mu_2[A] &\leqslant 0 \quad \text{if} \quad t \in [T_1, T_2] \end{aligned} \tag{4.5}$$

where $T_1 = \sqrt{5} - 2$, $T_2 = \sqrt{5}$.

Recall that for a linear equation, the region M containing the solutions $\underline{y}(t)$ and $\underline{\tilde{y}}(t)$ can be taken to be $\mathbb{R}^2$. We can therefore choose $\underline{\tilde{y}}(t) \equiv \underline{0}$, and it follows from (4.3) or (4.4) that contractivity is then equivalent to $\|\underline{y}(t)\|$ being monotonic decreasing. From the exact solution (3.2) of (3.1) we can compute

$$Y := \|\underline{y}(t)\|_2^2 / \|\underline{y}(0)\|_2^2$$

for a range of A_1 and A_2. Some plots of Y against t are shown in Figure 2.

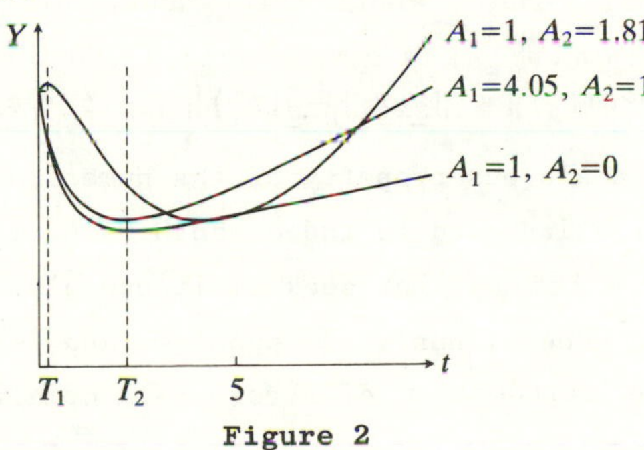

Figure 2

The values $A_1 = 1$, $A_2 = 0$ represent an arbitrary choice and we see from Fig. 2 that Y is then indeed monotonic decreasing for $t \in [T_1, T_2]$ but it is not monotonic increasing for all $t \in [0, T_1]$ or $t \in [T_2, \infty)$. However, a search of the parameter space (A_1, A_2) reveals

(i) For all A_1, A_2, Y is monotonic decreasing for $t \in [T_1, T_2]$.

(ii) For $A_1 = 1$, $A_2 = 1.81$, Y is monotonic increasing for $t \in [0, T_1]$.

(iii) For $A_1 = 4.05$, $A_2 = 1$, Y is monotonic increasing for $t \in [T_2, \infty)$.

Thus $[T_1, T_2]$ is the only interval of t for which Y is monotomic decreasing for all values of A_1 and A_2, exactly as predicted by the logarithmic norm (equation (4.5)).

We have thus demonstrated that for Example 1 the logarithmic norm — in notable contrast to the spectrum of the Jacobian — gives reliable and sharp information on the qualitative behaviour of the exact solutions. (In the case of nonlinear examples the information on contractivity will be equally reliable, but not always as sharp.)

Returning to the general nonlinear problem, we recall that the solutions $\underline{y}(t)$ and $\underline{\tilde{y}}(t)$ must lie in the convex region M containing the exact solution of (2.1). This means that we may not talk of solutions decaying, but only of solutions being contractive,

$$\|\underline{y}(t_2) - \underline{\tilde{y}}(t_2)\| \leqslant \|\underline{y}(t_1) - \underline{\tilde{y}}(t_1)\|, \quad t_0 \leqslant t_1 \leqslant t_2 \leqslant T$$

and demand an analogous property of the numerical solution. This is not a restriction, and is indeed both natural and desirable — natural because it does not seek to impose global qualitative behaviour on a general nonlinear problem, and desirable, since the inevitable introduction of errors in a numerical solution can be thought of as equivalent to jumping onto a neighbouring solution curve, and contractivity of the numerical solution is then precisely the property we would wish the method to have.

We are thus led to a new breed of stability definition with the following general syntax :

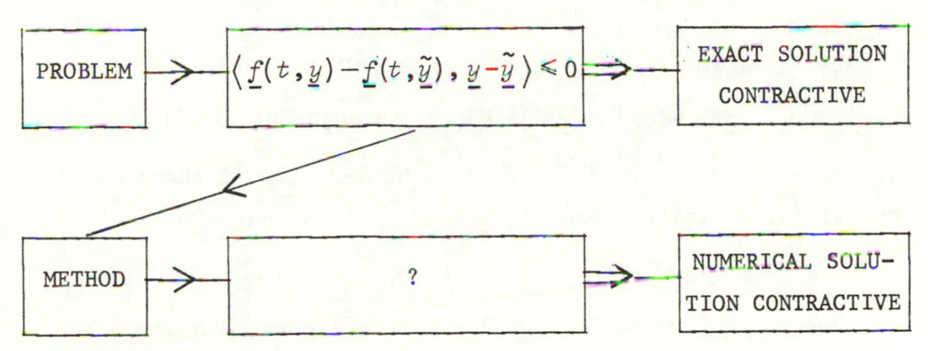

By contractivity of the numerical solution, we mean that

$$\|\underline{y}_{n+1}-\underline{\tilde{y}}_{n+1}\| \leq \|\underline{y}_n-\underline{\tilde{y}}_n\|, \text{ for a one-step method, and}$$

$$\|\underline{Y}_{n+1}-\underline{\tilde{Y}}_{n+1}\| \leq \|\underline{Y}_n-\underline{\tilde{Y}}_n\|, \text{ for a } k\text{-step method,}$$

where

$$\underline{Y}_n = [\underline{y}_{n+k-1}, \underline{y}_{n+k-2}, \cdots, \underline{y}_n]^T \in \mathbb{R}^{km}. \tag{4.6}$$

The remainder of this paper will be mostly concerned with replacing the query in the middle box of the lower line by the appropriate condition.

5. G-STABILITY

The earliest work on conditions for a method to be contractive was due to Dahlquist (1975a,b) and dealt with a variant of linear multistep methods. We assume the problem (2.1) is in autonomous form and is dissipative, i.e.

$$\underline{\dot{y}} = \underline{f}(\underline{y}), \ \underline{y}(t_0) = \underline{y}_0, \ \langle \underline{f}(\underline{y})-\underline{f}(\underline{\tilde{y}}), \underline{y}-\underline{\tilde{y}} \rangle \leq 0 .$$

Consider the linear multistep method

$$\rho(E)\,\underline{y}_n = h\sigma(E)\,\underline{f}_n \tag{5.1}$$

where

$$\rho(E) = \sum_{j=0}^{k} \alpha_j E^j, \quad \sigma(E) = \sum_{j=0}^{k} \beta_j E^j$$

and E is the forward shift operator. The *one-leg-twin* of (5.1) is defined by

$$\rho(E)\, \underline{y}_n = h\underline{f}\,(\sigma(E)\, \underline{y}_n)\,. \tag{5.2}$$

A "scalar" form of the vector of back values defined by (4.6) is

$$\underline{X}_n := [x_{n+k-1}\,,\; x_{n+k-2}\,, \cdots, x_n] \in \mathbb{R}^k.$$

The one-leg method (5.2) is said to be *G-stable* if there exists a positive definite symmetric matrix G such that

$$\underline{X}_1^T G \underline{X}_1 - \underline{X}_0^T G \underline{X}_0 \leqslant 2\sigma(E)\, x_0 \cdot \rho(E)\, x_0\,.$$

The "vector equivalent" of $\underline{X}^T G \underline{X}$ defines a norm, the *G-norm*:

$$\|\underline{Z}_n\|_G^2 := \sum_{i,j=1}^{k} g_{ij} \langle \underline{z}_{n+k-i}\,,\; \underline{z}_{n+k-j} \rangle$$

where

$$\underline{Z}_n = [\underline{z}_{n+k-1}\,,\; \underline{z}_{n+k-2}\,, \cdots, \underline{z}_n] \in \mathbb{R}^{km}\,.$$

The syntax for G-stability is then

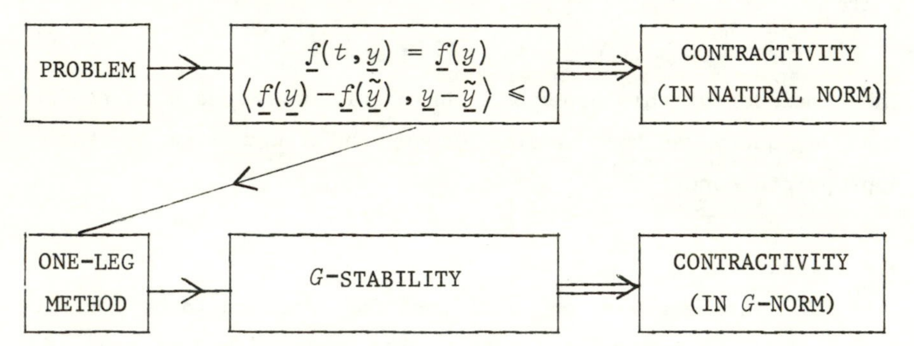

This definition differs from those that will follow later in this paper in that the norm in which the exact solution is contractive (the norm corresponding to the inner product in the middle box of the first line) is essentially unrelated to the G-norm in which the numerical solution is contractive. It was established early on in this development that G-stability implied A-stability, but Dahlquist (1978) eventually established the remarkable result that G-stability is equivalent to A-stability. Thus in one sense, G-stability fails to help us in our quest for better stability definitions. G-stability was nonetheless a landmark in stability theory and has left useful results on error bounds in terms of the G-norm (Dahlquist, 1975a,b).

Strictly speaking, G-stability has been defined for one-leg methods (5.2). However, if $\{\underline{y}_n\}$ satisfies (5.2) then $\{\sigma(E)\underline{y}_n\}$ satisfies the multistep method (5.1) (Dahlquist, 1978). Hence, G-stability implies boundedness — but not always contractivity — in the standard multistep framework.

6. RUNGE-KUTTA METHODS

Most of the development of nonlinear stability theory has been in the context of Runge-Kutta methods. The general s-stage RK method is defined by

$$\begin{aligned} \underline{k}_i &= \underline{f}(t_n + c_i h\,,\ \underline{y}_n + h \sum_{j=1}^{s} a_{ij}\underline{k}_j), \quad i = 1,2,\cdots,s \\ \underline{y}_{n+1} &= \underline{y}_n + h \sum_{j=1}^{s} b_j \underline{k}_j\,, \end{aligned} \tag{6.1}$$

with the corresponding Butcher array

$$\begin{array}{c|c} \underline{c} & A \\ \hline & \underline{b}^T \end{array} \qquad \begin{aligned} \underline{c} &= [c_1\,,\ c_2\,,\cdots,c_s]^T \\ \underline{b} &= [b_1\,,\ b_2\,,\cdots,b_s]^T \\ A &= [a_{ij}] \end{aligned}$$

Broad categories of RK methods can be defined as follows:

TYPE	STRUCTURE	A	WORK/STEP
EXPLICIT (CLASSICAL)	$a_{ij}=0$, $i\geqslant j$	STRICTLY LOWER TRIANGULAR	
SEMI-IMPLICIT	$a_{ij}=0$, $i>j$	LOWER TRIANGULAR	s SYSTEMS OF DIMENSION m
SINGLY DIAGONALLY IMPLICIT (SDIRK)	$a_{ij}=0$, $i>j$ $a_{ii}=a$	LOWER TRIANGULAR CONSTANT DIAGONAL	s SYSTEMS OF DIMENSION m, COMMON LU DECOMPOSITION
SINGLY IMPLICIT (SIRK)	—	—	As for SDIRK
IMPLICIT	—	FULL	ONE SYSTEM OF DIMENSION sm

For a description of SDIRK methods see Nørsett (1974) and Alexander (1977). For SIRK methods (Burrage (1978a,b, 1982), the matrix A is not necessarily lower triangular, but an implementation deriving from the work of Butcher (1976) results in the work per step being comparable with that for SDIRK methods. Not all authors concur with the above use of the terms "semi-implicit" and "singly diagonally implicit".

When $\underline{f}(t,\underline{y})=\underline{f}(t)$, an RK method reduces to a quadrature formula, and this has led to RK methods being named according to the quadrature formulae to which they reduce, the main groups being Gauss-Legendre, Radau and Lobatto. An RK method applied to the scalar test equation $\dot{y}=\lambda y$ yields the difference equation $y_{n+1}=R(h\lambda)y_n$, where $R(\cdot)$, a rational function, is the *stability function* of the RK method. Clearly

$$h\lambda\in R_A \Leftrightarrow |R(h\lambda)|<1,$$

where R_A is the region of absolute stability. In the following table we summarise the properties of several well-known RK methods of quadrature type :

s- STAGE RK METHOD	ORDER	STABILITY FUNCTION
GAUSS	$2s$	(s,s) - PADÉ
RADAU IA RADAU IIA	$2s-1$	$(s-1,s)$ - PADÉ
LOBATTO IIIA LOBATTO IIIB	$2s-2$	$(s-1,s-1)$ - PADÉ
LOBATTO IIIC		$(s-2,s)$ - PADÉ

All of the above methods are implicit; from the last column, it follows from a well-known result that all are A-stable. In Radau quadrature the abscissae include one of the two end-points of the interval of integration, and hence there are two possible classes; the suffix A refers to the Radau methods of Ehle (1969) as opposed to those of Butcher (1964). The suffix III in the Lobatto methods refers to their membership of a class defined by Butcher (1964); the IIIA and IIIB methods are due to Ehle (1969), and the IIIC to Chipman (1971). The coefficients of low order methods appearing in the above table can be found in Dekker and Verwer (1984).

In (6.1), the second argument in the right hand side of the equation defining $\underline{k}_i$ is an approximation to

$$\int_0^{c_i} \underline{f}(t_n + h\tau)\,d\tau\,,$$

and if each such approximation has order s, then the s-stage RK method is called a *collocation method*. (Wright (1970), Burrage (1978b)).

Finally, we mention two further definitions relevant to RK methods:

An RK method is *nonconfluent* if $c_i = c_j \Rightarrow i = j$.

An RK method is *irreducible* if there does not exist another RK method of lower stage number which gives identical numerical results.

(A precise definition of irreducibility is rather cumbersome, and there exists more than one definition; we use the term in the sense defined by Hundsdorfer and Spijker (1981). The reader is referred to Dekker and Verwer (1984) for a full discussion.)

7. *B*-, *BN*-, ALGEBRAIC AND *AN*-STABILITY

All of the stability definitions in this section apply to RK methods. The first is *B-stability* (Butcher (1975)), and assumes the equation in (2.1) is in autonomous form, $\dot{\underline{y}} = \underline{f}(\underline{y})$. Its syntax is

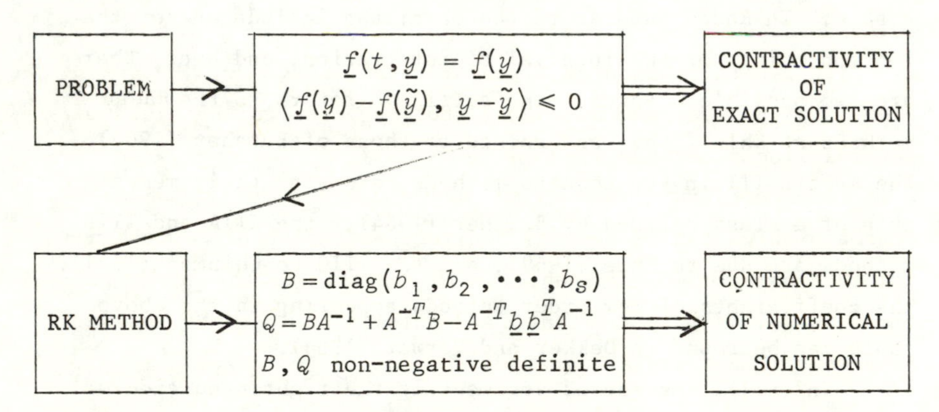

The definition of *B*-stability is given by the rightmost box on the lower line, and the box to its left gives a sufficient condition for *B*-stability; it is a difficult condition to apply in practice.

BN-stability (Burrage and Butcher (1979), Crouziex (1979)) extends the definition to cover a problem in non-autonomous form.

Our next definition, *algebraic stability* (Burrage and Butcher (1979), Crouziex (1979)) is much easier to apply in practice, and applies to the non-autonomous case. Its syntax is as follows:

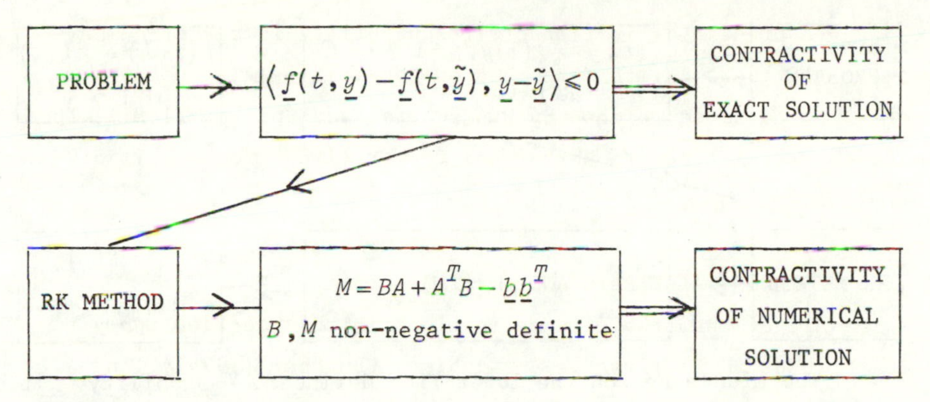

This time, it is the middle box of the lower line which gives the definition of algebraic stability, the box to its right being a consequence.

(A generalization of algebraic stability, (k,l)-*algebraic stability* (Burrage and Butcher (1980)) applies to the class of general linear methods.)

AN-stability (Hundsdorfer and Spijker (1981)) takes an approach which can be seen as a direct extension of A-stability. The test equation for A-stability is essentially the scalar equation $\dot{y} = \lambda y$, $\lambda \in \mathbb{C}$ a constant, and this is replaced by the variable coefficient scalar equation

$$\dot{y} = \lambda(t)y, \quad \lambda(t) \in \mathbb{C} \tag{7.1}$$

whose general solution satisfies

$$y(t+h) = \kappa y(t), \quad \kappa = \exp\left(\int_t^{t+h} \lambda(\tau)\,d\tau\right). \tag{7.2}$$

Applying an RK method to (7.1) yields

$$y_{n+1} = K(\psi)\, y_n, \quad K(\psi) = 1 + \underline{b}^T \psi (I - A\psi)^{-1} \underline{e} \tag{7.3}$$

where

$$\psi = \mathrm{diag}(\xi_1, \xi_2, \cdots, \xi_s), \quad \xi_i = h\lambda(t_n + hc_i), \quad \underline{e} = [1, 1, \cdots 1]^T.$$

Requiring analogous behaviour for (7.1) and (7.3) leads to the following syntax for AN-stability:

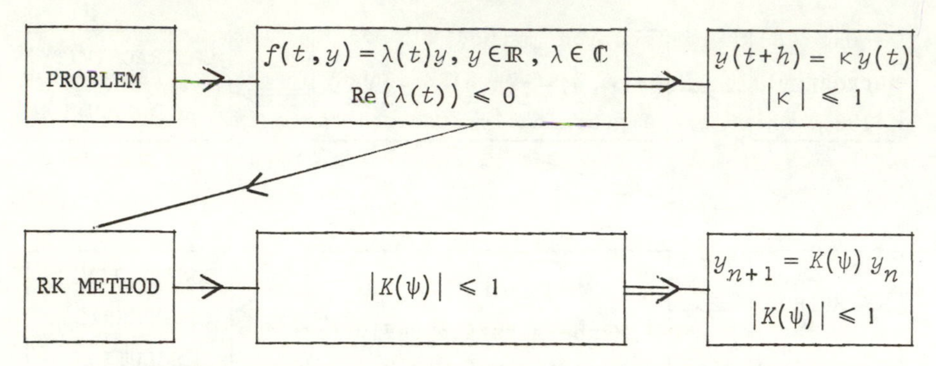

The middle box on the lower line defines *AN*-stability. It should be noted that, unlike *A*-stability, the test equation for *AN*-stability cannot be generalized in a natural way from a scalar to a vector equation.

Not surprisingly, *B*-, *BN*-, algebraic, *AN*- and *A*-stability are inter-related. For any RK method the following holds:

$$\text{Algebraic stability} \Rightarrow \begin{cases} BN\text{-stability} \Rightarrow B\text{-stability} \\ \qquad\qquad\qquad\qquad \Downarrow \\ AN\text{-stability} \Rightarrow A\text{-stability} \end{cases}$$

For RK methods which are either nonconfluent or irreducible the following holds:

$$\left.\begin{array}{l} \text{Algebraic stabilty} \\ \Updownarrow \\ BN\text{-stability} \\ \Updownarrow \\ AN\text{-stability} \end{array}\right\} \Rightarrow B\text{-stability} \Rightarrow A\text{-stability}$$

Of the RK-methods of quadrature type quoted earlier (which one recalls are all *A*-stable), the Gauss, Radau IA, Radau IIA and Lobatto IIIC methods are all algebraically stable; the Lobatto IIIA and IIIB are not.

It is possible to achieve algebraic stability with less than full implicitness. Thus there exist 2- and 3-stage SDIRK methods of orders 3 and 4 respectively due to Nϕrsett (1974), a

2-stage SIRK collocation method of order 3 (Nørsett (1976), Burrage (1978a)) and a 3-stage SIRK method of order 4 (Burrage (1982)) which are algebraically stable. For a full description of these methods, the reader is referred to Dekker and Verwer (1984).

8. CONCLUSIONS

In this paper we have restricted our attention to the development of nonlinear theory, and have not addressed the large and difficult question of the implementation of methods possessing these stability properties. The question of existence and uniqueness of solutions of the algebraic equations defining the new function value at each step, and the sensitivity of the internal stages of RK methods to error give rise to further definitions of *BS*- and *BSI*-stability (Frank, Schneid and Ueberhuber (1981, 1982)) but these are not definitions that fall within our general framework.

The author finds it very difficult to draw clear conclusions on the importance of the development of nonlinear stability theory to the practical problem of developing even more robust and efficient algorithms for solving stiff *IVP*s than we have at present. There can be no doubt that the absolute stability theory we depended on ten years ago was theoretically inadequate; nor can it be denied that the development of nonlinear stability is an attractive piece of mathematics which fully and convincingly overcomes the theoretical deficiencies of absolute stability — we could not have asked for better. Yet, several factors have prevented nonlinear stability theory from having as substantial an impact as we might have expected on the way in which we actually solve stiff *IVP*s in practice. Firstly, existing algorithms do not explicitly test for stability, but rely on good error estimates to pick up potential instability by detecting error growth; but stability is a key factor in the selection of methods on which these algorithms are based. Secondly, as mentioned earlier, absolute stability works well enough most of the time; it is surprisingly difficult to find counter-examples like Example 1. Lastly, the Newton iterations, inescapable in implicit RK methods, have to be handled carefully if algebraic stability is not to be lost, and this makes for expensive implementations. There have been some implementations of algebraically stable methods, but they do not seem yet to be threatening *BDF*—based methods as standard routines for large and difficult stiff systems. Whether this will happen in the future remains to be seen.

REFERENCES

Alexander, R. (1977), "Diagonally implicit Runge-Kutta methods for stiff *ODE*s", *SIAM J. Numer. Anal.*, *14*, pp.1006-1021.

Burrage, K. (1978a), "A special family of Runge-Kutta methods for solving stiff differential equations", *BIT*, *18*, pp.22-41.

Burrage, K. (1978b), "High order algebraically stable Runge-Kutta methods", *BIT*, *18*, pp.373-383.

Burrage, K. (1981), "Iteration schemes for singly-implicit Runge-Kutta methods", Computer Science Report No.24, University of Auckland.

Burrage, K. (1982), "Efficiently implementable algebraically stable Runge-Kutta methods", *SIAM J. Numer. Anal.*, *19*, pp.245-258.

Burrage, K. and Butcher, J.C. (1979), "Stability criteria for implicit Runge-Kutta methods", *SIAM J. Numer. Anal.*, *16*, pp.46-57.

Burrage, K. and Butcher, J.C. (1980), "Nonlinear stability of a general class of differential equation methods", *BIT*, *20*, pp.185-203.

Butcher, J.C. (1964), "Integration processes based on Radau quadrature formulas", *Math. Comp.*, *18*, pp.233-244.

Butcher, J.C. (1975), "A stability property of implicit Runge-Kutta methods", *BIT*, *15*, pp.358-361.

Butcher, J.C. (1976), "On the implementation of implicit Runge-Kutta methods", *BIT*, *16*, pp.237-240.

Chipman, F.H. (1971), "*A*-stable Runge-Kutta processes", *BIT*, *11*, pp.384-388.

Crouzeix, M. (1979), "Sur la B-stabilite des methodes de Runge-Kutta", *Numer. Math.*, *32*, pp.75-82.

Dahlquist, G. (1959), "Stability and error bounds in the numerical integration of ordinary differential equations", Trans. Royal. Inst. of Technology No.130, Stockholm.

Dahlquist, G. (1975a), "On stability and error analysis for stiff nonlinear problems", Report No TRITA-NA-7508, Dept. of Information Processing and Computer Science, Royal Inst. of Technology, Stockholm.

Dahlquist, G. (1975b), "Error analysis for a class of methods for stiff nonlinear initial problems", Report No TRITA-NA-7511, Dept. of Numerical Analysis and Computing Science, Royal Inst. of Technology, Stockholm.

Dahlquist, G. (1978), "*G*-stability is equivalent to *A*-stability", *BIT*, *18*, pp.384-401

Dahlquist, G. and Jeltsch, R. (1979), "Generalized disks of contractivity for explicit and implicit Runge-Kutta methods", Report No TRITA-NA-7906, Dept. of Numerical Analysis and Computing Science, Royal Inst. of Technology, Stockholm.

Dekker, K. and Verwer, J.G. (1984), *Stability of Runge-Kutta Methods for Stiff Nonlinear Equations*, North Holland (Amsterdam).

Ehle, B.L. (1969), "On Padé approximations to the exponential function and *A*-stable methods for the numerical solution of initial value problems", Univ. of Waterloo Dept. Applied Analysis and Computer Science Research Report No CSSR 2010.

Frank, R., Schneid, J. and Ueberhuber, C.W. (1981), "*B*-convergence of Runge-Kutta methods", Bericht nr 48/81, Institut fur Numerische Mathematik, Tech. Univ. Wien.

Frank, R., Schneid, J. and Ueberhuber, C.W. (1982), "Stability properties of implicit Runge-Kutta methods", Bericht Nr 52/82, Institut fur Numerische Mathematik, Tech. Univ. Wien.

Hirsch, M.W. and Smale, S. (1974), *Differential Equations, Dynamical Systems, and Linear Algebra*, Academic Press (New York).

Hundsdorfer, W.H. and Spijker, M. (1981), "A note on *B*-stability of Runge-Kutta methods", *Numer. Math.*, *36*, pp.319-331.

Lambert, J.D. (1973), *Computational Methods in Ordinary Differential Equations*, John Wiley & Sons (New York — London).

Nφrsett, S.P. (1974), "Semi-explicit Runge Kutta methods", Report Mathematics and Computation No 6/74, Dept. of Mathematics, University of Trondheim.

Nφrsett, S.P. (1976), "Runge-Kutta methods with a multiple real eigenvalue only", *BIT*, *16*, pp.388-393.

Wright, K. (1970), "Some relationships between implicit Runge-Kutta, collocation and Lanczos τ-methods, and their stability properties", *BIT*, *10*, pp.217-227.

J.D. Lambert
Department of Mathematical Sciences
University of Dundee
Dundee DD1 4HN
Scotland

15 Stiff ODE Initial Value Problems and their Solution

A. R. CURTIS

ABSTRACT

Numerical solution of stiff initial value problems in ordinary differential equations (ODEs) is discussed in the light of experience on a large number of practical problems, many of them large and very stiff, arising mainly from mass action kinetics, often with diffusion and/or advection.

An outline of the numerical method used is given, followed by a discussion of problem characteristics and their impact on choice and implementation of method. Special attention is given to exploitation of sparsity in large problems, need for relative accuracy control, effect of conservation laws and effect of discontinuities. A simple example is used to explain how stiffness arises, and to link it with cancellation in derivative evaluation and ill-conditioning of corrector equations.

More detailed discussions are given of techniques for error control, corrector convergence testing, and discontinuity handling, based on experience on demanding problems.

1. INTRODUCTION

We discuss in this paper the numerical solution of stiff ordinary differential equation (ODE) initial value problems of the form

$$\underline{y}' = \frac{dy}{dt} = \underline{f}(t,\underline{y}) , \qquad \underline{y}(0) = \underline{y}_0 , \tag{1.1}$$

where in general $\underline{y}, \underline{f}, \underline{y}_0$ are vectors of dimension m (but the scalar case is included for $m = 1$). Stiffness is usually discussed in terms of the eigenvalues of the Jacobian matrix

$$J = \partial \underline{f} / \partial \underline{y} \tag{1.2}$$

of the system, although this is not necessarily the best approach. We shall try to throw some light on how it may arise, but roughly speaking we assume that any eigenvalues of J with positive or zero real parts are very small (in relation to the time steps to be taken when solving the problem), and at least some eigenvalues have negative real parts large enough to cause difficulty unless a method suitable for stiff problems is used.

In our experience, mild stiffness is comparatively rare; the (non-dimensional) product of integration stepsize with the most negative eigenvalue often becomes enormous, so the question of fine distinctions between stiff and non-stiff problems rarely arises.

In Section 2 we outline the numerical method we use, and in Section 3 discuss characteristics of practical problems from one extensive area of applications, to highlight some features which are relevant to the design or choice of numerical method, and which may not be obvious. In Section 4 we discuss how stiffness arises in such problems, and relate it to heavy cancellation in evaluation of the function $\underline{f}(t, \underline{y})$ in (1.1).

In Sections 5 and 6 we discuss the design of error testing and of corrector convergence testing respectively, in the light of problem characteristics. In Section 7 we discuss the handling of discontinuities, which are common in many applications.

2. OUTLINE OF NUMERICAL METHOD

The numerical method we describe is incorporated into the Harwell Subroutine Library (Hopper, 1984) package DC03, and in the FACSIMILE (Curtis and Sweetenham, 1985) program. Like DIFSUB (Gear, 1971) and GEAR (Gear and Hindmarsh, 1974), it uses a so-called 'fixed-step' variable-order backward differentiation formula (BDF) method, using orders up to 4 normally, this limit being changeable, up to a maximum of 6, under user control. While I do not claim here that this is the only method suitable for very stiff problems, I do believe that some practical problems are such that non-BDF methods would have great difficulty with them; while the expected advantages of true variable-step BDF methods such as EPISODE (Byrne and Hindmarsh, 1975) do not seem to have been achieved in practice. For our present purposes, the method can best be outlined as follows.

At time t_{n-1} we have a (vector) local approximating polynomial $\underline{u}_{n-1}(t)$, of degree K (the order of the method), to $\underline{y}(t)$:

$$\underline{y}(t) \simeq \underline{u}_{n-1}(t) . \tag{2.1}$$

The process of taking an integration step, of length h, from t_{n-1} to t_n consists of constructing a corresponding approximating polynomial at t_n of the form

$$\underline{u}_n(t) = \underline{u}_{n-1}(t) + \underline{c}Q_K((t-t_n)/h) \tag{2.2}$$

where

$$Q_K(\theta) = (1+\theta)(1+\theta/2)\cdots(1+\theta/K) \tag{2.3}$$

is a fixed polynomial of degree K which vanishes at $\theta = -1, -2, \cdots, -K$, and where the correction $\underline{c}$ is chosen to make $\underline{u}_n(t)$ satisfy the differential equations (1.1) at $t = t_n$. That is, noting that $Q_K(0) = 1$ and writing

$$\underline{y}^p = \underline{u}_{n-1}(t_n) \tag{2.4}$$

$$\underline{f}^p = \underline{u}'_{n-1}(t_n) \tag{2.5}$$

$$\beta = Q'_K(0) = 1 + 1/2 + \cdots + 1/K , \tag{2.6}$$

$\underline{c}$ is chosen to satisfy the (generally nonlinear) equations

$$\underline{f}^p + \beta\underline{c}/h = \underline{f}(t_n, \underline{y}^p + \underline{c}) . \tag{2.7}$$

The "predictor-corrector" difference $\underline{c}$ is used to generate a Milne-type error estimate for stepsize control.

Then the value $\underline{y}_n$ of the computed solution at t_n is taken to be $\underline{u}_n(t_n)$. Because $Q_K(\theta)$ vanishes when $\theta = -1$, $\underline{u}_n(t)$ can be used for interpolation within the step just completed. If the order K is to be changed for the next step, an appropriate change to $\underline{u}_n(t)$ is made between steps. Fairly simple changes are made: on reducing the order by 1, the highest degree term is deleted; on increasing it, a highest degree term derived from the error estimate is added. It has to be remembered that these simple changes destroy the interpolation property, so they should not be made until completion of any output required within the step just completed.

The nonlinear equations (2.7) are solved iteratively at each step. It is necessary to use quasi-Newton iteration, because direct functional iteration diverges on stiff problems. A Newton iteration would have the form

$$(J - \beta I/h)\delta\underline{c} = \underline{r} = \underline{f}^p + \beta\underline{c}/h - \underline{f}(t_n, \underline{y}^p + \underline{c}) \qquad (2.8)$$

where $\underline{c}$ is the correction so far, and $\delta\underline{c}$ is a new increment to be added to it. Quasi-Newton iterations are used, in the sense that J is replaced by an approximation to the Jacobian matrix, computed by finite differencing and usually at an earlier time step; it is only recomputed if corrector convergence fails. An additional point worth remembering is that only a small number of corrector iterations, typically 1 to 3, are done; if a preset limit which is often 3 is reached without convergence, convergence failure action is taken. We have in fact found it worthwhile to increase the limit beyond 3 on large problems, where re-evaluation of J is relatively expensive, but we still set an absolute upper limit of 8.

3. SOME BASIC PROBLEM CHARACTERISTICS

Many of our stiff problems arise from modelling mass action kinetics of one sort or another, sometimes in conjunction with diffusion and/or advection by known solvent flow patterns. For those familiar with the stiffness which can arise from parabolic partial differential equations, it is important to emphasise that this is not usually the main source of stiffness in these problems, being relatively mild compared with that due to the mass action kinetics. We discuss below how stiffness does arise, but first we wish to point out some other characteristics of these problems which have important effects on the performance of numerical methods.

First, many reaction kinetics systems modelled are quite large, for example up to 100-200 concentration components (up to several thousand, if spatial inhomogeneity is also being modelled). An immediate consequence of this is the need to exploit any sparsity which is generally present in the Jacobian matrix of the system. Large savings in computer time are frequently achievable in this way. Also, it is unrealistic to expect the problem originator to code the evaluation of this matrix, so it must be evaluated either by finite differencing or by some analytical method; the latter usually assumes a restricted problem structure. The cost of analysing the sparsity pattern of the Newton iteration matrix is relatively high, so we do it only once.

Many of these concentrations are of unstable radicals with very short lifetime (i.e. large removal coefficients – we shall see below how this leads to stiffness). Even in a single problem, concentrations of different species differ by many orders of magnitude from each other, and a single concentration may vary, exponentially or cyclically, through many orders of magnitude during the integration. Because different people prefer different units (e.g. moles or molecules per unit volume, the latter in cubic metres or cubic centimetres), the range from problem to problem is even greater.

The 'owner' of the problem has no way of knowing in advance how large most components are expected to be, in order to suggest scaling factors or error weights. Moreover, in case of cyclic behaviour small values may be critically important in determining the duration, for example, of relatively inactive phases which determine the period. Note also that many components may start from zero initial values, but their growth must still be tracked accurately to get a good solution. In small test problems, by doing enough analysis it is often possible to predict the approximate magnitudes of the solution components, but this is quite impracticable on large problems.

Thus it is essential that accuracy control should be on some kind of relative basis for each component. To mention just one well-known test problem (Curtis, 1978), it is plain nonsense to do absolute accuracy testing (i.e. relative to 1) on a problem whose largest solution component is of order 10^{-3}, and the other three are of order 10^{-10} or smaller — and these components are even smaller where their accuracy needs to be controlled to give good global accuracy.

The existence of conservation laws in mass action kinetics is another feature relevant to the design of numerical methods. Each conservation law, which may express the constancy of the total amount of one element in a chemical reaction system, implies the existence of a constant vector $\underline{a}$ (whose elements are small non-negative integers) such that

$$\underline{a}^T \underline{y}(t) = \text{constant}\,, \tag{3.1}$$

from which it follows that the equations (1.1) satisfy

$$\underline{a}^T \underline{f}(t,\underline{y}) = 0\,. \tag{3.2}$$

It follows from (1.2) that

$$\underline{a}^T J = \underline{0}^T \tag{3.3}$$

and the same is true for finite-difference approximations to J (if, as usual, the same increment is used in calculating all the elements in one column). Therefore J and approximations to it have a deficiency in rank equal to the number of independent conservation laws. We shall see later that this is important in very stiff problems. The presence of diffusion and/or advection does not affect the argument if the operators are discretised in a conservative way, since they only move material from place to place without creating or destroying it.

Many of our problems, as reported by Curtis (1983), have almost a continuum of negative eigenvalues of J, so that fixed partitioning into stiff and non-stiff subsystems will not be successful because of the need for a wide range of step sizes. Re-partitioning from time to time during a run is sometimes proposed; the analysis needed to do this for a general system is comparable with that for analysing the sparsity pattern of the Newton matrix, so doing it more than once in a run is a high price to pay. This is especially the case since, at least at the larger step sizes, the dimension of the stiff subsystem would rarely be small compared with that of the full system, so that the gain to be expected from partitioning would be small.

A further characteristic of many problems is discontinuity of time derivatives. In a pollution model, for example, there is frequently an initial phase with no pollutant input, to establish a steady state. Then the rate of pollutant input will increase discontinuously to a nonzero value. If photochemistry is being modelled, important reactions will be switched on and off to simulate lamp state in the laboratory, or sunrise and sunset in nature. A similar effect occurs in modelling radiation chemistry in the coolant of a nuclear reactor; as the coolant enters or leaves the core, input of ionising radiation is switched on or off. To be a useful modelling tool, a numerical method must be capable of handling such discontinuities successfully, and if possible efficiently.

It is often thought that solutions of stiff problems have the character of decreasing exponentials, and that numerical methods specially tailored for such behaviour should be used. This is contrary to much experience; there may well be phases of (increasing or decreasing) exponential behaviour, but these are rarely the stiffest phases, which tend to occur either when initial exponential transients have died out, or before the feedstock exhaustion which initiates decaying exponential behaviour. During the transient itself, the step size remains small in order to follow it accurately.

There is also considerable justified concern over the vulnerability of BDF methods in problems with slowly decaying oscillatory complementary functions (eigenvalues near the imaginary axis). This does not normally happen in reaction kinetics problems; however, if instability in choice of stepsize is experienced, it may be due to this cause, and then restricting the maximum order to 2 should be effective; even a limit of 3 may be. If the solution itself has important oscillatory components, stepsizes will necessarily be restricted in order to follow them accurately, so the problem is less likely to become stiff. However, in many such problems the use of stiff methods is not greatly inefficient; the main exception to this is computation of gravitational orbits, where very high accuracy is often needed over many cycles.

4. ORIGIN AND NATURE OF STIFFNESS

We can illustrate how stiffness arises by considering a single concentration component in a mass action kinetics model. It satisfies a differential equation of the form

$$y' = P - Ry \qquad (4.1)$$

where its production rate P and removal coefficient R depend on time, in general, through the values of other concentration components (also explicitly, in some problems) but — and this is the key point — P and R usually do not depend strongly on y itself.

If the values of P and R vary sufficiently on a time scale of order $1/R$, then any significant difference between y and $z(t) = P/R$ tends to decay exponentially with coefficient $1/R$. More accurately, we can show that after such exponential decay we have asymptotically in $1/R$

$$y = z(t) - z'(t)/R + O(R^{-2}) . \tag{4.2}$$

Thus y is slowly varying in relation to the $1/R$ time scale, with the result that (subject to the behaviour of other solution components) an efficient integration method can take time steps h for which hR is extremely large (commonly up to 10^{15}, once up to 10^{23}, in our experience).

Two conclusions follow from this. First, there is very strong cancellation between the production and removal terms in the differential equation, since error control will not allow $|hy'|$ to greatly exceed y, so that

$$y' = P - Ry = O(y/h) = O(P/hR) . \tag{4.3}$$

Thus about $\log(hR)$ significant figures are lost through cancellation. Clearly, the numerical method must be extremely rugged in face of such severe cancellation in derivative evaluation.

Secondly, it is easy to see that J must have eigenvalues with negative real parts at least of order $-R$. But as explained above, because of conservation laws J itself is often singular, so it is only the relatively extremely small multiple β/h of the unit matrix which makes the quasi-Newton matrix

$$J - \beta I/h \tag{4.4}$$

in (2.8) non-singular. This matrix thus has condition number of order hR, but the situation is saved because both the predicted and the corrected solution vectors satisfy the conservation laws to high accuracy. Thus the corresponding eigenvalues are almost absent from the residuals. Where different methods or orders or stepsizes, or even different linear algebra, are used for different solution components, as is sometimes recommended, the second term in (4.4) is no longer a multiple of the unit matrix. In that case, the eigenvalues of the Jacobian and Newton matrices no longer coincide, and this safeguard is lost. I have no idea how serious the effects of this would be on very stiff problems with conservation laws.

Once again, because of the ill-conditioning, the need for utmost ruggedness of the numerical methods is clear. Moreover, it is often the case with ill-conditioned non-symmetric matrices that their eigenvector system is very skew (i.e. far from orthogonal), and we shall see later that this possibility should be taken into account in testing for convergence of corrector iterations.

In order to ensure adequate ruggedness of numerical methods, it is necessary to develop them using test problems which achieve the high degree of stiffness occurring in practical applications. Very few test problems in the literature are stiff enough for this, but some can be made stiffer by running them for much longer. The value of $h\|J\|$, for some convenient norm, can be monitored; if it does not approach the reciprocal of a rounding error on the computer in use, then either the problem is not sufficiently demanding or the code is not sufficiently rugged – if frequent irregular step size changes are being made, probably the latter. The cost of doing this is not excessive, since the work needed to multiply the problem time by a fixed factor (e.g. 10) is often roughly constant, once the stiff phase is reached. Thus the total computing resources needed are roughly proportional to the logarithm of the problem time.

Of course, if such rigorous testing has been done, any failure of the code in use is likely to be on a large, demanding problem, making debugging difficult; but meanwhile it will have been succeeding on many other demanding problems on which it would have failed with less realistic testing.

5. ERROR TESTING AND STEP SIZE CONTROL

We have indicated that some sort of relative accuracy for each component must be aimed at in practice; clearly simple relative error testing may get into difficulty if a solution component passes through zero, or underflows, or — much more commonly — simply starts from an initial value of zero. We can avoid these difficulties by holding a vector $\underline{w}$ of comparison values which are never allowed to become zero, but which are otherwise set, at the start of each integration step, close to the absolute values of the solution components. Then if $\underline{e}$ is the vector of error estimates and τ the tolerance, we control h so that for each component

$$|e_i| \leqslant \tau w_i \tag{5.1}$$

instead of

$$|e_i| \leqslant \tau |y_i| \, . \tag{5.2}$$

One can think of a number of ways of defining such comparison values; for example the choice (for each component)

$$w_i = \max(|y_i|, |hy'_i|) \tag{5.3}$$

might deal with the case of a component passing through zero, but not with the frequent case where a component starts from zero initial value with zero initial slope. Once a component has become sufficiently large, we use instead at the start of step n the updating formula (for each component)

$$w_{i,n} = \sigma w_{i,n-1} + (1-\sigma)|y_{i,n-1}| \tag{5.4}$$

where σ is chosen small enough to allow a $\underline{w}$ component to decrease exponentially at least as fast as the corresponding $\underline{y}$ component. We can estimate a suitable value for σ as follows: if $y = y_0 \exp(-\alpha t)$, relative accuracy testing with tolerance τ should restrict h so that

$$\alpha h \leqslant \mu = (\tau / c_{K+1})^{1/(K+1)} \tag{5.5}$$

when using order K, where c_{K+1} is a known constant for each K. It follows that the factor $\exp(-\alpha h)$ by which y decreases in one step cannot be smaller than $\exp(-\mu)$; we actually set $\sigma = \exp(-2\mu)$, for safety, whenever τ or K changes.

It would be desirable to use (5.4) on a decreasing component nearly down to underflow level, if necessary, to avoid a loss of control which can allow the component to go negative with, frequently, consequent numerical troubles. However, in many problems this is too expensive; it can be estimated to need several hundred integration steps starting from a value of order unity, even at relaxed tolerance on a computer whose underflow level is not particularly small. For this reason we put a floor under the decrease of each comparison value, set at a fixed multiple (10^{-10} by default) of the maximum value so far taken by the corresponding solution component. If this leads to numerical difficulty from solution components going negative when they shouldn't, this multiple may be decreased, or even set to zero to switch off the effect.

It is necessary to be able to cope with solution components whose initial values, and often initial slopes also, are zero. We first choose a small tentative initial step size, by the criterion that for each nonzero solution component $|hy'_i/y_i|$ does not exceed the error tolerance. For any zero component whose initial slope y'_i is nonzero we use (5.3), noting that if h should be reduced on the first step this comparison value must be scaled appropriately.

If the initial slope is also zero, we set w_i to a very large number which we do not update, to switch off testing on the component until it has become nonzero (actually, until it is at least 10^{-30}, although this value can be changed). Testing on the component is then introduced gradually, by resetting w_i to a value larger than the component by a factor of 10^{10}, and leaving the updating formula (5.4) to adapt it to the increasing component value. Although this is rather artificial, something has to be done about this very common situation, and the method works very successfully in practice.

6. CORRECTOR ITERATIONS AND CONVERGENCE TESTING

It has been suggested that one can avoid an unnecessary solution of the corrector equations (2.8) by testing the multiple $h\underline{r}/\beta$ of the residual vector, instead of the correction vector $\delta\underline{c}$, for convergence. The basis for this is that the eigenvalues of the Newton matrix (4.4) are either close to $-\beta/h$ or have even more negative real parts; from this it is incorrectly inferred that the norm of $h\underline{r}/\beta$ must (approximately) be at least as large as that of $\delta\underline{c}$.

This conclusion is false because of possible skewness of the eigensystem of J. Consider a 2×2 matrix B with right eigenvectors $\underline{u}=(1,0)^T$, with eigenvalue 10, and $\underline{v}=(1,\varepsilon)^T$, with eigenvalue 2. If $\underline{r}$ is the unit vector $(\underline{v}-\underline{u})/\varepsilon=(0,1)^T$, the solution of the equations

$$B\delta\underline{c}=\underline{r} \tag{6.1}$$

is given by

$$\delta\underline{c}=(0.5\underline{v}-0.1\underline{u})/\varepsilon=(0.4/\varepsilon,0.5)^T \tag{6.2}$$

whose norm exceeds that of $\underline{r}$ by a factor at least $0.4/\varepsilon$, which we can make as large as we like by choice of ε, even though both eigenvalues of B exceed unity.

With eigenvalues of J covering a wide range, and with possible skewness of its eigenvectors, there is simply no good metric in which to assess the residual vector $\underline{r}$. On the other hand, the correction vector increment $\delta\underline{c}$ can be assessed relative to the components of $\underline{y}$, or in practice to those of the error comparison vector $\underline{w}$. Moreover, there is a natural stopping criterion, at some small multiple (which we take as 0.1) of the error tolerance τ.

Even apart from this, over-pessimistic decisions can result from testing $h\underline{r}/\beta$ instead of $\delta\underline{c}$, as well as over-optimistic ones indicated by the above example. If $\delta\underline{c}$ contains a quite insignificant component of an eigenvector of J corresponding to a numerically very large eigenvalue, the corresponding component can completely dominate $\underline{r}$.

Since we have to solve the corrector equations (2.8) in order to test $\delta\underline{c}$, we always add this increment to $\underline{c}$. We can then avoid even evaluating the residuals for the next iteration if we can predict that the convergence test will be passed. We do this by predicting the rate of convergence, based on earlier experience, and applying a safety factor in testing the predicted value.

Experience with large and difficult problems has led us to be fairly conservative in corrector convergence testing, except for predicting convergence. It is not safe to test only the norms of successive increments $\delta\underline{c}$ for convergence, since we have found that such a test can be passed (based on a few iterations) although small components are diverging. Accordingly, we insist that each component must show convergence by a factor not exceeding 0.5 per iteration (or all must be small enough on the first iteration).

We have found that the rate of convergence frequently varies considerably from one iteration to the next, but is reasonably consistent between corresponding iterations on successive steps. Therefore, for predicting convergence we build up experience of convergence factors for individual iterations, updating them from step to step with the maximum ratio over all components (above rounding error level) of $|\delta c_i|$ on the current iteration to its value on the previous one. We discard these empirical convergence factors whenever the Newton matrix changes (because of a change in h or recomputation of J), and then force a minimum of two corrector iterations for one step, to get at least one convergence rate estimate. We also use these factors, ϕ say, (if less than 0.5, so that convergence is a possibility) in pessimistic estimates $\delta\underline{c}/(1-\phi)$ of the error in $\underline{c}$, taking the sum of a geometric series whose first term is the current correction.

7. HANDLING DISCONTINUITIES

As pointed out above, discontinuities in a derivative, often the first, of the solution will certainly be coded by many users of general purpose software for stiff ODE problems, so it is necessary to be able to handle them. An imposed jump discontinuity in a solution component itself is clearly beyond the scope of an ODE code, which must be allowed to keep solution components under its own control, so the user must code a restart of the integration procedure if he makes such a change.

However, in the case of a discontinuity in a derivative, say the qth derivative $\underline{y}^{(q)}(t)$ where $1 \leqslant q \leqslant K$, one possible approach is to leave the error testing to reduce the stepsize until the discontinuity can be crossed without triggering further stepsize reduction. Two questions arise with this approach:

(a) Will it work, or will the stepsize be reduced indefinitely?

(b) Will the actual errors after the discontinuity be small enough?

The first question is easily answered: if there is no safety factor in the stepsize reduction following failure of the error test, there is a possibility that the test will never be passed. However, such a safety factor is invariably built in: if for example it aims to produce an error estimate equal to half the tolerance, one can estimate worst cases (for a large discontinuity at $q=1$) ranging from about 5 failures and stepsize reductions at order 1 to about 20 at order 5. Assuming that discontinuities are relatively infrequent, this may be quite acceptable.

The above answer is unfortunately not the whole story, as is shown by the following case history. In a problem with reaction rate coefficients depending on temperature, which was one of the variables, the originator programmed a long initial phase at constant temperature to reach a steady state, followed by a phase during which the temperature fell at a steady rate. When we attempted to use the above method, the stepsize in the initial phase had become very large, and the code therefore attempted a large step across the discontinuity, which it would have reduced following failure of the error test. However, it was never allowed to get as far as this, because the first corrector iteration produced a negative temperature, which caused failure in taking its square root for the rate coefficient evaluation. We had therefore to use the more elaborate method described below.

In spite of this unfortunate experience on one problem, we have found that leaving the code to deal with discontinuities as best it can is successful, and not unduly expensive, on a surprisingly wide range of problems. This includes both hard discontinuities in $\underline{f}$ $(q=1)$ and the milder but still severe near discontinuities which arise in modelling sunrise and sunset in photochemical problems. We allow up to 5 failures at fixed order on a step, then reduce the order (if $K>1$) and try a further 5 times, trying a restart after 5 failures at order 1.

To study our second question, about accuracy, is very difficult in general. A detailed study of a special case, where the solution is exactly a polynomial of degree $K+1$, where many steps have been taken at constant h, and where the discontinuity is passed on the next (shorter) step, suggests that the actual error on that step is unlikely to exceed about $(K+1)\tau$, where τ is the error tolerance. This is quite a satisfactory result as far as it goes, especially in a stiff system where errors will subsequently die out.

The above case is too special to be completely reassuring; however, some limited numerical tests were done, and did not show excessive errors, even including errors of interpolation within steps, near a discontinuity. The tests were all on scalar problems of the form

$$y' = s' + \lambda(s-y) \qquad y(0)=1 \tag{7.1}$$

where

$$s(t) = e^{t} + \alpha(t-1)_{+} \tag{7.2}$$

for various values of λ, α and the error tolerance. Thus $s(t)$ has a jump discontinuity of size α in its first derivative at $t=1$, giving $q=1$.

Values of λ of order 10 were mainly used, giving scarcely any stiffness, since it was found that errors became smaller when stiffness was increased. In a few tests, a discontinuity in λ was also forced at $t=1$. Values of α in the range $(-10^6, 10^6)$ were used, and in all cases the actual global relative error $(y-s)/s$ was compared with the error tolerance, which was either 10^{-6} or 10^{-3}. Although no attempt was made to cover the parameter ranges closely, it was felt that reasonably representative results were obtained. Except near $s=0$ for negative α, the worst relative errors were less than 4 times, and generally less than twice, the tolerance.

It is often unwise to ignore a discontinuity in this way. In that case, the following method is recommended; it does unfortunately demand a little more work on the part of the user. A switching function $g(t,\underline{y})$ is constructed, which changes sign at the discontinuity. Two versions (in practice, usually one version with a switch) of the functions $\underline{f}(t,\underline{y})$ are coded, for use before and after the discontinuity. Integration is carried on using the 'before' version, and at the end of each integration step the switching function is tested. When it is found to have changed sign, it is known that the discontinuity should have occurred during the step just completed. Inverse interpolation is carried out, using the current approximating polynomial $\underline{u}_n(t)$ to the solution, to find the point at which the switching function is zero. Then a switch is made to the 'after' coding for $\underline{f}(t,\underline{y})$ and the integration is restarted (as for a new problem) from that point.

This method is highly robust and accurate, subject only to the 'before' version of $\underline{f}(t,\underline{y})$ being definable past the switching point as a smooth function, which is usually the case. On the few problems where we have tried both methods, there has usually been little to choose between them for efficiency — partly because our code allows rapid increase of stepsize after a restart.

REFERENCES

Byrne, G.D. and Hindmarsh, A.C. (1975), "A polyalgorithm for the numerical solution of ordinary differential equations", *ACM Trans. Math. Software*, *1*, pp.71-96.

Curtis, A.R. (1978) "Solution of large, stiff initial value problems — the state of the art", in *Numerical Software — Needs and Availability*, ed. Jacobs, D.A.H., Academic Press (London) pp.257-278.

Curtis, A.R. (1983), "Jacobian matrix properties and their impact on choice of software for stiff ODE systems", *I.M.A. J. Numer. Anal.*, *3*, pp.397-415.

Curtis, A.R. and Sweetenham, W.P. (1985) "FACSIMILE Release *H* User's Manual", Report AERE-R 11771.

Gear, C.W. (1971), "Algorithm 407, DIFSUB for solution of ordinary differential equations", *Comm. ACM*, *14*, pp.185-190.

Hindmarsh, A.C. (1974), "GEAR: Ordinary differential equation system solver", Report UCID-3001, Rev. 3, Lawrence Livermore Laboratory, University of California, Livermore.

Hopper M.J. (1984), "Harwell Subroutine Library: A catalogue of subroutines (1984)", Report AERE-R 9185 (5th Edition).

A.R. Curtis
Computer Science and Systems Division
A E R E Harwell
Oxon OX11 ORA
England

16 Order Stars and Stability

G. WANNER

ABSTRACT

Order stars, developed in 1978 in collaboration with E. Hairer and S.P, Nϕrsett, have become a successful tool for proving the non-existence of numerical methods under certain stability and order requirements. An example is Dahlquist's theorem on the non-existence of A-stable multistep methods of order 3 (1963). This theorem has subsequently led to the conjecture of Daniel and Moore (1970) that no A-stable q-stage multistep Runge-Kutta method (and q-derivative multistep method) exists of order $2q+1$.

For one-step methods A-stability amounts to the study of rational approximations $R(z)$ to $\exp(z)$, $z \in \mathbb{C}$. For some methods $R(z)$ is a Padé approximation and the question is, will R then result in an A-stable method. This question (the "Ehle conjecture") has originally led to the idea of order stars in 1978.

The same technique then allowed to give nice proofs of Dahlquist's theorem as well as the Daniel-Moore conjecture and the theorems of R. Jeltsch and O. Nevanlinna (1979-1981) on the comparison of stability domains for explicit and implicit methods.

Much work (A. Iserles, R. Jeltsch and others) has since then been devoted to the application of these ideas to obtain order bounds for stable difference schemes for hyperbolic and parabolic equations. Also the first Dahlquist barrier has been handled in this manner.

1. INTRODUCTION

We have discovered the order stars, because sometime in 1977 somewhere in California a secretary posted a letter of application for a position in Geneva into the wrong mail box. The letter arrived by surface mail three months too late and the position could be offered to Syvert Nørsett. This was the beginning of a fantastic collaboration; from January to June 1978 we were sitting in the same office and discussing every day order stars, the sandwich paper, and perturbed collocation.

The starting point of our subject, initiated in 1951 by Dahlquist and Rutishauser, is the explication of instability phenomena occurring in the numerical integration of differential equations nowadays called *stiff*. An example is the following

$$u_i'(t) = 1 + u_i^2 v_i - 4.4\, u_i + \frac{\alpha}{(\Delta x)^2}(u_{i-1} - 2u_i + u_{i+1}) \tag{1.1}$$

$$v_i'(t) = 3.4\, u_i - u_i^2 v_i + \frac{\alpha}{(\Delta x)^2}(v_{i-1} - 2v_i + v_{i+1})$$

$$u_0(t) = u_2(t), \quad u_{61}(t) = u_{59}(t) \qquad (t > 0)$$

$$v_0(t) = v_2(t), \quad v_{61}(t) = v_{59}(t) \qquad (t > 0)$$

$$u_i(0) = 1, \; v_i(0) = \min(1+7x,\, 12-7x), \; x = \frac{i-1}{59}, \; (i = 1, \cdots, 60),$$

$$\alpha = \frac{1}{20}, \qquad \Delta x = \frac{1}{59}$$

which arises from a chemical reaction (the "brusselator") with a diffusion term. We have integrated this system for 60 steps of integration with two different methods: An explicit *RK* method (Dormand and Prince) and an implicit method (the Rosenbrock code ROW4A). The results are shown in Figure 1 and show a significant difference between the explicit and implicit codes: In order to explain this instability, one linearizes in the neighbourhood of an equilibrium point the differential equation

$$\underline{y}' = \underline{f}(\underline{y}) \approx \frac{\partial \underline{f}}{\partial \underline{y}}(\underline{y}_0)(\underline{y} - \underline{y}_0)$$

and transforms the matrix $\frac{\partial \underline{f}}{\partial \underline{y}}(\underline{y}_0)$ to diagonal form

$$T^{-1} \frac{\partial \underline{f}}{\partial \underline{y}}(\underline{y}_0)\, T = \operatorname{diag}(\lambda_1, \cdots, \lambda_n).$$

The differential equation then is reduced to

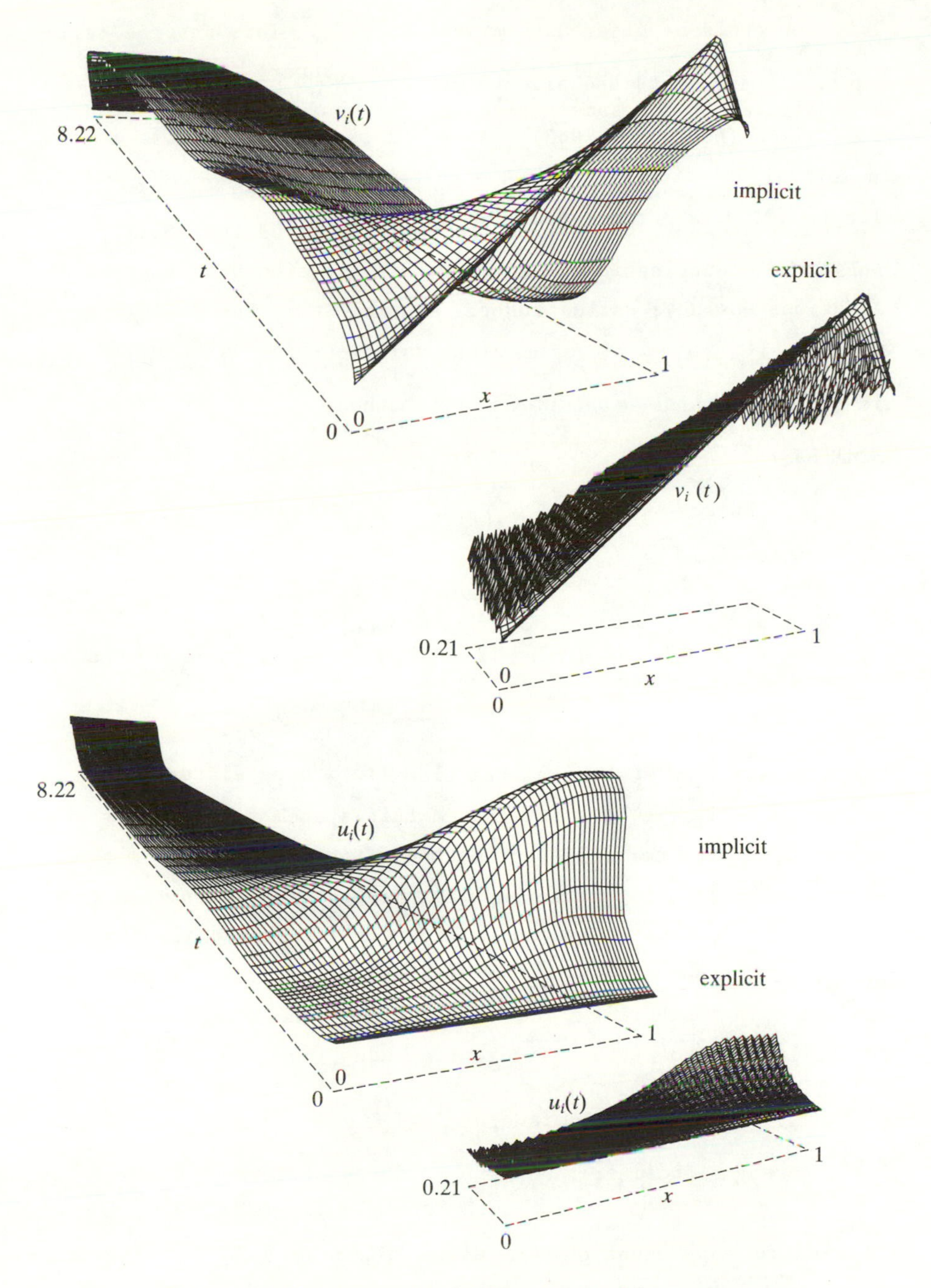

FIG. 1 Solutions of problem (1.1) in perspective projection (60 steps each)

$$y' = \lambda y \tag{1.2}$$

where λ represents the eigenvalues of $\frac{\partial f}{\partial \underline{y}}(\underline{y}_0)$.

DEFINITION (Dahlquist 1963): A method is called A-*stable* if the numerical solution of (1.2) is always stable for arbitrary step-length $h > 0$ and eigenvalues λ with $\operatorname{Re}\lambda \leqslant 0$.

DEFINITION: The set

$$S = \left\{ z \in \mathbb{C}: z = h\lambda \text{ produces stable numerical solutions for } y' = \lambda y \right\}$$

is called the *stability domain* of a method.

EXAMPLES:

1. Euler

$$y_{n+1} = y_n + hf(y_n)$$

$$y_{n+1} = (1+z)y_n$$

2. Backward Euler

$$y_{n+1} = y_n + hf(y_{n+1})$$

$$y_{n+1} = \frac{1}{1-z} y_n$$

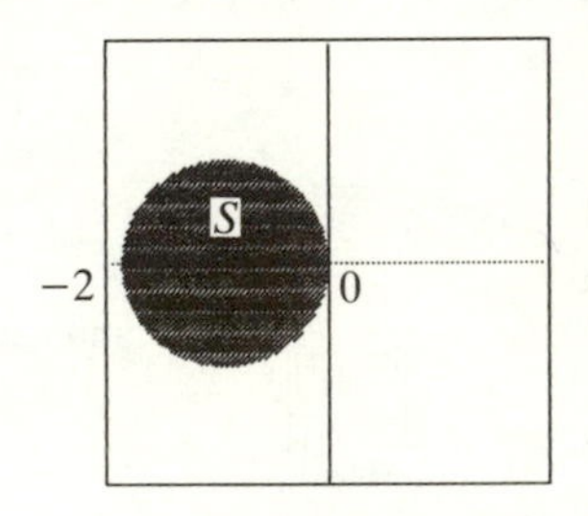

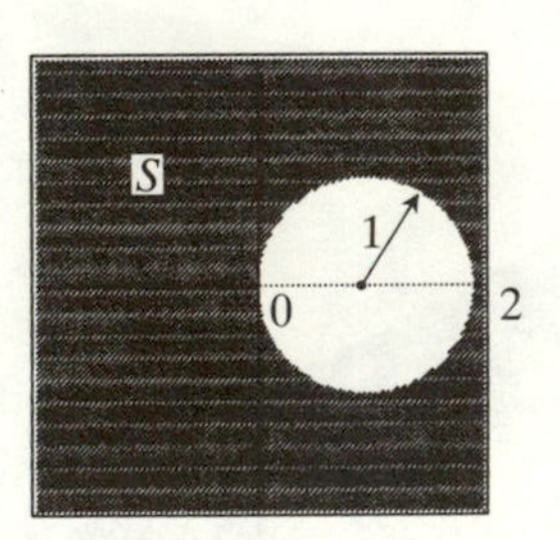

EXAMPLE 3:

$$\text{BDF:} \quad \nabla y_{n+1} + \frac{1}{2}\nabla^2 y_{n+1} + \cdots + \frac{1}{p}\nabla^p y_{n+1} = hf(y_{n+1}).$$

For $p = 2$ we obtain

$$\frac{3}{2} y_{n+1} - 2y_n + \frac{1}{2} y_{n-1} = hf(y_{n+1}) = zy_{n+1} \tag{1.3}$$

and

$$\left(\frac{3}{2} - z\right) y_{n+1} - 2y_n + \frac{1}{2} y_{n-1} = 0.$$

This difference equation is, as usual, solved by setting $y_n = C_1 r_1^n + C_2 r_2^n$, where r_1 and r_2 are the solutions of the corresponding characteristic equation

$$\left(\tfrac{3}{2} - z\right) r^2 - 2r + \tfrac{1}{2} = 0\,. \qquad (1.4)$$

The two roots of this equation are holomorphic functions of the complex variable z. Their moduli are represented in Figure 2. The stability domain S is the set of z for which $r_1(z)$ and $r_2(z)$ are $\leqslant 1$ in modulus. One can observe a branch point at $z = -\frac{1}{2}$ and a pole of *one* of the two roots at $z = \frac{3}{2}$. Comparing (1.4) and (1.3) it can be understood *that this pole has its origin in the implicit stage of formula (1.3)*.

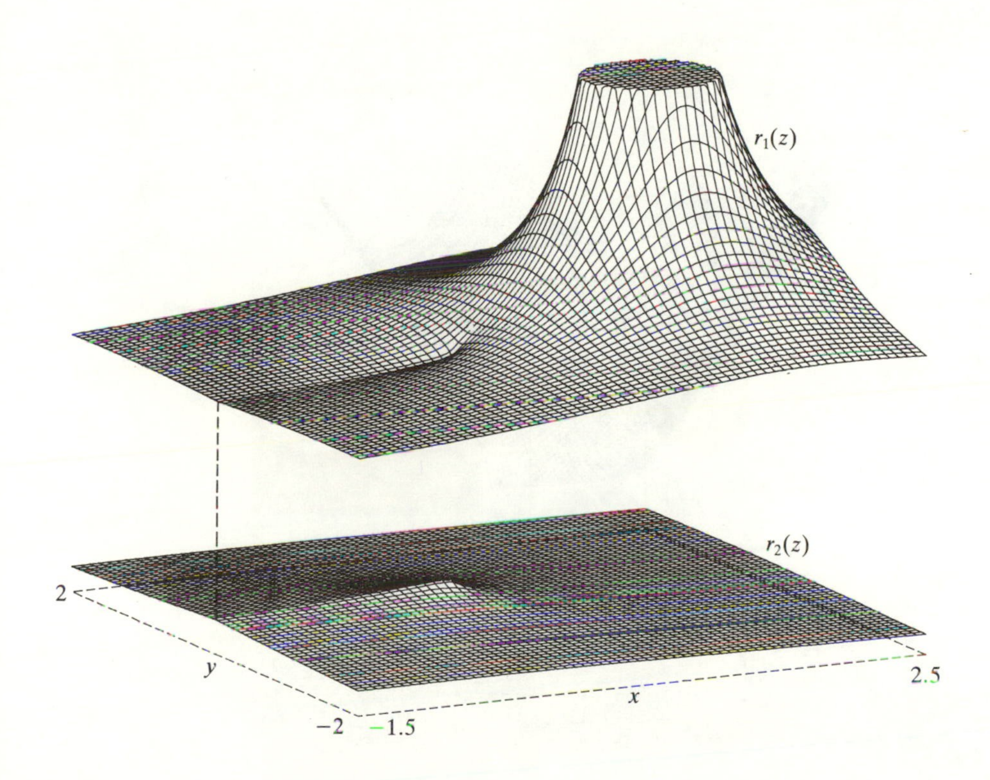

FIG. 2 The two roots of equation (1.4) for the implicit BDF_2 formula

EXAMPLE 4: The explicit Adams method of order 2

$$y_{n+1} = y_n + h\left[\frac{3}{2} f_n - \frac{1}{2} f_{n-1}\right]$$

leads to

$$r^2 - \left(\frac{2+3z}{2}\right) r + \frac{z}{2} = 0$$

whose roots are plotted in Figure 3. This time there is no finite pole, because there is no implicit stage in the formula. Instead, the *one explicit* function evaluation per step size gives rise to *one pole at infinity* of one of the roots.

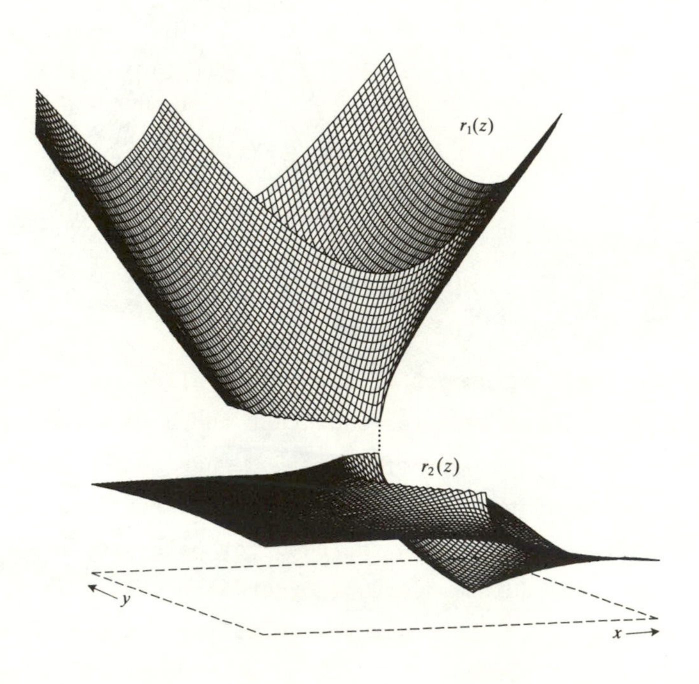

FIG. 3 **Characteristic roots for the second order explicit Adams method**

EXAMPLE 5: The method of Runge of order 2

$$g_1 = y_0 + hf(y_0)$$

$$y_1 = y_0 + \frac{h}{2}\left(f(y_0) + f(g_1)\right)$$

leads to

$$r(z) = \left(1 + z + \frac{z^2}{2!}\right). \tag{1.6}$$

Here, the *two* explicit function evaluations per step give rise to a pole of *order* 2 at infinity.

EXAMPLE 6: The implicit RK method of Gauss-Kuntzmann-Butcher of order 4

$$g_1 = y_0 + h\left(\tfrac{1}{4}f(g_1) + \left(\tfrac{1}{4} - \frac{\sqrt{3}}{6}\right)f(g_2)\right)$$

$$g_2 = y_0 + h\left(\left(\tfrac{1}{4} + \frac{\sqrt{3}}{6}\right)f(g_1) + \tfrac{1}{4}f(g_2)\right)$$

$$y_1 = y_0 + h\left(\tfrac{1}{2}f(g_1) + \tfrac{1}{2}f(g_2)\right)$$

gives rise to

$$r(z) = \frac{1 + \frac{z}{2} + \frac{z^2}{12}}{1 - \frac{z}{2} + \frac{z^2}{12}}\ . \tag{1.7}$$

Here, the *two implicit stages* of the method allow *two finite poles* of $r(z)$. These examples show that there is a correspondence between *numerical work of a method* and the number and position of *poles* of the corresponding characteristic function. The following theory will show the need for poles to obtain order and stability. The quintessence of order star theory is thus the need of numerical work for order and stability and the impossibility of "miracle" methods.

2. THE PADÉ TABLE

The stability functions of the RK methods seen here (and, in fact, of most of the high-order methods) are Padé approximations to the exponential function:

	0	1	2	3
0	1	$1+z$ Euler	$1+z+\frac{z^2}{2!}$ ERK 2	$1+z+\frac{z^2}{2!}+\frac{z^3}{3!}$ ERK 3
1	$\frac{1}{1-z}$ Backw. Euler	$\frac{1+\frac{z}{2}}{1-\frac{z}{2}}$ Trapezoidal R.	$\frac{1+\frac{2}{3}z+\frac{2.1}{3.2}\frac{z^2}{2!}}{1-\frac{1}{3}z}$	...
2	$\frac{1}{1-z+\frac{z^2}{2!}}$ Chipman	$\frac{1+\frac{1}{3}z}{1-\frac{2}{3}z+\frac{2.1}{3.2}\frac{z^2}{2!}}$ Ehle IIA	$\frac{1+\frac{z}{2}+\frac{z^2}{12}}{1-\frac{z}{2}+\frac{z^2}{12}}$ IRK 4	...
3	$\frac{1}{1-z+\frac{z^2}{2!}-\frac{z^3}{3!}}$	...		

Many scattered results with partially very complicated proofs were known before 1978 about the stability properties of these approximations. These are summarized in the following table:

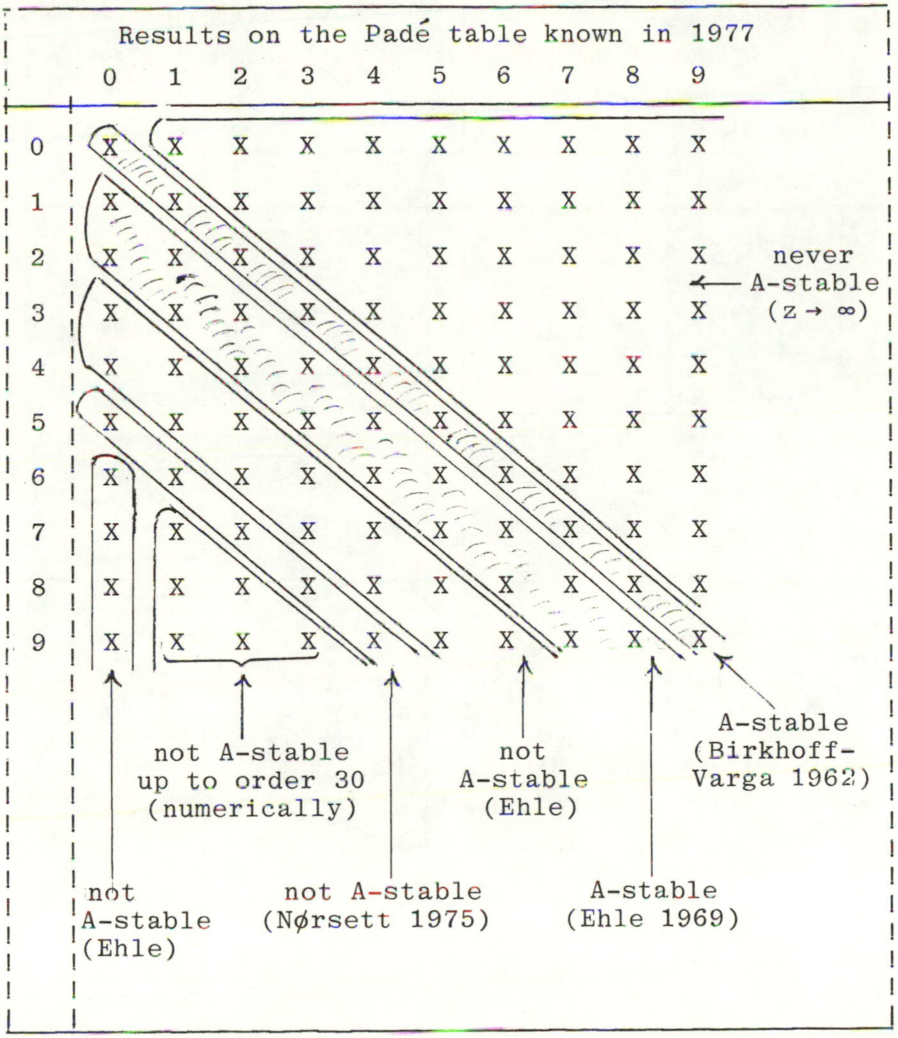

Results on the Padé table known in 1977

	0	1	2	3	4	5	6	7	8	9
0	X	X	X	X	X	X	X	X	X	X
1	X	X	X	X	X	X	X	X	X	X
2	X	X	X	X	X	X	X	X	X	X
3	X	X	X	X	X	X	X	X	X	X
4	X	X	X	X	X	X	X	X	X	X
5	X	X	X	X	X	X	X	X	X	X
6	X	X	X	X	X	X	X	X	X	X
7	X	X	X	X	X	X	X	X	X	X
8	X	X	X	X	X	X	X	X	X	X
9	X	X	X	X	X	X	X	X	X	X

never A-stable ($z \to \infty$)

not A-stable up to order 30 (numerically)

not A-stable (Ehle)

A-stable (Birkhoff-Varga 1962)

not A-stable (Ehle)

not A-stable (Nørsett 1975)

A-stable (Ehle 1969)

An open problem (Ehle's conjecture) was the question if *all* approximations below the second subdiagonal are not A-stable. A study of the stability domains of these Padé approximations (see Figure 4) does not give much insight. The breakthrough came with the following idea: One observes that in the neighbourhood of the origin the lines of equal height of $|r(z)|$ are vertical, because they approximate those of the exponential function for

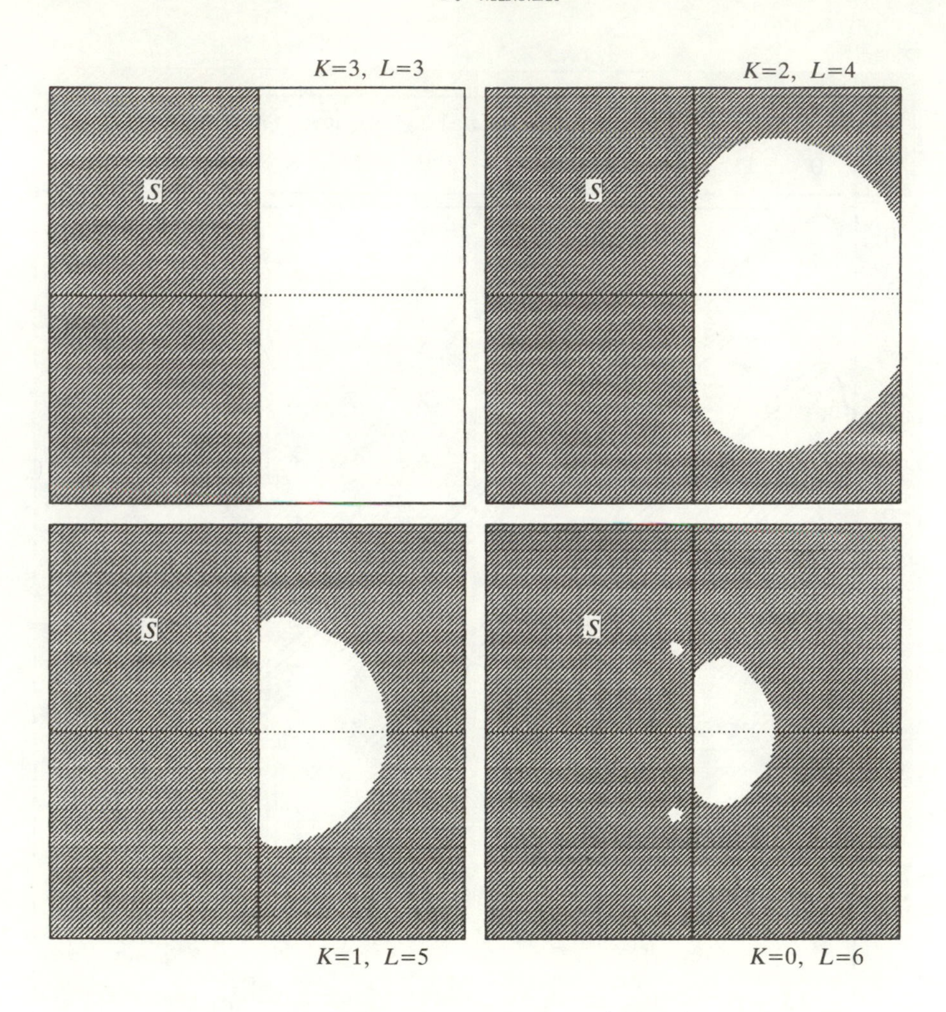

FIG. 4 **Stability domains for Padé approximations.**
K = **number of zeros,** L = **number of poles**

which $|e^z| = e^x$ where $z = x + iy$. More information might therefore be expected from the set

$$A = \{z : |r(z)| > |e^z|\} = \{z : |q(z)| > 1\}, \quad q(z) = \frac{r(z)}{e^z} \qquad (2.1)$$

which we call *the order star*. Computer plots of these sets for some Padé approximations are presented in Figure 5.

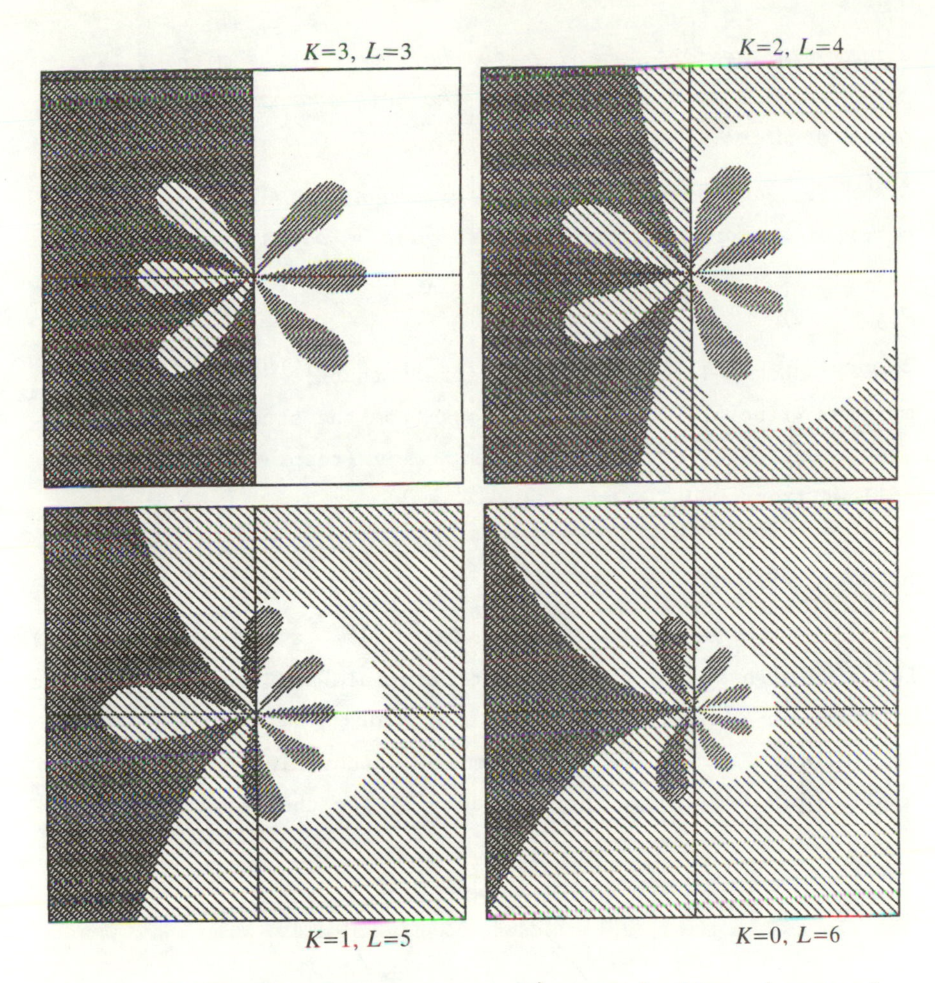

FIG. 5 Order stars for Padé approximations (compared to stability domains). The shaded area denotes A

The most important properties are summarized in the following Lemmas:

LEMMA 1. On the imaginary axis A *and* S *are complementary. Therefore A-stability implies that* $A \cap i\mathbb{R} = \emptyset$. *If there is no pole of* r *in* $\mathbb{C}^-$, *this condition is also sufficient.*

LEMMA 2. As $z \to \infty$, *exactly two branches of the boundary go to infinity.*

LEMMA 3. As $z \to 0$, *the order star is star-like with* $p+1$

equiangular sectors, where p *is the order of the method. The positive real axis is "white" in the neighbourhood of* 0 *if the error constant* C *is positive.*

LEMMA 4. Each bounded connected component of A *which contains* m *"order-sectors" must contain at least* m *poles of* $r(z)$*. The same is true for the complement of* A *and zeros of* $r(z)$*. (See Figure 6.)*

PROOFS: Lemma 1 follows from the fact that $|e^{iy}| = 1$ and the maximum principle. Lemma 2 follows from the strong increase (resp. decrease) of $|e^z|$ when $\text{Re}\, z \to +\infty$ (resp. $-\infty$). Lemma 3 follows from

$$e^z = r(z) + Cz^{p+1} + O(z^{p+2}) \qquad z \to 0 \tag{2.2}$$

or, by division,

$$\frac{r(z)}{e^z} = 1 - Cz^{p+1} + O(z^{p+2}) \qquad z \to 0\,. \tag{2.2'}$$

Lemma 4 follows from the argument principle and the fact that the argument of $q(z) = r(z)/e^z$ *decreases* along the boundary ∂A. This latter fact can be seen from the Cauchy-Riemann equations applied to $\log(q(z))$. For more details see the original paper from 1978.

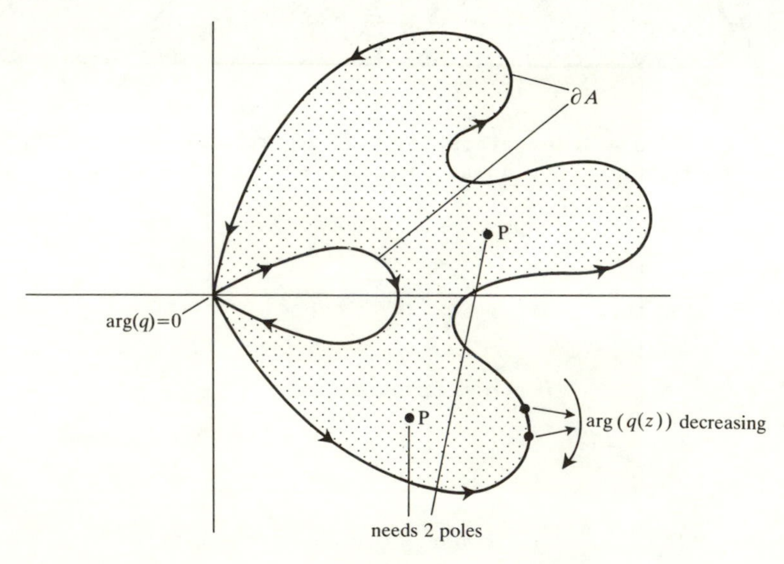

FIG. 6 Illustration for Lemma 4

3. A-STABLE LINEAR MULTISTEP METHODS

The famous theorem of Dahlquist (1963) states that there is no A-stable linear multistep method of order 3. Moreover, the smallest error constant of all A-stable methods of order 2 is that of the trapezoidal rule.

For the *proof* of this theorem, one considers the order star (2.1) as a subset of the Riemann surface (see Figure 2). There will be a star with $p+1$ rays on the *principal* sheet which is the holomorphic continuation of the root $r(0)=1$. For A-stability, the order star is not allowed to cross the imaginary axis on *any* of the sheets. If now the method would be of order 3, there would be 4 sectors, half of them in $\mathbb{C}^+$. This part of the order star must then, by Lemma 4, contain at least two poles (Figure 7). Further, all A-stable methods of order 2 must have a negative order constant (Lemma 3 and Figure 8). For the proof of the *second part* we need a new idea: Instead of comparing $r(z)$ to the true solution e^z, we compare it to the trapezoidal rule

$$r_2(z) = \frac{1+z/2}{1-z/2} \tag{3.1}$$

for which

$$e^z = r_2(z) - \frac{1}{12} z^3 + O(z^4). \tag{3.2}$$

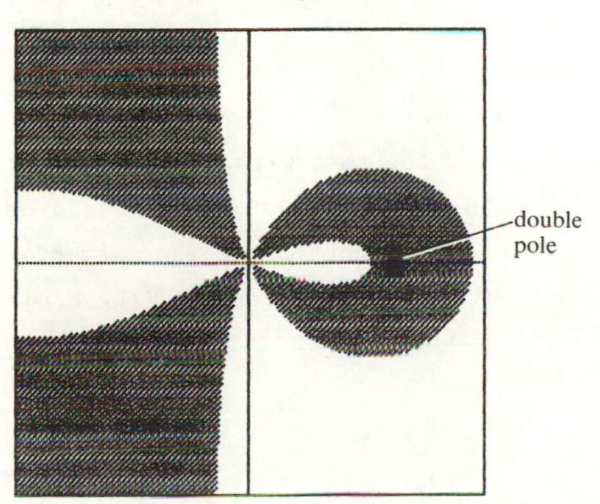

FIG. 7 A-stable method of order 3 (two stage DIRK method, $\gamma = 0.7786751$)

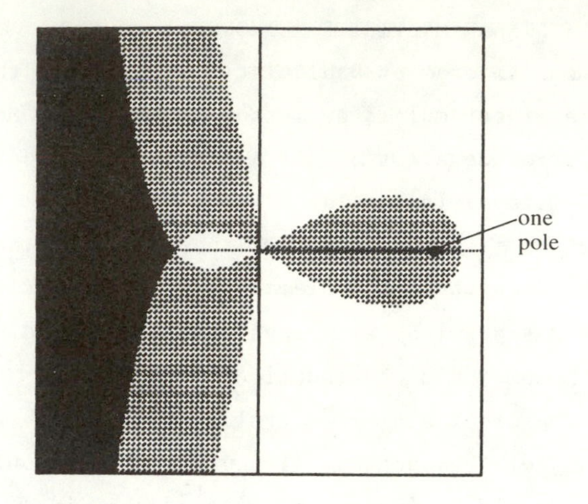

FIG. 8 A-stable linear multistep method of order 2 (BDF_2)

If we subtract (2.2) from (3.2) we obtain

$$\frac{r(z)}{r_2(z)} = 1 - \left(C + \frac{1}{12}\right) z^3 + O(z^4). \tag{3.3}$$

Thus the "comparison" order star (instead of (2.1))

$$B = \left\{ z : \left| \frac{r(z)}{r_2(z)} \right| > 1 \right\} \tag{3.4}$$

will have three sectors whose colour determines the sign of $C + \frac{1}{12}$. Its *poles* are the poles of $r(z)$ and the zeros of $r_2(z)$. The latter, however, are known to be in $\mathbb{C}^-$. Thus, if we restrict our investigation to $\mathbb{C}^+$, the same arguments as above remain valid and Lemma 3 together with Figure 8 gives $C < -\frac{1}{12}$. □

Just by adding more sectors to the above pictures and by replacing $r_2(z)$ by the diagonal Padé approximations, the above proofs lead to many more general results (the conjecture of Daniel and Moore).

4. THE THEOREMS OF JELTSCH AND NEVANLINNA

R. Jeltsch proved in 1978 that no explicit RK method with s stages can exist whose stability domain contains entirely the stability circle with radius s of s Euler steps with stepsize

h/s. Therefore, Euler's method could be seen as "the most stable" of all methods. The same result was obtained for Euler's method compared to linear multistep methods (O. Nevanlinna). With the help of the order star theory, the *converse* became true too: *Any explicit method is "more stable" than any other*, and vice versa, *under the hypothesis that the two methods require the same numerical work* (= i.e. have the same number of poles at infinity). More precisely:

THEOREM: If S_1 and S_2 are two stability domains of two explicit methods $r_1(z)$ and $r_2(z)$ of order at least 1 with the same number of poles at infinity, then

$$S_1 \not\supset S_2 \quad \text{and} \quad S_2 \not\supset S_1 .$$

The *proof* is essentially the same as the proof of the theorem of Dahlquist ("method B" (3.4)) with the following correspondences:

Dahlquist's Theorem		Jeltsch-Nevanlinnna's Theorem
$r(z)$	$\longleftrightarrow$	$r_1(z)$
$\frac{1+z/2}{1-z/2}$	$\longleftrightarrow$	$r_2(z)$
imag. axis	$\longleftrightarrow$	boundary of S_2
$\mathbb{C}^-$	$\longleftrightarrow$	interior of S_2
$\mathbb{C}^+$	$\longleftrightarrow$	exterior of S_2
$r(z)$ A-stable	$\longleftrightarrow$	$S_1 \supset S_2$.

Since the quotient $r_1(z)/r_2(z)$ has *no* pole left outside S_2, the methods cannot be of order 1. There is, however, still a hypothesis left: The proof only works if $r_2(z)$ is a univalent function (for example a one-step method). For multistep methods one needs the hypothesis *that the instability is produced exclusively by the principal root* which is then used in the comparison order star B (3.4) and only defined *outside* S_2. This is called "property C" by Jeltsch and Nevanlinna. That such a condition is necessary is clear, since a completely unstable linear multistep formula will, of course, never have the property of the theorem: Linear multistep methods, however, whose root locus curve is simply closed, *have* property $\mathbb{C}$.

EXAMPLE: The stability domains of the $Adams_2$ method $r_A(z)$ defined in (1.5) and of the RK_2 method $r_{RK}(z)$ (see (1.6)) are compared in Figure 9.

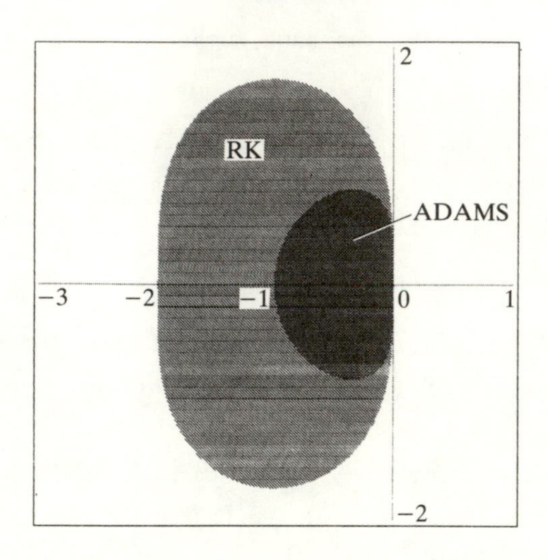

FIG. 9 Stability domain of $Adams_2$ **compared to** RK_2

RK_2 turns out to contain entirely the stability domain of Adams. But this comparison is not fair since RK_2 needs twice as much work per step. We therefore compare *one* step of RK_2 with stepsize $2h$ to *two* steps of $Adams_2$ with stepsize h, i.e. we define

$$B = \left\{ z : \left| \frac{r_A^2(z)}{r_{RK}(2z)} \right| > 1 \right\} . \tag{4.1}$$

Now the stability domain of RK is "shrinked" by a factor 2 ("scaled stability domain") and *neither of the two methods is more stable than the other one* (see Figure 10).

For more details and various generalizations (also to implicit methods) we refer to the papers of Jeltsch-Nevanlinna.

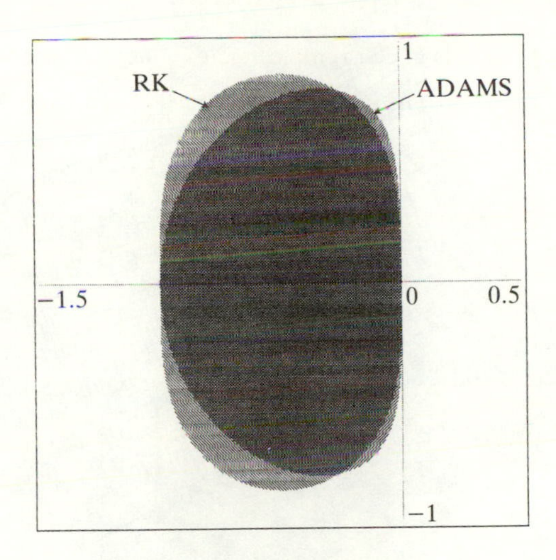

FIG. 10 Scaled stability domains

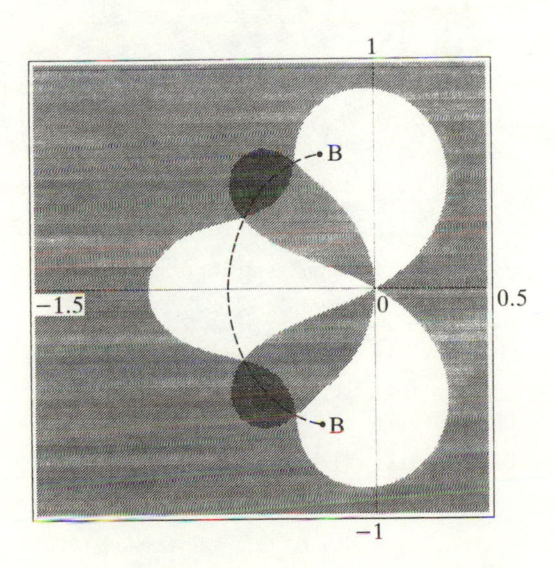

FIG. 11 The order star B (4.1)

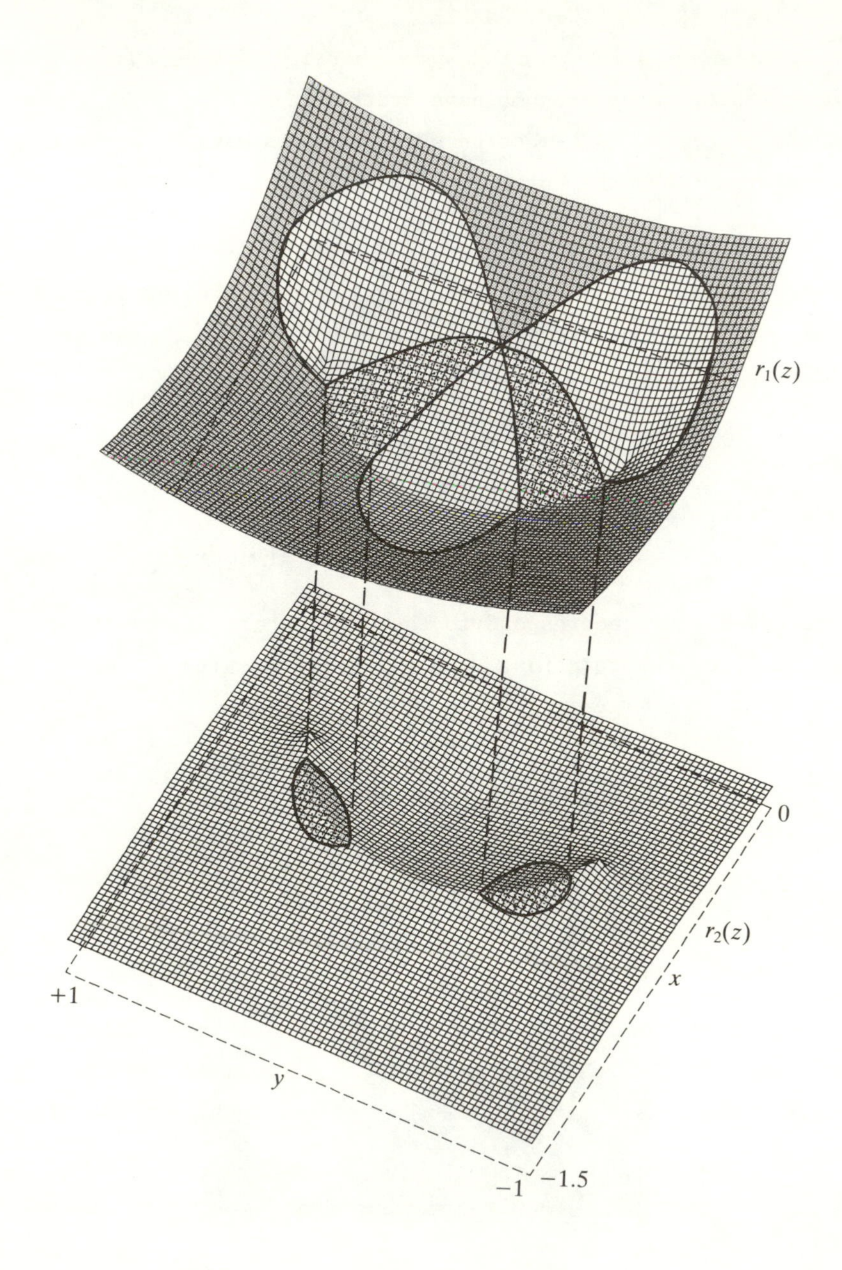

FIG. 12 The order star B of (4.1) on the Riemann surface for $r_A(z)$

5. THE FIRST DAHLQUIST BARRIER

The so-called "first Dahlquist barrier" states that a stable linear k-step method cannot have order higher than $k+2$. Stable methods of order $k+2$ must be symmetric and have an error constant larger than the optimal-order method with ρ-polynomial

$$\rho_0(\zeta) = (\zeta-1)(\zeta+1)^{k-1} . \qquad (5.1)$$

Iserles and Nørsett (1984) have first succeeded to prove this theorem by order stars. The crucial idea was to use the order condition

$$S(z) = \frac{\sigma(e^z)}{\rho(e^z)} - \frac{1}{z} = -Cz^{p-1} + O(z^p) \qquad (5.2)$$

and to define the "order star of the second kind"

$$C = \left\{ z \in \mathbb{C} : \operatorname{Re} S(z) > 0 \right\} . \qquad (5.3)$$

These order stars correspond to the order stars of the first kind (2.1) by the relation $S(z) = \log q(z)$:

First kind	$\longleftrightarrow$	Second kind
$\lvert q(z) \rvert > 1$	$\longleftrightarrow$	$\operatorname{Re} S(z) > 0$
$\arg q(z)$ decreases (see proof of Lemma 4)	$\longleftrightarrow$	$\operatorname{Im} S(z)$ decreases
use of quotient in (2.1) and (3.4)	$\longleftrightarrow$	use of difference in (5.2)
bounded connected component contains a pole	$\longleftrightarrow$	boundary of connected component passes through a pole

Corresponding to the proof of Dahlquist's 1963-theorem, the order star (5.3) with $S(z)$ as in (5.2) only gives the order result without an estimate of the error constant. If one uses instead

$$S(z) = \frac{\sigma(e^z)}{\rho(e^z)} - \frac{\sigma_0(e^z)}{\rho_0(e^z)} \qquad (5.4)$$

where ρ_0 and σ_0 are the polynomials of the "optimal" method, the proof gives both results and simplifies considerably (E. Hairer, unpublished seminar notes).

6. ORDER BOUNDS FOR HYPERBOLIC SCHEMES

Important work has been done by Iserles (1982), Iserles and Strang (1983), Iserles and Williamson (1983) on order bounds for difference schemes for hyperbolic equations. This has been taken up by Jeltsch and Strack (1985) who have simplified the proofs. Also, by using order stars which "compare" the difference scheme to be studied with an "optimal" scheme instead of the exact solution, they obtain, as well, estimates for the error constants. A survey on all these results is given in Jeltsch (1985). There is no use of going into further details here, since the paper of Jeltsch is so nicely written.

ACKNOWLEDGEMENTS: Many thanks to E. Hairer and A. Iserles for their careful reading of the manuscript.

REFERENCES

Butcher, J.C. (1964), "Implicit Runge-Kutta processes", *Math. Comp.*, *18*, pp.50-64.

Daniel, J.W. and Moore, R.E. (1970), *Computation and Theory in Ordinary Differential Equations*, Freeman (San Francisco).

Dahlquist, G. (1963), "A special stability problem for linear multistep methods", *BIT*, *3*, pp.27-43.

Dormand, J.R. and Prince, P.J. (1980), "A family of embedded Runge-Kutta formulae", *J. Comp. Appl. Math.*, *6*, pp.19-26.

Ehle, B.L. (1969), "On Padé approximations to the exponential function and A-stable methods for the numerical solution of initial value problems", Ph.D. Thesis, Department of Applied Analysis and Computer Science, Univ. of Waterloo, Canada.

Hairer, E. (1985-86), Unpublished seminar notes.

Iserles, A. (1982), "Order stars and a saturation theorem for first order hyperbolics", *IMA J. Numer. Anal.*, *2*, pp.49-61.

Iserles, A. and Nørsett, S.P. (1984), "A proof of the first Dahlquist barrier by order stars", *BIT*, *24*, pp.529-537.

Iserles, A. and Strang, G. (1983), "The optimal accuracy of difference schemes", *Trans. Amer. Math. Soc.*, *277*, pp.779-803.

Iserles, A. and Williamson, R.A. (1984), "Stability and accuracy of semidiscretized finite difference methods", *IMA J. Numer. Anal.*, *4*, pp.289-307.

Jeltsch, R. (1985), "Stability and accuracy of difference schemes for hyperbolic problems", *J. Comput. Appl. Math.*, *12 & 13*, pp.91-108.

Jeltsch, R. and Nevanlinna, O. (1981), "Stability of explicit time discretizations for solving initial value problems", *Numer. Math.*, *37*, pp.61-91.
Jeltsch, R. and Nevanlinna, O. (1982), "Stability and accuracy of time discretizations for initial value problems", *Numer. Math.*, *40*, pp.245-296.
Jeltsch, R. and Strack, K.G. (1985), "Accuracy bound for semi-discretizations of hyperbolic problems", *Math. Comp.*, *45*, pp.365-376.
Strang, G. and Iserles, A. (1983), "Barriers to stability", *SIAM J. Numer. Anal.*, *20*, pp.1251-1257.
Wanner, G., Hairer, E. and Nørsett, S.P. (1978), "Order stars and stability theorems", *BIT*, *18*, pp.475-489.

G. Wanner
Section de Mathématiques
Université de Genève
C.P. 240
CH-1211 Genéve 24
Switzerland

17 The State of the Art in the Numerical Treatment of Integral Equations

CHRISTOPHER T. H. BAKER

ABSTRACT

We provide a perspective on the current status of integral equations and their numerical treatment, giving particular emphasis to Volterra equations. We note the extensive work in *boundary integral techniques*, the interest in the *smoothness of solutions* and its effect on practical schemes, problems raised by *singular integral equations*, and the increasing attention paid to *evolutionary problems* (Volterra and Abel equations) both in the analysis of qualitative behaviour, like stability, and the construction of new methods. *Multigrid methods* have been devised for Fredholm equations, *new formulae* and *fast methods* have been found for Volterra equations, *fractional integration* methods have been constructed for Abel equations. The treatment of Fredholm equations is evocative of results in numerical linear algebra, whilst that of Volterra problems relates to the treatment of differential equations. (Stimulating results arise where integral equations increase these horizons.) Our presentation provides background to the other contributions on integral equations.

1. INTRODUCTION

Abel's work on integral equations dates to more than a century ago. Integral equations provide a tool for solving practical problems, but the analysts and functional analysts soon found the underlying structure of interest. Some of the early theory (involving Fredholm determinants) is related to quadrature methods; the practical approach of Nyström, using quadrature, dates to 1930. In 1953, Fox and Goodwin addressed systematically the numerical solution of integral equations. The subject did not appear in the first state-of-the-art meeting but was recognized by Ben Noble's contribution on the second such occasion. Then, the majority of the citations were to research papers; also referenced were the book of Anselone (1971), the book (then forthcoming) by the present writer, and two symposia procedings: Anselone (editor, 1964), and Delves and Walsh (editors, 1974). A decade later, the reader can now refer to Albrecht and Collatz (editors, 1980), Anderssen *et al.* (editors, 1980), Atkinson (1976), Baker (1977), Baker and Miller (editors, 1982), Chatelin (1983), Delves and Mohamed (1985), Golberg (editor, 1979), Groetsch (1984), Hämmerlin and Hoffmann (editors, 1985), Ivanov (1976), Jaswon and Symm (1977), Linz (1985), te Riele (editor, 1980), etc. and, soon, Brunner and van der Houwen (to appear, 1986) with its extensive references. With this explosion of interest, any writer is selective in what he includes in an article of this type. We aim to give references which lead to further reading, and wish to convey a *perspective*. Though our motivation is that of practical application (we refer to Lonseth (1977), the bibliography of Noble (1971) and the books of, for example, Anderssen *et al.* (1980), Burton (1983), Corduneanu (1973), Jaswon and Symm (1977), Miller (1971), etc., for source material)), we write in general terms. It is dangerous to isolate numerical methods from the problems to which they apply, since the aim is one of *simulation*. However, a global perspective serves to introduce the two other contributions on integral equations, and we shall highlight some areas of advance.

2. INTEGRAL EQUATIONS

An integral equation is a functional equation in which the unknown function appears as part of an integrand. Thus, Abel's classical equation of the second kind assumes the form

$$y(t) - \int_0^t (t-s)^{-\frac{1}{2}} y(s)\,dx = g(t) \qquad (t \geqslant 0) \tag{2.1}$$

where g is given and the problem involves the determination of y. We can classify integral equations as linear or nonlinear; other classifications are now fairly standard in texts: equations of Fredholm type, of Volterra type, of Abel type, weakly-singular equations, Wiener-Hopf equations, Cauchy principal-value equations, Urysohn equations, Hammerstein equations, equations of the first kind, of the second kind, etc., etc. Some of the terminology arose historically and may cause confusion. Thus, the equation for y, of the form

$$y(x) - \int_0^1 H(x,s,y(s))\,ds = g(x) \qquad (0 \leqslant x \leqslant 1), \tag{2.2}$$

with "kernel" $H(x,s,u)$ and inhomogeneous term $g(x)$ being prescribed and smooth, is commonly called a *nonlinear Fredholm equation* (since the limits of integration are fixed). However, the Fredholm theory of integral equations is a theory applicable to linear integral equations exemplified by

$$y(x) - \lambda \int_0^1 K(x,s)\,y(s)\,ds = g(x) \tag{2.3}$$

and in this case $K(x,s)$ is called the kernel. (The known value λ is introduced as a convenient parameter.) We reserve until later our remarks on Volterra equations, which some consider as a subset of Fredholm equations. (We regard (2.2) as a Volterra equation if $H(x,s,u)$ vanishes for $s > x$, the range of integration then reducing to $[0,x]$.) The classical linear integral equations of Fredholm type are (a) eqn. (2.3), which is a Fredholm equation of the second kind; (b) the eigenvalue problem

$$\int_0^1 K(x, s)\, y(s)\, ds = \kappa y(x) \qquad (0 \leqslant x \leqslant 1), \qquad (2.4)$$

(where we are required to find scalar κ and corresponding nontrivial y), and (c) the Fredholm equation of the first kind

$$\int_0^1 K(x, s)\, y(s)\, ds = g(x) \qquad (0 \leqslant x \leqslant 1). \qquad (2.5)$$

Under reasonable assumptions, the total eigensystem of (2.4) forms a countable set $\{\kappa_r, y_r\}$, which may be empty. Easy examples show that (2.5) may have no solution. It is convenient to adopt the notation

$$K\phi(x) = \int_0^1 K(x, s)\, \phi(s)\, ds \qquad (2.6)$$

so that (2.3) – (2.5) can be written in shorthand form as

$$(I - \lambda K)\, y = g\,; \quad Ky = \kappa y\,; \quad Ky = g\,. \qquad (2.3'), (2.4'), (2.5')$$

Let us assume the functions H, g, K, to be "smooth". Then there is some inter-relation between eqns (2.2) – (2.5). First, Newton's method for (2.2) involves the repeated solution of equations of the form (2.3) with $\lambda = 1$, and $K(x, s)$ of the form $(\partial/\partial u)\, H(x, s, u(s))$. Secondly, the Fredholm theory establishes a condition for a unique solution of (2.3) in terms of eigensolutions of (2.4). Thirdly, the solution of (2.5) can be discussed, when $K(x, s) = \overline{K(s, x)}$ (the Hermitian-symmetric case), in terms of the total eigensystem satisfying (2.4) (and the theory can be extended to the nonsymmetric case, using the singular values and singular functions of $K(x, s)$). All these results are standard in the literature (Smithies, 1952); they are a source of comfort because practical methods for the solution of (2.2)–(2.5) involve their replacement by algebraic equations for which corresponding results apply. *We issue a caveat that "natural" methods of discretizing (2.5) may not succeed in practice.*

3. NUMERICAL ANALYSIS OF FREDHOLM EQUATIONS

The processes by which discrete equations are obtained can be summarized as: (a) quadrature techniques and (b) expansion methods, such as collocation, Galerkin, and discrete Galerkin methods. We identify some essential features if we concentrate on linear equations. In quadrature methods, one replaces the integral (involving the solution y) by a finite sum involving the approximate solution $y_m(t)$ at m distinct abscissae $s_i = s_i^{(m)}$ and obtains a system of algebraic equations by forcing equality at these absicissae. Thus, if K, g are smooth, we can employ a standard quadrature

$$\int_0^1 \phi(s)\,ds = \sum_1^m W_j \phi(s_j)$$

to derive from (2.3) equations of the form

$$y_m(s_i) - \lambda \sum_1^m W_j K(s_i, s_j)\, y_m(s_j) = g(s_i) \qquad (i = 2, 3, \cdots, m).$$

(If the kernel K is weakly-singular, we use an appropriate "generalized" quadrature.) For expansion methods one seeks a solution of the form

$$y_m(x) = \sum_1^m \alpha_j \phi_j(x) ;$$

the functions $\{\phi_j\}$ are assumed linearly independent. Such a function will yield a residual $r_m(\{\alpha_j\}; x)$ when substituted in the integral equation, and the values $\alpha_j = \alpha_j^{(m)}$ are chosen so that $r_m(\{\alpha_j\}; x)$ satisfies certain constraints. For collocation, we force r_m to vanish at m collocation points, whilst in Galerkin methods we require r_m to be orthogonal to the space spanned by $\{\phi_j\}$, and so on. Thus, the Galerkin equations for (2.3) have the form

$$\sum_{j=1}^{m} [(\phi_j, \phi_k) - \lambda(K\phi_j, \phi_k)]\, \alpha_j = (g, \phi_k) \qquad (k = 1, 2, \cdots, m)$$

where (,) is an inner-product.

The notation (2.6) has been introduced as a convenience, but if the class of functions ϕ for which $K\phi$ is defined forms a linear function space X then K defines a linear operator on X. With suitable assumptions, (a) X is a Banach space and (b) K is a compact operator ((a) or (b) may hold separately). Since matrices represent operators on a finite-dimensional space (which can be normed as a Banach space), and these operators are compact, the analyst can unify the theory for well-behaved integral equations and matrix theory into a theory of compact operators. A linear integral equation (2.3) or (2.4) can be regarded as "nice" if it fits into this framework. In many cases, numerical methods for (2.3) can then be analyzed by viewing them as the replacement of $(I-\lambda K)\, y = g$ by a perturbed equation $(I-\lambda K_m)\, y_m = g_m$ where $g_m, y_m \in X$ and $K_m \in (X \to X)$, and m denotes a parameter of discretization.

Suppose $(I-\lambda K)$ is invertible; $(I-\lambda K)^{-1} = (I + \lambda R_\lambda)$ where R_λ is an integral operator whose kernel is called the resolvent kernel. The unique true solution y can be shown to solve an equation of the form $(I-\lambda K_m)\, y = g_m + \tau$ so we can deduce, *provided* $I-\lambda K_m$ is invertible, that

$$y - y_m = (I - \lambda K_m)^{-1}\, \tau$$

and

$$\|y - y_m\| \leqslant \|(I - \lambda K_m)^{-1}\|\, \|\tau\| .$$

Thus, we can establish convergence in norm if we show (i) that $\|\tau\| \to 0$, (ii) that $I - \lambda K_m$ is invertible for m sufficiently large, and (iii) that $\|(I-\lambda K_m)^{-1}\|$ is $O(1)$ as $m \to \infty$. In some cases, the theory of (ii) and (iii) parallels that familiar from the error analysis of numerical linear algebra, because $\|K-K_m\| \to 0$ as $m \to \infty$. However, this result is not always valid; then, one seeks to modify the theory by showing that

$$\|(K-K_m)\phi\| \to 0$$

as $m \to \infty$ for $\phi \in X$ and deducing (assuming appropriate K_m) that

$$\|(K-K_m)K_m\| \to 0$$

as $m \to \infty$. The latter is sufficient to establish the required results for $(I-\lambda K_m)$. We can refer to Anselone (1971), who indicates extensions to nonlinear equations, cf. Atkinson (1976), Baker (1977), etc. The preceding is a thumbnail sketch of the proof of convergence of many numerical methods for the linear equation (2.3). The theory for (2.4) can be pursued along related lines, because the eigenvalues of K are the values $\kappa = \lambda^{-1}$ such that $(I-\lambda K)^{-1}$ does not exist. However, the eigenvalue problem is simpler yet richer than this might imply: many properties analogous to those obtaining for the eigenvalue problem of matrices can be established. See Baker (1977), Chatelin (1981),(1983). The algebraic problems can be formulated, in the above cases, so that the conditioning mirrors that in the integral equation.

Note: Wendland (these proceedings) emphasizes that compactness of the operators is not relevant in an alternative viewpoint, for methods where a Gårding inequality applies.

To give a state-of-the-art perspective, we identify some recent areas of advance. Our sketch indicates the proof of convergence in the norm of X. However, *superconvergence* can occur: for instance, there may be a set of points $\{t_i^{(m)}\}$ such that

$$\sup_i |y(t_i^{(m)}) - y_m(t_i^{(m)})|$$

tends to zero faster than $\|y-y_m\|$. Further, some post-processing of y_m may yield benefits: (a) if y_m is obtained from an expansion method applied to (2.3), one step of a *Neumann iteration* yields from y_m the new approximation $\mathring{y}_m := g + \lambda K y_m$ and $\mathring{y}_m$ frequently compares favourably with y_m; if y_m is the Galerkin approximation, then $\mathring{y}_m$ is called the *iterated Galerkin approximation*; see Sloan (1982). (For (2.4), we define $\mathring{y}_m := \kappa_m^{-1} K y_m$ where κ_m approximates κ; cf. Chatelin (1981).) (b) *Extrapolation* and *deferred correction* can be applied with the quadrature method. Now the equations arising from discretization of Fredholm equations are not sparse. Accordingly, iterative techniques and in particular *multigrid methods* have found application. We refer to Hemker and Schippers (1981), Schippers (1982). All of our earlier observations relate to "nice" Fredholm equations but there are other important cases where advances are being made: *equations on an infinite interval* (e.g. $y(x) - \int_0^\infty K(x,s)\,y(s)\,ds = g(x)$), and the superconvergence result of Sloan and Spence (1986), for *first-kind* equations see Davies *et al.* (1985a,b); *boundary integral methods* (Wendland, 1982), for a case of non-compact operators see Chandler and Graham (1985); and the important class of *Cauchy principal-value equations* (see Anderssen *et al.* (1980), Elliott (1982), etc.).

When we turn to (2.5), we encounter some practical difficulty, and the possibility of stating a controversial view. The equation arises in practice, and (though it is often ill-posed) there are claims of success with the problem. The analogue of the theory of methods for (2.3) corresponds to the replacement of $Ky = g$ by $K_m y_m = g_m$, but if the operators are compact, K^{-1} and K_m^{-1} do not exist on X, and the construction of methods with an error analysis is thwarted. The theoretical difficulty represents practical reality. The algebraic equations obtained by discretizing (2.5) in a natural way are frequently ill-conditioned (or singular). The smoother the kernel, the more severe is the problem. As a consequence of ill-conditioning, least-squares techniques or linear-programming techniques frequently find an application to (2.5) when $K(x,s)$ is smooth. Many such approaches amount to imposing constraints upon the permissible "solutions". Since (2.5) may possess no solution in X (the classical theory (Smithies, 1952) treats the equation in $X = L^2[0,1]$), the search for pseudo-solutions is not inappropriate. Least-squares techniques for algebraic equations $Au = b$ are closely related to the construction of $u = A^+b$ where A^+ is the generalized inverse of A. We can write

$$A^+ = \lim(\alpha^2 I + A^*A)^{-1}A^*$$

(the limit being as $\alpha \to 0$). This suggests the search (in a wholly analogous way, as a pseudo-solution of (2.5)), for the function K^+g. It is hoped that, for small α, K^+g can be approximated by $(\alpha^2 I + K^*K)^{-1}K^*g$ $(\alpha \neq 0)$. (If one works in $L^2[0,1]$, which seems restrictive, K^* is the integral operator with kernel $\overline{K(s,x)}$). For this and other approaches see Davies (1983), Groetsch (1977, 1984), Bates and Wahba (1982), etc. The *optimum choice* of the regularization parameter α, under various assumptions, has occupied many. There is some danger of choosing α to fulfill an earlier unexpressed assumption as to the nature of y: at issue, we believe, is a fundamental need for the correct mathematical modelling of the problem. Note that if $K(x,s)$ has a weak singularity, ill-conditioning may be scarcely noticeable. For, a condition number of the matrix involved in the discretized problem can be represented by the ratio of its largest to least singular values, which approximate singular values of K; if $K(x,s)$ is weakly-singular, the singular values of K decay slowly to zero and ill-conditioning is slow to manifest itself.

4. VOLTERRA EQUATIONS (MISE-EN-SCÈNE)

It is commonplace for time-dependent phenomena to be

modelled by an initial-value problem for (a system of) differential equations $y'(t) = f(t,y(t))$ $(t \geq 0)$, $y(0) = y_0$. Many such phenomena are better modelled by relating $y'(t)$ not only to the current value $y(t)$ but to values $\{y(s) \mid s \leq t\}$. In particular we have delay-differential equations of the form

$$y'(t) = f(t,y(t)) + \sum_{i=0}^{M} f_i(t,y(t-\tau_i)) \qquad (t \geq 0)$$

where the lags τ_i are state-dependent, say $\tau_i = \tau_i(t,y(t)) \geq 0$; initial conditions required depend on these lags. It is a small generalization to introduce the integro-differential equation

$$y'(t) = f(t,y(t)) + \int_{-\infty}^{t} H^{\#}(t,s,y(s))\,ds$$

which also has applications; here, $y'(t)$ depends upon $y(s)$ for $-\infty < s \leq t$. If we suppose $y(s)$ to be prescribed for non-positive s: $y(s) = \phi(s)$ for $-\infty < s \leq 0$, we obtain a Volterra integro-differential equation

$$y'(t) = f(t,y(t)) + \int_{0}^{t} H^{\#}(t,s,y(s))\,ds + d(t) \quad (t \geq 0) \quad (4.1)$$

with $y(0) = \phi(0)$. Integration yields

$$y(t) = \int_{0}^{t} H(t,s,y(s))\,ds + g(t) \qquad (t \geq 0) \quad (4.2)$$

where the kernel H is

$$H(t,s,y(s)) = \int_{0}^{t} H^{\#}(\tau,s,y(s))\,d\tau + f(s,y(s))$$

and $g(t)$ is a "known forcing term". Equation (4.2) is a Volterra integral equation of the second kind, and equations of this form arise directly in applications. (The generalized Abel equation corresponds to a weakly-singular kernel

$$H(t,s,u) = M(t,s,u)/(t-s)^{1-\nu}$$

with $0 < \nu < 1$.) We assume smooth H, g in (4.2) unless specific reference is made to Abel-type kernels. Note that (4.2) is said to be in *canonical form* if g vanishes. We review some features to which we shall return later.

Suppose that $H(t,s,u)$ is finitely decomposable, i.e.

$$H(t,s,u) = \sum A_i(t)\, B_i(s,u),$$

the sum being over $i = 1, 2, \cdots, N$. Then the solution of (4.3) can be written in the form

$$y(t) = \sum \alpha_i(t)\, A_i(t) + g(t)$$

where the functions α_i satisfy the system of differential equations

$$\alpha_i'(t) = B_i(t, \sum \alpha_l(t)\, A_l(t) + g(t)) \tag{4.3}$$

and the initial conditions $\alpha_i(0) = 0$. Bownds (1982) based a code for the general case (4.2) on this result. There is much reminiscent of initial-value problems for ordinary differential equations in the study of eqn (4.2), but there are some areas where (4.2) reflects a richer structure.

Note that (4.1) is a special case of the integro-differential equation

$$y'(t) = G\left(t, y(t), \int_0^t H(t,s,y(s))\, ds\right) \qquad (t \geqslant 0) \tag{4.4}$$

with prescribed $y(0)$. Equation (4.4) can be written as a system of coupled Volterra equations; if we analyze (4.2) where y, H, g, are regarded as vector-valued functions, we obtain results for (4.4). Conditions have to be imposed upon H in order to establish the existence and (where appropriate) the unicity of the solution y; the theory is simplest where the kernel H is linear in y. Usually, the term "a solution $y(\cdot)$" is taken to be synonymous with "a continuous solution $y(\cdot)$", or "a differentiable solution $y(\cdot)$". The theory becomes more rewarding if one requires additional properties. Thus, from an abstract viewpoint, one may seek y in a suitable function space X. Some examples are:

(a) $X = L^p[0,\infty)$, the functions y with $\left\{\int_0^\infty |y(t)|^p\, dt\right\}^{1/p}$ finite,

(b) $X = C_l[0,\infty)$, the functions continuous for $t \geqslant 0$ such that $\lim y(t)$ as $t \to \infty$ exists and equals l, $|l| < \infty$. If one is interested in the boundedness of solutions, one may take $X = B[0,\infty)$, the space of functions bounded on $[0,\infty)$, or the space (c) $X = BC[0,\infty) = B \cap C\,[0,\infty)$ where (d) $X = C[0,\infty)$ denotes the functions continuous at every $t \geqslant 0$. Under suitable smoothness conditions on H, g, a solution of (4.2) in a suitable space X will lie in $C[0,\infty)$. But the conditions that y lie in the space (b) or the space (c), for example, require the imposition of conditions upon H, g that guarantee that a solution y has a particular type of behaviour or asymptotic behaviour. Such results can often be related to "stability".

A popular approach to (4.2) is to consider the "natural iteration":

$$y_{n+1}(t) = \int_0^t H(t,s,y_n(s))\,ds + g(t) \qquad (t \geqslant 0),$$

with a suitably chosen $y_0(t)$, e.g. $y_0(t) = g(t)$. This iteration can be used to establish the existence (and the behaviour) of y. Suppose $H(t,s,u)$ is continuous and bounded for $0 \leqslant s \leqslant t \leqslant T$, $|u|$ finite, and it satisfies a Lipschitz condition

$$|H(t,s,u) - H(t,s,v)| \leqslant \Lambda |u-v|$$

uniformly for $0 \leqslant s \leqslant t \leqslant T$. If, further, $g \in C\,[0,T]$, then eqn (4.2) has a unique continuous solution on $[0,T]$ with

$$y(t) = \lim_{n\to\infty} y_n(t)\,.$$

The preceding can be generalized to cover the weakly-singular kernel where

$$|H(t,s,u) - H(t,s,v)| \leqslant \Lambda |u-v|\,|t-s|^{-\alpha} \qquad (0 \leqslant \alpha < 1).$$

In the case of a general linear integral equation

$$y(t) = \lambda \int_0^t K(t,s)\,y(s)\,ds + g(t) \qquad (t \geqslant 0), \qquad (4.5)$$

$$y(t) = g(t) + \lambda \int_0^t R_\lambda(t,s)\,g(s)\,ds \qquad (t \geqslant 0) \qquad (4.6)$$

where

$$R_\lambda(t,s) = K(s,t) + \lambda K^{(2)}(s,t) + \lambda^2 K^{(3)}(s,t) + \cdots$$

and

$$K^{(n)}(s,t) = \int_0^t K(t,\sigma)\, K^{(n-1)}(\sigma,s)\, d\sigma, \quad K^{(1)} = K.$$

The resolvent R_λ is related to K by

$$R_\lambda(t,s) = \lambda \int_0^t K(t,\sigma)\, R_\lambda(\sigma,s)\, d\sigma + K(s,t) \quad (t \geqslant 0). \qquad (4.7)$$

As a special case, referred to later, consider

$$y(t) = \int_0^t \lambda y(s)\, ds + g(t) \qquad (t \geqslant 0);$$

if $g(t)$ is continuous, this equation has a unique continuous solution

$$y(t) = \lambda \int_0^t \exp\{\lambda(t-s)\}\, g(s)\, ds + g(t) \qquad (t \geqslant 0).$$

In the case that $K(t,s)$ has a weak singularity, of the type found in *Abel's equation*, the successive "iterated kernels" $K^{(n)}$ become smoother but the resolvent inherits a weak singularity and it follows that when g is smooth, *the solution y may be expected to have a badly-behaved derivative*.

In the case of a convolution equation, with $K(t,s) = k(t-s)$, viz.

$$y(t) = \lambda \int_0^t k(t-s)\, y(s)\, ds + g(t) \qquad (t \geqslant 0), \qquad (4.8)$$

the resolvent R_λ is a convolution kernel $r_\lambda(t-s)$; the function r_λ is the solution of (4.8) corresponding to the choice $g(t) = k(t)$. We can use $*$ to denote a convolution, so that (4.8) can be expressed as $y = \lambda k * y + g$; then $r_\lambda = \lambda k * r_\lambda + k$. The study of (4.8) by Laplace transform techniques may be suggested (the technique applies, indeed, for Abel's equation of the second kind). Denote by $L\{\phi\}$ the Laplace transform of ϕ:

$$L\{\phi\} := \int_0^\infty \exp(-t\sigma)\, \phi(\sigma)\, d\sigma,$$

provided the integral converges. (Under careful assumptions) we have $L\{k*y\} = L\{k\} L\{y\}$; we deduce that $L\{y\}[1 - \lambda L\{k\}] = L\{g\}$, and that $y = L^{-1}\{L\{g\}/(1-\lambda L\{g\})\}$ and one would be tempted to write $y = \rho_\lambda * g$ where $L\{\rho_\lambda\} = [1-\lambda L\{k\}]^{-1}$. This is possible only if one admits that ρ_λ could be a generalized function; if one writes $\rho_\lambda = \delta + \lambda r_\lambda$ where δ is the Dirac delta function, we recover the form $y = g + \lambda r_\lambda * g$. See Doetsch (1974) for rigour. Where results for the behaviour of a resolvent r_λ can be found, it is possible in a number of cases to deduce properties of the solution y. Suppose that $r_\lambda \in C \cap L^1[0,\infty)$. If $g \in BC[0,\infty)$ in (3.14), then clearly $y \in BC[0,\infty)$ (we have $\sup|y(t)| \leqslant \sup|g(t)| \times (1+|\lambda|\,\|r_\lambda\|_1)$, etc.). Again (provided $r_\lambda \in C \cap L^1[0,\infty)$), if $g \in C_0[0,\infty)$, then $y \in C_0[0,\infty)$; alternatively, if $g \in C \cap L^1[0,\infty)$ then $y \in C \cap L^1[0,\infty)$. The review of Tsalyuk (1979) summarizes some similar results. A classical theorem of Paley and Wiener (cf. Miller, 1971) establishes a criterion ensuring that r_λ be in $L^1[0,\infty)$ when $k \in L^1[0,\infty)$ (if, in addition, $k \in C[0,\infty)$ then $r_\lambda \in C[0,\infty)$).

THEOREM (Paley and Wiener): Suppose that $k \in L^1[0,\infty)$ and the function r_λ satisfies $r_\lambda - \lambda k * r_\lambda = k$. Then $r_\lambda \in L^1[0,\infty)$ if and only if

$$\lambda \int_0^\infty \exp(-wt)\, k(t)\, dt \neq 1$$

for $\mathrm{Re}(w) \geqslant 0$.

An extension to systems of equations is stated in Lubich (1983a). If $k \in C \cap L^1[0,\infty)$ is given, the Paley-Wiener theorem gives a condition to be satisfied by λ, but one which is in general difficult to check. For Hermitian positive-definite $k \in C \cap L^1[0,\infty)$ it transpires that the condition is simple: merely that $\mathrm{Re}(\lambda) < 0$. Though there are various equivalent definitions, we can say that a function $k \in C \cap L^1[0,\infty)$ is Hermitian positive-definite if, on defining $k(t) = \overline{k(-t)}$ for $t<0$ (the Hermitian extension of k to negative argument), we have

$$\textstyle\sum_i \sum_j k(t_i - t_j)\, z_i \bar{z}_j \geqslant 0$$

for all terminating sequences $\{z_i\}$ and any choice of $\{t_i\}$.

(*Note* that a doubly-infinite sequence $\{a_l\}_{-\infty}^{\infty}$ is called Hermitian positive definite if $(a_l = \bar{a}_{-l})$ and

$$\sum_i \sum_j a_{i-j} z_i \bar{z}_j \geq 0$$

for all terminating sequences $\{z_i\}$, and we see from the above that for every $h > 0$, the Hermitian extension of k defines a sequence $\{k(ih)\}_{-\infty}^{\infty}$ which is Hermitian positive definite.)

Positive-definite functions occur in particular as the characteristic functions of probability theory; these arise frequently as kernels in the renewal equation. If k is completely monotone it is Hermitian positive-definite. Observe that the Hermitian extension of k may yield a function which is not smooth at 0 (for smoothness we need $k'(0) = 0$ etc.).

5. NUMERICAL METHODS FOR VOLTERRA EQUATIONS

Classical numerical methods for Volterra equations of the second kind (which we call Volterra methods) fall into the following classes: (a) discretization methods based on quadrature and Runge-Kutta formulae, and (b) expansion methods (considered in these proceedings by H. Brunner). For equations of Abel type, considered later, the Volterra discretization methods have to be modified and smoothness of the solution is an important consideration. The solution $y(t)$ of a Volterra equation may be required "for $t \in [0, \infty)$" and the problem is an evolutionary one; the numerical methods generally reflect these features since they involve a step-by-step (or a block-by-block) process which can be executed indefinitely. Thus, whereas the numerical solution of Fredholm equations involves the solution of a large number of simultaneous algebraic equations, the numerical solution of Volterra equations generally involves the repeated solution of equations involving either one or (a block of) a few unknown values. For the Volterra integro-differential equation, discretization methods can be constructed which combine the classical techniques for differential equations and those for Volterra equations. Note that if we write an initial-value problem for a differential equation $y'(t) = f(t, y(t))$ in integral equation form

$$y(t) = \int_0^t f(s, y(s))\, ds + y(0)$$

and apply a Volterra method to this equation of the second kind, we obtain a class of methods (which may be novel) for initial-value problems. It is an important feature, with many consequences, that when Volterra methods are applied in this fashion, they can be reduced to *familiar* methods for the initial-value problem.

Recent advances have taken a number of forms. First, a better understanding of the *theoretical foundations* of classical methods has emerged. Secondly, the *practical implementation* of these classical methods has been re-examined and developed. Thirdly, *new methods* and *new formulae* (generally modifications of classical formulae) have appeared. We should not overlook the *adaptive approach* of Jones and McKee (1985). Perhaps surprisingly, the insight into Volterra methods has also yielded results for first kind equations and weakly-singular Abel equations.

5.1 *Quadrature Methods.* Consider the Volterra equation of the second kind

$$y(t) - \int_0^t H(t, s, y(t))\, ds = g(t) \qquad (t \geqslant 0). \tag{5.1}$$

Recall that some authors consider the *canonical form*, in which $g(t) \equiv 0$. For convenience, we shall examine quadrature formulae for the computation of approximations y_r to $y(rh)$ $(r = 1, 2, 3, \ldots)$ where h is fixed. Equation (5.1) yields $y(0) = g(0)$, and we assign y_0 this value. If we set $t = rh$ in (5.1), we see the need to discretize integrals of the form

$$\int_0^t H(t, s, y(s))\, ds.$$

Now a family of quadrature rules for integrating over successive intervals $[0,rh]$, each of the form

$$\int_0^{rh} \phi(s)\,ds = h\sum_0^r \omega_{rj}\,\phi(jh) \tag{5.2}$$

gives approximations

$$\int_0^{rh} \Phi(rh,s)\,ds = h\sum_0^r \omega_{rj}\,\Phi(rh,jh)\ .$$

Setting $\Phi(rh,s) = H(rh,s,y(s))$ leads us to discretize (5.1) with the set of equations

$$y_r - h\sum_0^r \omega_{rj} H(rh,jh,y_j) = g(rh) \qquad (r = 1,2,3,\ldots), \tag{5.3}$$

$y_0 = y(0) = g(0)$. Equations (5.3) can certainly be solved in a step-by-step manner if (as we assume) H satisfies a Lipschitz condition and h is sufficiently small. (With $\omega_{rr} \neq 0$ the equations are implicit, but a natural iteration,

$$y_r^{(k+1)} = h\omega_{rr}\,H(rh,rh,y_r^{(k)}) + h\sum_0^{r-1} \omega_{rj}\,H(rh,jh,y_j) + g(rh),$$

with a suitable predicted value $y_r^{(0)}$, can be employed. Newton's method may be preferred when h is large.) The success of the method is now determined by the choice of rules (5.2). As with a $\{\rho,\sigma\}$ linear multistep (LMM) formulae for ordinary differential equations, we have to consider (a) convergence, and the order of convergence, and (b) stability and related questions for (5.3). The $\{\rho,\sigma\}$ LMM formulae have to satisfy (Dahlquist's) zero-stability criterion: the analogous condition for (5.2) is that the weights ω_{rj} should be uniformly bounded. Then convergence of the values y_r follows if the rules (5.2) are themselves (in an appropriate sense) convergent, and the order of accuracy of the approximations is determined by that of the truncation errors

$$\tau_r := y(rh) - h\sum_0^r \omega_{rj}\,H(rh,jh,y(jh)) - g(rh) \quad (r = 1,2,3,\ldots).$$

We wish to convey the impression that a sensible choice of quadrature rules applied to an equation of the second kind will give approximations which converge on a finite range $[0, T]$, as $h \to 0$. Unfortunately, this property is inadequate as a model for the usefulness of the numerical method.

Initially, quadrature schemes (5.2) were formed in an *ad hoc* fashion using familiar rules; a summary of such formulae appeared in Baker (1977), Baker and Keech (1978). In particular, the choice $\omega_{rr} = \omega_{r0} = \frac{1}{2}$, otherwise $\omega_{rj} = 1$, gives a tableau of weights ω_{rj} corresponding to the repeated trapezium rule:

	$j = 0$	1	2	3	4
$r = 1$	$\frac{1}{2}$	$\frac{1}{2}$			
2	$\frac{1}{2}$	1	$\frac{1}{2}$		
3	$\frac{1}{2}$	1	1	$\frac{1}{2}$	
4	$\frac{1}{2}$	1	1	1	$\frac{1}{2}$
...	...	...	...	...	...

(5.4)

We pause to note the consequence of applying this choice to

$$y(t) = \lambda \int_0^t y(s)\, ds + 1.$$

Differentiating this basic equation yields $y'(t) = \lambda y(t)$; differencing the corresponding equations (5.3) for $\{y_r\}$ (obtained using (5.4)), we obtain the relation satisfied when the trapezium rule is applied as a LMM to $y'(t) = \lambda y(t)$. Further, the rule (5.4) applied to the canonical equation with a finitely separable kernel is equivalent to applying the LMM to the equivalent system (set $g \equiv 0$ in eqn. (4.3)) of ordinary differential equations. Consider the application to the basic equation of other rules (5.2): the weights of the Gregory rule of given order, of which the repeated trapezium rule is the simplest case, can be employed if r is sufficiently large. The use of the weights of a fixed Gregory rule is equivalent to employing an Adams-Moulton formulae (with certain starting values) for

$y' = \lambda y$. The rules mentioned above give weights ω_{rj} which can be applied safely, in practice, with some safeguards. So can the choice

	$j=0$	1	2	3	4	5	......		
1	$\frac{1}{2}$	$\frac{1}{2}$							(trapezium rule)
2	$\frac{1}{3}$	$\frac{4}{3}$	$\frac{1}{3}$						(Simpson's rule)
3	$\frac{1}{3}$	$\frac{4}{3}$	$\frac{1}{3}+\frac{1}{2}$	$\frac{1}{2}$					(Simpson's rule + trapezium rule)
$r=4$	$\frac{1}{3}$	$\frac{4}{3}$	$\frac{2}{3}$	$\frac{4}{3}$	$\frac{1}{3}$				(repeated Simpson's rule, etc.)
5	$\frac{1}{3}$	$\frac{4}{3}$	$\frac{2}{3}$	$\frac{4}{3}$	$\frac{1}{3}+\frac{1}{2}$	$\frac{1}{2}$			
6	$\frac{1}{3}$	$\frac{4}{3}$	$\frac{2}{3}$	$\frac{4}{3}$	$\frac{2}{3}$	$\frac{4}{3}$	$\frac{1}{3}$		
7	$\frac{1}{3}$	$\frac{4}{3}$	$\frac{2}{3}$	$\frac{4}{3}$	$\frac{2}{3}$	$\frac{4}{3}$	$\frac{1}{3}+\frac{1}{2}$	$\frac{1}{2}$	(5.5)
...	...	...	...	...	...	...	...	...	

As a variant of the family of rules (5.5) we can consider

r									
1	$\frac{1}{2}$	$\frac{1}{2}$							(trapezium rule)
2	$\frac{1}{3}$	$\frac{4}{3}$	$\frac{1}{3}$						(Simpson's rule)
3	$\frac{1}{2}$	$\frac{1}{2}+\frac{1}{3}$	$\frac{4}{3}$	$\frac{1}{3}$					(trapezium rule + Simpson's rule)
$r=4$	$\frac{1}{3}$	$\frac{4}{3}$	$\frac{2}{3}$	$\frac{4}{3}$	$\frac{1}{3}$				(repeated Simpson's rule, etc.)
5	$\frac{1}{2}$	$\frac{1}{2}+\frac{1}{3}$	$\frac{4}{3}$	$\frac{2}{3}$	$\frac{4}{3}$	$\frac{1}{3}$			
6	$\frac{1}{3}$	$\frac{4}{3}$	$\frac{2}{3}$	$\frac{4}{3}$	$\frac{2}{3}$	$\frac{4}{3}$	$\frac{1}{3}$		
7	$\frac{1}{2}$	$\frac{1}{2}+\frac{1}{3}$	$\frac{4}{3}$	$\frac{2}{3}$	$\frac{4}{3}$	$\frac{2}{3}$	$\frac{4}{3}$	$\frac{1}{3}$	(5.6)
...	...	...	...	...	...	...	...	...	

Although the theory shows (5.6) to provide a scheme converging with the same order of accuracy as (5.5) it is almost universally disastrous in practice. The problem can be appreciated by analysing the application of the methods to our basic equation; (5.6) reduces to the application of Simpson's rule, which has as its stability region a section of the imaginary axis and has undesirable "growth parameters". However, (5.5) reduces to the application of a cyclic LMM based on the trapezium rule and Simpson's rule; the cyclic combination has a nontrivial stability region. (It was once speculated that the repetition property

$$\omega_{r+1,j} = \omega_{r,j} \qquad (r_0 \leqslant j \leqslant r-r_1)$$

was required in order to achieve an acceptable scheme; it is now appreciated that this is not so.)

Other classical *ad hoc* rules are equivalent to using linear multistep methods or cyclic LMMs, and such rules (5.2) are called *reducible* or *cyclically reducible*, respectively. A rule (5.2) is $\{\rho,\sigma\}$-reducible where

$$\begin{aligned} \rho(\mu) &= \alpha_0\,\mu^k + \alpha_1\mu^{k-1} + \cdots + \alpha_k\,, \\ \sigma(\mu) &= \beta_0\,\mu^k + \beta_1\mu^{k-1} + \cdots + \beta_k\,, \end{aligned} \tag{5.7}$$

when (with the conventions $\alpha_l = 0$, $\beta_l = 0$, for $l \notin \{0,1,\cdots,k\}$, $\omega_{n,j} = 0$ when $j > n$):

$$\textstyle\sum_l \alpha_l \omega_{n-l,j} = \beta_{n-j}\ . \tag{5.8}$$

This approach promotes (Wolkenfelt (1981), van der Houwen *et al.* (1981, 1985)) the use of LMMs which have favourable properties to generate the weights in (5.2) using the given $\{\rho,\sigma\}$. Indeed, if a zero-stable $\{\rho,\sigma\}$ LMM is applied to the equation $y'(t) = \phi(t)$ with starting values of the form (5.2), the subsequent approximations to

$$y(rh) = \int_0^{rh} \phi(s)\,ds$$

are of the form

$$h\sum_0^r \omega_{rj}\,\phi(jh)\ ,$$

and the triangular matrix of weights ω_{rj} has the property that it is a bordered isoclinal matrix, so that $\omega_{rj} = w_{r-j}$ if $r_0 \leqslant j \leqslant r$ (compare the weights displayed in (5.4) to (5.6)); the weights are uniformly bounded as a consequence of zero-stability. Note that, as a consequence of (5.8),

$$w_0 + w_1\zeta + w_2\zeta^2 + \cdots = \sigma(\zeta^{-1})/\rho(\zeta^{-1})\ . \tag{5.9}$$

(Thus, $\rho(\mu) = \mu - 1$, $\sigma(\mu) = \frac{1}{2}(\mu + 1)$ gives $w_0 = \frac{1}{2}$, $w_l = 1$ $l \neq 0$, in (5.4).) The backward differential formulae (BDF), of order not exceeding 6, generate (with suitable starting rules) a family of rules (5.2) having some desirable stability properties. A slight complication is the generation (Baker and Derakhshan, 1986) of these "BDF-quadrature" weights; the classical *ad hoc* rules had a repetition property which ensured that the weights of the r-th rule (5.2) could be written down readily. Actually, the BDF-weights have an *asymptotic* repetition factor and correctly rounded values do repeat. We refer to Wolkenfelt (1981), which incorporates further interesting work.

5.2 *Runge-Kutta Methods*. Runge-Kutta (RK) methods fall into the following classes: (a) extended RK methods and (b) mixed quadrature RK methods, which can employ either (i) Pouzet-RK (PRK) formulae or (ii) Beltyukov-RK formulae. The latter formulae have now been generalized (for Volterra-RK formulae, see Brunner, Hairer and Nørsett (1982); for the extension of some of this work to *Abel equations* see Lubich (1983b)). Those attracted by theoretical aspects will find an interest in Volterra Runge-Kutta trees and their relation to the order of methods. The extended PRK formulae are simplest because they are directly related to ("classical") RK formulae for ordinary differential equations, and have the order of accuracy expected from their role in differential equations. Considering the application of PRK formulae to (5.1), we require Runge-Kutta parameters of a Butcher tableau $\begin{array}{c|c} \underline{c} & A \\ \hline & \underline{b}^T \end{array}$ which correspond to the family of quadrature rules

$$\int_0^{c_r} \phi(s)\,ds \simeq \sum_{s=0}^{p-1} A_{rs}\,\phi(c_s)$$
$$\int_0^{1} \phi(s)\,ds \simeq \sum_{s=0}^{p-1} b_s\,\phi(c_s). \qquad (5.10)$$

It is convenient to regard b_s as an alias for $A_{p,s}$, with $c_p = 1$. Then, on writing $t_{n,r}$ for $nh + c_r h$, the PRK formulae are expressible in the form

$$Y_{n,r} - h \sum_{s=0}^{p-1} A_{rs} H(t_{n,r}, t_{n,s}, Y_{n,s}) = \Gamma_{n,r} \quad (r = 0, 1, \cdots, p) \tag{5.11}$$

where

$$\Gamma_{n,r} = g(t_{n,r}) + h \sum_{j=0}^{n-1} \sum_{s=0}^{p-1} b_s H(t_{n,r}, t_{j,s}, Y_{j,s}). \tag{5.12}$$

The values $Y_{n,r}$ approximate $y(t_{n,r})$ and we take $y((n+1)h)$ to be approximated by $Y_{n,p} \equiv y_{n+1}$. In the mixed quadrature-PRK methods, formula (5.12) is amended to read

$$\Gamma_{n,r} = g(t_{n,r}) + h \sum_{j=0}^{n} \omega_{nj} H(t_{n,r}, jh, y_j), \quad (y_0 = y(0);\ y_l = y_{l-1,p},\ l \geqslant 1). \tag{5.13}$$

These mixed methods combine features of quadrature methods and RK formulae with a degree of economy and can have good local accuracy, but can lose favourable stability properties (Baker and Wilkinson (1980)). The mixed methods have been modified to improve their stability properties (for *modified methods* and other variants of methods, see Wolkenfelt (1981)); *however*, Baker (1985) showed that it is possible to construct implicit mixed methods with some good stability properties ($A(\alpha)$-stability).

There are numerous ways for us to motivate the various RK formulae. First, consider the canonical form $(g(t) \equiv 0)$ of (4.2) with a finitely decomposable kernel; its solution is determined (§4) by the solution of a system of differential equations. The classical RK method applied to this system is equivalent to the PRK method applied directly to the integral equation. Secondly, we can regard $y(t)$ in (4.2) as the section $Y(t,t)$ of the function $Y(t,s)$ such that $Y(t,0) = g(t)$, and

$$(\partial/\partial s) Y(t,s) = H(t, s, Y(s,s)),$$

and extend classical RK methods formally to solve the latter partial differential equation. Alternatively, we can regard a classical RK tableau as defining rules (5.10) which can be used to discretize the integral equation in a fashion analogous to the quadrature methods. Finally, the RK methods can be considered to be one-step incremental methods which advance the solution y from $t=0$ to $t=h$; a process of shifting the origin, by modifying $g(t)$ to incorporate "past" contributions from the solution, allows the determination of $y_r = y(rh)$ $(r = 1, 2, \cdots)$. The latter viewpoint is used in the analysis of a class of general methods by Hairer, Lubich and Nørsett (1983) and we indicate the perspective briefly in terms of PRK methods. Equations (5.11) define a set of values $Y_{n,r}(\Gamma)$, for $r = 0, 1, \cdots, p-1$ in terms of the values $\Gamma_{n,r}$ which correspond to approximate values of the history term

$$\gamma(t, \tau) := g(\tau) + \int_0^t H(\tau, s, y(s))\, ds$$

with $t = nh$, $\tau \in \{nh + c_r h\}$. Then $y_{n+1} = \Gamma_{n,p} + h\Phi_n$ where the incremental function

$$\Phi_h \equiv \Phi_n(\{Y_{n,r}(\Gamma) \mid r = 0, 1, \cdots, p-1\})$$

is

$$\sum_{s=0}^{p-1} b_s H((n+1)h,\ (n+c_s)h,\ Y_{n,s}(\Gamma)).$$

The perspective allows us to separate (i) the effect of approximating $\gamma(t, \tau)$ by values of $\Gamma_{n,r}$ and (ii) the local error which corresponds to one step using an increment Φ defined using the values $Y_{n,r}(\gamma)$ obtained in terms of the *exact* history term γ. The viewpoint assigns to y_{n+1} (alias $Y_{n,p}$) a special role amongst the values $Y_{n,r}$ which is appropriate to its higher-order accuracy.

5.3 *Discretization of Abel-type Equations*. The generalized Abel equation of the second kind reads

$$y(t) - \int_0^t (t-s)^{\nu-1} H(t,s,y(s))\,ds = g(t) \quad (t \geqslant 0). \tag{5.14}$$

Here, $\nu \in (0,1]$, and H is assumed smooth; $\nu = 1$ is the Volterra case. To discretize the integral using quadrature rules when $\nu \neq 1$ requires analogues of (5.2), which are accurate for appropriate Φ and are of the type

$$\int_0^{rh} (rh-s)^{\nu-1} \Phi(s)\,ds \simeq h^{\nu} \sum_0^r \omega_{rj}^{(\nu)} \Phi(jh) \quad (0<\nu<1). \tag{5.15}$$

The resulting quadrature methods for (5.14) have the form of (5.3) but with $h\omega_{rj}$ replaced by $h^{\nu}\omega_{rj}^{(\nu)}$. Note that if implicit methods $(\omega_{rr}^{(\nu)} \neq 0)$ are to be used, the natural iteration analogous to that described for the non-singular equation requires h^{ν} to be small, with a stricter constraint on the size of h. The construction of (5.15) merits some attention. If $H(t,s,y(s))$ is a smooth function of s, then it is reasonable to employ formulae (5.15) which are constructed to be accurate for smooth Φ. Amongst this class of formulae are generalizations of the interpolatory rules which form the basis for the choices in (5.2). But frequently (the linear Abel equation providing an example) a smooth function g gives rise to a non-smooth function y, so such formulae may then be suitable only if a small or graded stepsize h is employed. Perhaps more intriguing are the fractional integration formulae recently discussed by Lubich (1986) and intended for use in the present context. The array of weights $\omega_{rj}^{(\nu)}$ forms a bordered isoclinal matrix (familiar from the case $\nu = 1$) with $\omega_{rj}^{(\nu)} = w_{r-j}^{(\nu)}$ for $r_0 \leqslant j \leqslant r$: the values $w_{r-j}^{(\nu)}$ are the coefficients (cf. Baker and Derakhshan, 1986) of the formal power series

$$w_0^{(\nu)} + w_1^{(\nu)}\zeta + w_2^{(\nu)}\zeta^2 + \cdots = \Gamma(\nu)\,\{\sigma(\zeta^{-1})/\rho(\zeta^{-1})\}^{\nu}, \tag{5.16}$$

and the theory shows that the remaining weights can be chosen to cope with some of the expected bad behaviour in the true solution.

5.4 *Remarks on First Kind Equations.* We should not omit some reference to work on first-kind equations. The simplest equation

$$\int_0^t y(s)\,ds = g(t)$$

has a solution $g'(t)$ if $g(0)=0$ and g' exists; it is ill-posed as an equation in $C[0,\infty)$; the ill-posedness of Abel's classical first kind equation is milder, since fractional derivatives are involved. Some insight can be obtained by relating the discretization of the equation to numerical differentiation, see in particular Taylor (1976). Note that the trapezium rule (though convergent) has poor stability properties; the mid-point rule is safer. For BDF methods, refer to Wolkenfelt (1981). For extensions to Abel's equation of the first kind see Eggermont (1981), and most recently, the relevant work of Lubich (to appear, 1986).

6. RECENT ADVANCES IN THE STABILITY ANALYSIS OF VOLTERRA METHODS

By implication, methods for the treatment of Volterra and Abel equations of the second kind can be extended to the treatment of integro-differential equations. Questions concerning the choice of numerical schemes require us to consider a classification of various types of Volterra equation. Convolution equations (4.8), either of Volterra or of Abel type, appear prominently, but not exclusively, in applications (the renewal equation is of this form).

An area of advance has been the study of numerical stability. The basic equation

$$y(t) = \lambda \int_0^t y(s)\,ds + g(t) \tag{6.1}$$

(the convolution equation for which results are most readily obtained) is evidently rather simple. It is therefore important to question the validity of insight obtained by studying this case. First, let us note that the manipulative techniques used in analyzing methods for this equation can sometimes be extended to equations with more general separable kernels. Secondly, the use of localizing assumptions, in which a general equation is modelled locally by the basic equation with a value λ reflecting the partial derivative $(\partial/\partial v)H(t,s,v)$ (at, say, $t=s$) is open to the same objections as in the corresponding study of differential equation, but it provides a good insight. Recent advances in the theory have provided some guide, in precise terms, into the extent to which the basic equation gives genuine insight for the more general convolution case, displayed in eqn (4.8). This is the linear form of

$$y(t) - \lambda \int_0^t k(t-s)\,\Phi(y(s))\,ds = g(t) \qquad (t \geqslant 0).$$

The nonlinear version has been studied by Lubich (1985), and is subsumed in the discussion of Nevanlinna (1976), but most work has been related to the linear case $(\phi(v)=v)$ to which we restrict our attention. A convolution equation (linear or nonlinear) is special, because the choice of a fixed stepsize h in a numerical method which preserves the convolution structure allows the use of FFT techniques; cf. Hairer, Lubich and Schlichte (1985), Baker and Derakhshan (1985). Suppose, then, that

$$y(t) - \lambda \int_0^t k(t-s)\,y(s)\,ds = g(t) \qquad (t \geqslant 0). \qquad (6.2)$$

With * to denote convolution, $y-\lambda k*y=g$, and if the function g is perturbed by the addition of a function δg, then y changes to $y+\delta y$, where $\delta y-\lambda k*\delta y=\delta g$. This equation is, of course, of the same form as the original equation (6.2) for the function y. In consequence, *if we consider linear equations*, the question of analytic stability (namely, how does the solution respond to perturbations in g) is the same as the question of qualitative behaviour (namely, how does a solution y behave when g is of a particular form). For the basic equation, with

$k(t) = 1$ for $t \geqslant 0$, the question is already answered: If

$$y(t) - \lambda \int_0^t y(s)\,ds = g(t)$$

then

$$y(t) = g(t) + \lambda \int_0^t \exp\{\lambda(t-s)\}\, g(s)\,ds$$

(the resolvent kernel corresponding to the case $g(t) \equiv 1$) and we deduce that if g is bounded and continuous $(g \in BC[0,\infty))$ then $y(t)$ has the same property provided that $\mathrm{Re}(\lambda) \leqslant 0$. On the other hand, if g is continuous and tends to zero $(g \in C_0[0,\infty))$ then $y \in C_0[0,\infty)$ if $\mathrm{Re}(\lambda) < 0$. With a slight imprecision in the terminology, the basic equation is commonly called "stable" if $\mathrm{Re}(\lambda) \leqslant 0$ and "asymptotically stable" if $\mathrm{Re}(\lambda) < 0$. Note that the criteria depend upon λ, not on g; however, the *concepts* of stability, blurred in our definition, depend upon the class of admissible functions g (or of perturbations δg). *One seeks conditions under which stability (and qualitative behaviour) is preserved when numerical methods are applied to the equation.* This perspective merits emphasis. In the case of the basic equation, conditions for numerical stability can be expressed succinctly using the concepts of a simple von Neumann polynomial and a Schur polynomial which arise in the theory of numerical methods for ODEs. As a sample of results, we have: (i) A $\{\rho,\sigma\}$-reducible quadrature rule applied to the basic equation is stable (g bounded implies that the approximate solution is bounded) if the characteristic polynomial $\rho(\mu) - \lambda h \sigma(\mu)$ is simple von Neumann and is "strictly" (or "asymptotically") stable (g bounded with limit zero implies the same of the approximate solution) if the characteristic polynomial is a Schur polynomial. (ii) If the amplification factor $R(\lambda h)$ associated with a RK tableau, satisfies $|R(\lambda h)| \leqslant 1$ then the corresponding PRK method is stable. ($R(\lambda h) := 1 + \lambda h \underline{b}^T (I - \lambda h A)^{-1}\, [1, 1, \cdots, 1]^T$ is familiar from ODE theory.) (iii) A mixed quadrature-PRK method combining the above parameters is stable if the characteristic polynomial $\rho(\mu) - \lambda h R(\lambda h)\,\sigma(\mu)$ is simple von Neumann.

The fact that conditions (i) and (ii) are familiar from our study of ODEs means that we can borrow the concept of a region of numerical stability in the complex λh-plane as the set S of λh for which the scheme is numerically stable. We remind the reader that a strongly-stable $\{\rho,\sigma\}$-LMM has a stability region which includes a maximal disk D touching the imaginary axis at the origin and with centre on the negative real line; this disk is called the "stability disk" of the method, and degenerates to the whole left-half plane if the method is A-stable.

In the terms we have employed, the analytical stability of the basic equation (6.1) depends upon $\mathrm{Re}(\lambda)$. In order to extend the analysis to a general class of equation, one looks for an appropriate class of kernels k. For the case $k \in L^1[0,\infty)$, the Paley-Wiener theorem provides the necessary insight and we find what is needed in the following result:

THEOREM (Analytic stability for $\mathrm{Re}(\lambda) < 0$): The Hermitian positive-definite functions form (in $C \cap L^1[0,\infty)$) the largest class of kernels k for which $g \in BC[0,\infty)$ implies $y \in BC[0,\infty)$ whenever $\mathrm{Re}(\lambda) < 0$.

Now consider the discrete versions of the convolution integral equation, namely

$$y_n - \lambda \sum_{j=0}^{n} a_{n-j} y_j = d_n \qquad (n \in Z_+). \tag{6.3}$$

We assume that the sequence $\sum_0^\infty |a_l|$ converges (i.e. $\{a_0, a_1, a_2, \cdots\} \in l_1$), then a discrete version of the Paley-Wiener theorem applies.

THEOREM (Discrete Paley-Wiener): Suppose $\{a_0, a_1, a_2, \cdots\} \in l_1$ and $\{d_i\}$ is arbitrary with $\sup|d_i| < \infty$. Then $\sup|y_i| < \infty$ for all such $\{d_i\}$ if and only if $\lambda\sum_0^\infty a_r \zeta^r \neq 1$ for $|\zeta| \leqslant 1$.

This theorem is of interest, because the use of a zero-stable $\{\rho,\sigma\}$-reducible quadrature method applied to eqn (6.2) gives a system of discrete convolution equations in which $a_l = \omega_l k_l$ $(k_l \equiv k(lh))$ and $\sup|\omega_l| < \infty$ provided certain starting values are incorporated in the values d_i. The values a_l satisfy the conditions assumed *with the exception* that the condition that $\sum_0^\infty |a_l| < \infty$ (equivalently, that $\sum_0^\infty |k_l|$ converges) may not hold, and has to be added as an hypothesis. To proceed further, we wish to adapt the notion of a Hermitian positive-definite doubly-infinite sequence (encountered in Section 4) to the semi-infinite sequence $\{a_0, a_1, a_2, \cdots\}$. We say that $\{a_i\}_0^\infty$ is positive definite if the extension $\{\hat{a}_i\}_{-\infty}^\infty$ which is defined by $\hat{a}_i = a_i$ if $i>0$, $\hat{a}_i = \bar{a}_{-i}$ if $i<0$ and $\hat{a}_0 = a_0 + \bar{a}_0$ is Hermitian positive definite. We then require three observations: (a) if the strongly-stable $\{\rho,\sigma\}$-reducible LMM method has a stability disk D with radius $1/(2c)$, the associated $\{\rho,\sigma\}$-reducible quadrature is such that $\{w_0 + c, w_1, w_2, \cdots\}$ is positive-definite, (b) the termwise product of two positive-definite sequences is positive-definite, and (c) a bounded sequence $\{a_i\}$ is positive-definite if and only if

$$\operatorname{Re} \sum_0^\infty a_r \zeta^r \geqslant 0 \quad \text{for } |\zeta| < 1.$$

Result (a) is due to O. Nevanlinna (cf. Nevanlinna (1976)); (b) is due to Schur, being equivalent to a result concerning the Schur product of positive-definite matrices; (c) is associated with Toeplitz and Carathéodory. We now state the following (Lubich, 1983a):

THEOREM: Suppose that k is Hermitian positive-definite, normalized (without loss of generality) so that $k(0) = 1$, and that (for every h) $\{k(0), k(h), k(2h), \cdots\} \in l_1$. If $\{\rho,\sigma\}$ is strongly-stable, with stability disk D, then the $\{\rho,\sigma\}$-reducible quadrature method is numerically stable when $\lambda h \in D$.

If $\{\rho,\sigma\}$ is A-stable, the quadrature method with $\lambda h \in D$ simulates the analytic stability for $\mathrm{Re}(\lambda) < 0$. In general, D is a proper subset of the stability region S associated with the basic equation ($\lambda h \in S$ if and only if $\rho(\mu) - \lambda h \sigma(\mu)$ is a Schur polynomial). However, Figures 1 and 2 (prepared for the author by his student Mr Neville Ford) indicate that for the Gregory quadrature rules the set $S-D$ can be quite small whilst the corresponding set for BDF formulae is rather larger!

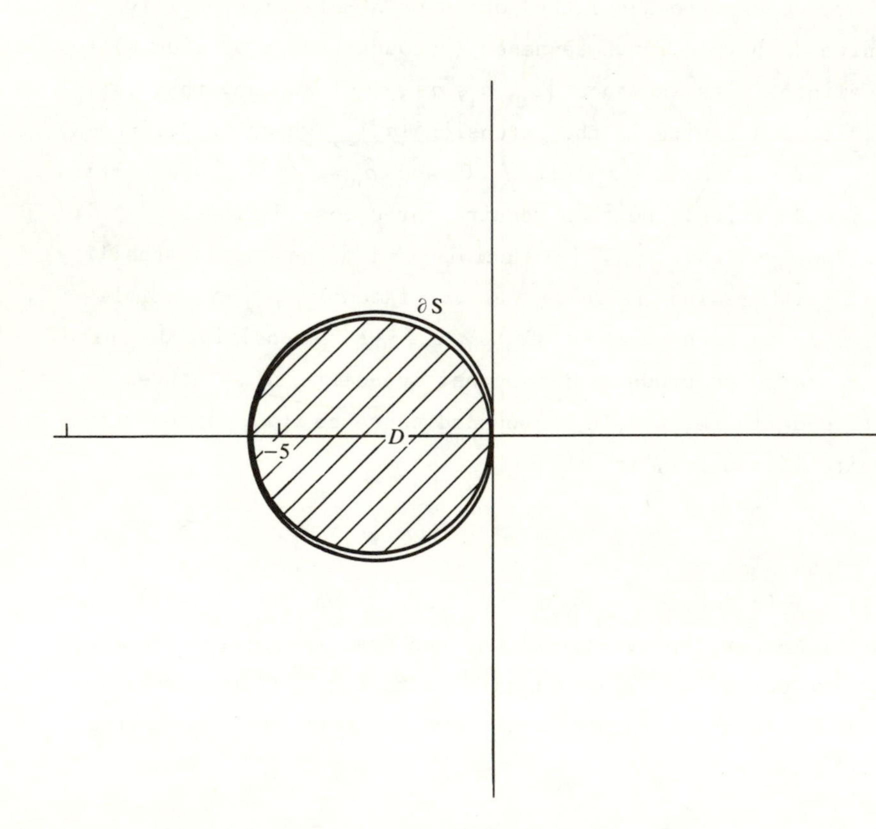

FIG. 1 Stability disk D included in the stability region S of the Gregory rules corresponding to the 2-step Adams-Moulton LMM

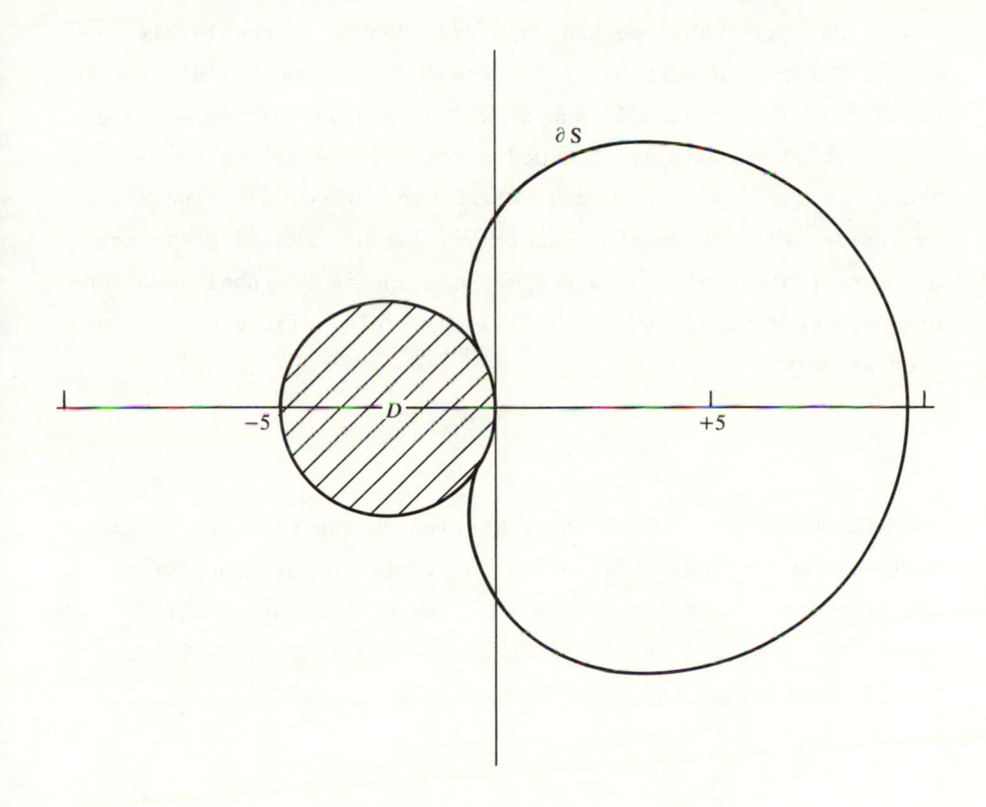

FIG. 2 Stability disk D included in the stability region S of the BDF quadrature rules corresponding to the 2-step BDF LMM

The figures show that, for the Gregory quadrature rules, the basic equation yields quite a reliable model for equations with positive-definite k.

Now, since A-stability of $\{\rho,\sigma\}$ is achievable only for order at most 2, we examine further avenues. Lubich (1983a) discusses $A(\alpha)$-stable methods applied when k is completely monotone. Hairer and Lubich (1984) extend the approach indicated to the extended-PRK methods and show that appropriate (algebraically-stable) methods of high order correctly simulate analytical behaviour for $\mathrm{Re}(\lambda) < 0$, and retain high order if the Hermitian extension of k is smooth. Lubich (1986) has extended the arguments to fractional quadrature methods applied to Abel equations of the second kind. (Note that (5.16) yields discrete convolution structure.)

ACKNOWLEDGEMENTS: The author acknowledges the kindness of many authors who have kept him abreast of their research and writings, the numerous fruitful conversations and collaboration with a number of them, and the interest of his own students particularly M.S. Derakhshan and N.J. Ford.

REFERENCES

Space prevents us from giving an exhaustive list of references; however, Brunner and van der Houwen (1986) have 50 pages of citations. The theses (proefschrift) of Schippers and of Wolkenfelt contain numerous papers by these authors.

Albrecht, J. and Collatz, L. (eds) (1980), *Numerical Treatment of Integral Equations*, Birkhäuser (Basel).

Anderssen, R.S., De Hoog, F.R. and Lukas, M.A. (eds) (1980), *The Application and Numerical Solution of Integral Equations*, Sijthoff and Noordhoff International (Alphen an de Rijn).

Anselone, P.M. (ed.) (1964), *Nonlinear Integral Equations*, University of Wisconsin Press (Madison).

Anselone, P.M. (1971), *Collectively Compact Operator Approximation Theory*, Prentice-Hall (Englewood Cliffs).

Atkinson, K.E. (1976), *A Survey of Numerical Methods for the Solution of Fredholm Integral Equations of the Second Kind*, SIAM Publication (Philadelphia).

Baker, C.T.H. (1977), *The Numerical Treatment of Integral Equations*, Clarendon Press (Oxford) (reprinted 1978).

Baker, C.T.H. (1985), "Concerning A(α)-stable mixed Volterra Runge-Kutta methods", in *Constructive Methods for the Practical Treatment of Integral Equations*, eds. Hämmerlin, G. and Hoffman, K.-H., Birkhäuser (Basel), pp.53-67.

Baker, C.T.H. and Derakhshan, M.S. (1985), "Preliminary report on a fast algorithm for the numerical solution (with error estimates) of a Volterra equation of the second kind having a convolution kernel", Numer. Anal. Tech. Rep. 106, University of Manchester.

Baker, C.T.H. and Derakhshan, M.S. (1986), "The use of NAG FFT routines in the construction of functions of power series used in fractional quadrature rules", Numer. Anal. Tech. Rep. 115, University of Manchester.

Baker, C.T.H. and Keech, M.S. (1978), "Stability regions in the numerical treatment of Volterra integral equations", *SIAM J. Numer. Anal. 15*, pp.394-417.

Baker, C.T.H. and Miller, G.F. (eds) (1982), *The Treatment of Integral Equations by Numerical Methods*, Academic Press, (London).

Baker, C.T.H. and Wilkinson, J.C. (1980), "Stability analysis of Runge-Kutta methods applied to a basic Volterra integral equation", *J. Austral. Math. Soc. Ser. B 22*, pp.515-538.

Bates, D.M. and Wahba G. (1982), "Computational methods for generalized cross-validation", in *The Treatment of Integral Equations by Numerical Methods*, eds. Baker, C.T.H. and Miller G.F., Academic Press (London), pp.283-296.

Bownds, J.M. (1982), "Comments on the performance of a FORTRAN subroutine for certain Volterra equations", in *The Treatment of Integral Equations by Numerical Methods*, eds. Baker, C.T.H. and Miller, G.F., Academic Press (London), pp.163-167.

Brunner, H., Hairer, E. and Nørsett, S.P. (1982), "Runge-Kutta theory for Volterra integral equations of the second kind", *Math. Comp. 39*, pp.147-163.

Brunner, H. and Nørsett, S.P. (1981), "Superconvergence of collocation methods for Volterra and Abel questions of the second kind", *Numer. Math. 36*, pp.347-358.

Brunner, H. and van der Houwen, P.J. (1986), *The Numerical Solution of Volterra Equations*, CWI Monographs, North-Holland (Amsterdam).

Burton, T.A. (1983), *Volterra Integral and Differential Equations*, Academic Press (New York).

Chandler, G.A. (1979), "Superconvergence of numerical solutions to second kind integral equations", Ph.D. Thesis, ANU Canberra.

Chandler, G.A. and Graham, I.G. (1985), "Uniform convergence of Galerkin solutions to noncompact integral operator equations", Tech. Rep. CMA, ANU Canberra.

Chatelin, F. (1981), "The spectral approximation of linear operators with applications to the computation of eigenelements of differential and integral operators", *SIAM Rev.* *23*, pp.495-522.

Chatelin, F. (1983), *Spectral Approximation of Linear Operators*, Academic Press (New York).

Corduneanu, C. (1973), *Integral Equations and Stability of Feedback Systems*, Academic Press (New York).

Cuppen, J.J.M. (1983), "A numerical solution of the inverse problem of electrocardiography", Academisch Proefschrift, Amsterdam.

Davies, A.R. (1983), "On a constrained Fourier extrapolation method for numerical deconvolution", in *Improperly Posed Problems and their Numerical Treatment*, eds. Hämmerlin G. and Hoffman K.-H., Birkhäuser (Basel), pp.65-80.

Davies, A.R. and Anderssen, R.S. (1985a), "Optimization in the regularization of ill-posed problems", Rep. CMA R30 85, ANU Canberra (to appear in *J. Austral. Math. Ser. B*)

Davies, A.R. and Anderssen, R.S. (1985b), "Improved estimates of statistical regularization parameters in Fourier differentiation and smoothing", Rep. CMA R01 85 ANU Canberra (to appear in *Numer. Math.*).

Davies, A.R., Iqbal, M., Maleknejad, K. and Redshaw, T.C. (1983), "A comparison of statistical and Fourier extrapolation methods for numerical deconvolution", in *Numerical Treatment of Inverse Problems in Differential and Integral Equations*, eds. Deuflhard, P. and Hairer, E., Birkhäuser (Basel), pp.320-334.

Delves, L.M. and Mohamed, J.L. (1985), *Computational Methods for Integral Equations*, Cambridge University Press (Cambridge).

Delves, L.M. and Walsh, J. (eds) (1974), *Numerical Solution of Integral Equations*, Clarendon Press (Oxford).

Deuflhard, P. and Hairer, E. (eds) (1983), *Numerical Treatment of Inverse Problems in Differential and Integral Equations*, Birkhäuser (Basel).

Dixon, J. and McKee, S. (1985), "A unified approach to convergence analysis of discretization methods for Volterra type", *IMA J. Numer. Anal.* *4*, pp.99-107.

Doetsch, G. (1974), *Introduction to the Theory and Applications of the Laplace Transformation*, Springer (Berlin).

Elliot, D. (1982), "The numerical treatment of singular integral equations — a review", in *The Treatment of Integral Equations by Numerical Methods*, eds. Baker, C.T.H. and Miller, G.F., Academic Press (London), pp.297-312.

Eggermont, P.P.B. (1981), "A new analysis of the trapezoidal discretization method for the numerical solution of Abel-type integral equations", *J. Integral Eqns.* *3*, pp.317-332.

Fox, L. and Goodwin, E.T. (1953), "The numerical solution of non-singular linear integral equations", *Phil. Trans. Roy. Soc. London Ser. A*, *245*, pp.501-534.

Golberg, M.A. (ed) (1979), *Solution Methods for Integral Equations*, Plenum Press (New York).

Graham, I.G. (1982), "Galerkin methods for second kind integral equations with weakly singular convolution kernels", *Math. Comp.* *39*, pp.519-533.

Groetsch, C.W. (1977), *Generalized Inverses of Linear Operators: Representation and Approximation*, Marcel Dekker (New York).

Groetsch, C.W. (1984), *The Theory of Tikhonov Regularization for Fredholm Equations of the First Kind*, Pitman (Boston).

Hairer, E., Lubich, Ch. and Nørsett, S.P. (1983), "Order of convergence of one-step methods for Volterra integral equations of the second kind", *SIAM J. Numer. Anal.* *20*, pp.569-579.

Hairer, E. and Lubich, Ch. (1984), "On the stability of Volterra-Runge-Kutta methods", *SIAM J. Numer. Anal.* *21*, pp.123-135.

Hairer, E., Lubich, Ch. and Schlichte, M. (1985), "Fast numerical solution of nonlinear Volterra convolution equations", *SIAM J. Sci. Stat. Comp.* *6*, pp.532-541.

Hämmerlin, G. and Hoffman K-H. (eds) (1983), *Improperly Posed Problems and their Numerical Treatment*, Birkhäuser (Basel).

Hämmerlin, G. and Hoffman, K-H. (eds) (1985), *Constructive Methods for the Practical Treatment of Integral Equations*, Birkhäuser (Basel).

Hemker, P.W. and Schippers, H. (1981), "Multiple grid methods for the solution of Fredholm integral equation of the second kind", *Math. Comp.* *36*, pp.215-232.

Hock, W. (1981), "An extrapolation method with stepsize control for nonlinear Volterra integral equations", *Numer. Math.* *38*, pp.155-178.

van der Houwen, P.J. and Blom, J.G. (1985), "Stability results for discrete Volterra equations: numerical experiments", in *Constructive Methods for the Practical Treatment of Integral Equations*, eds. Hammerlin, G. and Hoffman, K-H., Birkhäuser (Basel), pp.166-178.

van der Houwen, P.J. and te Riele, H.J.J. (1981), "Backward differentiation type formulas for Volterra integral equations of the second kind", *Numer. Math.* *37*, pp.205-217.

van der Houwen, P.J. and te Riele, H.J.J. (1985), "Linear multistep methods for Volterra integral and integro-differential equations", *Math. Comp.* *45*, pp.439-461.

Ivanov, V.V. (1976), *The Theory of Approximate Methods and their Application to the Numerical Solution of Singular Integral Equations*, Noordhoff (Leyden).

Jaswon, M.A. and Symm, G.T. (1977), *Integral Equation Methods in Potential Theory and Elastodynamics*, Academic Press (London).

Jones, H.M. and McKee, S. (1985), "Variable step size predictor-corrector schemes for second kind Volterra integral equations", *Math. Comp.* *44*, pp.391-404.

Linz, P. (1985), *Analytical and Numerical Methods for Volterra Equations*, SIAM Publications (Philadelphia).

Lonseth, A.T. (1977), "Sources and applications of integral equations", *SIAM Rev. 19*, pp.241-278.

Lubich, Ch. (1983a), "On the stability of linear multistep methods for Volterra convolution equations", *IMA J. Numer. Anal. 3*, pp.439-465.

Lubich, Ch. (1983b), "Runge-Kutta theory for Volterra and Abel integral equations of the second kind", *Math. Comp. 41*, pp.87-102.

Lubich, Ch. (1985), "On the numerical solution of Volterra equations with unbounded nonlinearity", *J. Integral Eqns. 10*, pp.175-183.

Lubich, Ch. (1986), "Zur numerische analysis von differential- und integral-gleichungen", Habiliationschrift, Universität Innsbruck.

Lubich, Ch. (to appear) "Fractional linear multistep methods for Abel-Volterra integral equations of the first kind", (Submitted to *IMA J. Numer. Anal.*)

Miller, R.K. (1971), *Nonlinear Volterra Integral Equations*, Benjamin (Menlo Park).

Nevanlinna, O. (1976), "Positive quadratures for Volterra equations", *Computing 16*, pp.349-357.

Noble, B. (1971), "A bibliography on 'Methods for solving integral equations' ", MRC Tech. Rep. 1176 and 1177, Univ. of Wisconsin, Madison.

Pouzet, P. (1962) "Étude, en vue de leur traitement numérique d'équations intégrales et intégro-différentielles du type de Volterra pour des problemes de conditions initiales", Thesis, University of Strasbourg.

te Riele, H.J.J. (ed.) (1980), *Colloquium Numerical Treatment of Integral Equations*, Mathematisch Centrum (Amsterdam).

Schippers, H. (1982), "Multiple grid methods for equations of the second kind with applications in fluid mechanics", Academisch Proefschrift, Amsterdam.

Sloan, I.H. (1982), "Superconvergence and the Galerkin method for integral equations of the second kind", in *The Treatment of Integral Equations by Numerical Methods*, eds. Baker, C.T.H. and Miller, G.F., Academic Press (London), pp.197-208.

Sloan, I.H. and Spence, A. (1986), "A projection method for integral equations on the half-line", *IMA J. Numer. Anal. 6*, pp.153-172.

Smithies, F. (1952), *Integral Equations*, Cambridge University Press (Cambridge).

Taylor, P.J. (1976), "The solution of Volterra integral equations of the first kind using inverted differentiation formulas", *BIT 16*, pp.416-425.

Tsalyuk, Z.B. (1979), "Volterra integral equations", *J. Soviet Math. 12*, pp.715-758.

Wendland, W.L. (1982), "On applications and the convergence of boundary integral methods", in *The Treatment of Integral Equations by Numerical Methods*, eds. Baker, C.T.H. and Miller, G.F., Academic Press (London), pp.463-476.

Wolkenfelt, P.H.M. (1981), "The numerical analysis of reducible quadrature methods for Volterra integral and integro-differential equations", Academisch Proefschrift, Amsterdam.

Christopher T.H. Baker
Department of Mathematics
The Victoria University of Manchester
Oxford Road
Manchester M13 9PL
England

18 Strongly Elliptic Boundary Integral Equations

W. L. WENDLAND

ABSTRACT

Strong ellipticity for boundary integral equations on smooth closed surfaces or curves is defined in the framework of pseudo-differential operators. The variational formulation provides a unifying analysis of Galerkin and collocation methods. For these versions of boundary elements and spline approximations we give a survey on asymptotic convergence results for families of boundary elements with diminishing mesh width. The class of integral equations contains the Fredholm integral equations of the second kind, some of the first kind and also Cauchy singular and hyper-singular integral equations.

1. INTRODUCTION

During the last decade, the numerical treatment of boundary iintegral equations in the form of boundary element methods has become a rather popular and powerful technique for engineering computations of boundary value problems, in addition to finite difference and finite element methods. The classical mathematical analysis, including that of numerical methods for integral equations, is usually based on the classification of integral equations, such as Fredholm integral equations of the second kind, of the first kind and Cauchy singular integral equations. Then each class is treated separately. In the boundary element methods, however, all these types of integral equations and even some more general classes are coming up in just one problem and often also in a coupled form. The engineer will go ahead and design a numerical method for his boundary element software; he will not bother whether his equations can be found at best in different parts of monographs on the numerical solution of integral equations. Hence, there is a strong desire for an appropriate unifying mathematical framework which provides the theory and fundamental principles for the justification and the error analysis of the boundary element methods. In fact, for elliptic boundary value problems which model time harmonic and steady state phenomena, variational methods provide the desired frame. In addition to existence and uniqueness, they also give the formulation of Ritz-Galerkin methods, and in the corresponding weak formulation they lead to the more general Galerkin-Petrov methods including collocation. In particular, with finite boundary elements they lead to the decisive properties for the convergence of the corresponding numerical methods. In the linear theory, the fundamental property of the associated quadratic form is *strong ellipticity* in the form of Gårding's inequality. For a continuous invertible linear operator in a Hilbert space and *all* Galerkin methods, the strong ellipticity is also necessary for their convergence (Vainikko, 1965).

The concept of strong ellipticity is commonly well understood for differential operators. For our boundary integral operators, however, one needs the more sophisticated theory of pseudodifferential operators. The latter has been far and successfully developed within the last 20 years. Here the boundary integral equations can be understood as to be standard pseudodifferential operators on the compact boundary manifold. Thus the Gårding inequality is equivalent to the positive definiteness of the principal symbol of the pseudodifferential operator. For particularly chosen boundary integral equations, the Gårding inequality can also be obtained from the coerciveness of the original strongly elliptic boundary value problem. Since the

Gårding inequality implies stability and convergence of *every* Galerkin approximation, this is particularly true for the boundary element Galerkin methods. However, finite elements including the boundary elements have also strong local approximation properties which might be the reason for corresponding versions of the localization principle (Prössdorf, 1984; Arnold and Wendland, 1985) that leads to a slight relaxation of the positive definiteness of the principal symbol. For the convergence of Galerkin as well as of collocation boundary element methods we require positive definiteness of the principal symbol only up to multiplication by regular functions or matrices thereof, this being our definition of *strong ellipticity* (Stephan and Wendland, 1976). Recently it has turned out that this version of strong ellipticity is for two-dimensional boundary value problems with boundary integral equations on the boundary curve and spline Galerkin or spline collocation not only sufficient but also necessary for stability and convergence (Prössdorf and Schmidt, 1981a; Schmidt, 1986).

This survey is in principle restricted to these concepts. Of course, there are far more problems coming up in theoretical and numerical analysis of boundary element methods, for a corresponding survey, see e.g., (Wendland, 1985a).

In Chapter 2 one finds the so-called "direct method" or "the method of Green's formula" which goes back to Sobolev (1936) and has been developed in (Calderon, 1963; Hörmander, 1966; Seeley, 1966), see also (Dieudonné, 1978). This method has been rediscovered in different contexts, e.g. (Brebbia *et al*, 1978, Herrera, 1984, 1985). Following Constabel and Wendland (1986) we obtain boundary integral equations in variational form satisfying a coerciveness inequality if the underlying boundary value problems of the interior and exterior are coercive elliptic, since the associated bilinear form of the boundary integral equations corresponds to the energy of the desired fields. A similar approach has been used by Feng (1980, 1983) and by

Feng and Yu (1983) with the Green function of the Dirichlet problem instead of the fundamental solution of $\mathbb{R}^n$. Hence the practical use of Feng's method is restricted to special domains in constrast to most of the engineering boundary element methods.

In Chapter 3 we present the general mapping properties of boundary integral and pseudodifferential operators and their representation in terms of Hadamard finite part integrals, differential operators and Cauchy singular integral operators.

Chapter 4 is devoted to the variational formulation of the boundary integral equations and to the property of strong ellipticity.

In Chapter 5 we give a survey on general and boundary element Galerkin methods including asymptotic convergence results of optimal order and strong ellipticity. As a consequence, we show some superapproximation results for iterated solutions of Fredholm integral equations of the second kind.

In Chapter 6 we collect the asymptotic convergence results for Galerkin-Petrov and collocation methods. For two-dimensional boundary value problems, i.e. the boundary integral equations on curves, we give a survey on the asymptotic convergence of naive spline collocation and Schmidt's ε-collocation. These results are rather complete whereas in three-dimensional problems with boundary integral equations on surfaces we give a review of collocation only for Fredholm integral equations of the second kind. For the general case not much is presently known.

We omit the presentation of the influence of numerical integration which leads to fully discretized algorithms. Corresponding references can be found in (Nedelec, 1976; Schwab and Wendland, 1985; Wendland, 1981, 1983, 1985a, 1985b).

2. MOTIVATION AND THE DIRECT METHOD

Since the class of boundary integral equations is generated by boundary value problems let us here indicate the transition from a regular elliptic boundary value problem to one of the possible boundary integral equations by the so-called "direct method" following (Costabel and Wendland, 1985), involving equations for the Cauchy data on the boundary. This method goes back to Sobolev, Calderón, Seeley and Hörmander and rests on the jump relations of multiple layer potentials.

Let Ω_1 be a bounded domain in $\mathbb{R}^n$ $(n \geq 2)$ whose boundary Γ is sufficiently smooth; the exterior domain is denoted by $\Omega_2 = \mathbb{R}^n \setminus \overline{\Omega_1}$ and their disjoint union by $\Omega = \overline{\Omega_1} \dot{\cup} \overline{\Omega_2}$. Then a function in $C^\infty(\Omega)$ means a *pair* of functions given on $\overline{\Omega_1}$ and $\overline{\Omega_2}$ whose limits on Γ may be different, i.e. u may jump across Γ. The boundary integral equations will be formulated for a regular elliptic boundary value problem in Ω_1 (or in Ω_2) where

$$
\begin{aligned}
&P^{(2m)}u = f \quad \text{in} \quad \Omega_1 \subset \mathbb{R}^n \text{ (or, alternatively, in } \Omega_2), \\
&R(\gamma u) = g \quad \text{on} \quad \Gamma .
\end{aligned}
\tag{2.1}
$$

(For exterior problems in Ω_2 append suitable additional radiation conditions at infinity.)

Here $P^{(2m)}$ denotes an elliptic $2m$-th order linear partial differential operator (or an $N \times N$ system of such), $R = ((R_{jk}))$, $j = 0, \cdots, m-1$, $k = 0, \cdots, 2m-1$, is a matrix of tangential differential operators of orders $\mu_j - k$, $0 \leq \mu_j \leq 2m-1$ (or $N \times N$ matrices of such). u is the desired unknown function (or an N-vector of functions) and

$$\gamma u = (u, \partial_n u, \cdots, \partial_n^{2m-1} u)^T \big|_\Gamma$$

denotes the column of Cauchy data where ∂_n is the normal derivative corresponding to the exterior unit normal vector $\vec{n}$ on Γ. If the differential equation in (2.1) is considered in the distributional sense then the possible jumps of the Cauchy data across Γ for $u \in C^\infty(\overline{\Omega_1} \dot{\cup} \overline{\Omega_2})$ will generate corresponding boundary distributions and the differential equation yields for any test function $\phi \in C_0^\infty(\mathbb{R}^n)$:

$$\int_{\mathbb{R}^n} \phi(y)\, P^{(2m)} u(y)\, dy = \int_{\mathbb{R}^n} \phi(y)\, f(y)\, dy \tag{2.2}$$
$$+ \sum_{k=0}^{2m-1} \sum_{l=0}^{2m-1-k} \int_\Gamma (\partial_n'^k \phi(y))\, P_{k+l+1} [\partial_n^l u_2(y) - \partial_n^l u_1(y)]\, do(y).$$

This is the *second Green formula* (see (Dieudonné, 1978, Section 23.48.13.4)) where $\partial_n'^k$ denotes the differential operator transposed to ∂_n^k, i.e. for all $\phi, \psi \in C_0^\infty(\mathbb{R}^n)$ we have

$$\int_\Gamma (\partial_n^k \psi)\, \phi\, do(y) = \int_\Gamma \psi (\partial_n'^k \phi)\, do(y).$$

The operators P_j in (2.2) are defined by the decomposition of $P^{(2m)}$ along Γ with respect to local coordinates in the normal and tangential directions via

$$P^{(2m)} \phi = \sum_{j=0}^{2m} P_j \partial_n^j \phi .$$

As is usual for the boundary integral methods, let us assume that the *fundamental solution* $G(x,y)$ of $P^{(2m)}$ is known, i.e. for every distribution ψ with compact support it is true that

$$\int_{\mathbb{R}^n} G(x,y)\, P_y^{(2m)} \Psi(y)\, dy = P_x^{(2m)} \int_{\mathbb{R}^n} (G,x,y)\, \Psi(y)\, dy$$
$$= \Psi(x) . \tag{2.3}$$

Hence, (2.2) with $\phi(y) = G(x,y)$ yields the *Green representation formula*

$$u_j(x) = \int_{\Omega_j} G(x,y)\, f(y)\, dy +$$
$$+ (-1)^j \sum_{\substack{k+l+1 \leqslant 2m \\ k,l \geqslant 0}} \int_\Gamma (\partial_{n(y)}'^k G(x,y) P_{k+l+1}(y) \partial_n^l u_j(y)\, do(y) \tag{2.4}$$

for $x \in \Omega_j$ and $j = 1$ or 2

(provided G satisfies suitable radiation conditions for exterior problems). If we take the Cauchy data at Γ from both sides in (2.4) then we obtain the relation

$$\partial_n^i u_j(x) = \partial_{n(x)}^i \int_{\Omega_j} G(x,y)\, f(y)\, dy$$
$$+ (-1)^j \sum_{\substack{k+l+1 \leq 2m \\ k,l \geq 0}} \partial_{n(x)}^i \int_\Gamma (\partial_{n(y)}^{\prime k} G(x,y)) \cdot P_{k+l+1}(y)\, \partial_n^l u_j(y)\, do(y)\,, \tag{2.5}$$

where $\Omega_j \ni x \to \Gamma$, $j = 1,2$, $i = 0,\cdots,2m-1$ on Γ.

Note that the right hand sides in (2.5) are defined by the limit values from Ω_1 or Ω_2, respectively. If they are expressed by *direct* integrals then, in general, one needs jump relations for potentials. In short, we write (2.5) as

$$\gamma u_j = \gamma \mathbb{G} f_j + (-1)^j K_j P \gamma u_j \text{ on } \Gamma. \tag{2.6}$$

For $f_j = 0$ and $j = 1$ or 2, the corresponding operators in (2.6) define the *Calderon projections*. In (2.1), however, the boundary conditions do not define *all* the Cauchy data. Therefore we require that R *possesses complementary boundary operators* $S = ((S_{il}))$, $i = 0,\cdots,m-1$; $l = 0,\cdots,2m-1$ such that $M = \begin{pmatrix} R \\ S \end{pmatrix}$ *becomes* an *invertible* square matrix of tangential boundary operators. If we augment the given boundary data g in (2.1) by yet unknown functions w then the Cauchy data can be expressed by g and w and vice versa:

$$\begin{pmatrix} g \\ w \end{pmatrix} = M\gamma u_j = \begin{pmatrix} R \\ S \end{pmatrix} \gamma u_j\,, \quad \gamma u_j = M^{-1} \begin{pmatrix} g \\ w \end{pmatrix}, \quad j = 1,2. \tag{2.7}$$

Inserting (2.7) into (2.6) implies a system of $2m \cdot N$ linear equations for the $m \cdot N$ unknown components of w. Hence we have a whole variety of boundary integral equations for w. If we choose the first set of equations from (2.6) we find the system of *boundary integral equations of the first kind on* Γ :

$$Aw := -RK_j PM^{-1}\begin{pmatrix}0\\w\end{pmatrix} = g - (-1)^j RK_j\begin{pmatrix}g\\0\end{pmatrix} - R\gamma\mathbb{G} f_j \,. \quad (2.8)$$

From the choice of the second set we obtain the system of

boundary integral equations of the second kind on Γ ;

$$\tilde{A}w := w - (-1)^j SK_j PM^{-1}\begin{pmatrix}0\\w\end{pmatrix} = (-1)^j SK_j PM^{-1}\begin{pmatrix}g\\0\end{pmatrix} + S\gamma\mathbb{G} f_j \,. \quad (2.9)$$

The operators A and $\tilde{A}$ are defined by compositions of tangential differential operators in M^{-1} and P, of potentials (see (2.5)) and of boundary operators. Note that for the explicit formulations the jump relations of potentials are to be incorporated. A more careful analysis as in (Chazarain and Piriou, 1982, Chapter 5) shows that A in (2.8) is a continuous linear mapping of the boundary Sobolev space V^s into W^s for any real s where

$$V^s = \prod_{j=0}^{m-1} H^{-m+\mu_j+s+\frac{1}{2}}(\Gamma;\mathbb{C}^N), \quad W^s = \prod_{j=0}^{m-1} H^{m-\mu_j+s-\frac{1}{2}}(\Gamma,\mathbb{C}^N).$$

These spaces are associated with the boundary value problem (2.1). Correspondingly, $\tilde{A}$ in (2.9) turns out to be continuous from V^s into itself.

For a regular C^∞ manifold Γ and $\sigma \geqslant 0$ the Sobolev space $H^\sigma(\Gamma)$ can be defined by the closure of $C^\infty(\Gamma)$ with respect to the norm $\|\cdot\|_\sigma$ defined by

$$\|f\|_\sigma^2 = \sum_{|l|\leqslant[\sigma]} \sum_{j=1}^{M} \int_{\Gamma_j} |\partial^l f|^2 \, do \, +$$

$$+ \sum_{|l|=[\sigma]} \sum_{j=1}^{M} \int_{\Gamma_j}\int_{\Gamma_j} \frac{|\partial^l f(x) - \partial^l f(y)|^2}{|x-y|^{n-1+2(\sigma-[\sigma])}} \, do_x \, do_y$$

where ∂^l denote all covariant derivatives and

$$\bigcup_{j=1}^{M} \Gamma_j$$

is a finite covering of Γ by regular charts. $[\cdot]$ denotes the

Gaussian bracket. For integer σ skip the last term. For $\sigma<0$ use the duality with respect to $L_2(\Gamma)$,

$$\|f\|_\sigma = \sup_{\|\phi\|_{-\sigma}\leqslant 1} \left|\langle f,\phi\rangle_{L_2(\Gamma)}\right| = \sup_{\|\phi\|_{-\sigma}\leqslant 1} \left|\int_\Gamma f\bar{\phi}\,do\right| .$$

For $n=2$ one can use a global chart for the closed curve Γ which corresponds to a 1-period parameter representation. With respect to this representation all functions become 1-periodic and may be represented by their Fourier series. Then the Sobolev norms can be introduced via

$$\|f\|_\sigma^2 = |\hat{f}(0)|^2 + \sum_{k\neq 0} |k|^{2\sigma} |\hat{f}(k)|^2 ,$$

where $\hat{f}(k)$ denotes the k-th Fourier coefficient.

Both types of equations (2.8) and (2.9) are common in practical applications. For instance, in the traction problem of linear elasticity $P^{(2)}$ with $m=1$ is the differential operator of the Navier system, u with $N=n=2$ or 3 is the displacement and R is the traction operator on Γ. Here $S=(1,0)$ with 1 the identity. w is the desired boundary displacement, (2.4) is the Betti formula and (2.5) the Somigliana identity. Equation (2.9) becomes a Cauchy singular integral equation of the second kind (Mikhlin and Prössdorf, 1986) whereas the equation (2.8) has a hypersingular integral operator of the first kind (Nedelec, 1982b). However, this operator is closely related to the energy of the elastic field (Hsiao *et al*, 1984; Hsiao and Wendland, to appear; McDonald *et al*, 1974; Mustoe *et al*, 1982; Nedelec, 1978b, 1982a, 1982b).

Besides elasticity, equations of the form (2.8), (2.9) are also used in thermoelasticity, electrostatics and electromagnetic fields, viscous flows and seepage problems, geodesy and many further applications. We shall give some references in Section 3.

Besides elasticity, also in the general case the operator A in (2.8) is closely related to the energy of the field.

THEOREM 1 (Costabel and Wendland, 1986): *Let the original boundary value problem provide an energy bilinear form* $\phi(u,v)$ *having the following properties:*

For every compact $K \subset \mathbb{R}^n$ *there exists a constant* c_K *such that for all* $u, v \in C_0^\infty(K)$

$$|\Phi(u,v)| \leq c_K \|u\|_{H^m(\mathbb{R}^n)} \|v\|_{H^m(\mathbb{R}^n)} , \tag{2.10}$$

$$\mathrm{Re}\,\phi(u,u) = \mathrm{Re}\Big\{ \int_{\Omega_1 \cup \Omega_2} (P^{(2m)} u)^T \bar{u}\, dx + \int_\Gamma \{(R\gamma u_1)^T \overline{S\gamma u_1} - (R\gamma u_2)^T \overline{S\gamma u_2}\}\, do \Big\}, \tag{2.11}$$

there exist positive constants λ, ε *and* $c \geq 0$ *such that for every* $u \in C_0^\infty(\overline{\Omega_1} \dot{\cup} \overline{\Omega_2})$ *with* $R\gamma u_1 = R\gamma u_2$ *the Gårding inequality holds:*

$$\mathrm{Re}\,\Phi(u,u) \geq \lambda \|u\|^2_{H^m(\overline{\Omega_1} \dot{\cup} \overline{\Omega_2})} - c\|u\|^2_{H^{m-\varepsilon}}. \tag{2.12}$$

Then assumptions (2.10) – (2.12) *imply coercivity of the first kind integral operator* A *on* Γ, *i.e. there exists a positive constant* β *and* $c \geq 0$ *such that for all* $w \in V^0(\Gamma)$

$$\mathrm{Re}\,\langle Aw, w\rangle_{L_2(\Gamma)} \geq \beta \|w\|^2_{V^0(\Gamma)} - c\|w\|^2_{V^{-\varepsilon}(\Gamma)} . \tag{2.13}$$

Consequently, in this case and in view of the Rellich imbedding theorem, the operator A can be written as

$$A = D - C \tag{2.14}$$

where D is a positive definite operator and C is compact. This decomposition is the decisive property for the convergence of boundary element methods. On the other hand, Theorem 1 also gives a mathematical justification of the Levin-Schwinger principles (Hönl *et al*, 1961, p.391 ff.) and also gives a modern interpretation of variational methods in combination with boundary integral equations as have been used by C. Gauss (1839, V p.232 ff.), G Fichera (1961), Fichera and Ricci (1976) and Ricci (1974).

3. GENERAL BOUNDARY INTEGRAL EQUATIONS AND SOME APPLICATIONS

Besides the boundary integral equations obtained by the direct method as in Section 2, far more kinds of boundary integral equations are known from various applications and from classical and modern potential theory (Colton and Kress, 1983; Hartmann, 1981; Kral, 1980; Kupradze *et al*, 1979; Müller, 1969; Muskhelishvili, 1963; Zabreyko *et al*, 1975). In all cases the solution to (2.1) is sought in the form of boundary potentials

$$u(x) = \sum_{k=0}^{2m-1} \int_{\Gamma} \left(\partial^k_{n(y)} G(x,y)\right) B_k w(y)\, do(y) \quad \text{for} \quad x \notin \Gamma \tag{3.1}$$

with tangential differential operators B_k and yet unknown boundary charges w. Then the boundary conditions of (2.1) applied to the *ansatz* (3.1) yield in general (with appropriate jump relations of potentials) boundary integral equations of the form

$$\begin{aligned} Aw(x) = B_1^* w(x) + \sum_{j,k=0}^{M} B^*_{2jk}\, \partial^j_{n(x)} \cdot \\ \cdot \int_{\Gamma\setminus\{x\}} \left(\partial^k_{n(y)} G(x,y)\right) B_k w(y)\, \chi(|x-y|)\, do(y) + \\ + \int_{\Gamma} k(x,y)\, w(y)\, do(y)\,, \end{aligned} \tag{3.2}$$

where B_1^* and B^*_{2jk} are also tangential differential operators, where $\chi(\rho)$ is a C_0^∞ cut-off function being identically 1 near the origin and where the remainder term has a sufficiently smooth kernel $k(x,y)$.

Again, the boundary integral operators are defined by compositions of differential operators, the trace operator on Γ and boundary potentials defined by the fundamental solution. For $\Gamma \in C^\infty$ we therefore have

THEOREM 2 (Dieudonné, 1978, Section 23.53; Chazarain and Piriou, 1982, Chapter 5; Eskin, 1980, Section 23.21): *The boundary integral operators in boundary integral equations to regular*

elliptic boundary value problems are classical pseudo-differential operators of integer orders.

This characterization has several consequences. First let us consider the local representation of operators A in (3.2). To this end consider the local representation of Γ with respect to tangential coordinates about $x \in \Gamma$:

$$y \in \Gamma : y = x + F(x,t)\vec{n}(x) + \sum_{j=1}^{n-1} t_j \vec{e}_j(x) \,, \quad t = (t_1, \cdots, t_{n-1}) \,. \tag{3.3}$$

$\vec{n}(x)$ denotes the exterior normal vector and $\vec{e}_j(x)$ a basis of tangential unit vectors to Γ and x. For $\Gamma \in C^\infty$ the functions in (3.3) are also C^∞ and this representation holds in some neighbourhood of x. Since Γ is compact, there exists a global positive constant γ such that (3.3) holds with $|t| < \gamma$. Let with (3.3)

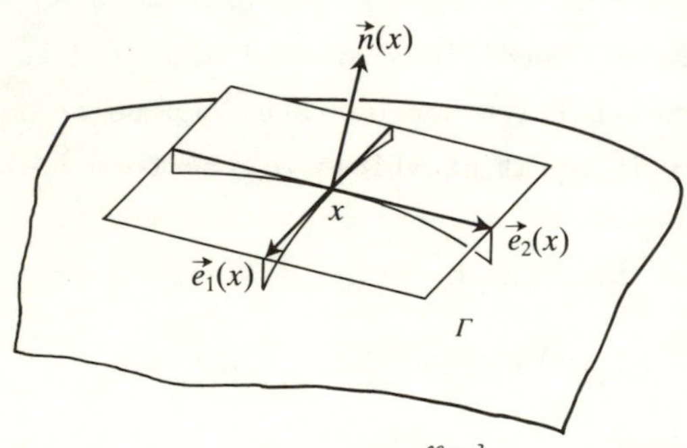

$$w(x;t) := w\Big(x + F\vec{n}(x) + \sum_{j=1}^{n-1} t_j \vec{e}_j(x)\Big) = w(y) \,, \quad \partial_t^l = \partial_{t_1}^{l_1} \cdots \partial_{t_{n-1}}^{l_{n-1}}$$

and let $\chi(\rho) = 0$ for $\rho \geqslant \frac{1}{2}\gamma$.

With the parameter representation (3.3) all the quantities in (3.2) can be expressed in terms of t (many of the necessary explicit relations can be found in (Martensen, 1968, p.75 ff.)) and the boundary integral operators finally take the form

$$Aw(x) = \sum_{|l| \leqslant p} b_l(x)\, \partial_t^l w(x;t) \Big|_{t=0}$$

$$+ \sum_{0 \leqslant \beta} \text{finite part} \int_{0<|t|} |t|^{\beta-p-n+1} f_{1\beta}\Big(x; \frac{t}{|t|}\Big) \chi(|t|)\, w(x;t)\, dt +$$

$$+ \sum_{\max\{0,-p-n+1\} \leqslant \beta} \int_{0<|t|} |t|^{\beta} \ln |t| \, f_{2\beta}\left(x; \frac{t}{|t|}\right) \chi(|t|) w(x;t) dt$$

$$+ \int_{\Gamma} k(x,y) w(y) do(y) \,. \qquad (3.4)$$

The integrals in (3.4) are understood as Hadamard finite part integrals, they also correspond to the representation of distributions (see (Eskin, 1980, Chapter I, Section 3)). $p \in \mathbb{Z}$ is the order of the operator. For negative p skip the first sum. The functions $f_{j\beta}$, $j=1,2$, depend in C^{∞} on the observation point x but also on $\frac{t}{|t|}$ tracing the $n-1$-dimensional unit sphere. They correspond to the "characteristics" in the case of Cauchy singular integrals, where $p=0$ (Mikhlin and Prössdorf, 1986) With polar coordinates, in case $p \geqslant 0$, (3.4) can be integrated by parts. Then one finds a regularized representation consisting of differentiations and at most Cauchy principal value singular integrals,

$$Aw(x) = \sum_{|l| \leqslant p} \Big\{ b_l^*(x) \, \partial_t^l w(x;t) \big|_{t=0} +$$

$$+ \sum_{0 \leqslant \beta} \text{p.v.} \int_{|t|>0} |t|^{\beta-n+1} f_{1l\beta}^*\left(x; \frac{t}{|t|}\right) \chi(|t|) \, \partial_t^l w(x;t) \, dt$$

$$+ \sum_{\max\{0-p-n+1\} \leqslant \beta} \int_{|t|>0} |t|^{\beta} \ln|t| f_{2l\beta}^*\left(x; \frac{t}{|t|}\right) \cdot$$

$$\cdot \chi(|t|) \, \partial_t^l w(x;t) \, dt \Big\}$$

$$+ \int_{\Gamma} k^*(x,y) \, w(y) \, do(y) \,. \qquad (3.5)$$

The above representations show clearly that the boundary integral equations will be classical Fredholm integral equations of the second kind with weakly singular kernels only in very special cases whereas in general we must accept rather general integro-differential operators. For the numerical treatment, in particular for boundary element methods, this fact implies many difficulties. For example, numerical integration for the above integrals is by no means satisfactorily developed yet (see e.g. (Schwab and Wendland, 1985)).

The analysis of the above classical pseudo-differential operators has been developed very extensively during the last decades by the use of the Fourier transform (see e.g. (Hörmander, 1966; Chazarain and Piriou, 1982; Eskin, 1980; Petersen, 1983; Taylor, 1981; Treves, 1980)). To this end consider the above operators for a moment in a fixed chart on Γ where all functions depend on the tangential local coordinates. Then Fourier transform and its reciprocity formula for $w \in C_0^\infty$ read as

$$w(x;t) = \int\limits_{\mathbb{R}^{n-1}} \int\limits_{\mathbb{R}^{n-1}} e^{2\pi i(t-\tau)\cdot\xi}\, w(x;\tau)\, d\tau\, d\xi$$

and the operator A takes the classical form of a *pseudo-differential operator*

$$Aw(x) = \int\limits_{\mathbb{R}^{n-1}} \int\limits_{\mathbb{R}^{n-1}} e^{-2\pi i\xi\cdot\tau}\, a(x;\xi)\, \chi(|\tau|)\, w(x;\tau)\, d\tau\, d\xi + \\ + \int\limits_{\Gamma} k^{**}(x,y)\, w(y)\, do(y), \tag{3.6}$$

where k^{**} is a sufficiently smooth kernel and where the Fourier transformed kernels of (3.5), i.e.

$$a(x;\xi) = \sum_{|l|\leq p} b_l(x)(i\xi)^l + \text{finite part} \sum_{0<|t|} e^{2\pi i t\cdot\xi} \cdot \\ \cdot \Big\{ \sum_{0\leq\beta} |t|^{\beta-p-n+1} f_{1\beta}\Big(x;\frac{t}{|t|}\Big) \\ + |t|^{\beta} \ln|t|\, f_{2\beta}\Big(x;\frac{t}{|t|}\Big)\Big\}\, \chi(|t|)\, dt$$

define a representation of the symbol associated with A. Since the integro-differential operators here are classical pseudo-differential operators of integer order p, the symbol a admits an asymptotic expansion as follows.

By $S^j_{1,0}$ let us denote the Kohn-Nirenberg class of symbols $q(x,\xi)$ in $C^\infty(\Gamma \times \mathbb{R}^n)$ with the property that for any multi-indices k, l there exist constants c_{lk} such that

$$|\partial_x^k \partial_\xi^l q(x,\xi)| \leq c_{lk}(1+|\xi|)^{j-|l|}.$$

Then a admits an asymptotic expansion

$$a(x;\xi) \simeq \sum_{0\leq\beta} a_\beta(x;\xi) \text{ with } a_\beta \in S^{p-\beta}_{1,0} ,$$

where a_β is positive homogeneous, i.e.

$$a_\beta(x;\lambda\xi) = \lambda^{p-\beta} a_\beta(x;\xi) \quad \text{for } \lambda \geq 1 \quad \text{and} \quad |\xi| \geq 1$$

and where

$$a(x;\xi) - \sum_{\beta=0}^{N} a_\beta(x;\xi) \in S^{p-N-1}_{1,0}$$

for any integer $N \geq 0$. The first term $a_0(x;\xi)$ is called the *principal symbol* of A in (3.5), (3.6).

In the two-dimensional case $n=2$, the curve Γ can be given by a 1-periodic *global* parameter representation

$$y = X(t) .$$

For systems of curves introduce corresponding vector valued functions (Arnold and Wendland, 1983). Here Fourier expansions yield for the representation (3.6) the form

$$Aw(\tau) = \sum_{l\in\mathbb{Z}} a(\tau,l)\, \hat{w}(l)\, e^{2\pi i l\tau} + \int_0^1 k(\tau,t)\, w(t)\, dt \qquad (3.7)$$

where the principal symbol a_0 is characterized by its homogeneity

$$a_0(\tau,l) = |l|^p a_0(\tau, \operatorname{sign} l) \quad \text{for } l \neq 0. \qquad (3.8)$$

The specific operators which have been obtained by the direct method in the equations (2.8) are given by the matrix $A = ((A_{jk}))$ where A_{jk} are ($N \times N$ matrix) pseudo-differential operators of orders $\mu_{j-1} + \mu_{k-1} + 1 \dot{-} 2m$ on Γ, $j,k = 1,\cdots,m$. Correspondingly, in $\tilde{A} = ((\tilde{A}_{jk}))$ of (2.9) we find ($N \times N$ matrix) pseudo-differential operators of orders $\mu_{k-1} - \mu_{j-1}$ (see (Costabel and Wendland, 1986)).

For illustration let us classify some of the common operators in applications. Some of the corresponding principal symbols can be found in (Wendland, 1985b). Many explicit applications can also be found in (Brebbia *et al*, 1984) and (Wendland, 1983).

Operators of order 0 *which become Fredholm integral equation operators of the second kind with compact operators*: Flow problems in ideal flows and corresponding panel methods (Martensen, 1959; Wagner and Urban, 1985), viscous flow problems (Hebeker, 1986a, Hsiao and Kress, 1985), Darcy flows (Bischoff, 1982; Liggett and Liu, 1983), acoustics and electromagnetic fields (Colton and Kress, 1983; Goldstein, 1979; Gregoire *et al*, 1974; Ha Duong, 1979; Kleinman and Roach, 1974; Kupradze, 1956; Müller, 1969; Poggio and Miller, 1973; Schenck, 1968; Wendland, 1983).

Operators of order 0 *which define Cauchy singular integral equations*: Elasticity and thermoelasticity (Cruse, 1973; Hsiao and Wendland, to appear; Jentsch, 1974, 1975; Kupradze, 1965; Kupradze *et al*, 1979; Maul, 1976; Rieder, 1962; Rizzo, 1967), geodesy (Stock, 1985), subsonic compressible flows treated by the Prandtl-Glauert transform (Maskew, 1982) and Stokes flows (Hebeker, 1984).

Operators of order -1 *which define Fredholm integral equations of the first kind*: For $n=2$ the method goes back to Fichera (1961) and has been extended to rather general situations in (Costabel, 1984). Examples are Symm's integral equation of conformal mapping (Christiansen, 1981; Gaier, 1976; Hayes *et al*, 1972; Hoidn, 1983; Hsiao *et al*, 1984; Symm, 1963, 1967; Wendland, 1983); ideal flows (Djaoua, 1981), seepage problems (Bischoff, 1982; Liggett and Liu, 1983), viscous flow problems and Stokes flows (Fischer, 1980, 1982; Fischer *et al*, 1985; Fischer and Rosenberger, 1985; Hsiao *et al*, 1980, 1984; Hsiao and MacCamy, 1973; Hsiao and Wendland, 1977; Zhu, 1982): electrostatics and electromagnetic problems (Bendali, 1982; Costabel and Stephan, 1985; Le Roux, 1974a, 1974b; MacCamy and Stephan, 1983; Nedelec, 1978a; Nedelec and Planchard, 1973; Stephan, 1983); acoustics (Agranovich, 1977; Djaoua, 1975; Filippi, 1977a. 1977b; Giroire, 1978; Giroire and Necelec, 1978; Stephan, 1983; Wendland, 1983); elasticity (Bolteus and Tullberg, 1981; Brebbia, 1978; Brebbia *et al*, 1984; Christiansen and Hansen, 1975; Cruse, 1973; Hartmann, 1981; Hsiao *et al*, 1984; Hsiao and Wendland, 1977, to appear; Jaswon and Symm, 1977; Kuhn and Möhrmann, 1983; Nedelec, 1977b; Niwa *et al*, 1982; Rizzo, 1967); plate bending (Hartmann, 1981; Hartmann and Zotemantel, 1986; Hsiao *et al*, 1984; Hsiao and MacCamy, 1973) and torsion problems (Jaswon and Symm, 1977; Mehlhorn, 1970).

Operators of order 1 *which define hypersingular equations*: Acoustics and electrostatics (Agranovich, 1977; Aziz *et al*, 1980; Costabel and Stephan, 1983; Durand, 1983; Filippi, 1977b; Giroire, 1978; Giroire and Nedelec, 1978; Ha Duong, 1979; Nedelec, 1982a; Stephan, 1983; Wendland, 1983; Wolfe, 1980); elasticity (Bamberger, 1983; Bamberger and Ha Duong, 1985; Hartmann, in preparation; Hsiao and Wendland, to appear; Jaswon, 1981; Mustoe *et al*, 1982; Nedelec, 1980, 1982a, 1982b; Watson, 1979, 1982) and Stokes flows (Hebeker, 1986b).

4. VARIATIONAL FORM AND STRONG ELLIPTICITY

Similar to direct finite difference or finite element approximations of the boundary value problem (2.1), all approximation methods for boundary integral equations and boundary element methods and the corresponding analysis are based on the *weak* and *the variational formulation of boundary integral equations.* As usual, a distribution $v \in H^{\sigma}(\Gamma)$ with some real σ is called a *weak solution* of equation

$$Av = f \quad \text{iff} \quad (Av, \phi)_{L_2(\Gamma)} = (f, \phi)_{L_2(\Gamma)} \tag{4.1}$$

for all test functions $\phi \in C^{\infty}(\Gamma)$. Since the boundary integral operators are considered to be systems of pseudo-differential operators we have the following continuity property.

LEMMA 3 (Hörmander, 1966): *For a pseudo-differential operator* A *of real order* β *on* Γ *and any real Sobolev index* σ *we have continuity of* $A : H^{\sigma}(\Gamma) \to H^{\sigma-\beta}(\Gamma)$:

$$\|Av\|_{H^{\rho}(\Gamma)} \leq c(A,\Gamma,\sigma)\|v\|_{H^{\sigma}(\Gamma)} \quad \text{with} \quad \rho = \sigma - \beta\,. \tag{4.2}$$

Correspondingly, if $A = ((A_{lk}))$ is given by a matrix of pseudo-differential operators A_{lk} of orders $\alpha_{lk} \in \mathbb{R}$, $l,k = 1,\cdots,m$ then these orders and additional real indices $\sigma_1,\cdots,\sigma_m$ and $\rho_1,\cdots,\rho_m$ can always be chosen such that

$$\sigma_k = \alpha_{lk} + \rho_l \quad \text{for } l,k = 1,\cdots,m. \tag{4.3}$$

(Note that α_{lk} can be increased by adding zero terms to the symbol's expansion.) Then (4.1) also holds with

$$H^{\rho} := \prod_{l=1}^{m} H^{\rho_l}(\Gamma) \quad \text{and} \quad H^{\sigma}, \text{ correspondingly.}$$

Of course, (4.1) also holds if every A_{lk} by itself is a $N \times N$ matrix of equal-order pseudo-differential operators.

A weak solution v is called a *variational solution* iff $v \in H^{\sigma}(\Gamma)$ and

$$\begin{aligned}(Av,w)_{H^{(\rho+\sigma)/2}(\Gamma)} &:= \sum_{l,k=1}^{m} (A_{lk}v_k, w_l)_{H^{(\rho_l+\sigma_l)/2}(\Gamma)} \\ &= \sum_{l=1}^{m} (f_l, w_l)_{H^{(\rho_l+\sigma_l)/2}(\Gamma)} \\ &=: (f,w)_{H^{(\rho+\sigma)/2}(\Gamma)}\end{aligned} \tag{4.4}$$

for all $w \in H^{\sigma}(\Gamma)$. Note that the bilinear form on the left hand side of (4.4) extends continuously to $H^{\sigma-\alpha}(\Gamma) \times H^{\sigma+\alpha}(\Gamma)$ for any $\alpha \in \mathbb{R}$ with $\sigma \pm \alpha = (\sigma_1 \pm \alpha,\cdots,\sigma_n \pm \alpha)$ by duality. As for the class of operators in Theorem 1, also here we want to know whether for the general case also a Gårding inequality is valid or not. However, the corresponding concept requires a slight generalization for applicability. To this end let Λ^α denote the α-th order Bessel operator having the principal symbol $|\xi|^\alpha$ for $|\xi| \geq 1$ and defining ismorphisms $\Lambda^\alpha : H^\sigma(\Gamma) \rightarrow H^{\sigma-\alpha}(\Gamma)$. For $n=2$ e.g. define using the Fourier series

$$\Lambda^\alpha u(\tau) = \hat{u}(0) + \sum_{k \neq 0} |k|^\alpha \hat{u}(k)\, e^{2\pi i k \tau}.$$

In accordance with (Stephan and Wendland, 1976) we call the pseudo-differential operator A *strongly elliptic* iff to the principal symbol matrix $a_0(x,\xi)$ there exist a C^∞ matrix-valued function $\Theta(x) = ((\Theta_{jl}))$ and a constant $\gamma_0 > 0$ such that for all $x \in \Gamma$, $\zeta \in \mathbb{C}^{N \cdot m}$ and $|\xi| = 1$

$$\operatorname{Re} \zeta^T (\Theta(x)\, a_0(x;\zeta))\bar{\zeta} \geq \gamma_0 |\zeta|^2 . \tag{4.5}$$

THEOREM 4 (Hörmander, 1966): *If A is a strongly elliptic pseudo-differential operator with orders* (4.3) *then there exists a constant $\gamma_1 > 0$ and a linear compact operator $C : H^\sigma \rightarrow H^\rho$ such that the Gårding inequality holds in the form*

$$\operatorname{Re} ((\Lambda^{-\rho}\Theta\Lambda^\rho) Aw, w)_{H^{(\rho+\sigma)/2}(\Gamma)} \geq \gamma_1(\rho,\sigma) \|w\|^2_{H^\sigma(\Gamma)} - \operatorname{Re}(Cw, w)_{H^{(\rho+\sigma)/2}(\Gamma)} \tag{4.6}$$

where

$$\Lambda^{-\rho}\Theta\Lambda^\rho := \left((\Lambda^{-\rho_j}\Theta_{jl}\Lambda^{\rho_l}) \right).$$

Note that here we need the rather elaborate theory of pseudo-differential operators whereas Theorem 1 is based on the coercivity of the original boundary value problems. As a consequence of Theorem 4 it is true that

THEOREM 5: *If A is a strongly elliptic pseudo-differential operator with* (4.3) *then the classical Fredholm alternative is valid for the equation*

$$Av = f \in H^\rho(\Gamma)$$

for $v \in H^{\sigma}(\Gamma)$.

For the proof apply Nikolski's theorem (Nikolski, 1943) to

$$\Lambda^{-\rho}\,\Theta\,\Lambda^{\rho} A = D - C \tag{4.7}$$

in $H^{(\rho+\sigma)/2}$, since D is positive definite due to (4.6), then use ellipticity of equations (4.7) (Treves, 1980, Chapter III). The property (4.5) provides a criterion for the validity of Gårding's inequality for arbitrary boundary integral operators. However, the explicit global construction of $\Theta(x)$ to given $a_0(x;\xi)$ is in general by no means trivial. In case $n=2$, however, a very simple criterion by Prössdorf, Schmidt and Rathsfeld is available.

THEOREM 6 (Prössdorf and Rathsfeld, 1985; Prössdorf and Schmidt, 1981b): *For* $n=2$ *the pseudodifferential operator* A *with* (4.3) *on the closed curve* Γ *is strongly elliptic if and only if*

$$\det\left(\mu a_0(x,1) + (1-\mu) a_0(x,-1)\right) \neq 0$$

for all $x \in \Gamma$ *and all real* $\mu \in [0,1]$.

For the proof consider the operator

$$\Lambda^{\rho} A \Lambda^{\sigma} = \left((\Lambda^{\rho_l} A_{lk} \Lambda^{\sigma_k}) \right)$$

whose principal part is a system of Cauchy singular integral operators, for which the result has been shown in (Prössdorf and Rathsfeld, 1985; Prössdorf and Schmidt, 1981b).

5. GALERKIN METHODS

The variational formulation of the boundary integral equations (4.4) is best suited for Galerkin's method. Let us introduce a family of finite dimensional nested subspaces

$$H_h \subset H^{\sigma} = \prod_{k=1}^{m} H^{\sigma_k}(\Gamma) \tag{5.1}$$

where h denotes a parameter associated to the dimension M of H_h by $h = cM^{-(n-1)}$. Here and in the sequel c will denote a generic constant independent of h. Then the well-known Galerkin procedure for (4.4) is to find $w_h \in H_h$ by solving the quadratic finite system of linear equations that are equivalent to

$$(u_h, Aw_h)_{H^{(\rho+\sigma)/2}(\Gamma)} = (u_h, f)_{H^{(\rho+\sigma)/2}(\Gamma)} \quad \text{for all } u_h \in H_h. \tag{5.2}$$

We assume that the bilinear form in (4.4) is continuous on H^σ.

If the finite dimensional equations (5.2), are inverse stable then the approximate solutions w_h converge quasi optimally to the desired solution w.

LEMMA 6: (Cea's lemma (Cea, 1964). See also (Mikhlin, 1962).) *If the stability condition* (Babuška, 1970)

$$\sup_{u_h \in H_h} \frac{|(u_h, Av_h)_{H^{(\rho+\sigma)/2}(\Gamma)}|}{\|u_h\|_\sigma} \geq \gamma \|v_h\|_\sigma \tag{5.3}$$

holds for all $v_h \in H_h$ *where* $\gamma > 0$ *then*

$$\|w - w_h\|_\sigma \leq c \inf_{u_h \in H_h} \|w - u_h\|_\sigma . \tag{5.4}$$

The proof is straightforward, combining continuity of A, inverse stability and unique solvability of (5.2). In view of (5.4) we shall require an approximation property of H_h,

$$\lim_{h \to 0} \inf_{u_h \in H_h} \|w - u_h\|_\sigma = 0 . \tag{D}$$

For actual operators it remains to show (5.3) which usually requires detailed analysis involving the approximating spaces H_h and properties of A.

THEOREM 7: (Stephan and Wendland, 1976). *Let* $A: H^\sigma \to H^\rho$ *be bijective and continuous. Then there exist* $h_0 > 0$ *and* $c > 0$ *such that the Galerkin equations* (5.2) *are uniquely solvable for every* $0 < h \leq h_0$ *and the approximate solutions* w_h *satisfy* (5.4), *provided one of the additional properties holds* :

(i) A *satisfies* (4.6) *with* $\Theta_{jl} = \delta_{jl}$ *and* $C = 0$.

(ii) A *satisfies* (4.6) *with* $\Theta_{jl} = \delta_{jl}$ *and* H_h *satisfies* (*D*).

(iii) *A satisfies (4.6) and H_h satisfies in addition to (D) the following approximation property for any given $\phi \in C^\infty(\Gamma)$:*
There exist $\delta > 0$, $c_\phi \geq 0$ such that

$$\|\phi v_h - P_h \phi v_h\|_\sigma \leq c_\phi h^\delta \|v_h\| \qquad (C)$$

for all $v_h \in H_h$ where P_h denotes the $H^{(\rho+\sigma)/2}$ projection of H^σ onto H_h.

(The property (C) underlies the discrete localization principle (Arnold and Wendland, 1985; Prössdorf, 1984).)

Sketch of the proof: Set $D = \Lambda^{-\rho}\Theta\Lambda^\rho A + C$.
Then D is in all three cases a H^σ elliptic bilinear form due to (4.6). Now it remains to construct for any given $v_h \in H_h$ some $u_h \in H_h$ such that

$$\left|(u_h, Av_h)_{H^{(\rho+\sigma)/2}(\Gamma)}\right| \geq \gamma \|v_h\|_\sigma \|u_h\|_\sigma \qquad (5.5)$$

holds.

(i) Take $u_h = v_h$. Then

$$|(v_h, Av_h)| \geq \mathrm{Re}\,(v_h, Dv_h) \geq \gamma_0 \|v_h\|_\sigma^2$$

yields (5.3). Note that the mapping G_h defined by solving (5.2), i.e.

$$G_h w := w_h = (P_h D P_h)^{-1} P_h D w$$

is a stable projection onto H_h, $\|G_h w\|_\sigma \leq c\|w\|_\sigma$ where c is independent of h.

(ii) Take $u_h = [I - G_h D^{-1} C]\, v_h$.
Since C is compact and $\lim_{h\to 0} G_h v = v$ for every $v \in H^\sigma$ due to assumption (D) we find that $G_h D^{-1} C$ converges to $D^{-1} C$ in the operator norm associated with $H^\sigma \to H^\sigma$. (See (Anselone, 1971; Bruhn and Wendland, 1967, Hilfssatz 3).) Hence $[I - G_h D^{-1} C]^{-1}$ exists for almost all h and converges to $A^{-1} D$ in the same operator norm. This yields stability and (5.5).

(iii) Take $u_h = P_h \Lambda^\rho \Theta^* \Lambda^{-1} [I - G_h D^{-1} C]\, v_h$.

Then $\|u_h\|_\sigma \leqslant c\,\|v_h\|_\sigma$. Inserting u_h into the bilinear form gives with the compactness of $(\Lambda^\rho \Theta^* \Lambda^{-\rho} - \Theta^*)$ (Taylor, 1981, Theorem 5.1), the above arguments and with (C)

$$\begin{aligned}
\mathrm{Re}(u_h, Av_h)_{(\rho+\sigma)/2} &= \mathrm{Re}((P_h - I)(\Lambda^\rho \Theta^* \Lambda^{-\rho} - \Theta^*)[\cdots]v_h, Av_h) \\
&\quad + \mathrm{Re}((P_h - I)\,\Theta^*[\cdots]v_h, Av_h) \\
&\quad + \mathrm{Re}(\Lambda^\rho \Theta^* \Lambda^{-\rho}[\cdots]v_h, Av_h) \\
&\geqslant -c\|(P_h - I)(\Lambda^\rho \Theta^* \Lambda^{-\rho} - \Theta^*)\|\,\|v_h\|_\sigma^2 \\
&\quad - ch^\delta \|v_h\|_\sigma^2 + \mathrm{Re}([\cdots]v_h, D[I - G_h D^{-1} C]v_h) \\
&\geqslant \gamma_0 \|v_h\|_\sigma^2 - (o(1) + ch^\delta)\|v_h\|_\sigma^2 \\
&\geqslant \gamma\,\|v_h\|_\sigma\,\|u_h\|_\sigma
\end{aligned}$$

where $\gamma > 0$ for h small enough, i.e. (5.5).

5.1 *Galerkin Boundary Element Methods*

Let us specialize the family H_h by choosing a regular $S_h^{d,r}$ family of boundary finite elements in the sense of Babuška and Aziz (1972), $r<d$. Further let us restrict n to 3 or 2. For $n=3$, $S_h^{d,r}$ can be defined by using triangulations of the parameter domains associated with the charts of a finite atlas for Γ. On the triangulations we either use piecewise polynomial finite elements which are transplanted onto Γ by the use of the corresponding chart's application. Or we also approximate Γ by Lagrangian elements and define isoparametric boundary elements similar to shell elements as introduced by Nedelec (1976). For $n=2$ we use the global parameter representation and introduce 1-dimensional 1-periodic piecewise $(d-1)$-degree polynomial splines on a family of partitions $\{t_0 = 0 < t_1 < \cdots < t_M = 1\}$ of the unit interval, $h = \max_{j=1,\cdots,M}(t_j - t_{j-1})$. Again with the regular application of the parameter representation we transplant the splines onto Γ or, as in three dimensions, we also approximate Γ by a spline interpolation of the parameter representation (Le Roux, 1974a).

Then $H_h = S_h^{d,r}$ provides the following properties:

Approximation property: $S_h^{d,r} \subset H^r(\Gamma)$. *For any* $\tau \leq r$, $\tau \leq \beta \leq d$ *there exists* $c > 0$ *and for every* $v \in H^\beta(\Gamma)$ *there exists a family* $v_h \in S_h^{d,r} = H_h$ *such that*

$$\|v - v_h\|_\tau \leq c h^{\beta-\tau} \|v\|_\beta . \tag{A}$$

Since C^∞ is dense in all $H^\rho(\Gamma)$, (*A*) implies (*D*) for $r \geq \sigma$. (If $\sigma = (\sigma_1, \cdots, \sigma_m)$ then take m $S_h^{d,r}$ systems.) Note that in general the trial functions in H_h do not need to be continuous.

If the triangulations or partitions, respectively, are regular then the $S_h^{d,r}$ systems also provide the

Inverse assumption: *For* $\tau \leq \beta \leq r$ *there exist constants* M *such that for all* $v_h \in H_h = S_h^{d,r}$,

$$\|v_h\|_\beta \leq M h^{\tau-\beta} \|v_h\|_\tau . \tag{I}$$

LEMMA 8 (Arnold and Wendland, 1985; Nitsche and Schatz, 1972; Strang, 1972): *Properties* (*A*) *and* (*I*) *imply* (*C*).

Clearly, these properties provide with Theorem 7 and Cea's lemma error estimates of optimal order by using (*A*) for estimating the right hand side in (5.4). Since we apply Galerkin's method to strongly elliptic problems in scales of Sobolev spaces, these results can be further extended by using the Aubin-Nitsche duality arguments. For simplicity let us now assume that in (4.3) there holds in particular

$$\alpha_{lk} = \sigma_k + \sigma_l , \quad \sigma_k = -\rho_k = \tfrac{1}{2}\alpha_{kk}, \qquad k = 1, \cdots, m , \tag{5.6}$$

and, hence, the scalar products in (5.2) are the usual L_2-scalar product. (The more general case can be handled correspondingly.)

THEOREM 9 (Arnold and Wendland, 1985; Hsiao and Wendland, 1981; Nedelec, 1977b; Stephan and Wendland, 1976). *Let* A *be given as a bijective system of strongly elliptic pseudodifferential operators whose orders satisfy* (5.6). *Let the boundary element systems provide* (*A*) *and* (*I*) *and let* $-d \leq \sigma_k \leq r$. *Then there exists* $h_0 > 0$ *such that for all* $0 < h \leq h_0$ *the Galerkin equa-*

tions (5.2) *are uniquely solvable and the Galerkin solutions* w_h *satisfy the optimal order error estimates*

$$\|w-w_h\|_\tau \leqslant ch^{\beta-\tau}\|w\|_\beta \tag{5.7}$$

for $\alpha_{kk}-d \leqslant \tau \leqslant \beta \leqslant d$, $\tau \leqslant r$, $-r \leqslant \beta$, $k=1,\cdots,m$.

Remarks: These estimates can be extended to componentwise estimates with different degree boundary elements (Wendland, 1983), but we omit here these tedious details.

In connection with the inverse assumption (I), Theorem 9 yields also asymptotic estimates for the conditioning of the discrete equations (Wendland, 1985a, Lemma 6.1).

For $n=2$ and splines, the index $\tau < d-\frac{1}{2}$ does not need to be an integer (Arnold and Wendland, 1981).

Since the boundary finite elements also provide local approximation properties the following result by Braun can also be shown for the above class of problems.

THEOREM 10 (Braun, 1985): *Let* A *and the boundary element system satisfy the assumptions of Theorem* 9. *Let* $\Gamma_0 \subset \Gamma_1 \subset \Gamma$ *be compact subsets with* $\Gamma_0 \subset \mathring{\Gamma}_1$ *where the latter denotes the open interior. Then we also have local superapproximation of the Galerkin solution, i.e.*

$$\|w-w_h\|_{H^\tau(\Gamma_0)} \leqslant ch^{\beta-\tau}\left\{\|w\|_{H^\beta(\Gamma_1)} + \|w\|_{H^\tau(\Gamma)}\right\}.$$

For practical computations pointwise error bounds are often preferable.

THEOREM 11 (Rannacher and Wendland, 1985): *Let* A *and the boundary element systems satisfy the assumptions of Theorem* 9 *and let in addition* $\sigma_k \leqslant 0$. *On each element* T *of the triangulation or partition, respectively, in addition there holds for all* $v_h \in H_h$

$$\|v_h\|_{H^d(K)} \leqslant c\|v_h\|_{H^{d-1}(K)}$$

and

$$\|v_h\|_{H^j(K)} \leqslant ch^{l-j}\|v_h\|_{H^l(K)} \quad \text{for } 0 \leqslant l \leqslant j \leqslant d-1$$

and $l, j, d \in \mathbb{Z}$.

Then the Galerkin solutions w_h *satisfy the pointwise error estimates*

$$|w(x) - w_h(x)| \leq c h^d \left(\log \frac{1}{h}\right)^{\frac{n}{2}-1} \|w\|_{W^{d,\infty}(\Gamma)}.$$

With the approximate boundary charge w_h the original potential field $u(x)$ can be computed via (3.1). Away from Γ one finds superconvergence since the kernels in (3.1) are real analytic for $x \neq y$ for real analytic coefficients of $P^{(2m)}$ and since the tangential operators B_k can be integrated by parts. Then

$$u_h(x) = \int_\Gamma \left\{ \sum_{k=0}^{2m-1} B_k' \partial_{n(y)}^k G(x,y) \right\} w_h(y)\, do(y)$$

defines the approximate field where B_k' denotes the operator adjoint to B_k.

LEMMA 12: *Let all assumptions of Theorem* 9 *be satisfied. Then we have for the potential fields*

$$|\partial^l u(x) - \partial^l u_h(x)| \leq c(l, \text{distance } (x,\Gamma))\, h^{\beta + d - 2 \max\{\sigma_k\}} \|w\|_{H^\beta(\Gamma)}.$$

5.2 *Convergence and Strong Ellipticity*

As we have seen, strong ellipticity is a rather powerful *sufficient* condition for optimal order convergent Galerkin methods, and Theorem 7 is rather general being valid not only for boundary element methods but also in case (i) and (ii) for global approximations as for Fourier series spectral methods in case $n = 2$. Condition (*C*), however, seems to be connected with H_h having a basis with "small supports" as *B*-splines.

If we require convergence in H^σ for all Galerkin methods with *all* possible $H_h \to H^\sigma$ then A must be coercive satisfying (4.6) with $\Theta = 1$ due to a result by Vainikko (1965), i.e. for A being a pseudodifferential operator, A must be strongly elliptic with $\Theta = 1$.

On the other hand, if P_h and A commute modulo

collectively compact operators then the corresponding Galerkin method converges already for bijective $A: H^{\sigma}(\Gamma) \to H^{\sigma}(\Gamma)$. For $n=2$, on a curve this is the case for the Fourier series (spectral) method applied to elliptic equations with a principal symbol $a_0(x;\xi) = a_0(\xi)$, i.e. with a principal part having "constant coefficients". For singular integral equations ($a=0$) with variable coefficients one needs besides ellipticity and bijectivity of A also that all the right and the left factorization indices vanish (Gohberg and Fel'dman, 1974; Lamp *et al*, 1984; Prössdorf and Silberman, 1977). Thus, in these cases the commutator property is a rather strong connection between A and H_h.

For $n=2$ and the spline Galerkin method there is almost no relation between A and H_h due to the following theorem by Schmidt which goes back to (Prössdorf and Schmidt, 1981a).

THEOREM 13 (Schmidt, 1984): *Let A be one given pseudodifferential operator of order α on a simple closed curve Γ and let the spline Galerkin method with regular piecewise linears $r = 1 = d-1$ converge with optimal order. Then A is strongly elliptic.*

Hence, for two-dimensional boundary element methods *of optimal order convergence* strong ellipticity of A is even *necessary*.

5.3 *Iterated Smoothing and Superconvergence for Fredholm Integral Equations of the Second Kind*

In the special case $\alpha_{jk} = 0$ the equations (4.7) with strongly elliptic A are *equivalent* to Fredholm integral equations of the form

$$v + Qv + \frac{\gamma C v}{||D||^2} = \frac{\gamma}{||D||^2} f \quad \text{in } H^{\sigma}(\Gamma) \tag{5.8}$$

where

$$Q = \left\{ \frac{\gamma}{||D||^2} D - I \right\}$$

is a contraction with

$$||Q||^2 \leqslant 1 - \frac{\gamma^2}{||D||^2} < 1$$

(see (Ciarlet, 1978, p.9; Stephan and Wendland, 1982, Satz 1.3)).

Fredholm integral equations of this form arise e.g. in the classical boundary integral equation method for the Laplacian and the Helmholtz equation if Γ is nonsmooth, see e.g. (McLean, 1986; Wendland, 1981) and references therein.

If $Q=0$ in (5.8) then the compactness of C can be used for improving the quality of an approximation (Brakhage, 1960) including the order of convergence (Wendland, 1968). To this end let us define by

$$v_{(l+1)} := -Cv_{(l)} + f \quad \text{with} \quad v_{(0)} := v_h ,$$

$l = 0, \cdots, \rho - 1$, the ρ-times iterated approximation.

THEOREM 14 (Wendland, 1983, Theorem 3.2): *Let* $C: H^{\tau} \to H^{\tau+\kappa}$ *with* $\kappa > 0$ *be continuous where* $-d \leqslant \tau \leqslant \frac{n}{2} + \varepsilon - \kappa - \frac{1}{2}$, *i.e. in particular a pseudodifferential operator of order* $-\kappa$. *Let* $v_{(\rho)}$ *be the* ρ*-times iterated approximation with* $\rho > (\frac{n}{2} + d)\,\kappa^{-1}$. *Then we have pointwise superconvergence*

$$\|v - v_{(\rho)}\|_{L_\infty(\Gamma)} \leqslant ch^{2d} \|v\|_{H^d(\Gamma)} . \tag{5.9}$$

For the proof, the super approximation result (5.7) with $\beta = d = -\tau$ can be exploited in connection with smoothing of C^{ρ}. (5.9) is well known for classical Fredholm integral equations of the second kind (Brakhage, 1960; Chandler, 1979; Chatelin, 1983; Graham *et al*, 1985; Sloan and Thomée, 1986).

5.4 *Cauchy Singular Integral Equations*

Note that the results of Theorems 9 – 13 apply in particular to the spline Galerkin approximation of systems of Cauchy singular integral equations on a closed curve Γ without self-intersection,

$$Aw = b(x)\,w(x) + c(x)\,\frac{1}{\pi i} \int_\Gamma \frac{w(s)}{y(s) - (x_1 + ix_2)}\,dy(s) + \int_\Gamma k(x, y)\,w(s)\,ds_y = f(x) \tag{5.10}$$

where $y(s) = X_1(s) + iX_2(s)$ denotes the complex parameter representation of Γ. Here $\alpha_{jk} = 0$, $a_0(x,\xi) = b(x) + c(x)\frac{\xi}{|\xi|}$ for $|\xi| \geq 1$, and the strong ellipticity condition (4.8) reads as

$$\det(b(x) + \lambda c(x)) \neq 0 \text{ for all } x \in \Gamma \text{ and } \lambda \in [-1,1]. \quad (5.11)$$

Theorem 9 then assures for $\beta = d = -\tau$, i.e. piecewise polynomials of order $d-1$ the asymptotic convergence

$$\|w - w_h\|_{-d} \leq ch^{2d} \|w\|_d$$

in the $H^{-d}(\Gamma)$ norm and Theorem 11 gives pointwise convergence

$$|w(x) - w_h(x)| \leq ch^d \|w\|_{W^{d,\infty}(\Gamma)}.$$

Schmidt's result, Theorem 13 implies that for the *optimal order convergence* strong ellipticity (5.11) is even necessary. The example of a divergent spline collocation applied to the *non*-strongly elliptic equation (5.10) with $b = 0$, $c = 1$ in (Ivanov, 1976, p.204 ff.) indicates that also the spline Galerkin method may completely fail if strong ellipticity (5.11) is violated.

6. GALERKIN-PETROV METHODS, COLLOCATION AND SPLINE COLLOCATION ON CURVES

6.1 *The Galerkin-Petrov Method*

Many of the discretization methods including collocation can be seen as special cases of the rather general *Galerkin-Petrov method*: Find a trial function $w_h \in H_h$ such that for all test functions $u_h \in T_h$

$$(u_h, Aw_h)_{L_2(\Gamma)} = (u_h, f)_{L_2(\Gamma)}. \quad (6.1)$$

Here H_h denotes the trial space and T_h denotes the test space with $\dim H_h = M = \dim T_h$, whereas in general H_h and T_h can be different.

As in Lemma 6, also here stability implies convergence.

LEMMA 14 (Babuška, 1970): *If* $H_h \subset H^\sigma(\Gamma)$, *if*

$$|(u_h, Aw_h)_{L_2(\Gamma)}| \leq c\|u_h\|_{T_h} \|w_h\|_\sigma$$

for all $u_h \in T_h$ *and* $w_h \in H_h$ *and if*

$$\sup_{u_h \in T_h} \frac{|(u_h, Av_h)_{L_2(\Gamma)}|}{\|u_h\|_{T_h}} \geq \gamma \|v_h\|_\sigma \qquad (6.2)$$

for all $v_h \in H_h$ *where* $\gamma > 0$, *then*

$$\|w - w_h\|_\sigma \leq c \inf_{v_h \in H_h} \|w - v_h\|_\sigma .$$

The general Galerkin-Petrov method can be used for a large variety of problems; our strongly elliptic boundary integral equations form only a rather restricted subclass. Here in case $n = 2$ the choice of spline trial spaces H_h and trigonometric polynomial test spaces T_h yields an exponentially convergent scheme (Arnold, 1983). In (Ruotsalainen and Saranen, 1986) the trial space H_h consists of Dirac functions and the test space T_h of splines. In the least squares methods, $T_h = AH_h$.

6.2 *Point Collocation*

For point collocation methods, T_h is the space of Dirac functionals associated with M appropriately chosen collocation points $x_l \in \Gamma$, $l = 1, \cdots, M$, $\Xi := \{x_1, \cdots, x_M\}$. Then the collocation procedure for (4.1) is to find the vector coefficients γ_j of the approximate solution

$$w_h(x) = \sum_{m=1}^{M} \gamma_m \mu_m(x) , \quad x \in \Gamma \qquad (6.3)$$

where μ_m denotes a basis of H_h (for fixed h), by solving the collocation equations

$$\sum_{m=1}^{M} (A\mu_m)(x_l)\, \gamma_m = f(x_l) , \quad x_l \in \Xi . \qquad (6.4)$$

But contrary to Galerkin's procedure, for the collocation (6.4) convergence results for our class of equations are known as yet only in the case of $n = 2$, i.e. for one-dimensional boundary integral equations, whereas for higher dimensions convergence has been shown only in the special case of Fredholm integral equations of the second kind (5.8) under additional restrictions on Q.

Now let us consider first the two-dimensional case $n = 2$ where Γ is a closed smooth nonintersecting curve and all functions on Γ are considered to be 1-periodic depending on the parameter t. For H_h given by *global* trigonometric polynomials and for Cauchy singular integral equations, the point collocation has been analyzed in (Prössdorf and Silbermann, 1977).

6.3 *Naive Spline Collocation on Curves*

In engineering boundary element methods almost all the codes (Mackerle and Andersson, 1984) are based on collocation methods. For two-dimensional problems these boundary element methods are based on $(d-1)$-degree 1-periodic polynomial splines $H_h = S_h^{d,r}$ as in Section 5.1. For the error analysis presented here, however, we require in addition that the splines are $(d-2)$-times continuously differentiable. Then with any $r < d - \frac{1}{2}$ the splines provide the approximation property (A) and for quasi-uniform partitions also the inverse assumption (I) (Arnold and Wendland, 1983). For odd splines, i.e. when $d-1$ is odd, we choose the break points as $\Xi = \{x_l = X(t_l)\}$, whereas for even $d-1$, the collocation points are chosen corresponding to midpoints $\Xi = \{x_l = X((t_{l+1} + t_l)/2)\}$.

These choices correspond to spline interpolation – we therefore call the corresponding collocation (6.4) *naive collocation* which provides the following convergence results.

THEOREM 15: *Let A be an invertible strongly elliptic pseudo-differential operator of order* α *(or a system of such). Then the naive spline collocation method converges asymptotically with optimal order, i.e. there exist positive constants* h_0, c *such that* (6.3) *is uniquely solvable for all* h *with* $0 < h \leq h_0$ *and there holds the asymptotic error estimate*

$$\|w - w_h\|_\tau \leq c h^{\beta-\tau} \|w\|_\beta \qquad (6.5)$$

provided

(i) $(d-1)$ *is odd*, $\alpha < d-1$, $\alpha \leqslant \tau \leqslant (\alpha + d)/2 \leqslant \beta \leqslant d$ (see (Arnold and Wendland, 1983), *or*

(ii) *the family of partitions is quasiuniform*, $\alpha < d-1$, $\alpha \leqslant \tau < d - \frac{1}{2}$, $\tau \leqslant \beta \leqslant d$, $\alpha + d \leqslant 2\beta$ (see (Prössdorf and Schmidt, 1981b) *for* $\alpha = 0$ *and* $d = 2$, (Arnold and Wendland, 1985; Saranen and Wendland, 1985; Schmidt, 1984, 1985, 1986; Prössdorf and Rathsfeld, 1985) *for the general case*) *or*

(iii) *the family of partitions is quasiuniform*, $(d-1)$ *is even* $d-1 \leqslant \alpha < d - \frac{1}{2}$, $\tau < \beta$ <u>or</u> $\alpha + \frac{1}{2} < \tau = \beta$ (see (Arnold and Wendland, 1985)).

Remarks to the proof: The proofs given in the referred papers are rather different and partly also rather tedious. However, at least for two cases we are able to recover these proofs as special cases of Lemma 14 with $\tau = \sigma$ by constructing appropriate test elements $u_h \in T_h$ to given $v_h \in H_h$. In the remaining cases the Aubin-Nitsche duality arguments and localization principles in connection with Korn's trick and Fourier expansions provide the basic tools of the proofs.

(i) In this case $d \geqslant 2$ is even. For $v_h \in H_h$ the distribution $(d/dt)^d v_h$ defines an appropriate linear combination of Dirac functionals on the partition

$$(-1)^{d/2} \left(\frac{d}{dt}\right)^d v_h = \sum_{l=1}^{M} c_l(v_h)\, \delta(t_l - t) \qquad (6.6)$$

where $c_l(v_h) \in \mathbb{R}^N$.

Further, for any continuous $\psi(t)$ we define the operator π_h by

$$\pi_h \psi := \sum_{l=1}^{M} \psi(t_l)\, \tfrac{1}{2}\,(t_{l+1} - t_{l-1})\, \delta(t_l - t).$$

Then to any $v_h \in H_h$ we choose for showing (6.2),

$$u_h(t) := \sum_{l=1}^{M} \Theta^*(t_l)\, c_l(v_h)\, \delta(t_l - t) + \pi_h \Theta^* v_h$$
$$- \pi_h A^{*-1} C^* \Big\{ \sum_{l=1}^{M} c_l(v_h)\, \delta(t_l - \cdot) + v_h \Big\} \tag{6.7}$$

where $D = \Theta A + C$.

T_h will be equipped with the $H^{(\alpha-d)/2}(\Gamma)$ norm, since $T_h \subset H^{(\alpha-d)/2}(\Gamma)$ due to $(\alpha-d)/2 < -\frac{1}{2}$. Then integration by parts and application of the arguments in (Arnold and Wendland, 1983) eventually yield the estimate

$$\left|(u_h, Av_h)_{L_2(\Gamma)}\right| \geqslant \mathrm{Re}\,(v_h, Dv_h)_{H^{d/2}(\Gamma)} - O(h^\delta \|v_h\|_\sigma^2)$$
$$\geqslant \gamma \|v_h\|_\sigma^2 - O(h^\delta \|v_h\|_\sigma^2)$$

where $\sigma = (\alpha + d)/2$ and $\delta > 0$. From (6.6) one also has

$$\|u_h\|_{(\alpha-d)/2} \leqslant c\, \|v_h\|_\sigma$$

which implies (6.2) for $h > 0$ being sufficiently small.

(ii) In this case let us consider only even $(d-1)$, uniform partitions, the special case of $\Theta = 1$ and a principal symbol $a_0(x, \xi)$ of A which does not depend on x as in (Saranen and Wendland, 1985). Let π be the characteristic function of $[0,1)$ and let $\pi^d := \pi * \pi * \cdots * \pi = (*\pi)^d$ be the d-fold convolution of π. Then let the B-splines

$$\mu_k := \pi^d\,((t/h) - k + 1)\,,$$

and their 1-periodic extensions form a basis of H_h. Correspondingly, we define

$$\tilde{\mu}_k := \pi^{d+1}((t/h) - k + 3/2)\,.$$

Then for any trial function

$$v_h = \sum_{k=1}^{M} \gamma_k \mu_k$$

we choose the Dirac functionals

$$(-1)^{(d+1)/2}\left(\frac{d}{dt}\right)^{(d+1)} \sum_{k=1}^{M} \tfrac{1}{4} h\gamma_k |\sin \pi k h/2|^{-1} \tilde{\mu}_k$$

$$= \sum_{l=1}^{M} c_l(v_h)\, \delta\left(t_l + \frac{h}{2} - t\right).$$

The test distribution $u_h \in T_h$ can be chosen as in (6.7) where $\delta(t_l - t)$ needs everywhere to be replaced by $\delta(t_l - t + h/2)$.

For the remaining cases see the references given in Theorem 15.

Remarks: Note that the convergence results of Theorem 15 apply in particular to Cauchy singular integral equations (5.10) where $\alpha = 0$. Here (iii) provides optimal order convergence of the "panel method" with piecewise constants, i.e. $d = 1$ and naive collocation. Theorem 15 can be extended appropriately to strongly elliptic systems with varying orders α_{lk} satisfying (5.6) or only (4.3). The details, however, are yet to be done.

In connection with the inverse assumption (I) we are also able to give asymptotic estimates for the conditioning of collocation equations (6.3) (Wendland, 1985a, Lemma 6.1).

As for the spline Galerkin method in Theorem 13 also for naive spline collocation the strong ellipticity is not only sufficient for optimal order convergence but also necessary.

THEOREM 16 (Prössdorf and Schmidt, 1981b; Schmidt, 1986): *Let A be a given pseudodifferential operator of order α on a simple closed curve Γ and let the naive spline collocation method converge with optimal order. Then A is strongly elliptic.*

Remark: The results of Theorem 15 and 16 correct and replenish the erroneous proposition by Gabdulhaev (1970), (Fenyö and Stolle, 1984, Satz 5, p.298) which are only valid if the inverses of (6.3) are uniformly bounded. But the latter is just the crucial result in the above analysis.

6.4 *Schmidt's ε-Collocation* (Schmidt, 1986)

Ivanov (1976, p.204 ff.) shows that naive spline collocation with piecewise linears diverges for the *non* strongly elliptic Hilbert transform, i.e. (5.10) with $b=0$ and $c=1$. However, in this case collocation at the midpoints of the subintervals becomes convergent (Schmidt, 1983). Correspondingly, with the appropriate choice of the collocation points *depending on the principal part of* A spline collocation converges with optimal order for a larger class than strongly elliptic operators. Following (Schmidt, 1986) we consider only *equidistant* subdivisions and define here

$$\Xi = \{x_l = X(t_l + \varepsilon h)\}$$

where the fixed constant $\varepsilon \in [0,1)$ will be chosen in connection with A. (For $\varepsilon = 0$ we have the break points and for $\varepsilon = \frac{1}{2}$ the midpoints.)

THEOREM 17 (Schmidt, 1986):

(i) *Let* $\varepsilon = 0$ *and* $d > \alpha + 1$ *or* $\varepsilon = \frac{1}{2}$ *and* $d > \alpha + \frac{1}{2}$, $\alpha \in \mathbb{Z}$. *Then the ε-collocation on equidistant subdivisions converges with optimal order* (6.5) *if and only if*

$$\mu a_0(x,1) + (-1)^{d+2\varepsilon}(1-\mu)\, a_0(x,-1) \neq 0$$

for all $\mu \in [0,1]$ *and* $x \in \Gamma$.

(ii) *Let* $\varepsilon \in (0,1)$, $d < \alpha + \frac{1}{2}$, $\alpha \in \mathbb{Z}$. *Then the ε-collocation on equidistant subdivisions converges with optimal order* (6.5) *if and only if*

$$a_0(x,1) \neq 0\,, \quad a_0(x,-1) \neq 0$$

and

$$a_0(x,1) \sum_{k=0}^{\infty} (k+\tau)^{\alpha-d}\, e^{2\pi i k\varepsilon} + (-1)^d a_0(x,-1) \sum_{k=1}^{\infty} (k+\tau)^{\alpha-d}\, e^{-2\pi i k\varepsilon} \neq 0$$

for all $\tau \in (0,1)$ *and all* $x \in \Gamma$.

Remark: These results on ε-collocation show that the operator A and the spline and test spaces are to be related properly in order to obtain convergence of optimal order — the consistency of the method is by far not sufficient for its convergence. These results have been confirmed for equations with Cauchy singular integral operators A in (Niessner and Ribault, 1986). Also note that ε corresponds to Brunner's collocation parameter c_j (see these Proceedings and the monograph (Chandler and Graham, 1986)).

6.4 *Boundary Element Collocation in Higher Dimensions* n

For higher dimensional problems where Γ is a surface, i.e. an $(n-1)$-dimensional manifold in $\mathbb{R}^n$, boundary element collocation has been analyzed so far only for Fredholm integral equations of the second kind of the form

$$Aw := (I+Q+C)\,w = f \quad \text{on} \quad \Gamma \tag{6.8}$$

which corresponds to (5.8). Here the results for point collocation (6.4) are obtained in the topology of uniform convergence, and they are based on the concepts of discrete perturbations of the identity by contractions and collectively compact operator families (Anselone, 1971; Baker, 1977; Chatelin, 1983; McLean, 1986). Surveys on the results for boundary element methods are given in (Wendland, 1983, 1985b). Here $X_0 = C^0(\Gamma) \subset X = W^0_\infty(\Gamma)$ if $d = 1$, and $X_0 = X = C^0(\Gamma)$ if $d \geq 1$ equipped with the corresponding norms. Let I_h denote the Lagrangian interpolation operator associated with H_h and Ξ. Then (6.4) reads as

$$w_h + I_h Q w_h + I_h C w_h = I_h f \quad \text{on} \quad \Gamma\,.$$

THEOREM 18 (Anselone, 1971; Bruhn and Wendland, 1967; Stummel, 1971): *Suppose* $g \in X_0$, $Q : X_0 \to X_0$ *is continuous*, $C : X_0 \to X$ *is compact*, $\lim\limits_{h\to 0} \|I_h g - g\|_X = 0$ *for any* $g \in X_0$,

$$\|I_h Q\|_{X,X} \leq q < 1 \textit{ uniformly for all } h > 0,$$

$$\lim_{h \to 0} \|I_h Cg - Cg\|_X = 0 \quad \textit{for every } g \in X_0$$

and suppose that

$$\bigcup_{0<h} \{w = (I_h - I)\,C\phi \mid \ \phi \in X \ \text{and} \ \|\phi\| \leq 1\}$$

is collectively compact in X_0 (Anselone, 1971). *If* A *is bijective then there exist positive constants* h_0, c_1, c_2 *such that* (6.4) *is uniquely solvable for all* $h < h_0$ *and we have asymptotic estimates*

$$\|w_h\|_X \leq c_1 \|g\|_X$$

and

$$\|w - w_h\|_X \leq c_2 \{\|(I_h - I)\,Qw\|_X + \|(I_h - I)\,Cw\|_X\}.$$

Remarks: These estimates imply convergence of the panel methods as in (Wendland, 1968; Kleinman and Wendland, 1977; Sloan and Thomée, 1986; McLean. 1986; Hebeker, 1986a; Nowak, 1986). Further smoothing by iteration again yields higher order approximations, see (Graham *et al*, 1985; Joe, 1986).

7. CONCLUDING REMARKS

As has been indicated in the previous chapters, the concept of strongly elliptic boundary integral or pseudodifferential equations provides the basic properties for the efficiency and asymptotic convergence of their numerical treatment, in particular of the popular boundary element methods.

In most practical applications, however, the boundaries have discontinuous curvatures, corners and edges and these geometrical discontinuities generate corresponding singularities of the desired solutions of the original boundary value problems as well as of the desired boundary charges. The singularities require on the one hand modifications of the corresponding analysis, on the other hand they pollute the accuracy of numerical computations if they are not incorporated into the trial and test spaces appropriately. The latter is usually performed by augmenting the trial space with singular functions or by using graded meshes.

For equations on curves, i.e. for two-dimensional boundary value problems, the extended analysis is either based on the functional analysis of continuous functions with maximum norm and its dual space, or on L_2 in connection with arguments as in Theorem 18 and for Fredholm integral equations of the second kind, or on the Mellin transform with the origin at the geometrical discontinuity and appropriate weighted Sobolev spaces. These results are based on the local regularity of the solutions (Grisvard, 1985) and are now rather complete.

For spline Galerkin methods applied to Fredholm integral equations of the second kind on curves with corners, see e.g. (Bruhn and Wendland, 1967; Chandler, 1984; Chandler and Graham, 1986; Costabel, to appear a; Costabel and Stephan, to appear; Djaoua, 1981; McLean, 1986; Raugel, 1978; Wendland, 1983) and for spline collocation (Atkinson and de Hoog, 1984; Bruhn and Wendland, 1967; Chandler and Graham, 1986; McLean, 1986; Papamichael and Hough, 1983; Ruland, 1978; Wendland, 1983).

The concept of strong ellipticity has been extended to corner problems by the use of the Mellin transform. Here most contributions have been made for Galerkin methods as in (Costabel, to appear, b; Costabel and Stephan, 1983, 1985, 1986b, to appear; Costabel *et al*, 1983; Fischer, 1982; Lamp *et al*, 1984; Stephan and Wendland, 1984, 1985; Weisel, 1979; Wendland *et al*, 1979) whereas spline collocation for more general strongly elliptic equations can be found in (Costabel and Stephan, 1986b).

For equations on surfaces in $\mathbb{R}^3$, i.e. for 3-dimensional boundary value problems one finds yet only few results either on surfaces with corners and edges for the boundary element collocation for Fredholm integral equations of the second kind (6.8) as in (Gaier, 1976; Giroire and Nedelec, 1978; Hebeker, 1986a; Wendland, 1968, 1983) or for boundary element Galerkin methods applied to strongly elliptic equations on bounded smooth pieces of surfaces as in crack and screen problems (Bamberger, 1983;

Bamberger and Ha Duong, 1985; Costabel and Stephan, 1986a; Durand, 1983; Stephan, 1984) or on general Lipschitz boundaries (Costabel, to appear a). This numerical analysis still needs to be completed.

In spite of the excellent numerical performance of boundary element methods there is their restriction to the cases where the fundamental solution $G(x,y)$ is known explicitly. Hence, inhomogeneous materials and nonlinear partial differential equations require additional domain integral equations and far more extensive numerical work. Another possibility is to combine finite element approximation in the subdomains of nonhomogeneity and nonlinearity with boundary element potentials in the homogeneous subdomains which seems to combine the best features of both methods (Bettes *et al*, 1979). Here the corresponding error analysis is available only for the case corresponding to Fredholm boundary integral equations of the second kind with a compact integral operator (Johnson and Nedelec, 1980) whereas for the case of general strongly elliptic problems there are known only some preliminary results (Wendland, 1986) indicating that perhaps even the corresponding algorithms will have to be appropriately modified.

REFERENCES

Agranovich, M.S. (1977), "Spectral properties of diffraction problems", in *The General Method of Natural Vibrations in Diffraction Theory*, eds. Voitovic, N.N., Katzenellenbaum, B.Z. and Sivov, A.N., (Russian) Izdat. Nauka (Moscow).

Anselone, P.M. (1971), *Collectively Compact Operator Approximation Theory*, Prentice Hall (Englewood Cliffs, N.J.).

Arnold, D.N. (1983), "A spline-trigonometric Galerkin method and an exponentially convergent boundary integral method", *Math. Comp.*, *41*, pp.383-397.

Arnold, D.N. and Wendland, W.L. (1983), "On the asymptotic convergence of collocation methods", *Math. Comp.*, *41*, pp.349-381.

Arnold, D.N. and Wendland, W.L. (1985), "The convergence of spline collocation for strongly elliptic equations on curves", *Numer. Math.*, *47*, pp.317-341.

Atkinson, K. and de Hoog, F. (1984), "The numerical solution of Laplace's equation on a wedge", *IMA J. Numer. Anal.*, *4*, pp.19-41.

Aziz, A.K., Dorr, M.R. and Kellog, R.B. (1980), "Calculation of electromagnetic scattering by a perfect conductor", Report TR 80-245, Naval Surface Weapons Center, Silver Spring, Maryland 20910.

Babuška, I. (1970), "Error bounds for finite element methods", *Numer. Math.*, *16*, pp.322-333.

Babuška, I. and Aziz, A.K. (1972), "Survey lectures on the mathematical foundations of the finite element method", in *The Mathematical Foundation of the Finite Element Method with Applications to Partial Differential Equations*, ed. Aziz, A.K., Academic Press (New York), pp.3-359.

Baker, C. (1977), *The Numerical Treatment of Integral Equations*, Clarendon Press (Oxford).

Bamberger, A. (1983), "Approximation de la diffraction d'ondes élastiques — une nouvelle approche (I)", Rapport Int. 91, Centre Math. Appl. Ecole Polytechnique, Palaiseau, France.

Bamberger, A. and Ha Duong, T. (1985), "Diffraction d'une onde acoustique par une paroi absorbante: Nouvelles equations integrales", Rapport Int. 121, Centre Math. Appl. Ecole Polytechnique, Palaiseau, France.

Bendali, A. (1982), "Numerical analysis of the exterior boundary-value problem for the time-harmonic Maxwell equations by a boundary finite element method", Rapport Int. 83, Centre Math. Appl. Ecole Polytechnique, Palaiseau, France.

Bettes, P., Kelly, D.W. and Zienkiewicz, O.C. (1979), "Marriage à la mode. The best of both worlds (Finite elements and boundary integrals)", in *Energy Methods in Finite Analysis*, eds. Glowinski, R., Rodin, E.Y. and Zienkiewicz, O.C., J. Wiley (Chichester).

Bischoff, H. (1982), "Die Feldintegralmethode zur Beschreibung der Grundwasserbewegung", in *Fünftes Wassertechnisches Seminar, Grundwasserbewirtschaftung, Grundwassermodelle, Grundwasseranreicherung*, Schriftenreihe WAR 16, Inst. Wasserversorgung, Abwasserbeseitigung und Raumplanung TH Darmstadt, Fed. Rep. Germany, pp.57-86.

Bolteus L. and Tullberg, O. (1981), "BEMSTAT — A new type of boundary element program for two-dimensional elasticity problems", in *Boundary Element Methods*, ed. Brebbia, C.A., Springer-Verlag (Berlin-Heidelberg-New York), pp.518-537.

Brakhage, H. (1960), "Über die numerische Behandlung von Integralgleichungen nach der Quadraturformelmethode", *Numer. Math.*, *2*, pp.183-196.

Braun, K. (1985), "Lokale Konvergenz der Ritz-Projektion auf finite Elemente für Pseudo-Differentialgleichungen negativer Ordnung", Doctoral Thesis, Universität Freiburg, Germany.

Brebbia, C.A. (1978), *The Boundary Element Methods for Engineers*, Pentech Press (London).

Brebbia, C.A., Telles, J.C.F. and Wrobel, L.C. (1984), *Boundary Element Techniques*, Springer-Verlag (Berlin).
Bruhn, G. and Wendland, W.L. (1967), "Über die näherungsweise Lösung von linearen Funktionalgleichungen", *Internat. Schriftenreihe Numer. Math.*, *7*, pp.136-164.
Brunner, H. and van der Houwen, P.J. (1986), *The Numerical Solution of Volterra Equations*, North Holland (Amsterdam), to appear.
Calderon, A.P. (1963), "Boundary value problems for elliptic equations", Outlines of the joint Soviet-American Symposium on Partial Differential Equations (Novosibirsk), pp.303-304.
Céa, J. (1964), "Approximation variationelle des problèmes aux limites", *Ann. Inst. Fourier (Grenoble)*, *14*, pp.345-444.
Chandler, G.A. (1979), "Superconvergence of numerical solutions to second kind integral equations", Ph.D. Thesis, Australian National University, Canberra.
Chandler, G.A. (1984), "Galerkin's method for boundary integral equations on polygonal domains", *J. Austral. Math. Soc. Ser. B*, *26*, pp.1-13.
Chandler, G.A. and Graham, I.G. (1986), "Product integration — collocation methods for non-compact integral operator equations", *Math. Comp.*, to appear.
Chatelin, F. (1983), *Spectral Approximation of Linear Operators*, Academic Press (New York).
Chazarain, J. and Piriou, A. (1982), *Introduction to the Theory of Linear Partial Differential Equations*, North-Holland (Amsterdam).
Christiansen, S. (1981), "Condition number of matrices derived from two classes of integral equations", *Math. Methods Appl. Sci.*, *3*, pp.364-392.
Christiansen, S. and Hansen, E. (1975), "A direct integral equation method for computing the hoop stress in plane isotropic sheets". *J. Elasticity*, *5*, pp.1-14.
Ciarlet, P.G. (1978), *The Finite Element Method for Elliptic Problems*, North-Holland (Amsterdam-New York-Oxford).
Colton, D. and Kress, R. (1983), *Integral Equation Methods in Scattering Theory*, John Wiley and Son (New York).
Costabel, M. (1984), "Starke Elliptizität von Randintegraloperatoren erster Art", Habilitationsschrift, Technische Hochschule Darmstadt, Germany.
Costabel, M., "Boundary integral operators on Lipschitz domains", to appear.
Costabel, M., "Boundary integral equations for mixed boundary value problems", to appear.
Costabel, M. and Stephan, E. (1985), "Boundary integral equations for mixed boundary value problems in polygonal domains and Galerkin approximation", in *Mathematical Models and Methods in Mechanics 1981*, eds. Fiszdon, W. and Wilmański, K., *Banach Center Publications*, *15*, PWN-Polish Scientific Publ. (Warsaw), pp.175-251.

Costabel, M. and Stephan, E. (1983), "The normal derivative of the double layer potential on polygons and Galerkin approximation", *Applicable Anal.*, *16*, pp.205-228.
Costabel, M. and Stephan, E. (1985), "A direct boundary integral equation method for transmission problems", *J. Math. Anal. Appl.*, *106*, pp.367-413.
Costabel, M. and Stephan, E. (1986), "A boundary element method for three-dimensional crack problems", in *Innovative Numerical Methods in Engineering*, eds. Shaw R.P. *et al*, Springer-Verlang (Berlin-Heidelberg-New York-Tokyo), pp.351-359.
Costabel, M. and Stephan, E., (1986), "On the convergence of collocation methods for boundary integral equations on polygons", *Math Comp.*, to appear (Preprint 982, Technical Univ. Darmstadt, Dept. Math., Germany).
Costabel, M., Stephan, E. and Wendland, W.L. (1983), "On boundary integral equations of the first kind for the bi-Laplacian in a polygonal plane domain", *Ann. Scuola Norm. Sup. Pisa, Cl. Sci.*,(*4*), *10*, pp.197-241
Costabel, M. and Wendland, W.L. (1986), "Strong ellipticity of boundary integral operators". *J. Reine Angew. Math.*, to appear. (Preprint 889, Technical Univ. Darmstadt, Dept. Math., Germany).
Cruse, T.A. (1973), "Applications of the boundary integral equation method to three-dimensional stress analysis", *Comput. and Structures*, *3*, pp.309-369.
Dieudonné, J. (1978), *Eléments d'Analyse, Vol. VIII*, Gauthier-Villars (Paris).
Djaoua, M. (1975), "Méthode d'éléments finis pour la résolution d'un problème extérieur dans R^3", Rapport Int. 3, Centre Math. Appl. Ecole Polytechnique, Palasseau, France.
Djaoua, M. (1981), "A method of calculation of lifting flows around two-dimensional corner-shaped bodies", *Math. Comp.*, *36*, pp.405-425.
Durand, M. (1983), "Layer potentials and boundary value problems for the Helmholtz equation in the complement of a thin obstacle", *Math. Methods Appl. Sci.*, *5*, pp.389-421.
Eskin, G.I. (1980), *Boundary Value Problems for Elliptic Pseudodifferential Equations*, AMS Transl. Math. Mon. 52, (Providence, Rhode Island).
Feng, K. (1980), "Differential vs. integral equations and finite vs. infinite elements", *Math. Numer. Sinica*, *2*, pp.100-105.
Feng, K. (1983), "Finite element method and natural boundary reduction", in *Proc. Intern. Congress Math.*, August 16-24, (Warsaw), pp.1439-1453.
Feng, K. and De Hao Yu (1983), "Canonical integral equations of elliptic boundary value problems and their numerical solutions", in *Proc. China – France Symp. on the Finite Element Method*, April 1982, Beijing, Gordon and Breach (New York), pp.211-215.

Fenyö, S. and Stolle, H.W. (1984), *Theorie und Praxis der linearen Integralgleichungen, Vol. 4*, Birkhäuser-Verlag (Basel-Boston-Stuttgart).

Fichera, G. (1961), "Linear elliptic equations of higher order in two independent variables and singular integral equations", in *Proc. Conf. Partial Differential Equations in Continuum Mechanics*, ed. Langer, R.E., University Press (Madison, Wisconsin).

Fichera, G. and Ricci, P. (1976), "The single layer potential approach of boundary value problems for elliptic equations", in *Lecture Notes in Mathematics*, *561*, Springer-Verlag (Berlin), pp.39-50.

Filippi, P. (1977), "Potentiels de couche pour les ondes mécaniques scalaires", *Rév. CETHEDEC*, *51*, pp.121-175.

Filippi, P. (1977), "Layer potentials and acoustic diffraction", *J. Sound Vibration*, *54*, pp.473-500.

Fischer, T. (1980), "Ein Verfahren zur Berechnung schleichender Umströmungen beliebiger dreidimensionaler Hindernisse mit Hilfe singulärer Störungsrechnung, Integralgleichungen erster Art und Randelementmethode", Diplom Thesis, Technische Hochschule Darmstadt, Germany.

Fischer, T. (1982), "An integral equation procedure for the exterior three-dimensional viscous flow", *Integral Equations Operator Theory*, *5*, pp.490-505.

Fischer, T., Hsiao, G.C. and Wendland, W.L. (1985), "Singular perturbations for the exterior three-dimensional slow viscous flow problem", *J. Math. Anal. Appl.*, *110*, pp.583-603.

Fischer, T. and Rosenberger, R. (1986), "A boundary integral method for the numerical computation of the forces exerted on a sphere in viscous incompressible flows near a plane wall", *Z. Angew. Math. Phys.*, to appear. (Preprint 872, Technical Univ. Darmstadt, Dept. Math., Germany 1985).

Gabdulhaev, B.G. (1970), "A direct method for solving integral equations", *Amer. Math. Soc. Transl. Ser. 2*, *91*, pp.213-223.

Gaier, D. (1976), "Integralgleichungen erster Art und konforme Abbildungen", *Math. Zeitschr.*, *147*, pp.113-139.

Gauss, C.F. (1839), "Allgemeine Lehrsätze in Beziehung auf die im verkehrten Verhältnisse des Quadrats der Entfernung wirkenden Anziehungs – und Abstoßungs-Kräfte", *Werke*, 2nd Edition, Göttingen, 1877, Vol. V, pp.194-242.

Giroire, I. (1978), "Integral equation methods for exterior problems for the Helmholtz equation", Rapport Int. 40, Centre Math. Appl., Ecole Polytechnique, Palaiseau, France.

Giroire, J. and Nedelec, J.C. (1978), "Numerical solution of an exterior Neumann problem using a double layer potential", *Math. Comp.*, *32*, pp.973-990.

Gohberg, I.C. and Fel'dman, I.A. (1974), *Convolution Equations and Projection Methods for their Solution*, AMS Transl. Math. Mon. (Providence, Rhode Island).

Goldstein, C. (1979), *Numerical methods for Helmholtz type equations in unbounded domains*, BNL – 26543, Brookhaven Lab., Brookhaven, N.Y.

Gregoire, J.P., Nedelec, J.C. and Planchard, J. (1974), "Problèmes relatifs à l'équations d'Helmholtz", *EDF Bull, Direction Etudes Rech. Ser. C Math. Inform.*, *2*, pp.15-32.

Graham, I.G., Joe, S. and Sloan, I.H. (1985), "Iterated Galerkin versus iterated collocation for integral equations of the second kind", *IMA J. Numer. Anal.*, *5*, pp.355-369.

Grisvard, P. (1985), *Elliptic Problems in Nonsmooth Domains*, Pitman (London).

Ha Duong, T. (1979), "La méthode de Schenck pour le résolution numérique de radiation acoustique", *EDF Bull. Direction Etudes Rech. Sér. C. Math. Inform.*, *12*, pp.15-50.

Ha Duong, T. (1980), "A finite element method for the double layer potential solutions of the Neumann exterior problem", *Math. Methods Appl. Sci.*, *2*, pp.191-208.

Hartmann, F. (1981), "Elastostatics", in *Progress in Boundary Element Methods, Vol. 1*, ed. Brebbia, C.A., Pentech Press (Plymouth), pp.84-167.

Hartmann F. "Kompatibilität auf dem Rand", in preparation.

Hartmann, F. and Zotemantel, R. (1986), "The direct boundary element method in plate bending", *Internat. J. Numer. Methods Engrg.*, to appear.

Hayes, J.K., Kahaner, D.K. and Kellner, R.G. (1972), "An improved method for numerical conformal mapping", *Math. Comp.*, *26*, pp.327-334.

Hebeker, F.K. (1984), "New hydrodynamical potentials and a free boundary value problem of the stationary Stokes equations", Preprint 636 A4 SFB 72, Universität Bonn, Germany.

Hebeker, F.K. (1986), "Efficient boundary element methods for three dimensional exterior viscous flows", *Numer. Meth. Partial Diff. Equations*, to appear.

Hebeker, F.K. (1986/87), "An integral equation of the first kind for a free boundary value problem of the stationary Stokes equations", *Math. Methods Appl. Sci.*, to appear.

Herrera, I. (1984), *Boundary Methods. An Algebraic Approach*, Pitman (London).

Herrera, I. (1985), "Unified approach to numerical methods, Part 1. Green's formulas for operators in discontinuous fields", *Numer. Meth. Partial Diff. Equations*, *1*, pp.12-37.

Hess, J.L. (1986), "Calculation of acoustic fields about arbitrary three-dimensional bodies by a method of surface source distributions based on certain wave number expansions", Report DAC 66901, McDonnell Douglas.

Hess, J.L. and Smith, A.M.O. (1967), "Calculation of potential flow about arbitrary bodies", in *Progress in Aeronautical Sciences*, *8*, ed. Kuchemann, D., Pergamon (Oxford), pp.1–138.

Honl, H., Maue, A.W. and Westpfahl, K. (1961), "Theorie der Beugung", in *Handbuch der Physik*, *25/1*, ed. Flügge, S., Springer-Verlag (Berlin).

Hörmander, L. (1966), "Pseudo-differential operators and non-elliptic boundary problems", *Ann. of Math.*, *83*, pp.129-209.

Hoidn, H.P. (1983), "Die Kollokationsmethode angewandt auf die Symmsche Integralgleichung", Doctoral Thesis, ETH Zürich, Switzerland.

Hsiao, G.C., Kopp, P. and Wendland, W.L. (1980), "A Galerkin collocation method for integral equations of the first kind", *Computing*, *25*, pp.89-130.

Hsiao, G.C., Kopp, P. and Wendland, W.L. (1984), "Some applications of a Galerkin collocation method for integral equations of the first kind", *Math. Methods Appl. Sci.*, *6*, pp.280-325.

Hsiao, G.C. and Kress, R. (1985), "A modified boundary integral equation method for the two-dimensional exterior Stokes problem", *Appl. Numer. Math.*, *1*, pp.77-93.

Hsiao, G.C. and MacCamy, R.C. (1973), "Solution of boundary value problems by integral equations of the first kind", *SIAM Rev.*, *15*, pp.687-705.

Hsiao, G.C. and Wendland, W.L. (1977), "A finite element method for some integral equations of the first kind", *J. Math. Anal. Appl.*, *58*, pp.449-481.

Hsiao, G.C. and Wendland, W.L. (1981), "The Aubin-Nitsche lemma for integral equations", *J. Integral Equations*, *3*, pp.299-315.

Hsiao, G.C. and Wendland, W.L. (1983), "On a boundary integral method for some exterior problems in elasticity", *Dokl. Acad. Nauk SSR*, Special issue ded. Academician V.D. Kupradze 80th birthday, to appear.

Ivanov, V.V. (1976), *The Theory of Approximate Methods and their Applications to the Numerical Solution of Singular Integral Equations*, Noordhoff Int. Publ. (Leyden).

Jaswon, M.A. (1981), "Some theoretical aspects of boundary integral equations", in *Boundary Element Methods*, ed. Brebbia, C.A., Springer-Verlag (Berlin-Heidelberg-New York), pp.399-411.

Jaswon, M.A. and Symm, G.T. (1977), *Integral Equation Methods in Potential Theory and Elastostatics*, Academic Press (London).

Jentsch, L. (1974), "Über stationäre thermoelastische Schwingungen in inhomogenen Körpern", *Math. Nachr.*, *64*, pp.171-231.

Jentsch, L. (1975), "Stationäre thermoelastische Schwingungen in stückweise homogenen Körpern infolge zeitlich periodischer Außentemperatur", *Math. Nachr.*, *69*, pp.15-37.

Joe, S. (1986), "Discrete collocation methods for second kind Fredholm integral equations", *SIAM J. Numer. Anal.*, to appear.

Johnson, J. and Nedelec, J.C. (1980), "On the coupling of boundary integral and finite element methods", *Math. Comp.*, *35*, pp.1063-1079.

Kleinman, R.L. and Roach, G.F. (1974), "Boundary integral equations for the three-dimensional Helmholtz equation", *SIAM Rev.*, *16*, pp.214-236.

Kleinman, R. and Wendland, W.L. (1977), "On Neumann's method for the exterior Neumann problem for the Helmholtz equation", *J. Math. Anal. Appl.*, *57*, pp.107-202.

Kral, J. (1980), *Integral Operators in Potential Theory*, *Lecture Notes in Mathematics 823*, Springer-Verlag (Berlin).

Kuhn, G. and Möhrmann, W. (1982), "Boundary element method in elastostatic theory and applications", *Appl. Math. Modelling*, *7*, pp.97-105.

Kupradze, V.D. (1956), *Randwertaufgaben der Schwingungstheorie und Integralgleichungen*, Dt. Verl. Wissenschaften (Berlin).

Kupradze, V.D. (1965), *Potential Methods in the Theory of Elasticity*, Israel Progr. Sci. Transl. (Jerusalem).

Kupradze, V.D., Gegelia, T.G., Basheleishvili, M.O. and Burchuladze, T.V. (1979), *Three-Dimensional Problems of the Mathematical Theory of Elasticity and Thermoelasticity*, North-Holland (Amsterdam).

Lamp, U., Schleicher, T., Stephan, E. and Wendland, W.L. (1984), "Galerkin collocation for an improved boundary element method for a plane mixed boundary value problem", *Computing*, *33*, pp.269-296.

Lamp, U., Schleicher, T. and Wendland, W.L. (1985), "The fast Fourier transform and the numerical solution of one-dimensional integral equations", *Numer. Math.*, *47*, pp.15-38.

Le Roux, M.N. (1974), "Résolution numérique du problème du potential dans le plan par une méthode variationelle d'éléments finis", Doctoral Thesis, University Rennes, France.

Le Roux, M.N. (1974), "Equations intégrales pour le problème du potential électrique dans le plan", *C.R. Akad. Sci. Paris*, *278*, *Sér. A*, pp.541-544.

Ligett, J.A. and Liu, P. (1983), *The Boundary Integral Equation Method for Porous Media Flow*, George Allen Unwin (London).

MacCamy, R.C. and Stephan, E. (1983), "A boundary element method for an exterior problem for three-dimensional Maxwell's equations", *Applicable Anal.*, *16*, pp.141-163.

Mackerle, J. and Andersson, T. (1984), "Boundary element software in engineering", *Advances in Eng. Software*, *6*, pp.66-102.

Martensen, E. (1959), "Berechnung der Druckverteilung an Gitterprofilen in ebener Potentialströmung mit einer Fredholmschen Integralgleichung", *Arch. Rational Mech. Anal.*, *3*, pp.235-270.

Martensen, E. (1968), *Potentialtheorie*, Teubner-Verlag (Stuttgart).

Maskew, B. (1982), "Prediction of subsonic aerodynamic characteristics: A case for low-order panel methods", *J. Aircraft*, *19*, pp.157-163.

Maul, J. (1976), "Eine einheitliche Methode zur Lösung der ebenen Aufgaben der linearen Elastostatik", Schriftenreihe ZIMM 24, Akad. Wiss. DDR, Berlin.

McDonald, B.H., Friedmann, M. and Wexler, A. (1974), "Solution of integral equations", *IEEE Trans. Microwave Theory Tech.*, pp.237-248.

McLean, W. (1986), "Boundary integral methods for the Laplace equation", Ph.D. Thesis, Australian National University, Canberra.

Mehlhorn, G. (1970), "Ein Beitrag zum Kipp-Problem bei Stahlbeton – und Spannbetonträgern", Doctoral Dissertation, D17, Technische Hochschule Darmstadt, Germany.

Mikhlin, S.G. (1962), *Variationsmethoden der Mathematischen Physik*, Akademie-Verlag (Berlin).

Mikhlin, S.G. and Prössdorf, S. (1980), *Singular Integral Operators*, Akademie-Verlag (Berlin).

Müller, C. (1969), *Foundations of the Mathematical Theory of Electromagnetic Waves*, Springer-Verlag (Berlin-Heidelberg-New York).

Muskhelishvili, N.L. (1963), *Some Basic Problems of the Mathematical Theory of Elasticity*, Noordhoff (Groningen).

Mustoe, G.W., Volait, F. and Zienkiewicz, O.C. (1982), "A symmetric direct boundary integral equation method for the two-dimensional elastostatics", *Res. Mechanica*, *4*, pp.57-82.

Nedelec, J.C. (1976), "Curved finite element methods for the solution of singular integral equations on surfaces in R^3", *Comput. Methods Appl. Mech. Engrg.*, *8*, pp.61-80.

Nedelec, J.C. (1977), "Méthodes d'éléments finis courbes pour la résolution des équations integrales singulières sur des surfaces de $\mathbb{R}^3$", *RAIRO Anal. Numer.*, *12*, pp.212-229.

Nedelec, J.C. (1977), *Approximation des Equations Intégrales en Mécanique et en Physique*, Lecture Notes, Centre de Mathématiques Appliquées, Ecole Polytechnique, Palaiseau, France.

Nedelec, J.C. (1978), "Approximation par potential de double cuche du problème de Neumann extérieur", *C.R. Acad. Sci. Paris, Sér. I. Math.* *280*, pp.616-619.

Nedelec, J.C. (1978), "La méthode des éléments finis appliquée aux équations integrales de la physique", *First Meeting AFCET-SMF on Applied Mathematics*, Palaiseau, Vol. 1, pp.181-190.

Nedelec, J.C. (1980), "Formulations variationelles de quelques équations intégrales faisant intervenir des parties finies", in *Innovative Numerical Analysis for the Engineering Sciences*, ed. Shaw, R., Univ. Press of Virginia (Charlottesville), pp.517-524.

Nedelec, J.C. (1982), "Integral equations involving non integrable kernels", in *Conferentie van Numeriek Wiskundigen*, 18-20 Oktover 1982, Zeist, Werlegemeenschap Numericke Wiskunde, Mathematical Centre, Amsterdam, Holland, pp.47-57.

Nedelec, J.C. (1982), "Integral equations with non integrable kernels", *Integral Equations Operator Theory*, *5*, pp.562-572.

Nedelec, J.C. and Planchard, J. (1973), "Une méthode variationelle d'éléments finis pour la résolution numérique d'un problème extérieur dans R^3", *RAIRO Anal. Numer.*, *7*, pp.105-129.

Niessner, H. and Ribault, H. (1986), "Condition of boundary integral equations arising from flow computations", *Comp. and Appl. Math.*, to appear.

Nikolski, S.M. (1943), "Linear equations in normed linear spaces", *Isvestija Akad. Nauk. SSSR*, *Ser. Matem.*, *7*, (Russian), pp.147-163.

Nitsche, J. and Schatz, A. (1972), "On local approximation properties of L_2-projections on spline subspaces", *Applicable Anal.*, *2*, pp.61-168.

Niwa, Y., Kobayashi, S. and Kitahara, M., "Applications of the boundary integral equation method to eigenvalue problems of elastodynamics", in *Boundary Element Methods in Engineering*, ed. Brebbia, C.A., Springer-Verlag, (Berlin-Heidelberg-New York), pp.297-311.

Nowak, P. (1986), "Higher order panel method for potential flow problems in 3-D", in *Notes on Numerical Fluid Mechanics*, eds. Hackbusch, W. and Witsch, K., Vieweg-Verlag (Braunschweig), to appear.

Papamichael, N. and Hough, D.M. (1983), "An integral equation method for the numerical conformal mapping of interior, exterior and double-connected domains", *Numer. Math.*, *41*, pp.287-307.

Petersen, B.E. (1983), *Introduction to the Fourier Transform and Pseudo-Differential Operators*, Pitman (London).

Poggio, A.J. and Miller, E.K. (1973), "Integral equation solutions of three-dimensional scattering problems", in *Computer Techniques for Electromagnetics*, ed. Mittra, R., Pergamon (Oxford).

Prössdorf, S. (1984), "Ein Lokalisierungsprinzip in der Theorie der Splineapproximationen und einige Anwendungen", *Math. Nachr.*, *119*, pp.239-255.

Prössdorf, S. and Rathsfeld, A. (1985), "On strongly elliptic singular integral operators with piecewise continuous coefficients", *Integral Equations Operator Theory*, *8*, pp.825-841.

Prössdorf, S. and Schmidt, G. (1981), "A finite element collocation method for singular integral equations", *Math. Nachr.*, *100*, pp.33-66.

Prössdorf, S. and Schmidt, G. (1981), "A finite element collocation method for systems of singular equations", Preprint P-MATH-26/81, Akad. d. Wiss. DDR, Inst. Math., DDR 1080 Berlin, Mohrenstr. 39.

Prössdorf, S. and Silbermann, B. (1977), *Projektionsverfahren und die näherungsweise Lösung singulärer Gleichungen*, Teubner (Leipzig).

Rannacher, R. and Wendland, W.L. (1985), "On the order of pointwise convergence of some boundary element methods, Part I. Operators of negative and zero order", *RAIRO Modél. Math. Anal. Numér.*, *19*, pp.65-88.

Raugel, G. (1978), "Résolution numérique de problèmes elliptiques dans des domaines avec coins", Doctoral Thesis, Ser. A, Univ. Rennes, France, pp.63-506.

Ricci, P. (1974), "Sui potenziale di simplice strato per le equazioni ellittiche di ordne superiore in due variabili", *Rend. Mat.*, *7*, pp.1-39.

Rieder, G. (1962), "Iterationsverfahren und Operatorgleichungen in der Elastizitätstheorie", *Abh. Braunschweig. Wiss. Ges.*, *14*, pp.109-443.

Rizzo, F.J. (1967), "An integral equation approach to boundary value problems of classical elastostatics", *Quart. Appl. Math.*, *25*, pp.83-95.

Ruland, C. (1986), "Ein Verfahren zur Lösung von $(\Delta + k^2)\,u = 0$ in Aussengebieten mit Ecken", *Applicable Anal.*, *7*, pp.69-79.

Ruotsalainen, K. and Saranen, J. (1986), "A dual method to the collocation method", *SIAM J. Numer. Anal.*, to appear.

Saranen, J. and Wendland, W.L. (1985), "On the asymptotic convergence of collocation methods with spline functions of even degree", *Math. Comp.*, *45*, pp.91-108.

Schenck, H.A. (1968), "Improved integral formulation for acoustic radiation problems", *J. Acoust. Soc. Amer.*, *44*, pp.41-58.

Schleicher, K.-T. (1983), "Die Randelementmethode für gemischte Randwertprobleme auf Eckengebieten unter Berücksichtigung von Singulärfunktionen", Diplom Thesis, Technische Hochschule Darmstadt, Germany.

Schmidt, G. (1983), "On spline collocation for singular integral equations", *Math. Nachr.*, *111*, pp.177-196.

Schmidt, G. (1984), "The convergence of Galerkin and collocation methods with splines for pseudodifferential equations on closed curves", *Z. Anal. Anwendungen*, *3*, pp.371-384.

Schmidt, G. (1985), "On spline collocation methods for boundary integral equations in the plane", *Math. Methods Appl. Sci.*, *7*, pp.74-89.

Schmidt, G. (1986), "On ε-collocation for pseudodifferential equations on a closed curve", *Math. Nachr.*, *126*, pp.183-196.

Schwab, C. and Wendland, W.L. (1985), "3-D BEM and numerical integration", in *Boundary Elements in Engineering VII*, eds. Brebbia, C.A. and Maier, G., Vol. II, Springer-Verlag (Berlin), pp.13.85 – 13.102.

Seeley, R.T. (1966), "Singular integrals and boundary value problems", *Amer. J. Math.*, *88*, pp.781-809.

Sloan, I. and Thomée, V. (1986), "Superconvergence of the Galerkin iterates for integral equations of the second kind", *J. Integral Equations*, to appear.

Sobolev, S. (1936), "Méthode nouvelle à résoudre le problème de Cauchy pour les équations linéaires hyperboliques normales", *Mat. Sb.*, *1*, pp.39-72.

Stephan, E. (1983), "Solution procedures for interface problems in acoustics and electromagnetics", in *Theoretical Acoustics and Numerical Treatments*, ed. Filippi, P., CISM Courses and Lectures No. 277, Springer-Verlag (Wien-New York), pp.291-348.

Stephan, E. (1984), "Boundary integral equations for mixed boundary value problems, screen and transmission problems in R^3", Habilitationsschrift, Technical Univ. Darmstadt (Preprint 848, Technical Univ. Darmstadt, Dept. Math., Germany).

Stephan, E. and Wendland, W.L. (1976), "Remarks to Galerkin and least squares methods with finite elements for general elliptic problems", *Manuscripta Geodaetica*, *1*, pp.93-123.

Stephan, E. and Wendland, W.L. (1982), *Mathematische Grundlagen der Finiten Element-Methoden*, Meth. Verf. Math. Physik 23, Peter Lang Verlag (Frankfurt/Main).

Stephan, E. and Wendland, W.L. (1984), "An augmented Galerkin procedure for the boundary integral method applied to two-dimensional screen and crack problems", *Applicable Anal.*, *18*, pp.183-220.

Stephan, E. and Wendland, W.L. (1985), "An augmented Galerkin procedure for the boundary integral method applied to mixed boundary value problems", *Appl. Numer. Math.*, *1*, pp.121-143.

Stock, B. (1985), "Über die Anwendung der Randelementmethode zur Lösung des linearen Molodenskiischen und verallgemeinerten Neumannschen geodätischen Randwertproblems", Doctoral Thesis D17, Technische Hochschule Darmstadt, Germany.

Strang, G. (1972), "Approximation in the finite element method", *Numer. Math.*, *19*, pp.113-137.

Stummel, F. (1971), "Diskrete Konvergenz linearer Operatoren I und II", *Math. Z.*, *120*, pp.231-264.

Symm, G.T. (1967), "Integral equations in potential theory, II", *Proc. Roy. Soc. London Ser. A*, *275*, pp.33-46.

Symm, G.T. (1967), "Numerical mapping of exterior domains", *Numer. Math.*, *10*, pp.437-445.

Taylor, M.E. (1981), *Pseudodifferential Operators*, Princeton Univ. Press (Princeton, N.Y.).

Treves, F. (1980), *Introduction to Pseudodifferential and Fourier Integral Operators I*, Plenum Press (New York, London).

Vainikko, G. (1965), "On the question of convergence of the Galerkin method", (Russian) *Tartu Rükl. Ül. Toimetised*, *177*, pp.148-152.

Wagner, S. and Urban, Ch. (1985), "Current activities in basic research work on panel methods in Germany", in *Colloquium on Fluid Mechanics*, Univ. Armed Forces, Munich, Germany.

Watson, J.O. (1979), "Advanced implementation of the boundary element method for two- and three-dimensional elastostatics", in *Developments in Boundary Element Methods*, eds. Banerjee, I.P.K. and Butterfields, R., Appl. Science Publ. Ltd. (London), pp.31-63.

Watson, J.O. (1982), "Hermitian cubic boundary elements for plane problems of fracture mechanics", *Res. Mechanica*, *4*, pp.23-43.

Weisel, J. (1979), "Lösung singulärer Variationsprobleme durch die Verfahren von Ritz und Galerkin mit finiten Elementen — Anwendungen in der konformen Abbildung", *Mitt. Math. Sem. Giessen*, *138*.

Wendland, W.L. (1968), "Die Behandlung von Randwertaufgaben im R_3 mit Hilfe von Einfach- und Doppelschichtpotentialen", *Numer. Math.*, *11*, pp.380-404.

Wendland, W.L. (1980), "On Galerkin collocation methods for integral equations of elliptic boundary value problems", *Internat. Schriftenreihe Numer. Math.*, *53*, pp.244-275.

Wendland, W.L. (1981), "Asymptotic accuracy and convergence", in *Progress in Boundary Element Methods*, *Vol. 1*, ed. Brebbia, C.A., Pentech Press (London-Plymouth), pp.289-313.

Wendland, W.L. (1983), "Boundary element methods and their asymptotic convergence", in *Theoretical Acoustics and Numerical Treatment*, ed. Filippi, P., CISM Courses and Lectures No. 277, Springer-Verlag (Wien-New York), pp.135-216.

Wendland, W.L. (1985), "On some mathematical aspects of boundary element methods for elliptic problems", in *The Mathematics of Finite Elements and Applications*, *V*, Mafelap 1984, ed. Whiteman, J.R., Academic Press (London), pp.193-227.

Wendland, W.L. (1985), "Asymptotic accuracy and convergence for point collocation methods", Chap. 9 of *Topics in Boundary Element Research*, *Vol. 2*, ed. Brebbia, C.A., Springer-Verlag (Berlin), pp.230-257.

Wendland, W.L. (1986), "On asymptotic error estimates for the combined boundary and finite element method", in *Innovative Numerical Methods in Engineering*, eds. Shaw, R. *et al*, Springer-Verlag (Berlin-Heidelberg-New York-Tokyo), pp.55-70.

Wendland, W.L., Stephan, E. and Hsiao, G.C. (1979), "On the integral equation method for the plane mixed boundary value problem of the Laplacian", *Math. Methods Appl. Sci.*, *1*, pp.265-321.

Wolfe, P. (1980), "An integral operator connected with the Helmholtz equation", *J. Funct. Anal.*, *36*, pp.105-113.

Zabreyko, P.P. *et al*, (1975), *Integral Equations — a Reference Text*, Noordhoff (Leyden).

Zhu, J. (1982), "Résolution par équations intégrales des problèmes de Stokes bi et tridimensionales", Doctoral Thesis, University Pierre et Marie Curie, Paris.

(see over for Name and Address)

W.L. Wendland
University Stuttgart
Math. Inst. A
Pfaffenwaldring 57
D-7000 Stuttgart 80
Germany FDR

19 Collocation Methods for One-Dimensional Fredholm and Volterra Integral Equations

H. BRUNNER

ABSTRACT

We survey recent advances in the numerical solution of second-kind Fredholm and Volterra integral equations by collocation methods in certain piecewise polynomial spaces. Particular attention is given to local and global superconvergence results for the collocation solution and its iterate and to integral equations possessing weakly singular kernels and nonsmooth solutions.

1. INTRODUCTION

The present survey is concerned with the state of the art in the numerical solution of one-dimensional second-kind integral equations of Fredholm type,

$$y(t) = g(t) + \lambda \int_0^T |t-s|^{-\alpha} K(t,s)\, y(s)\, ds\,,$$

$$t \in I := [0,T]\,, \quad 0 \leqslant \alpha < 1\,, \tag{1.1}$$

and of Volterra type,

$$y(t) = g(t) + \int_0^t (t-s)^{-\alpha} K(t,s)\, y(s)\, ds\,,$$

$$t \in I\,, \quad 0 \leqslant \alpha < 1\,, \tag{1.2}$$

in certain piecewise polynomial spaces. The paper will focus on collocation methods (Sections 2 and 3). However, in order to put the corresponding results in a somewhat wider perspective we use this introductory section to give a brief sketch of a number of other important developments which have occurred in the last ten years, and we conclude with a section on current software for integral equations (1.1) and (1.2).

The increasing interest in the numerical solution and the application of integral equations is reflected primarily in the number of conference proceedings which have appeared in recent years: the first of these, Delves and Walsh (1974), was followed by the proceedings edited by te Riele (1979a), Golberg (1979a), Albrecht and Collatz (1980), Anderssen, de Hoog and Lukas (1980), Baker and Miller (1982), and Hämmerlin and Hoffmann (1985) (compare also the special issue of the Journal of the Australian Mathematical Society, Sloan (1981)). While Noble (1977) exhibited the state of the art in the numerical analysis of integral equations some ten years ago, subsequent advances are described in the survey papers by te Riele (1979b), Golberg (1979b), Sloan (1980), Brunner (1982), and Baker (1982). As far as monographs are concerned we had, by the time of the last state-of-the-art conference a decade ago, the important books by Atkinson (1976a) and Baker (1977); within the last three years there appeared the ones by Chatelin (1983), Vainikko, Pedas and Uba (1984), Linz (1985), Delves and Mohamed (1985), and Brunner and van der Houwen (1986).

Recent advances in the theory and the numerical analysis of integral equations have led to their being used more frequently in various mathematical modelling processes. Of the numerous papers and books which reflect this we cite those by Lonseth (1977), Cushing (1977), Diekman (1979), Golberg (1979a), Anderssen, de Hoog and Lukas (1980), Graham (1980b), Anderssen and de Hoog (1982), Cameron and McKee (1983), Jakeman (1984), and Lodge, Renardy and Nohel (1986). Further sources of integral equations may be found in the references listed in their extensive bibliographies.

The last ten years have also witnessed the developments of

abstract frameworks for the analysis of numerical methods for integral equations, particularly for equations of Fredholm type. Here, we mention the contributions of Anselone and Lee (1976), Vainikko (1978), MacCamy and Weiss (1979), Chandler (1979), Graham (1980a), Chatelin (1983), Grigorieff (1983), Scott (1984), Dixon and McKee (1985), and Joe (1985a); in addition, consult Arnold and Wendland (1983) and Sloan and Thomée (1985) for further details and references.

We conclude this section with a brief description of some of the principal trends in the recent numerical analysis of one-dimensional integral equations of the forms (1.1) and (1.2).

(i) *Fredholm Integral Equations*: Here, research in the last decade has focused on Galerkin and collocation methods (and their iterated versions, introduced in Sloan (1976)) in spaces of continuous or discontinuous piecewise polynomials. Their local and global superconvergence properties are now well understood (see, e.g., Sloan (1976), Chandler (1979, 1980), Brunner (1981, 1984a), Chatelin and Lebbar (1981, 1984), Sloan (1982), Chatelin (1983), Spence and Thomas (1983), Sloan (1984), Joe (1985a,b,c,d), Graham, Joe and Sloan (1985), and Sloan and Thomée (1985) on equations with bounded kernels; analogous results for equations with weakly singular kernels may be found in Chandler (1979), Graham (1980a, 1981, 1982b), Schneider (1981), Vainikko and Uba (1981), Thomas (1981), Vainikko, Pedas and Uba (1984), and Joe (1985a)). Compare also Graham (1981) and Graham and Schneider (1985) for extensions of some of the above results to weakly singular Fredholm equations in $\mathbb{R}^n$.

(ii) *Volterra Integral Equations*: Systematic development and analysis of general classes of numerical methods for Volterra integral equations began only in the late 1970s (Noble's state-of-the-art paper of 1977 has a mere four pages dedicated to Volterra equations). For equations with bounded kernels the two main classes of methods are generalizations of analogous ones for initial-value problems in ODEs: they are the Volterra linear multistep methods of van der Houwen and te Riele (1985) (compare also Matthys (1976), McKee (1979), and Wolkenfelt (1982, 1983) for related ideas), and the Volterra-Runge-Kutta methods of Brunner, Hairer and Nørsett (1982) (see also van der Houwen and Blom (1978), van der Houwen (1980), van der Houwen, Wolkenfelt and Baker (1981), Duncan (1982), Hairer, Lubich and Nørsett (1983), Schlichte (1984), and Peluso and Piazza (1985)). Lubich (1983, 1985) introduced, respectively, Volterra-Runge-Kutta methods and fractional linear multistep methods for Volterra integral equations with weakly singular kernels.

Collocation methods for Volterra integral equations with bounded kernels were studied by Esser (1978) (Hermite-type collocation), Brunner (1980, 1981, 1984a), and Blom and Brunner (1985), while Logan (1976), Brunner and Nørsett (1981), Kershaw (1982), te Riele (1982), Brunner (1983), Brunner and te Riele (1984), Abdalkhani (1984), Brunner (1985a,b), and Dixon (1985) used collocation, based on polynomial splines or on special nonpolynomial splines, for Volterra equations with weakly singular kernels. Consult also Westreich and Cahlon (1980) and Cahlon (1981) for related methods.

2. COLLOCATION METHODS: REGULAR KERNELS AND SMOOTH SOLUTIONS

2.1 *The Approximating Polynomial Spline Spaces*

Let $I := [0,T]$ be the (compact) interval on which a given integral equation (1.1) or (1.2) is to be solved, and let $\{\Pi_N\}$ $(N \geq 1)$ denote a sequence of meshes on I, with Π_N: $0 = t_0 < t_1 < \cdots < t_N = T$ (for ease of notation we omit the superscript N in $t_n = t_n^{(N)}$ indicating the dependence of the mesh points t_n on N; the same convention will be used for other mesh-related quantities such as h). With a given mesh Π_N we associate:

$$h_n := t_{n+1} - t_n \; (n = 0, \cdots, N-1), \quad h := \max_n (h_n) \text{ (mesh diameter)},$$

$h' := \min_n (h_n)$; $\sigma_0 := [0, t_1]$, $\sigma_n := (t_n, t_{n+1}]$ $(n = 1, \cdots, N-1)$;

$Z_N := \{t_n : n = 1, \cdots, N-1\}$ (interior mesh points, or knots),

$\bar{Z}_N := Z_N \cup \{T\}$.

The following mesh types will play important roles in polynomial spline collocation (and Galerkin) methods:

(i) *Quasi-uniform meshes:* a mesh sequence $\{\Pi_N\}$ is called quasi-uniform if there exists a finite constant $\gamma \geq 1$, independent of N, such that $h/h' \leq \gamma$ for all $N \geq 1$ (recall that, by the above convention, h and h' depend on N). The special value $\gamma = 1$ yields the *uniform meshes*: $h_n = h = h' = TN^{-1}$ for all n. We note that the mesh diameter h of a given member of $\{\Pi_N\}$ satisfies

$$h \leq \gamma T N^{-1} \quad (= O(N^{-1})). \tag{2.1.1}$$

(ii) *Symmetrically graded meshes*: a mesh Π_N is called symmetrically graded if its points are given by

$$t_n := (2n/N)^r \, T/2, \quad \text{if } n = 0, \cdots, [N/2], \tag{2.1.2a}$$

and

$$t_n := T - t_{N-n}, \quad \text{if } n = [N/2]+1, \cdots, N. \tag{2.1.2b}$$

In the present context the *grading exponent* r is assumed to satisfy $r \geq 1$. For $r = 1$ we obtain the uniform mesh. For a general $r > 1$ we find that $h' = h_0 = 2^{r-1} T N^{-r}$, and

$$h < r T N^{-1} \quad (= O(N^{-1})); \tag{2.1.3}$$

i.e., the mesh diameter tends to zero as $N \to \infty$, a property shared by uniform mesh sequences. However, it is easily seen that a symmetrically graded mesh sequence with $r > 1$ is not quasi-uniform.

Mesh sequences of this type will be used when applying polynomial spline collocation methods to Fredholm integral equations (1.1) with $0 < \alpha < 1$ (cf. Section 3.1).

(iii) *Asymmetrically graded meshes*: a mesh Π_N will be called asymmetrically graded if its points are given by

$$t_n := (n/N)^r \cdot T, \quad n = 0, \cdots, N, \tag{2.1.4}$$

where the grading exponent is a real number $r \geqslant 1$. Here, we have $h' = h_0 = TNR^{-r}$, and the mesh diameter h satisfies again (2.1.3). However, an asymmetrically graded mesh sequence corresponding to $r > 1$ is not quasi-uniform.

Asymmetrically graded meshes will be of importance in connection with polynomial spline collocation methods for Volterra integral equations (1.2) with $0 < \alpha < 1$ (cf. Section 3.2).

We note that asymmetrically graded meshes were introduced by Rice in 1969 in the context of the optimal knot placement when approximating certain Hölder-continuous functions (such as $f(t) = t^{\beta}$, $0 < \beta < 1$, $t \in [0, 1]$) by polynomial spline functions. Symmetrically graded meshes were used, e.g., by Graham (1980a, 1982b), Schneider (1981), and Vainikko and Uba (1981).

For a given mesh Π_N and for given integers d and m, $-1 \leqslant d < m$, we define the space

$$S_m^{(d)}(Z_N) := \Big\{ u : u|_{\sigma_n} =: u_n \in \pi_m, \quad n = 0, \cdots, N-1; \\ u_{n-1}^{(j)}(t_n) = u_n^{(j)}(t_n) \text{ for } t_n \in Z_N \text{ and } j = 0, \cdots, d \Big\} \tag{2.1.5}$$

of (real) *polynomial spline functions* (or: piecewise polynomials) of degree m whose members possess the knots Z_N and lie in $C^d(I)$ if $d \geqslant 0$; if $d = -1$, then the members of $S_m^{(-1)}(Z_N)$ will in general have jump discontinuities at their knots. In (2.1.5), π_m denotes the space of (real) polynomials of degree not exceeding m.

In the following we shall use mostly the polynomial spline spaces corresponding to $d = 0$ (continuous piecewise polynomials) and $d = -1$ (discontinuous piecewise polynomials, with discontinuities at Z_N).

Note that $S_m^{(d)}(Z_N)$ is a finite-dimensional vector space, with

$$\dim S_m^{(d)}(Z_N) = N(m-d) + (d+1) . \tag{2.1.6}$$

A comprehensive treatment of polynomial spline functions, including the construction of suitable bases, may be found in Schumaker (1981).

2.2 *Fredholm Integral Equations of the Second Kind*

2.2.1 *Collocation and iterated collocation.* We consider first the Fredholm integral equation

$$y(t) = g(t) + \lambda \int_0^T K(t,s)\, y(s)\, ds , \qquad t \in I , \tag{2.2.1}$$

where $\lambda \neq 0$ is such that $1/\lambda$ is not an eigenvalue of the underlying Fredholm integral operator, and where g and K are real-valued and continuous functions. The unique solution $y \in C(I)$ of (2.2.1) will be approximated in the polynomial spline space $S_{m-1}^{(-1)}(Z_N)$, using collocation techniques (recall that the dimension of this space is Nm; cf. (2.1.6)). To be more precise, let $\{c_j\}$ be m real numbers satisfying $0 \leqslant c_1 < \cdots < c_m \leqslant 1$, and define the sets

$$X_n := \left\{ t_{n,j} := t_n + c_j h_n : \quad j = 1, \cdots, m \right\} \quad (n = 0, \cdots, N-1) , \tag{2.2.2a}$$

and

$$X(N) := \bigcup_{n=0}^{N-1} X_n . \tag{2.2.2b}$$

The $\{c_j\}$ will be referred to as *collocation parameters*, and $X(N)$ is the set of corresponding *collocation points*. Note that $|X(N)| \leqslant Nm$, with equality if and only if $c_m - c_1 < 1$.

We now seek an element $u \in S_{m-1}^{(-1)}(Z_N)$ which satisfies the given integral equation (2.2.1) on the set $X(N)$ ("collocation" on $X(N)$):

$$u(t) = g(t) + \lambda \int_0^T K(t,s)\, u(s)\, ds \quad \text{for all} \quad t \in X(N). \tag{2.2.3}$$

This *collocation equation* can be written as

$$u_n(t_{n,j}) = g(t_{n,j}) + \lambda \sum_{i=0}^{N-1} h_i \int_0^T K_{n,j}(t_i + \tau h_i)\, u_i(t_i + \tau h_i)\, d\tau \tag{2.2.4a}$$

$$(j = 1, \cdots, m;\ n = 0, \cdots, N-1),$$

where $K_{n,j}(\cdot) := K(t_{n,j}, \cdot)$, and where u_n is the restriction of u to the subinterval σ_n, given by

$$u_n(t_n + \tau h_n) = \sum_{l=1}^{m} L_l(\tau)\, U_{n,l}, \quad t_n + \tau h_n \in \sigma_n\,. \tag{2.2.4b}$$

Here

$$U_{n,l} := u_n(t_{n,l}) \quad \text{and} \quad L_l(\tau) := \sum_{k \neq l}^{m} (\tau - c_k)/(c_l - c_k).$$

Hence, in order to determine the collocation solution $u \in S_{m-1}^{(-1)}(Z_N)$ we have to solve a linear system in $\mathbb{R}^{Nm}$ for the values $(U_{0,1}, \cdots, U_{0,m}; \cdots; U_{N-1,1}, \cdots, U_{N-1,m})$; once these values have been found, u is completely determined by (2.2.4b). Note that if the collocation parameters $\{c_j\}$ are chosen such that

$$0 = c_1 < c_2 < \cdots < c_{m-1} < c_m = 1 \tag{2.2.5}$$

(implying that $|X(N)| = N(m-1) + 1 = \dim S_{m-1}^{(0)}(Z_N)$), then u will be continuous on I; i.e., we then have

$$u \in S_{m-1}^{(-1)}(Z_N) \cap C(I) = S_{m-1}^{(0)}(Z_N).$$

The *iterated collocation solution* u^I based on the above collocation approximation u is defined by

$$u^I(t) := g(t) + \lambda \int_0^T K(t,s)\, u(s)\, ds, \quad t \in I; \tag{2.2.6a}$$

in analogy to (2.2.4a) it can be written as

$$u^I(t) = g(t) + \lambda \sum_{i=0}^{N-1} h_i \int_0^1 K(t, t_i + \tau h_i)\, u_i(t_i + \tau h_i)\, d\tau, \quad t \in I. \tag{2.2.6b}$$

The iterate u^I has the following important properties:

(i) u^I coincides with the underlying collocation solution u at the collocation points: $u^I(t) = u(t)$ for all $t \in X(N)$.

(ii) If g and K in (2.2.1) satisfy $g \in C^q(I)$ and $K \in C^q(I \times I)$ for some $q \geqslant 0$, then u^I possesses the same degree of smoothness: $u^I \in C^q(I)$ (while u itself is, in general, not even continuous on I).

Iterated approximations to solutions of second-kind Fredholm integral equations were introduced by Sloan (1976), and it was shown that in the case of Galerkin approximations u^I is always more accurate than u (provided that N is sufficiently large). Compare also Sloan, Burn and Datyner (1975), and Sloan and Burn (1979) (where iterated polynomial collocation is studied). Further references will be given in the following section where we discuss the convergence properties of the collocation solution u and its iterate u^I.

2.2.2 *Local and global superconvergence.* We start with some notation. If f is piecewise continuous on I, let

$$\|f\|_\infty := \sup\{|f(t)| : t \in I\}.$$

For given I set $D := I \times I$.

Consider first the collocation solution u given by (2.2.4). If g and K in (2.2.1) satisfy $g \in C^m(I)$, $K \in C^m(D)$ for all $t \in I$, then

$$\|y-u\|_\infty = O(N^{-m}) \tag{2.2.7}$$

for any quasi-uniform mesh sequence $\{\Pi_N\}$. This result reflects the approximation power of the spline spaces $S_{m-1}^{(d)}(Z_N)$, $d \in \{-1, 0\}$, and it holds for any choice of the collocation parameters $\{c_j\}$.

As we shall see in Section 3.1, if the kernel $K(t,s)$ is replaced by the weakly singular kernel $|t-s|^{-\alpha} K(t,s)$, with $0 < \alpha < 1$ and $K \in C^m(D)$ for $t \in I$, then (2.2.7) will no longer hold: we only obtain $\|y-u\|_\infty = O(N^{-(1-\alpha)})$ on quasi-uniform $\{\Pi_N\}$, regardless of how we choose the degree of the approximating spline function. The optimal convergence order can be recovered if $\{\Pi_N\}$ is a symmetrically graded mesh sequence with suitable grading exponent r (depending on m and α).

Do there exist specific sets of collocation parameters $\{c_j\}$ for which we obtain

$$\max_{t \in \bar{Z}_N} |y(t)-u(t)| = O(N^{-p^*}), \tag{2.2.8a}$$

and

$$\|y-u^I\|_\infty = O(N^{-p^I}), \tag{2.2.8b}$$

with $p^* > m$, $p^I > m$? If so, then we say that u is *locally superconvergent* (on $\bar{Z}_N := Z_N \cup \{T\}$) and u^I is *globally superconvergent* (on I).

In order to see why and when u and its iterate u^I possess such superconvergence properties, and how the attainable orders p^* and p^I are related to the smoothness of g and K (and thus to that of the exact solution y), recall first that the solution of (2.2.1) is given by

$$y(t) = g(t) + \lambda \int_0^T R(t,s;\lambda)\, g(s)\, ds, \quad t \in I,$$

where $R(t,s;\lambda)$ denotes the resolvent kernel. It follows from the classical Fredholm theory that $R(t,s;\lambda)$ (as a function of t and s) inherits the smoothness of $K(t,s)$. Consider now (2.2.3): an alternative form of this collocation equation is

$$u(t) = g(t) - \delta(t) + \lambda \int_0^T K(t,s)\, u(s)\, ds, \quad t \in I,$$

where the residual δ vanishes at the collocation points $X(N)$ and is smooth on each subinterval σ_n (provided g and K are smooth functions). Since the *collocation error* $e := y-u$ satisfies the Fredholm integral equation

$$e(t) = \delta(t) + \lambda \int_0^T K(t,s)\, e(s)\, ds, \qquad t \in I,$$

it may be expressed in the form

$$e(t) = \delta(t) + \lambda \int_0^T R(t,s;\lambda)\, \delta(s)\, ds, \quad t \in I, \tag{2.2.9a}$$

or, equivalently, as

$$e(t) = \delta(t) + \lambda \sum_{i=1}^{N-1} h_i \int_0^1 R(t, t_i + \tau h_i; \lambda)\, \delta(t_i + \tau h_i)\, d\tau, \quad t \in I. \tag{2.2.9b}$$

By (2.2.1) and (2.2.6a), the *iterated collocation error* $e^I := y - u^I$ is related to e by

$$e^I(t) = \lambda \int_0^T K(t,s)\, e(s)\, ds, \quad t \in I.$$

Hence, using (2.2.9a) and the appropriate Fredholm identity for R and K, we obtain

$$e^I(t) = \lambda \int_0^T R(t, s; \lambda)\, \delta(s)\, ds, \quad t \in I \tag{2.2.10a}$$

(which closely resembles (2.2.9a), except that now the term $\delta(t)$ no longer occurs outside the integral). In analogy to (2.2.9b) we also write

$$e^I(t) = \lambda \sum_{i=0}^{N-1} h_i \int_0^1 R(t, t_i + \tau h_i; \lambda)\, \delta(t_i + \tau h_i)\, d\tau, \quad t \in I. \tag{2.2.10b}$$

Consider now the integrals in (2.2.9b) and (2.2.10b): according to remarks made earlier, their integrands are smooth if g and K are smooth functions. We set

$$\int_0^1 R(t, t_i + \tau h_i; \lambda)\, \delta(t_i + \tau h_i)\, d\tau = \sum_{l=1}^{m} w_l R(t, t_{i,l}; \lambda)\, \delta(t_{i,l}) + E_i(t), \quad t \in I;$$

i.e., we approximate these integrals by m-point interpolatory quadrature rules whose abscissae are the collocation points X_i (cf. (2.2.2a)) and which induce the quadrature errors $E_i(t)$. Since δ vanishes at all collocation points, (2.2.9b) and (2.2.10b) become, respectively,

$$e(t) = \delta(t) + \lambda \sum_{i=0}^{N-1} h_i E_i(t), \quad t \in I, \tag{2.2.11}$$

and

$$e^I(t) = \lambda \sum_{i=0}^{N-1} h_i E_i(t), \quad t \in I. \tag{2.2.12}$$

This shows that the convergence order of the iterated collocation error e^I is governed solely by that of the quadrature errors $E_i(t)$, and this order depends, on the one hand, on the choice of the collocation parameters $\{c_j\}$ and, on the other, on the degree of smoothness of the integrand. While the resolvent kernel inherits the smoothness of K, the smoothness of the residual δ (on σ_i) depends on that of K <u>and</u> of g (i.e., on that of the exact solution y of (2.2.1)). In contrast to this, the order of convergence of the collocation error e depends also on the order of δ: δ vanishes for $t \in X(N)$ but behaves globally like $\|\delta\|_\infty = O(N^{-m})$.

Thus, the following results on the attainable orders of local and global superconvergence hold for $u \in S_{m-1}^{(-1)}(Z_N)$ and its iterate u^I (compare, e.g., Chatelin and Lebbar (1981, 1984), Chatelin (1983), Brunner (1981, 1984a), and Joe (1985a,b,c)):

(i) If the collocation parameters $\{c_j\}$ are the *Gauss points* (zeros of the shifted Legendre polynomial $P_m(2s-1)$), then the collocation solution u is <u>not</u> locally superconvergent on $\bar{Z}_N$ (i.e., $p^* = m$ in (2.2.8a)); its iterate u^I, however, exhibits global superconvergence (on I) of order $p^I = 2m$, provided that $g \in C^{2m}(I)$ and $k \in C^{2m}(D)$ $(t \in I)$:

$$\|y - u^I\|_\infty = O(N^{-2m}). \tag{2.2.13}$$

(Note that by (2.2.11), u is locally superconvergent <u>on $X(N)$</u> of order $2m$.)

(ii) If the collocation parameters $\{c_j\}$ are the *Radau II points* (zeros of $P_{m-1}(2s-1) - P_m(2s-1)$), then u exhibits local superconvergence (on $\bar{Z}_N$) of order $p^* = 2m-1$. The corresponding iterated collocation solution u^I is globally superconvergent with the same order $p^I = p^*$.

(iii) If the $\{c_j\}$ are the *Lobatto points* (zeros of $s(s-1)P'_{m-1}(2s-1)$), then u (which is now an element of $S_{m-1}^{(0)}(Z_N)$, since we have $c_1 = 0$ and $c_m = 1$; cf. (2.2.5)) satisfies (2.2.8a) with $p^* = 2m-2$; for its iterate u^I we obtain the global superconvergence order $p^I = 2m-2$. In other words, if we employ *continuous* spline approximations of degree $m-1$, then the best possible orders of superconvergence are $p^* = 2m-2$ locally (for u) and $p^I = 2m-2$ globally (for u^I). The iterated collocation solution u^I can exhibit a higher order of global superconvergence only if the underlying collocation solution u is discontinuous at the knots Z_N.

In Chatelin and Lebbar (1981, 1984), Chatelin (1983), and Joe (1985a,b,c) the above superconvergence results are derived in a different way, using a functional-analytic setting. The most comprehensive analysis is due to Joe (1985a) where the setting is that of "fractional derivative spaces" (Nikolskii spaces). This author exhibits the precise relationship between the attainable order p^I in (2.2.8b) and the degree of smoothness of g and K; his analysis encompasses also the discrete collocation and iterated collocation solutions (cf. Section 2.2.3).

Chatelin and Lebbar (1981) showed that if K in (2.2.1) is a *Green's kernel*,

$$K(t,s) = \begin{cases} k_1(t,s) & \text{for } 0 \leqslant s \leqslant t \leqslant T, \\ k_2(t,s) & \text{for } 0 \leqslant t \leqslant s \leqslant T, \end{cases}$$

with k_1 and k_2 smooth on their triangular domains but with K possessing a low degree of smoothness on $s = t$, then the global superconvergence property (2.2.13) is lost: u^I now possesses only local superconvergence on $\bar{Z}_N$ of order $p^* = 2m$. (An even more extreme case is given by a Volterra kernel, where $k_2(t,s) \equiv 0$ and where K is not continuous on $s = t$; see Section 2.3.)

The study of superconvergence in the approximate solution of Fredholm integral equations has its origins in the early 1970s (we mention the influential paper by de Boor and Swartz (1973); see also Křížek and Neittaanmäki (1984) for a comprehensive survey of superconvergence results in ODEs, PDEs, and integral equations). The first superconvergence results in spline approximations for second-kind Fredholm equations concern Galerkin-type approximations; see Sloan (1976), Richter (1978), and Chandler (1979, 1980).

Early papers on spline collocation for Fredholm integral equations are by Houstis and Papatheodoru (1978) and Volk (1979). The mathematical foundation of spline collocation in $S_{m-1}^{(-1)}(Z_N)$ in the setting of the function space $L_\infty(I)$ is due to Atkinson, Graham and Sloan (1983). The paper by Graham, Joe and Sloan (1985) provides much insight into the relative merits of spline collocation and spline Galerkin methods (and their iterative variants), especially with regard to the necessary smoothness requirements on the kernel and the exact solution.

Finally, the reader interested in collocation approximations for higher-dimensional Fredholm integral equations is referred to the important paper by Arnold and Wendland (1983) and its extensive bibliography. In addition, see also Graham (1985).

2.2.3 *Discrete collocation solutions*. Usually the integrals occurring in the collocation equations (2.2.4a) and (2.2.6b) cannot be found analytically and will thus have to be evaluated by suitable quadrature rules, leading to a perturbation in u and its iterate u^I. We shall refer to the resulting approximations $\hat{u}$ (which is still in $S_{m-1}^{(-1)}(Z_N)$) and $\hat{u}^I$ as *discrete collocation solution* and *discrete iterated collocation solution*. It is clear that the order of convergence of $\hat{u}$ or $\hat{u}^I$ will not exceed that of the "exact" collocation solutions u or u^I. On the other hand, a bad choice of the quadrature rules will contaminate these orders, especially in the case of optimal global superconvergence in u^I (cf. (2.2.13)). If, however, g and K

possess only a low degree of smoothness, then there is generally no point in employing quadrature rules possessing a high degree of precision.

The question of how to select appropriate quadrature rules when discretizing the collocation equations (2.2.4a) and (2.2.6b) has been answered in detail in Joe (1985a,c). Here, we shall consider only a particular (though important) case, namely, discretization based on interpolatory m-point quadrature formulas with abscissas given by the collocation points. Setting $\hat{U}_{n,j} := \hat{u}_n(t_{n,j})$, the discretized version of (2.2.4a) is then given by

$$\hat{U}_{n,j} = g(t_{n,j}) + \lambda \sum_{i=0}^{N-1} h_i \sum_{l=1}^{m} w_l K_{n,j}(t_{i,l}) \hat{U}_{i,l} \quad (2.2.14)$$

$$(j = 1, \cdots, m;\; n = 0, \cdots, N-1).$$

The values $\hat{U}_{n,j}$ obtained as the solution of this linear system are then used in the discretized version of (2.2.6b),

$$\hat{u}^I(t) := g(t) + \lambda \sum_{i=0}^{N-1} h_i \sum_{l=1}^{m} w_l K(t, t_{i,l}) \hat{U}_{i,l}, \quad t \in I. \quad (2.2.15)$$

The quadrature weights $\{w_l\}$ are

$$w_l := \int_0^1 \prod_{k \neq l}^{m} (\tau - c_k)/(c_l - c_k)\, d\tau \quad (l = 1, \cdots, m).$$

Comparison of (2.2.14) and (2.2.15) also reveals that the postprocessing (2.2.15) of the discrete collocation solution $\hat{u}$ is very inexpensive; the major part of the computational effort is involved in the computation of $\hat{u}$.

It can be shown that, for sufficiently smooth g and K in (2.2.1), the local and global superconvergence results for u and u^I remain valid for the discrete collocation solution $\hat{u}$ and its iterate $\hat{u}^I$ (as given by (2.2.14) and (2.2.15)); compare Brunner (1984a) and Joe (1985a,c) for details.

An important earlier analysis of discretized collocation in a functional-analytic setting was given by Anselone and Lee

(1976) (see also Anselone and Krabs (1979) where this approach is extended to Fredholm integral equations (1.1) with $0 < \alpha < 1$). Discretized Galerkin methods have been studied by Spence and Thomas (1983), Joe (1985d), and Atkinson and Bogomolny (1985). The recent monograph by Delves and Mohamed (1985) describes various implementations of Galerkin-type methods for Fredholm integral equations ("fast" versions can frequently be obtained by employing appropriate FFT techniques).

As mentioned above, the calculation of the iterated collocation (or Galerkin) solution may be viewed as a *postprocessing* method. Another class of such methods is given by the *extrapolation methods* (see, e.g., the book by Marchuk and Shaidurov (1983)). An interesting (and, in our context, relevant) observation is due to Lin Qun and Liu Jiaquan (1980): if the kernel of (2.2.1) has a low degree of smoothness (e.g., if it is a Green's kernel), then a globally superconvergent approximation can be obtained by applying extrapolation techniques to the iterated collocation solution (which is, in this case, not globally superconvergent; cf. Chatelin and Lebbar (1981)).

2.3 *Volterra Integral Equations of the Second Kind*

2.3.1 *Collocation and iterated collocation.* If the Volterra integral equation

$$y(t) = g(t) + \int_0^t K(t,s)\,y(s)\,ds\,, \quad t \in T := [0,T]\,, \qquad (2.3.1)$$

is solved numerically by collocation in $S_{m-1}^{(-1)}(Z_N)$, using the collocation points $X(N)$ introduced in (2.2.2), then the collocation equation assumes the form

$$u_n(t_{n,j}) + h_n \int_0^{c_j} K_{n,j}(t_n + \tau h_n)\,u_n(t_n + \tau h_n)\,d\tau + \\ + \sum_{i=0}^{n-1} h_i \int_0^1 K_{n,j}(t_i + \tau h_i)\,u_i(t_i + \tau h_i)\,d\tau \qquad (2.3.2)$$

$$(j = 1, \cdots, m;\ n = 0, \cdots, N-1),$$

where $K_{n,j}(\cdot) := K(t_{n,j}, \cdot)$, and where the restriction u_n of $u \in S_{m-1}^{(-1)}(Z_N)$ to the subinterval σ_n is again given by (2.2.4b). Since the upper limit of integration in a Volterra integral equation is the variable t, (2.3.2) represents, in contrast to the analogous Fredholm collocation equation (2.2.4a), a linear (block-) *triangular* system in $\mathbb{R}^{Nm}$. In other words, the collocation solution u is now generated *recursively*, by computing successively the restrictions $u_0, u_1, \cdots, u_{N-1}$ to the subintervals $\sigma_0, \sigma_1, \cdots, \sigma_{N-1}$; i.e., we solve a sequence of N "small" linear systems in $\mathbb{R}^m$ (where, in most applications, $m \leqslant 4$).

If g and K in (2.3.1) are continuous then the collocation solution u can be forced to be a member of the smoother spline space $S_{m-1}^{(0)}(Z_N)$ by prescribing collocation parameters $\{c_j\}$ with $c_1 = 0$ and $c_m = 1$ (recall (2.2.5)).

The *iterated collocation solution* u^I corresponding to the above collocation solution u is defined by

$$u^I(t) := g(t) + \int_0^t K(t,s)\,u(s)\,ds\,, \quad t \in I\,; \qquad (2.3.3a)$$

for $t = t_n + v h_n \in \sigma_n$, we obtain the analogue of (2.2.6b),

$$\begin{aligned} u^I(t) = g(t) + h_n \int_0^v K(t, t_n + \tau h_n)\,u_n(t_n + \tau h_n)\,d\tau\, + \\ + \sum_{i=0}^{n-1} h_i \int_0^1 K(t, t_i + \tau h_i)\,u_i(t_i + \tau h_i)\,d\tau\,. \end{aligned} \qquad (2.3.3b)$$

As in the case of Fredholm integral equations, u^I coincides with the underlying collocation solution u at the collocation points $X(N)$. However, u^I now no longer inherits the smoothness of g and K: if these functions are q times continuously differentiable, then we have, in general, only $u^I \in C(I)$, with its derivative being discontinuous on Z_N, regardless of q. It will thus come as no surprise that the iterated collocation solution cannot exhibit global superconvergence of order $p^I > m$; on $\bar{Z}_N$

the attainable order of local superconvergence is $p^* = 2m$ (cf. (2.3.6) below). This is reminiscent of the result for Fredholm integral equations with Green's kernels (Section 2.2.2).

2.3.2 *Local superconvergence properties.* The superconvergence results for Fredholm collocation solutions may be described as follows: if p^* is the order of local superconvergence of u at the collocation points $X(N)$, then its iterate u^I possesses the order $p^I = p^*$ of global superconvergence on I (cf. (2.2.11) and (2.2.12)). For Volterra integral equations (2.3.1) the iterated collocation solution u^I given by (2.3.3b) cannot be globally superconvergent on I: the global superconvergence results for the Fredholm iterate become local superconvergence results (on $\bar{Z}_N$) for the Volterra iterate. Before outlining the reason for this contrast we give a summary of local superconvergence results for quasi-uniform mesh sequences.

Table 2.3.1 Local Superconvergence for Volterra Integral Equations

Collocation parameters $\{c_j\}$	Attainable order: $u \in S_{m-1}^{(-1)}(Z_N)$
$-\ 0 \leqslant c_1 < \cdots < c_m \leqslant 1$	$\lVert y-u \rVert_\infty = O(N^{-m})$
$-$ Gauss points	$\lVert y-u^I \rVert_\infty = O(N^{-m})$
	$\max_{t \in \bar{Z}_N} \lvert y(t) - u^I(t) \rvert = O(N^{-2m})$
$-$ Radau II points	$\lVert y-u^I \rVert_\infty = O(N^{-m})$
$(u^I(t) = u(t), \quad t \in \bar{Z}_N)$	$\max_{t \in \bar{Z}_N} \lvert y(t) - u(t) \rvert = O(N^{-(2m-1)})$
$-$ Lobatto points	$\lVert y-u^I \rVert_\infty = O(N^{-(2m-2)})$
$(u^I(t) = u(t), \quad t \in \bar{Z}_N)$	$\max_{t \in \bar{Z}_N} \lvert y(t) - u(t) \rvert = O(N^{-(2m-2)})$

Note that this list is not exhaustive: u has local superconvergence of order $p^* = 2m-2$ also if $c_1, \cdots, c_{m-1}$ are the $m-1$ Gauss points in $(0,1)$ and $c_m = 1$. While u is then in $S_{m-1}^{(-1)}(Z_N)$, the collocation solution for the Lobatto points (which has the same order p^*) is in $S_{m-1}^{(0)}(Z_N)$.

The non-existence of global superconvergence results for u^I is a consequence of the fact that, in contrast to (2.2.10a), the iterated collocation error $e^I := y - u^I$ satisfies now

$$e^I(t) = \int_0^t R(t,s)\,\delta(s)\,ds\,, \quad t \in I\,, \tag{2.3.4}$$

where R is the resolvent kernel of K in (2.3.1), and where the residual δ vanishes at the collocation points $X(N)$ and is smooth on each subinterval σ_n. If $t = t_n$ is in $\bar{Z}_N$, then (2.3.4) may be written as

$$e^I(t_n) = \sum_{i=0}^{n-1} h_i \int_0^1 R(t_n, t_i + \tau h_i)\,\delta(t_i + \tau h_i)\,d\tau\,, \tag{2.3.5}$$

and the argument following (2.2.10b) can then be used to derive the result

$$\max_{t_n \in \bar{Z}_N} |e^I(t_n)| = O(N^{-2m})\,, \tag{2.3.6}$$

provided the $\{c_j\}$ are the m Gauss points in $(0,1)$. If, however, t is not in $\bar{Z}_N$; i.e., $t = t_n + vh_n$, with $0 < v < 1$, then the analogue of (2.3.5) contains also an integral over the proper subinterval $[0,v]$ of $[0,1]$, and it is easy to see that this leads to $|e^I(t)| = O(N^{-m})$ only.

The above superconvergence results are due to Brunner (1980, 1981), Brunner and Nørsett (1981) (for the collocation solution u), and to Brunner (1984a) and Blom and Brunner (1985) (for u^I; the last paper extends the results to nonlinear Volterra integral equations of the second kind). Compare also Brunner and van der Houwen (1986) for a full account of spline collocation and iterated collocation methods for (2.3.1) and its nonlinear counterpart.

The first local superconvergence results (for the Radau II points and the Lobatto points) for certain discrete collocation solutions were established by de Hoog and Weiss (1975); they generalize analogous superconvergence results given by Axelsson in 1969 for a special Volterra equation equivalent to the initial-value problem for $y' = f(t,y)$.

2.3.3 *Discretized collocation methods.* In the discretization of the collocation equation (2.3.2) we are faced with a problem not encountered in Fredholm collocation equations (recall (2.2.4a)): since the first integral in (2.3.2) has the upper limit c_j (instead of 1), and since the Volterra kernel $K(t,s)$ is defined only on the triangle $S := \{(t,s): 0 \leqslant s \leqslant t \leqslant T\}$, we are forced to employ a separate quadrature rule for each of the (nonempty) subintervals $[0,c_j]$. Obvious choices are the m-point interpolatory rules based on the affine images of the collocation parameters $\{c_1,\cdots,c_m\}$ in $[0,c_j]$: here, the abscissas are given by $\{t_n + c_j c_l h_n : l = 1,\cdots,m\}$, and the quadrature weights are $w_{j,l} := c_j w_l$ (with $\{w_l\}$ as in (2.2.14)). This particular choice leads to the modified *implicit Volterra-Runge-Kutta method* of de Hoog and Weiss (1975),

$$\hat{U}_{n,j} = g(t_{n,j}) + c_j h_n \sum_{l=1}^{m} w_l K_{n,j}(t_n + c_j c_l h_n)\,\hat{u}_n(t_n + c_j c_l h_n) + $$

$$+ \sum_{i=0}^{n-1} h_i \sum_{l=1}^{m} w_l K_{n,j}(t_{i,l})\,\hat{U}_{i,l} \qquad (2.3.7)$$

$$(j = 1,\cdots,m;\; n = 0,\cdots,N-1)$$

where $\hat{U}_{n,j} := \hat{u}_n(t_{n,j})$, and where the discrete collocation solution $\hat{u} \in S^{(-1)}_{m-1}(Z_N)$ is given on the subinterval σ_n by (2.2.4b), with u_n replaced by $\hat{u}_n$.

As has been indicated in the previous section (cf. (2.3.6)), the discrete iterated collocation solution $\hat{u}^I(t)$ is of practical interest only if $t = t_n \in \bar{Z}_N$. In order to preserve the local superconvergence property (2.3.6) of $u^I(t_n)$, we choose the discretization

$$\hat{u}^I(t_n) := g(t_n) + \sum_{i=0}^{n-1} h_i \sum_{l=1}^{m} w_l K(t_n, t_{i,l}) \hat{U}_{i,l}, \quad t_n \in \bar{Z}_N, \tag{2.3.8}$$

where the values $\hat{U}_{i,l}$ are determined from the discretized collocation equation (2.3.7), corresponding to the Gauss collocation parameters. Equation (2.3.8) is the analogue of (2.2.15) except that here the values of t are restricted to the finite set $\bar{Z}_N$.

A detailed analysis of discrete collocation solutions and their iterates may be found in Brunner (1984a) and, especially, in Brunner and van der Houwen (1986). Blom and Brunner (1985) extended the above analysis to nonlinear Volterra integral equations; no such analysis exists as yet for nonlinear Fredholm integral equations.

2.3.4 *Additional remarks.* The local superconvergence results encountered in polynomial spline collocation for second-kind Volterra integral equations do not carry over to Volterra integral equations of the first kind (i.e., to (2.3.1) with $y(t)$ on its left-hand side replaced by 0). In fact, local superconvergence at the points $\bar{Z}_N$ cannot occur (Brunner (1979a,b)). However, Eggermont (1986) has shown that postprocessing of u based on certain local interpolation techniques can lead to a slightly higher convergence order at these points.

In contrast to this, collocation in $S_m^{(0)}(Z_N)$ for first-order Volterra integro-differential equations yields local superconvergence of order $p^* = 2m$ for u if collocation occurs at the Gauss points; u' has then only the order $p = m$ on $\bar{Z}_N$. If the collocation parameters are the Radau II points, then both u and u' exhibit local superconvergence of order $p^* = 2m-1$ (Brunner (1984b, 1986b)). Compare also the survey paper by Bellen (1985) for local superconvergence results in the numerical solution of functional differential equations (of which the Volterra integro-differential equations are a special case).

3. COLLOCATION METHODS: WEAKLY SINGULAR KERNELS AND NONSMOOTH SOLUTIONS

3.1 *Fredholm Integral Equations of the Second Kind*

3.1.1 *Collocation and iterated collocation.* We consider now the Fredholm integral equation (1.1) with $0<\alpha<1$. As before, g and K are assumed to be continuous functions, and λ is such that the homogeneous version of (1.1) has only the trivial solution.

The collocation approximation $u \in S_{m-1}^{(-1)}(Z_N)$ to the solution of (1.1) is defined by a collocation equation similar to (2.2.4a), except that here the integrals over $[0,1]$ are of the form

$$\Phi_{n,i}(u;t) := \int_0^1 |(t-t_i)/h_i-\tau|^{-\alpha} K(t,t_i+\tau h_i)\, u_i(t_i+\tau h_i)\,d\tau\,, \tag{3.1.1}$$

with $t=t_{n,j} \in X(N)$, and they are to be multiplied by $h_i^{1-\alpha}$ instead of h_i. Again, if $c_1=0$ and $c_m=1$, then u is in the continuous spline space $S_{m-1}^{(0)}(Z_N)$.

If we employ quasi-uniform mesh sequences $\{\Pi_N\}$, then it turns out that for any smooth g and K we can only have $\|y-u\|_\infty = O(N^{-(1-\alpha)})$, while the best result for the corresponding iterated collocation solution,

$$u^I(t) := g(t)+\lambda \sum_{i=0}^{N-1} h_i^{1-\alpha}\,\Phi_{n,i}(u;t)\,, \qquad t\in I, \tag{3.1.2}$$

is $\|y-u^I\|_\infty = O(N^{-2(1-\alpha)})$ (see Chandler (1979)). Similar results hold for spline Galerkin methods and their iterative variants (Graham (1980a, 1982b)).

The reason for this disappointingly low order of convergence lies in the behaviour of the exact solution y near the endpoints $t=0$ and $t=T$ of I: its derivatives are, in general, unbounded at these points (Richter (1976), Pitkäranta (1979), Chandler (1979), Schneider (1979), Graham (1980a, 1982a), Vainikko and Pedas (1981), and Vainikko, Pedas and Uba (1984)). In order to describe this singular behaviour more precisely we use a terminology due to Rice (1969): a function $f \in C(I)$ is said to be of type $(\beta, \mu, \{0, T\})$ $(0 < \beta < 1, \mu \in N)$ if (i) f is Hölder continuous on I; i.e., if $|f(t) - f(t')| \leq L|t-t'|^{\beta}$ for $t, t' \in I$, and (ii) f has continuous derivatives of order μ on the open interval $(0, T)$, with $|f^{(k)}(t)| \leq C_1 t^{\beta-k}$ on $0 < t \leq T/2$, and $|f^{(k)}(t)| \leq C_2(T-t)^{\beta-k}$ on $T/2 \leq t < T$ $(k = 1, \cdots, \mu)$. Schneider (1979) has shown that, if g and K are suitably smooth, the exact solution y is of type $(1-\alpha, m, \{0, T\})$. Moreover, if g is only of type $(\beta, m, \{0, T\})$ with $0 < \beta < 1$, then y is of type $(\gamma, m, \{0, T\})$, where $\gamma := \min(1-\alpha, \beta)$. Hence, the above convergence result for u reflects the approximation power of the spline spaces $S_{m-1}^{(d)}(Z_N)$ for functions of type $(1-\alpha, m, \{0, T\})$ when the knot sequence is quasi-uniform (Rice (1969), Schumaker (1981)).

If the collocation solution $u \in S_{m-1}^{(-1)}(Z_N)$ and its iterate are to exhibit a higher order of convergence, then one has to resort to *symmetrically graded mesh sequences* (cf. (2.1.2)). The following results then hold:

(i) (Vainikko and Uba (1981)). If the solution y is of type $(1-\alpha, m, \{0, T\})$, then the collocation solution $u \in S_{m-1}^{(d)}(Z_N)$ $(d \in \{-1, 0\})$ satisfies $\|y-u\|_{\infty} = O(N^{-m})$ if $\{\Pi_N\}$ is a symmetrically graded mesh sequence of the form (2.1.2) corresponding to the grading exponent $r = m/(1-\alpha)$. A lower value of r will lead to a decrease of this order.

(ii) (Schneider (1981); see also Chandler (1979)). If the mesh sequence is graded as in (i), then we have, in general, $\|y-u^I\|_{\infty} = O(N^{-m})$. However, if the collocation parameters $\{c_j\}$ satisfy the relation

$$\int_0^1 \prod_{j=1}^{m} (s - c_j)\, ds = 0\,, \tag{3.1.3}$$

and if the solution y is of type $(1-\alpha, m+1, \{0, T\})$, then the iterated collocation solution u^I defined by (3.1.2) converges slightly faster,

$$\|y-u^I\|_\infty = O(N^{-(m+1-\alpha)}) \tag{3.1.4}$$

provided that the grading exponent in (2.1.2) is given by $r=r^I := (m+2-\alpha)/(2(1-\alpha))$. (Analogous results hold if the weak singularity $|t-s|^{-\alpha}$ in (1.1) is replaced by $\log|t-s|$.)

We note that (3.1.3) is fulfilled if the $\{c_j\}$ are the Gauss points, the Radau II points, or (for m odd) the Lobatto points. The order of convergence shown in (3.1.4) is optimal for $0<\alpha<1$; it is also attained if the grading exponent r is chosen as in (i), $r=m/(1-\alpha)$, provided that $m\geqslant 2$ (we then have $r>r^I$). See also Graham (1982b) for similar results concerning Galerkin solutions and their iterates.

Schneider (1981) actually derived the result (3.1.4) for the product integration method (see Atkinson (1976a) for an introduction to such methods). As will be indicated in the next section, a product integration method may be viewed as a particular discretized version of the iterated collocation method (3.1.2), and it can be shown that (3.1.4) is also valid for the "exact" iterated collocation solution obtained by (3.1.2).

3.1.2 *Product integration and discretized iterated collocation.* For a given mesh Π_N on I, the Fredholm integral equation (1.1), $0<\alpha<1$, can be rewritten in a mesh-related form,

$$y(t) = g(t) + \lambda \sum_{i=0}^{N-1} h_i^{1-\alpha}\,\Phi_{n,i}(y;t), \quad t\in I, \tag{3.1.5}$$

with integrals $\Phi_{n,i}(y;t)$ given as in (3.1.1). Assume that these integrals are approximated by m-point product integration rules whose abscissas and weights are given by $\{t_{i,l} := t_i + c_l h_i : l=1,\cdots,m\}$ and

$$w_l^{(i)}(t;\alpha) := \int_0^1 |(t-t_i)/h_i - \tau|^{-\alpha}\, L_l(\tau)\,d\tau \quad (l=1,\cdots,m), \tag{3.1.6}$$

respectively. We thus define the *product integration solution* z to y by

$$z(t) := g(t) + \lambda \sum_{i=0}^{N-1} h_i^{1-\alpha} \sum_{l=1}^{m} w_l^{(i)}(t;\alpha)\, K(t, t_{i,l}) Z_{i,l}, \quad t \in I, \tag{3.1.7a}$$

where the values $Z_{i,l} := z(t_{i,l})$ are found by solving the equations

$$Z_{n,j} = g(t_{n,j}) + \lambda \sum_{i=0}^{N-1} h_i^{1-\alpha} \sum_{l=1}^{m} w_l^{(i)}(t_{n,j};\alpha)\, K(t_{n,j}, t_{i,l}) Z_{i,l} \tag{3.1.7b}$$

$$(j = 1, \cdots, m;\ n = 0, \cdots, N-1).$$

In other words, the product integration solution for (3.1.5) is obtained by approximating the *bounded* part of the integrand in $\Phi_{n,i}(y;t)$ by an interpolant in $S_{m-1}^{(-1)}(Z_N)$; this results in integrals (3.1.6) which can be found analytically. (Note that in a collocation method it is the solution itself which is approximated in $S_{m-1}^{(-1)}(Z_N)$.) After having solved the linear system (3.1.7b), (3.1.7a) serves as a natural interpolation formula.

Suppose now that the collocation equation for $u = S_{m-1}^{(-1)}(Z_N)$ (i.e., (2.2.4a) with integrals (3.1.1)) and the iterated equation (3.1.2) are discretized by m-point product quadrature, and call the resulting discrete solutions $\hat{u}$ and $\hat{u}^I$. It is then readily seen that the values $\hat{U}_{n,j} := \hat{u}(t_{n,j})$ solve (3.1.7b); hence, if h is sufficiently small, then $Z_{n,j} = \hat{U}_{n,j}$, and it follows from (3.1.7a) that we must have $z(t) = \hat{u}^I(t)$ for $t \in I$. As indicated at the end of the previous section the order of convergence of the product integration solution z is thus given by that of the discrete collocation solution $\hat{u}^I$.

We conclude with a remark on the limitation inherent in the practical use of (symmetrically) graded meshes. Recall from (2.1.2) that the smallest meshwidth is $h_0 (= h_{N-1}) = 2^{r-1} T N^{-r}$. If the grading exponent is given by $r = m/(1-\alpha)$ and if $T = 1, \alpha = 1/2$, then we obtain $h_0 = 2^{2m-1} N^{-2m}$. It is clear that even for moderate values of m and N the value of h_0 will become very small (for instance, $m=4$ (piecewise cubics) and $N=100$ yield $h_0 = 32 \cdot 10^{-16}$), and this may result in a severely ill-conditioned linear system for the values $\hat{U}_{n,j}$ ($=Z_{n,j}$ in (3.1.7a)). Hence, one will often have to look for alternatives to (discrete) collocation and iterated collocation on graded meshes when solving weakly singular Fredholm integral equations. The interested reader is referred to the papers by Sloan (1980b) and Anselone (1981), and to the relevant references listed in the latter paper.

3.2 *Volterra Integral Equations of the Second Kind*

3.2.1 *Collocation and iterated collocation.* For $g \in C^m(I)$ and $K \in C^m(S)$, the solution of the Volterra integral equation (1.2) with $0 < \alpha < 1$ is m times continuously differentiable only in the left-open interval $(0, T]$; at $t = 0$ its derivatives are unbounded. To be more precise, and using an obvious modification of the definition given in Section 3.1.1, the solution is of type $(1-\alpha, m, \{0\})$ (compare Brunner (1983, 1985b); the non-linear case is studied in Logan (1976) and, especially, in Lubich (1983)). As a consequence, the collocation solution $u = S_{m-1}^{(-1)}(Z_N)$ will converge like $\|y-u\|_\infty = O(N^{-(1-\alpha)})$ if the underlying mesh sequence is quasi-uniform, in complete analogy to the result for weakly singular Fredholm integral equations. The optimal order, $\|y-u\|_\infty = O(N^{-m})$, is recovered if one employs *asymmetrically graded meshes* of the form (2.1.4) whose grading exponent is $r = m/(1-\alpha)$ (see Brunner (1985a,b)).

As in Volterra equations with smooth solutions the iterated collocation solution u^I can be superconvergent only at the points $\bar{Z}_N$: the best order, $p^* = m+1-\alpha$ (which equals the best order of global superconvergence of the Fredholm iterate u^I; cf. (3.1.4)), is attained if the collocation parameters satisfy (3.1.3).

We note that the principal tool used in the convergence

analysis of collocation (and product integration) methods, the generalized discrete Gronwall lemma, is discussed in detail in Dixon and McKee (1983), Scott (1984), and Dixon (1985); see also Kershaw (1982).

3.2.2 *Discretized collocation and product integration methods.* The discretization of the collocation equations defining u and u^I can again be achieved by employing appropriate product integration techniques. Since most of the integrals are of the form (3.1.1) we may use the quadrature rules of Section 3.1.2. Thus, if the discrete iterated collocation solution is evaluated only for $t = \bar{Z}_N$, then we have the discretization

$$\hat{u}^I(t_n) := g(t_n) + \sum_{i=0}^{N-1} h_i^{1-\alpha} \sum_{l=1}^{m} w_l^{(i)}(t_n;\alpha)\, K(t_n, t_{i,l})\, \hat{U}_{i,l},$$

with quadrature weights as in (3.1.6). However, when computing the values $\hat{U}_{n,j}$ of the discrete collocation solution $\hat{u} \in S_{m-1}^{(-1)}(Z_N)$ at $t = t_{n,j}$ we also encounter integrals over $[0, c_j]$ where the integrand contains the singular term $(c_j - \tau)^{-\alpha}$. The foregoing product quadrature rules are readily adapted to deal with these integrals, along the lines of Section 2.3.3. We note that the use of m-point product quadrature rules yields the same orders of convergence for $\hat{u}$ and $\hat{u}^I$ as for the "exact" collocation solutions u and u^I.

It is clear that collocation on asymmetrically graded meshes is of limited practical value: since the collocation solution is computed recursively, starting on the subinterval σ_0 of length $h_0 = TN^{-r}$, with $r = m/(1-\alpha)$, large values of N will lead to a very small h_0, and the resulting rounding errors may contaminate the subsequent recursive process. On the other hand, it is striking that if (1.2) with $0 < \alpha < 1$ is solved by collocation on *uniform meshes*, then *near* $t = T$ the collocation solution exhibits nearly $O(N^{-m})$-convergence and is almost invariably more accurate than the solution corresponding to the correct mesh grading given by $r = m/(1-\alpha)$. This seems to be a consequence of the flexibility inherent in discontinuous splines $u \in S_{m-1}^{(-1)}(Z_N)$ to "correct" any starting inaccuracies encountered at the beginning of a recursive process.

We conclude this section by mentioning that in the last ten years various authors have proposed product integration techniques for solving weakly singular Volterra equations; see, e.g., Logan (1976), Westreich and Cahlon (1980), Cahlon (1981), Cameron and McKee (1983, 1984), Franco, McKee and Dixon (1983), and McKee and Stokes (1983). Kershaw (1982) presents an elegant convergence analysis for the discretized collocation method in $S_1^{(0)}(Z_N)$ (the product trapezoidal rule), and Dixon (1985) then shows that on uniform meshes one can obtain convergence of order $2-\alpha$ on a subset of $\bar{Z}_N$ bounded away from $t=0$.

3.2.3 *Alternative methods*. There are essentially two ways of generating high-order approximations to solutions of weakly singular Volterra integral equations on *uniform meshes*:

(i) Collocation in *nonpolynomial spline spaces* whose elements reflect the nonsmooth behaviour of the exact solution near $t=0$. This approach has been studied in te Riele (1982), Brunner (1983), and Brunner and te Riele (1984).

(ii) *Fractional linear multistep methods* based on discrete fractional differentiation (and generalizing the (ρ, σ)-reducible quadrature methods of Wolkenfelt (1982)) have recently been proposed by Lubich (1985); see also Zheludev (1982), Cameron and McKee (1985), and Hairer and Maass (1985) for related ideas. These methods use a special starting procedure (which may be viewed as a discretized nonpolynomial spline collocation method), and they appear to represent the most promising approach for the efficient solution of Volterra integral equations with weakly singular kernels.

4. REMARKS ON RELATED SOFTWARE

Since the appearance of Atkinson's Fredholm routines IESIMP

and IEGAUS (Atkinson (1976b)) there has been considerable activity in software development for Fredholm and Volterra integral equations. Surveys of the state of the art may be found in Delves and Mohamed (1985) (Chapter 10) and Brunner and van der Houwen (1986) (Chapter 9). Since neither the NAG library nor IMSL list any Volterra routines (IMSL has no integral equation routines), we mention some of the recent routines for (nonlinear) Volterra integral equations.

The routine COLVI2 by Blom and Brummer (1985, 1986) is based on discretized collocation and iterated collocation; it makes use of the corresponding local superconvergence properties (cf. Section 2.3.2). The discretized collocation method of de Hoog and Weiss (1975) has been implemented (for the Lobatto points) by Schlichte (1984). Explicit Volterra-Runge-Kutta methods are used in Tanfulla and Ribighini (1981) (embedded methods of order 4 and 5), Duncan (1982) (embedded methods of Verner of orders 5 and 6), Hairer, Lubich and Schlichte (1985) (4th-order method and FFT techniques for equations with convolution kernels), and Baker and Derakhshan (1985) (Fehlberg extension of the previous algorithm). Linear multistep methods form the basis of the routines in Hock (1981) (extrapolation), Kunkel (1982), and Williams and McKee (1985) (predictor-corrector methods). Finally, Bownds and Appelbaum (1985) employ kernel approximation techniques; the resulting system of ODEs is then solved by an ODE routine (compare also Bownds (1978, 1979, 1982), as well as Shampine (1985)).

Except for the routine VIESIN by Logan (1976) (product Simpson method) we are not aware of any other automatic programs for solving weakly singular Volterra integral equations of the form (1.2). (A routine based on fractional linear multistep methods is supposedly about to be published.) We mention, however, the package MATE by Jakeman (1980) designed for a somewhat different class of weakly singular Volterra integral equations where the given data are not exact (see also Jakeman (1984)).

The above remarks show that there is a definite need for the development of robust and efficient automatic routines for second-kind integral equations with weakly singular kernels (both of Volterra and Fredholm type), especially since such equations occur rather frequently in practical applications.

REFERENCES

Abdalkhani, J. (1984), "Collocation and Runge-Kutta methods for Volterra integral equations of the second kind with weakly singular kernels", *Congress. Numer. 42*, pp.87-100.

Albrecht, J. and Collatz, L. (1980), *Numerical Treatment of Integral Equations*, Internat. Ser. Numer. Math. 53, Birkhäuser Verlag (Basel-Boston).

Anderssen, R.S. and De Hoog, F.R. (1982), "Application and numerical solution of Abel-type integral equations", Math. Research Report No. 7-1982, Australian National University, Canberra.

Anderssen, R.S., De Hoog, F.R. and Lukas, M.A. (eds.) (1980), *The Application and Numerical Solution of Integral Equations*, Sijthoff and Noordhoff (Alphen/Rijn).

Anselone, P.M. (1981), "Singularity subtraction in the numerical solution of integral equations", *J. Austral. Math. Soc. Ser. B 22*, pp.408-418.

Anselone, P.M. and Krabs, W. (1979), "Approximate solution of weakly singular integral equations", *J. Integral. Eqns. 1*, pp.61-75.

Anselone, P.M. and Lee, J.W. (1976), "Double approximation methods for the solution of Fredholm integral equations", in *Numerische Methoden der Approximationstheorie, Bd. 3*, eds. Collatz, L. *et al.*, Internat. Ser. Numer. Math. 30, Birkhäuser Verlag (Basel), pp.9-34.

Arnold, D.N. and Wendland, W.L. (1983), "On the asymptotic convergence of collocation methods", *Math. Comp. 41*, pp.349-381.

Atkinson, K.E. (1976a), *A Survey of Numerical Methods for the Solution of Fredholm Integral Equations of the Second Kind*, SIAM (Philadelphia).

Atkinson, K.E. (1976b), "An automatic program for linear Fredholm integral equations of the second kind", *ACM Trans. Math. Software 2*, pp.154-171.

Atkinson, K.E. and Bogomolny, A. (1985), "The discrete Galerkin method for integral equations", preprint.

Atkinson, K.E., Graham, I. and Sloan, I.H. (1983), "Piecewise continuous collocation for integral equations". *SIAM J. Numer. Anal. 20*, pp.172-186.

Baker, C.T.H. (1977), *The Numerical Treatment of Integral Equations*, Clarendon Press (Oxford).

Baker, C.T.H. (1982), "An introduction to the numerical treatment of Volterra and Abel-type integral equations", in *Topics in Numerical Analysis*, ed. Turner, P.R., Lecture Notes in Math., 965, Springer-Verlag, (Berlin-Heidelberg-New York), pp.1-38.

Baker, C.T.H. and Derakhshan, M.S. (1985), "Preliminary report on a fast algorithm for the numerical solution (with error estimates) of a Volterra integral equation of the 2nd kind having a convolution kernel", Numer. Anal. Report No. 106, Dept of Mathematics, Universitv of Manchester.

Baker, C.T.H. and Miller, G.F. (eds.) (1982), *Treatment of Integral Equations by Numerical Methods*, Academic Press (London).

Bellen, A. (1985), "Constrained mesh methods for functional differential equations", in *Delay Equations, Approximation and Application*, eds. Meinardus, G. and Nürnberger, G., Internat. Ser. Numer. Math. 74, Birkhäuser Verlag (Basel-Boston), pp.52-70.

Blom, J.G. and Brunner, H. (1985), "The numerical solution of nonlinear Volterra integral equations of the second kind by collocation and iterated collocation methods", Report NM-R8522, Centre for Mathematics and Computer Science, Amsterdam (to appear in *SIAM J. Sci. Statist. Comput.*).

Blom, J.G. and Brunner, H. (1986), "Algorithm ***: Discretized collocation and iterated collocation for nonlinear Volterra integral equations of the second kind", Report NM-R8618, Centre for Mathematics and Computer Science, Amsterdam.

De Boor, C. and Swartz, B. (1973), "Collocation at Gaussian points", *SIAM J. Numer. Anal. 10*, pp.582-606.

Bownds, J.M. (1978), "On an initial-value method for quickly solving Volterra integral equations: a review", *J. Optimization Theory Appl. 24*, pp.133-151.

Bownds, J.M. (1979), "A combined recursive collocation and kernel approximation technique for certain singular Volterra integral equations", *J. Integral Eqns. 1*, pp.153-164.

Bownds, J.M. (1982), "Theory and performance of a subroutine for solving Volterra integral equations", *Computing 28*, pp.317-332.

Bownds, J.M. and Appelbaum, L. (1985), "Algorithm 627: A FORTRAN subroutine for solving Volterra integral equations", *ACM Trans. Math. Software 11*, pp.58-65.

Brunner, H. (1979a), "Superconvergence of collocation methods for Volterra integral equations of the first kind", *Computing 21*, pp.151-157.

Brunner, H. (1979b), "A note on collocation methods for Volterra integral equations of the first kind", *Computing 23*, pp.179-187.

Brunner, H. (1980), "Superconvergence in collocation and implicit Runge-Kutta methods for Volterra-type integral equations of the second kind", in *Numerical Treatment of Integral Equations*, eds. Albrecht, J. and Collatz, L., Internat. Ser. Numer. Math. 53, Birkhäuser Verlag (Basel-Boston), pp.54-72.

Brunner, H. (1981), "The application of the variation of constants formulas in the numerical analysis of integral and integro-differential equations", *Utilitas Math. 19*, pp.255-290.

Brunner, H. (1982), "A survey of recent advances in the numerical solution of Volterra integral and integro-differential equations", *J. Comput. Appl. Math. 8*, pp.213-229.

Brunner, H. (1983), "Nonpolynomial spline collocation for Volterra equations with weakly singular kernels", *SIAM J. Numer. Anal. 20*, pp.1106-1119.

Brunner, H. (1984a), "Iterated collocation methods and their discretizations for Volterra integral equations", *SIAM J. Numer. Anal. 21*, pp.1132-1145.

Brunner, H. (1984b), "Implicit Runge-Kutta methods of optimal order for Volterra integro-differential equations", *Math. Comp. 42*, pp.95-109.

Brunner, H. (1985a), "The numerical solution of weakly singular Volterra integral equations by collocation on graded meshes", *Math. Comp. 45*, pp.417-437.

Brunner, H. (1985b), "The approximate solution of Volterra equations with nonsmooth solutions", *Utilitas Math. 27*, pp.57-95.

Brunner, H. (1986), "High-order methods for the numerical solution of Volterra integro-differential equations", *J. Comput. Appl. Math., 15*, pp.301-309.

Brunner, H., Hairer, E. and Nørsett, S.P. (1982), "Runge-Kutta theory for Volterra integral equations of the second kind", *Math. Comp. 39*, pp.147-163.

Brunner, H. and van der Houwen, P.J. (1986), *The Numerical Solution of Volterra Equations*, CWI Monographs, Vol. 3, North-Holland (Amsterdam).

Brunner, H. and Nørsett, S.P. (1981), "Superconvergence of collocation methods for Volterra and Abel integral equations of the second kind", *Numer. Math. 36*, pp.347-358.

Brunner, H. and te Riele, H.J.J. (1984), "Volterra-type integral equations of the second kind with non-smooth solutions: high-order methods based on collocation techniques", *J. Integral Eqns. 6*, pp.187-203.

Cahlon, B. (1981), "Numerical solution of nonlinear Volterra integral equations", *J. Comput. Appl. Math. 7*, pp.121-128.

Cameron, R.F. and McKee, S. (1983), "High-accuracy product integration methods for the Abel equation, and their application to a problem in scattering theory", *Inter. J. Numer. Methods Engrg. 19*, pp.1527-1536,

Cameron, R.F. and McKee, S. (1984), "Product integration methods for second-kind Abel integral equations", *J. Comput. Appl. Math. 11*, pp.1-10.

Cameron, R.F. and McKee, S. (1985), "The analysis of product integration methods for Abel's equation using discrete fractional differentiation", *IMA J. Numer. Anal. 5*, pp.339-353.

Chandler, G.A. (1979), "Superconvergence of numerical methods to second kind integral equations", Ph.D. Thesis, Australian National University, Canberra.

Chandler, G.A. (1980), "Superconvergence for second kind integral equations", in *The Application and Numerical Solution of Integral Equations*, eds. Anderssen, R.S., De Hoog, F.R. and Lukas, M.A., Sijthoff and Noordhoff (Alphen/Rijn), pp.103-117.

Chatelin, F. (1983), *Spectral Approximation of Linear Operators*, Academic Press (New York).

Chatelin, F. and Lebbar, R. (1981), "The iterated projection solution for the Fredholm integral equation of second kind", *J. Austral. Math. Soc. Ser. B 22*, pp.439-451.

Chatelin, F. and Lebbar, R. (1984), "Superconvergence results for the iterated projection method applied to a Fredholm integral equation of the second kind and the corresponding eigenvalue problem", *J. Integral Eqns. 6*, pp.71-91.

Cushing, J.M. (1977), *Integrodifferential Equations and Delay Models in Population Dynamics*, *Lecture Notes in Biomath.*, *20*, Springer-Verlag (New York).

Delves, L.M. and Mohamed, J.L. (1985), *Computational Methods for Integral Equations*, Cambridge University Press (Cambridge).

Delves, L.M. and Walsh, J. (eds.) (1974), *Numerical Solution of Integral Equations*, Clarendon Press (Oxford).

Diekman, O. (1979), "Integral equations and population dynamics", in *Colloquium Numerical Treatment of Integral Equations*, ed. te Riele, H.J.J., MC Syllabus No. 41, Mathematisch Centrum (Amsterdam), pp.115-149.

Dixon, J. (1985), "On the order of the error in discretization methods for weakly singular second kind Volterra integral equations with nonsmooth solutions", *BIT 25*, pp.624-634.

Dixon J. and McKee, S. (1983), "Singular Gronwall inequalities", Numer. Anal. Report NA/83/44, Hertford College, University of Oxford.

Dixon, J. and McKee, S. (1985), "A unified approach to convergence analysis of discretization methods for Volterra-type equations", *IMA J. Numer. Anal. 5*, pp.41-57.

Duncan, R.P. (1982), "A Runge-Kutta method using variable stepsizes for Volterra integral equations of the 2nd kind", Tech. Report 157/82, Dept. of Comput. Science, University of Toronto.

Eggermont, P.P.B. (1986), "Improving the accuracy of collocation solutions of Volterra integral equations of the first kind by local interpolation", *Numer. Math. 48*, pp.263-279.

Esser, R. (1978), "Numerische Behandlung einer Volterraschen Integralgleichung", *Computing 19*, pp.269-284.

Franco, N.B., McKee, S. and Dixon, J. (1983), "A numerical solution of Lighthill's integral equation for the surface temperature of a projectile", *Mat. Appl. Comput. 2*, pp.257-271.

Golberg, M.A. (ed.) (1979a), *Solution Methods for Integral Equations*, Plenum Press (New York).

Golberg, M.A. (1979b), "A survey of numerical methods for integral equations", in *Solution Methods for Integral Equations*, ed. Goldberg, M.A., Plenum Press (New York), pp.1-58.

Graham, I.G. (1980a), "The numerical solution of Fredholm integral equations of the second kind", Ph.D. Thesis, University of New South Wales, Kensington.

Graham, I.G. (1980b) "Some application areas for Fredholm integral equations", in *The Application and Numerical Solution of Integral Equations*, eds. Anderssen, R.S., De Hoog, F.R. and Lukas, M.A., Sijthoff and Noordhoff (Alphen/Rijn), pp.75-102.

Graham, I.G. (1981), "Collocation methods for two dimensional weakly singular integral equations", *J. Austral. Math. Soc. Ser. B 22*, pp.456-473.

Graham, I.G. (1982a), "Singularity expansions for the solutions of second kind Fredholm integral equations with weakly singular kernels", *J. Integral Eqns. 4*, pp.1-30.

Graham, I.G. (1982b), "Galerkin methods for second kind integral equations with singularities", *Math. Comp. 39*, pp.519-533.

Graham, I.G. (1985), "Numerical methods for multidimensional integral equations", in *Computational Techniques and Applications: CTAC-83, Sydney*, eds. Noye, J. and Fletcher, C., North-Holland (Amsterdam), pp.335-351.

Graham, I.G., Joe, S. and Sloan, I.H. (1985), "Iterated Galerkin versus iterated collocation for integral equations of the second kind", *IMA J. Numer. Anal. 5*, pp.355-369.

Graham, I.G. and Schneider, C. (1985), "Product integration for weakly singular integral equations in $\mathbb{R}^m$", in *Constructive Methods for the Practical Treatment of Integral Equations*, eds. Hämmerlin, G. and Hoffman, K.-H., Internat. Ser. Numer. Math. 73, Birkhäuser Verlag (Basel-Boston), pp.156-165.

Grigorieff, R.D. (1983), "Differenzenapproximationen von Integrodifferentialgleichungen einer Veränderlichen", *Numer. Funct. Anal. Optim. 6*, pp.315-340.

Hairer, E., Lubich, Ch. and Nørsett, S.P. (1983), "Order of convergence of one-step methods for Volterra integral equations of the second kind", *SIAM J. Numer. Anal. 20*, pp.569-579.

Hairer, E., Lubich, Ch. and Schlichte, M. (1985), "Fast numerical solution of nonlinear Volterra convolution equations", *SIAM J. Sci. Statist. Comput. 6*, pp.532-541.

Hairer, E. and Maass, P. (1985), "Fractional multistep methods for the nonlinear Basset equation", Preprint No. 313, Sonderforschungsbereich 123, University of Heidelberg.

Hämmerlin, G. and Hoffman, K.-H. (eds.) (1985), *Constructive Methods for the Practical Treatment of Integral Equations*, Internat. Ser. Numer. Math. 73, Birkhäuser Verlag (Basel-Boston).

Hock, W. (1981), "An extrapolation method with step size control for nonlinear Volterra integral equations", *Numer. Math.* *38*, pp.155-178.

De Hoog, F. and Weiss, R. (1975), "Implicit Runge-Kutta methods for second kind Volterra integral equations", *Numer. Math.* *23*, pp.199-213.

Houstis, E.N. and Papatheodoru, T.S. (1978), "A collocation method for Fredholm integral equations of the second kind", *Math. Comp.* *32*, pp.159-173.

van der Houwen, P.J. (1980), "Convergence and stability results in Runge-Kutta type methods for Volterra integral equations of the second kind", *BIT 20*, pp.375-377.

van der Houwen, P.J. and Blom, J.G. (1978), "On the numerical solution of Volterra integral equations of the second kind. II. Runge-Kutta methods", Report NW 61/78, Mathematisch Centrum, Amsterdam.

van der Houwen, P.J. and te Riele, H.J.J. (1985), "Linear multistep methods for Volterra integral and integrodifferential equations", *Math. Comp.* *45*, pp.439-461. (Supplement: *Math Comp.* *45*, S21-S28).

van der Houwen, P.J., Wolkenfelt, P.H.M. and Baker, C.T.H. (1981), "Convergence and stability analysis for modified Runge-Kutta methods in the numerical treatment of second-kind Volterra integral equations", *IMA J. Numer. Anal.* *1*, pp.303-328.

Jakeman, A.J. (1980), "How, when and why to use Mate — A package of programs containing methods for Abel type equations and their linear functionals", *Mikroskopie 37*, pp.458-465.

Jakeman, A.J. (1984), "Numerical inversion of a second kind singular Volterra equation — the thin section equation of stereology", *Utilitas Math.* *26*, pp.193-213.

Joe, S. (1985a), "The numerical solution of second kind Fredholm integral equations", Ph.D. Thesis, University of New South Wales, Kensington.

Joe, S. (1985b), "Collocation methods using piecewise polynomials for second kind integral equations", *J. Comput. Appl. Math.* *12/13*, pp.391-400.

Joe, S. (1985c), "Discrete collocation methods for second kind Fredholm integral equations", *SIAM J. Numer. Anal.* *22*, pp.1167-1177.

Joe, S. (1985d), "Discrete Galerkin methods for Fredholm integral equations of the second kind", preprint.

Kershaw, D. (1982), "Some results for Abel-Volterra integral equations of the second kind", in *Treatment of Integral Equations by Numerical Methods*, eds. Baker, C.T.H. and Miller, G.F., Academic Press (London), pp.273-282.

Křížek, M. and Neittaanmäki, P. (1984), "On superconvergence techniques", Preprint 34, Dept. of Math., University of Jyväskylä, Finland.

Kunkel, P. (1982), "Ein adaptives Verfahren zur Lösung von Volterraschen Integralgleichungen zweiter Art", Diplomarbeit, Institut für Angew. Math., Universität Heidelberg.

Lin Qun and Liu Jiaquan (1980), "Extrapolation method for Fredholm equations with nonsmooth kernels", *Numer. Math.* *35*, pp.459-464.

Linz, P. (1985), *Analytical and Numerical Methods for Volterra Equations*, SIAM Studies in Appl. Math. Vol. 7, SIAM (Philadelphia).

Lodge, A.S., Renardy, M. and Nohel, J.A. (eds), (1986), *Viscoelasticity and Rheology*, Academic Press (New York).

Logan, J.E. (1976), "The approximate solution of Volterra integral equations of the second kind", Ph.D. Thesis, University of Iowa, Iowa City.

Lonseth, A.T. (1977), "Sources and applications of integral equations", *SIAM Rev.* *19*, pp.241-278.

Lubich, Ch. (1983), "Runge-Kutta theory for Volterra and Abel integral equations of the second kind", *Math. Comp.* *41*, pp.87-102.

Lubich, Ch. (1985), "Fractional linear multistep methods for Abel-Volterra integral equations of the second kind", *Math. Comp.* *45*, pp.463-469.

MacCamy, R.C. and Weiss, P. (1979), "Numerical approximations for Volterra integral equations", in *Volterra Equations*, eds. Londen S.-O. and Staffans, O.J., Lecture Notes in Math. 737, Springer-Verlag (Berlin-Heidelberg-New York), pp.173-191.

Marchuk, G.I. and Shaidurov, V.V. (1983), *Difference Methods and Their Extrapolations*, Springer-Verlag (New York).

Matthys, J. (1976), "A-stable linear multistep methods for Volterra integro-differential equations", *Numer. Math.* *27*, pp.85-94.

McKee, S. (1979), "Cyclic multistep methods for solving Volterra integro-differential equations", *SIAM J. Numer. Anal.* *16*, pp.106-114.

McKee, S. and Stokes, A. (1983), "Product integration methods for the nonlinear Basset equation", *SIAM J. Numer. Anal.* *20*, pp.143-160.

Noble, B. (1977), "The numerical solution of integral equations", in *The State of the Art in Numerical Analysis*, ed. Jacobs, D.A.H., Academic Press (London), pp.915-966.

Peluso, R. and Piazza, G. (1985), "Metodi di tipo Runge-Kutta per la risoluzione di equazioni integrali di Volterra di seconda specie fortemente stiff", *Boll. Un. Mat. Ital.* *B(6) 4*, pp.155-165.

Pitkäranta, J. (1979), "On the differential properties of solutions to Fredholm equations with weakly singular kernels", *J. Inst. Math. Appl.* *24*, pp.109-119.

Rice, J.R. (1969), "On the degree of convergence of nonlinear spline approximation", in *Approximation with Special Emphasis on Spline Functions*, ed. Schoenberg, I.J., Academic Press (New York), pp.349-365.

Richter, G.R. (1976), "On weakly singular Fredholm integral equations with displacement kernels", *J. Math. Anal. Appl.* *55*, pp.32-42.

Richter, G.R. (1978), "Superconvergence of piecewise polynomial Galerkin approximations for Fredholm integral equations of the second kind", *Numer. Math.* *31*, pp.63-70.

te Riele, H.J.J. (ed.) (1979a), *Colloquium Numerical Treatment of Integral Equations*, MC Syllabus 41, Mathematisch Centrum (Amsterdam).

te Riele, H.J.J. (1979b), "Introduction and global survey of numerical methods for integral equations", in *Colloquium Numerical Treatment of Integral Equations*, ed. te Riele, H.J.J., MC Syllabus 41, Mathematisch Centrum (Amsterdam) pp.3-25.

te Riele, H.J.J. (1982), "Collocation methods for weakly singular second-kind Volterra integral equations with nonsmooth solution", *IMA J. Numer. Anal.* *2*, pp.437-449.

Schlichte, M. (1984), "Anwendung eines impliziten Runge-Kutta-Verfahrens auf Volterra'sche Integralgleichungen zweiter Art mit Faltungskern", Diplomarbeit, Inst. für Angew, Math., Universität Heidelberg.

Schneider, C. (1979), "Regularity of the solution to a class of weakly singular Fredholm integral equations of the second kind". *Integral Eqns. Operator Theory* *2*, pp.62-68.

Schneider, C. (1981), "Product integration for weakly singular integral equations", *Math. Comp.* *36*, pp.207-213.

Schumaker, L.L. (1981), *Spline Functions: Basic Theory*, Wiley (New York).

Scott, J.A. (1984), "A unified analysis of discretization methods", D. Phil. Thesis, University of Oxford.

Shampine, L.F. (1985), "Solving Volterra integral equations with ODE codes", Preprint, Sandia Lab., Albuquerque.

Sloane, I.H. (1976), "Improvement by iteration for compact operator equations", *Math. Comp.* *30*, pp.758-764.

Sloan, I.H. (1980a), "A review of numerical methods for Fredholm equations of the second kind", in *The Application and Numerical Solution of Integral Equations*, eds. Anderssen, R.S., de Hoog, F.R. and Lukas, M.A., Sijthoff and Noordhoff (Alphen/Rijn), pp.51-74.

Sloan, I.H. (1980b), "The numerical solution of Fredholm equations of the second kind by polynomial interpolation", *J. Integral Eqns.* *2*, pp.265-279.

Sloan, I.H. (ed.) (1981), *Special Issue on the Numerical Solution of Integral Equations: J. Austral. Math. Soc. Ser. B 22, Part 4*.

Sloan, I.H. (1982), "Superconvergence and the Galerkin method for integral equations of the second kind", in *Treatment of Integral Equations by Numerical Methods*, eds. Baker, C.T.H. and Miller, G.F., Academic Press (London), pp.197-207.

Sloan, I.H. (1984), "Four variants of the Galerkin method for integral equations of the second kind", *IMA J. Numer. Anal.* *4*, pp.9-17.

Sloan, I.H. and Burns, B.J. (1979), "Collocation with polynomials for integral equations of the second kind: a new approach to the theory", *J. Integral Eqns.* *1*, pp.77-94.

Sloan, I.H., Burns, B.J. and Datyner, D. (1975), "A new approach to the numerical solution of integral equations", *J. Comput. Phys.* *18*, pp.92-105.

Sloan, I.H. and Thomée, V. (1985), "Superconvergence of Galerkin iterates for integral equations of the second kind", *J. Integral Eqns.* *9*, pp.1-23.

Spence, A. and Thomas, K.S. (1983), "On superconvergence properties of Galerkin's method for compact operator equations", *IMA J. Numer. Anal.* *3*, pp.253-271.

Tanfulla, M. and Ribighini, G. (1981), "Procedure di stima dell' errore nella risoluzione di equazioni integrali dì tipo Volterra", *Riv. Mat. Univ. Parma (4) 7*, pp.473-487.

Thomas, K.S. (1981), "Galerkin methods for singular integral equations", *Math. Comp.* *36*, pp.193-205.

Vainikko, G. (1978), "Approximative methods for nonlinear equations (two approaches to the convergence problem)", *Nonlinear Anal.* *2*, pp.647-687.

Vainikko, G. and Pedas, A. (1981), "The properties of solutions of weakly singular integral equations", *J. Austral. Math. Soc. Ser. B 22*, pp.419-430.

Vainikko, G., Pedas, A. and Uba, P. (1984), *Methods for Solving Weakly Singular Integral Equations* (Russian), Tartu. Gos. Univ. (Tartu).

Vainikko, G. and Uba, P. (1981), "A piecewise polynomial approximation to the solution of an integral equation with weakly singular kernel", *J. Austral. Math. Soc. Ser. B 22*, pp.431-438.

Volk, M. (1979), "Die numerische Behandlung Fredholm'scher Integralgleichungen zweiter Art mittels Splinefunktionen", Report HMI-B286, Hahn-Meitner-Institut, Berlin.

Westreich, D. and Cahlon, B. (1980), "Numerical solution of Volterra integral equations with continuous or discontinuous terms", *J. Inst. Math. Appl.* *26*, pp.175-186.

Williams, H.M. and McKee, S. (1985), "Variable stepsize predictor-corrector schemes for second kind Volterra integral equations", *Math. Comp.* *44*, pp.391-404.

Wolkenfelt, P.H.M. (1982), "The construction of reducible quadrature rules for Volterra integral and integro-differential equations", *IMA J. Numer. Anal.* *2*. pp.131-152.

Wolkenfelt, P.H.M. (1983), "Modified multilag methods for Volterra functional equations", *Math. Comp.* *40*, pp.301-316.

Zheludev, V.A. (1982), "Derivatives of fractional order and the numerical solution of a class of convolution equations", *Differential Eqns.* *18*, pp.1404-1413.

H. Brunner
Institut de Mathématiques
Université de Fribourg
Ch-1700 Fribourg
Switzerland

Present address:
Dept. of Mathematics and Statistics
Memorial University of Newfoundland
St. John's
Newfoundland A1C 5S7
Canada

20 Numerical Methods for Free and Moving Boundary Problems

COLIN W. CRYER

ABSTRACT

A survey is given of recent developments in numerical methods for free and moving boundary problems.

1. INTRODUCTION

We begin with a brief introduction to free boundary problems for readers who are not familiar with such problems.

In the theory of ordinary or partial differential equations one is interested in the solutions u of a given system of differential equations A on a domain $\Omega \subset R^n$:

$$Au = f, \text{ on } \Omega . \tag{1.1}$$

In applications we usually seek that solution u which satisfies a set of conditions on the boundary $\partial\Omega$ of Ω:

$$Bu = g, \text{ on } \partial\Omega . \tag{1.2}$$

If the conditions B can be chosen so that the solution u exists, is unique and depends continuously on the data f, g then the problem (1.1) is said to be well-posed. We will call the problem (1.1) – (1.2) a known-boundary problem (KBP) when we wish to emphasize the fact that the boundary $\partial\Omega$ is known.

There are many real-world problems in which the governing differential equation $Au = f$ is known but the domain Ω is not completely specified. For example, if appropriate assumptions are made about the physical properties of water and air, then the differential equations governing the flow inside and outside a falling raindrop are known, but the shape of the raindrop is not known *a priori*, so that the problem cannot be formulated as a KBP. Such problems arise very frequently, and indeed one can argue that in the real physical world no boundary is exactly known.

If Ω is not completely specified then it is necessary to provide additional information, over and above that required for a KBP, in order that the solution u be determined. At first sight it is not obvious in what form this additional information should be provided. However, on the basis of many examples, it has been found that it is physically and mathematically meaningful to provide the additional information in the form of an additional boundary condition $Cu=0$ along the unknown part Γ of $\partial\Omega$. Since Ω is not completely specified the problem is inherently nonlinear and we may expect multiple solutions. To ensure that the solution is unique, it is often necessary to impose some supplementary conditions which we denote by $Su=0$. In the case of a raindrop, for example, we must specify the volume V of the raindrop. Even this is not sufficient, because we could then have several raindrops with total volume V, so that we must also specify that Ω is simply-connected. In summary, an unknown-boundary problem (UBP) is a problem of the following type:

An open set $\Omega\subset R^n$ is partly given, part of its boundary, the unknown-boundary Γ, being unknown. It is required to find Γ and u such that

(a) $Au=f$, on Ω (the governing equations)
(b) $Bu=g$, on $\partial\Omega$ (the boundary conditions on all or part of $\partial\Omega$) (1.3)

(a) $Cu=f$, on Γ (additional boundary condition)
(b) $Su=0$ (supplementary conditions, possibly an empty set). (1.4)

A and B are such that, were Γ known, then equations (1.3) would constitute a well-posed KBP.

If A is a steady-state partial differential equation then the UBP is called a *free-boundary problem* (FBP) and the unknown-boundary Γ is known as the *free boundary* (FB). If A is an evolutionary partial differential equation then the UBP is called a *moving boundary problem* (MBP) and the unknown-boundary Γ is called the *moving boundary* (MB).

Many moving-boundary problems arise when considering phase-change problems, such as the melting of ice, and such problems are often called *Stefan* problems, since the work of Stefan in the 1890s on the melting of the polar ice is among the earliest work in this area.

We now give three simple examples of unknown boundary problems with analytic solutions.

EXAMPLE 1.1

The following is a model problem which is very useful for illustrative purposes but has no physical interpretation.

$$\Omega := (0,\tau) \subset R^1, \text{ where } \tau > 0 \text{ is not known,}$$
$$\Gamma := \{\tau\}$$
$$u \in C^2(\Omega)$$

where u and Γ satisfy:

Known-boundary problem:

$$\begin{aligned} &\text{(a) } -u_{xx} + (x-1) = 0, \text{ on } \Omega \quad \text{(governing equation)} \\ &\text{(b) } u(0) = 0, \quad u(\tau) = 0 \quad \text{(boundary conditions)} \end{aligned} \tag{1.5}$$

Unknown-boundary conditions:

$$\begin{aligned} &\text{(a) } u_x(\tau) = 0 \quad \text{(additional boundary condition)} \\ &\text{(b) } \emptyset \quad \text{(no supplementary conditions).} \end{aligned} \tag{1.6}$$

For fixed τ equations (1.5) have a unique solution $u(x;\tau)$ which may readily be found by integrating equation (1.5a) twice and imposing the boundary conditions (1.5b):

$$u(x;\tau) = x(x-\tau)\left(\frac{x+\tau}{6} - \frac{1}{2}\right).$$

Imposing the auxiliary boundary conditions (1.6a) we find that τ must satisfy the quadratic equation:

$$u_x(x;\tau)\Big|_{x=\tau} = x\left(\frac{x+\tau}{6} - \frac{1}{2}\right)\Big|_{x=\tau} = \tau\left(\frac{\tau}{3} - \frac{1}{2}\right) = 0,$$

so that $\tau = 3/2$. Hence the solution of this free-boundary problem is:

$$\Omega = (0, 3/2), \quad \Gamma = \{3/2\}, \quad u(x) = x(x-3/2)^2/6. \qquad (1.7)$$

EXAMPLE 1.2

Consider the two-dimensional flow of an ideal fluid past an air bubble at constant pressure, allowing for surface tension but neglecting gravity. Let u be the velocity potential, so that the velocity in the fluid is *grad* u. The mathematical formulation is:

Ω := domain in R^2 exterior to the bubble,

Γ := bubble surface, $u \in C^2(\Omega)$.

u and Γ must satisfy:

Known-boundary problem:

(a) $\nabla^2 u := u_{xx} + u_{yy} = 0$, on Ω (governing equation)

(b) (i) $grad\, u \to (1,0)$ as $x^2 + y^2 \to \infty$ (uniform flow at infinity) (1.8)

(ii) $\frac{\partial u}{\partial n} = 0$ on $\Gamma = \partial\Omega$ (streamline)

Unknown-boundary conditions:

(a) $|grad\ u|^2 = -2T\kappa$, on Γ (balance of forces)
where T is the surface tension (assumed constant) and κ is the curvature of Γ, (1.9)

(b) Ω simply connected (supplementary condition).

As is the case for many free-boundary problems with analytical solutions, a solution is most readily obtained using complex function theory. In 1955 McLeod showed that one possible solution is:

$$\Gamma := \{z = g(t): t = e^{i\theta}, \ 0 \leqslant \theta \leqslant 2\pi\}$$

where

$$g(t) := 2T\left(t - \frac{2}{3t} - \frac{1}{27\,t^3}\right). \tag{1.10}$$

We omit the expression for u.

EXAMPLE 1.3

Consider the following simple one-phase one-dimensional Stefan problem, first considered by Neumann in the 1860s. An insulated semi-infinite tube lying along the positive x-axis contains water at temperature 0°C, the freezing-point of water. At time $t=0$ the temperature at $x=0$ is lowered to -1°C, and this temperature is maintained for $t>0$. Starting at the left end of the tube, the water begins to freeze; the location of the ice-water unknown boundary is denoted by:

$$x = X(t)\,, \quad t \geqslant 0, \text{ with } \quad X(0) = 0\,.$$

The water temperature is assumed to remain constant at 0°C so that only the temperature in the ice needs to be determined (hence the term *one-phase*).

The mathematical formulation is:

$\Omega := \{(x,t)\colon 0 < x < X(t),\ 0 < t < \infty\} \subset R^2$ (the ice region)

$\Gamma := \{(X(t), t)\colon 0 \leqslant t < \infty\}$ (the water-ice region)

$u \in C^2(\Omega)$

where u is the temperature in the ice.

u and Γ must satisfy:

Known-boundary problem:

(a) $\dfrac{\partial u}{\partial t} = u_{xx}$, on Ω (governing equation, the heat equation)

(1.11)

(b) (i) $u(0,t) = -1$ (temperature at left end of tube)

(ii) $u(X(t),t) = 0$ on Γ (water-ice interface at freezing temperature)

Unknown-boundary conditions:

(a) $$\frac{\partial u}{\partial x} = L\,\frac{dX(t)}{dt}\,, \tag{1.12}$$

where L is the latent heat of water (a constant). This condition expresses the fact that L units of heat must be removed for one gram of water at 0°C to freeze.

The solution of the UBP is found with an approach which is very similar to that used to solve Example 1.1. One obtains a solution of the governing differential equation which satisfies one boundary condition and contains a parameter, and then chooses this parameter so as to satisfy the remaining boundary conditions. The solution is:

$$X(t) := 2\lambda\sqrt{t}$$

$$u(x,t) := -1 + \mathrm{erf}\,(x/2\sqrt{t})/\mathrm{erf}(\lambda)$$

where λ is the unique positive root of the equation

$$e^{-\lambda^2} = \sqrt{\pi} L \lambda\, \mathrm{erf}(\lambda)\ ,$$

and $\mathrm{erf}(x)$ denotes the error function.

The previous examples were chosen to illustrate different types of UBPs, with the proviso that the analytic solutions were known. There is an extensive literature on theoretical aspects of UBPs — existence, uniqueness, stability, qualitative properties of the UB and so forth. This literature is often of direct interest to the numerical analyst because specific UBPs are usually considered. At the risk of oversimplification, it may be said that a century of often brilliant effort by mathematicians has shown that the UBPs which arise in connection with physical problems are usually well-posed. As a general rule, the solution (u,Γ) is as smooth as the data will permit.

UBPs arise in a great variety of contexts, but the overwhelming majority of applications occur in continuum mechanics. In continuum mechanics an UBP arises in principle whenever there is a line or surface or interface on which the solution is discontinuous, and the UB corresponds to this line of discontinuity or surface of discontinuity or singular surface. The discontinuities can arise for many reasons:

- Two different materials may be present, for example air

and water.

- The material may occur in different phases, for example water and ice.
- The material properties may depend discontinuously upon certain variables. For example, the material may change from elastic to plastic when the stress reaches a critical value.
- There may be a jump in the pressure or velocity, as across a shock.

Within continuum mechanics UBPs occur most frequently in fluid mechanics (jets, bubbles, wakes, waves), porous flow (seepage through soil), and phase transition mechanics (melting and freezing).

The literature is scattered throughout many journals. The proceedings of the biennial international colloquia on free and moving boundary problems, the most recent of which is edited by Bossavit, Damlamian, and Fremond (1985), provide a record of current developments in the analysis, modelling, and numerical solution of UBPs, although the contributions in the engineering literature are not fully represented.

There are also a number of books which deal partly or exclusively with UBPs:

- Crank (1984) is an excellent general introduction at the intermediate level, which discusses most aspects of UBPs. Elliott and Ockendon (1982) is also a useful introduction at the same level. Friedman (1982) is an advanced text which gives a detailed description of recent theoretical developments.
- Diaz (1985) treats FBPs arising from nonlinear partial differential equations.
- Birkhoff and Zarantonello (1957) cover the extensive older literature on fluid mechanics FBPs, which is in danger of being forgotten. Monakhov (1983) considers the application of the theory of analytic functions to FBPs.
- Rubinstein (1971) is a classic text on the Stefan problem.
- Porous flow FBPs are discussed in many texts: recent works in which the theory of variational inequalities is applied include Baiocchi and Capelo (1984) and Chipot (1984).
- Panagiotopoulos (1985) considers applications in mechanics.

- Pironneau (1984) deals with optimal shape design, which has close connections with FBPs. Impulse control, another area with connections to UBPs, is the subject of Bensoussan and Lions (1984).
- Other texts which deal with one or other type of UBP include: Barbu (1984) — optimal control; Crowley and Elliott (1985) — crystal growth.

2. NUMERICAL METHODS

The numerical solution of the UBP (1.3)-(1.4) requires the solution of the KBP (1.3) and we may therefore expect all standard numerical methods for KBPs to be of use for solving UBPs. There is, however, an additional feature in that any given UBP can be manipulated in many ways before being solved numerically. In contrast to most other topics considered in this volume, the reformulation of the problem is generally considered as an integral part of the numerical method.

When solving an UBP numerically, a basic decision must therefore be made at the outset:

- Either the problem is solved in the original form, in which case the UB must be explicitly approximated throughout the calculation. This leads to *trial-free-boundary methods* for FBPs and *front-tracking methods* for MBPs.
- Or the problem is first manipulated so that the UB is eliminated at the cost of additional complications. In particular, the transformed problem must be nonlinear since the original UBP is nonlinear. The UB is then recovered at the end of the computation. There are several basic possibilities:
    1. *Front-fixing transformations*. The unknown domain Ω is mapped onto a known domain, at the cost of additional complexities in the differential equation and boundary conditions. This class of transformations includes the *isotherm migration method*.
    2. *Analytical transformation*. Using techniques such as conformal mapping it is sometimes possible to transform the problem radically, obtaining for example an equivalent integral equation.

3. *Fixed-domain transformation.* In certain cases it is possible to introduce a weak or generalised solution which is defined on a known domain and which implicitly contains information about the location of the unknown boundary. This class of transformations includes *variational inequalities* as well as the *enthalpy method* for Stefan problems.

The following sections consider these methods in greater detail and give examples of recent developments.

3. TRIAL-FREE-BOUNDARY METHODS

A trial-free-boundary method (TFBM) for the FBP

(a) $Au = f$, on Ω,

(b) $Bu = g$, on $\partial\Omega$;

(a) $Cu = f$, on Γ,

(b) $Su = 0$.

consists of the following procedure:

1. Choose an initial approximation to the FB Γ, and denote this approximation by $\Gamma^{(0)}$. Set $k=0$.

2. Let $\Omega^{(k)}$ correspond to $\Gamma^{(k)}$. Compute an approximation $u_h^{(k)}$ to the solution $u^{(k)}$ of:

$$\text{(a)}\quad Au^{(k)} = f, \text{ on } \Omega^{(k)}, \qquad \text{(b)}\quad Bu^{(k)} = g, \text{ on } \partial\Omega^{(k)}. \tag{3.3}$$

3. Obtain Γ_{new} by adjusting $\Gamma^{(k)}$ so that the conditions

$$\text{(a)}\quad Cu^{(k)} = f, \text{ on } \Gamma_{\text{new}}, \qquad \text{(b)}\quad Su^{(k)} = 0 \tag{3.4}$$

are satisfied approximately.

Set $k=k+1$, $\Gamma^{(k)} = \Gamma_{\text{new}}$.

4. Repeat steps 2 and 3 until the successive $\Gamma^{(k)}$ agree to the desired accuracy.

Within this general framework many variations are possible. The trial-free-boundary method has now been used for over

70 years and is in wide use, so that a great deal of experience has been accumulated, although, regrettably, rigorous convergence proofs are still lacking. The following comments are an attempt to summarise this experience, and indicate current practice.

1. The boundary conditions

$$\begin{aligned} Bu &= g, \text{ on } \partial\Omega \cap \Gamma \\ Cu &= f, \text{ on } \Gamma, \end{aligned} \tag{3.5}$$

may be replaced by equivalent conditions

$$\begin{aligned} B_1 u &= g_1, \text{ on } \partial\Omega \cap \Gamma \\ C_1 u &= f_1; \text{ on } \Gamma. \end{aligned} \tag{3.6}$$

For example, if $Bu = g$ takes the form of a Dirichlet condition $u = g$ and $Cu = f$ takes the form of a Neumann condition $\partial u/\partial n = 0$, then these conditions may obviously be interchanged. More complicated combinations are also possible.

There are various factors which influence the choice of the boundary conditions (3.6):

(a) *Ease of computation of* $u_h^{(k)}$. For example, if $u_h^{(k)}$ is being computed using finite elements, it may be easier to handle a natural boundary condition such as a normal derivative condition than to handle a Dirichlet condition.

(b) *Ease of computation of* Γ_{new}. It is for instance easier to satisfy equation (3.4a) if this is a Dirichlet condition than if it involves the derivatives of $u^{(k)}$.

(c) *Convergence properties of the algorithm*. This is of course the overriding consideration. The outcome of sixty years of experimentation up to 1976, as summarised by the present author, will be found in Crank (1984, section 8.2). More recently the trial-free-boundary method has been applied to many problems involving viscous and non-Newtonian fluids: see for example, Reinelt and Saffman (1985), Ryskin and Leal (1984a,b,c).

Recent discussions of the possible choices of equations (3.6) are given by Ryskin and Leal (1984, p.9), Tanner (1984, p.294). Tanner is concerned with the problem of the extrusion of a polymer from a die, and this work is particularly interesting in that for certain fluids it has not been possible to obtain convergence.

2. Any convenient numerical method may be used to compute the approximate solution $u_h^{(k)}$.

There is of course a great choice of possible methods, since in principle any numerical method for boundary value problems can be used. However, the fact that Ω is a general region means that it is advantageous to use a method which can easily handle general curved boundaries. In the last few years the most popular methods have been:

(a) *Integral equations*. For example Kelmanson (1983).

(b) *Finite-differences with boundary-fitted coordinates*. For example Reinelt and Saffman (1985), Christov and Volkov (1985).

(c) *Finite-elements*.

3. Various strategies may be used to compute Γ_{new}. Crank (1984, Section 8.2) gives a good summary.

It will be clear from the above that the trial-free-boundary method is both successful and complicated. The great strength of the method is that it is in principle applicable to every FBP. It is for this reason that the author believes that the method will continue to form an essential part of the toolkit of the numerical analyst.

4. FRONT-TRACKING

Front-tracking methods are the analogue for MBPs of trial-free-boundary methods for FBPs. The original coordinates are retained and the MB is explicitly approximated throughout the calculation.

It is not possible to discuss all the various versions of front-tracking which have been suggested in the literature. We merely mention some ideas which represent much current work, and refer the reader to the original papers for more details.

1. *The method of lines.* The method of lines has been extensively used by Meyer, whose paper (Meyer (1981)) provides a good introduction. The basic idea is to discretize the time variable. As a result, the partial differential equation is replaced by a sequence of one-dimensional FBPs at each time level.

2. *The method of Bonnerot and Jamet.* Bonnerot and Jamet (1979) use discontinuous finite elements to solve Stefan problems, and thus obtain a method with considerable flexibility.

3. *The method of Lynch.* Lynch (1982) has provided a general framework for deforming elements, that is finite elements which deform with time. This approach has been successfully used for many phase-change problems by Lynch and others.

5. FRONT-FIXING TRANSFORMATIONS

The idea underlying front-fixing transformations is that it is easier to solve a more complicated problem on a simple domain Ω. Therefore, a mapping is introduced which maps the unknown domain Ω onto a fixed domain. There is of course an unlimited number of possible transformations, but those known to the author fall roughly into three categories:

1. *Simple Analytic Transformations*

These are transformations in which specific properties of the problem in hand make it possible to choose a transformation such that the transformed problem is relatively simple. Examples include:

(a) *The 'ϕ-ψ transformation'*. For FBPs arising in ideal two-dimensional flow the stream-function ψ and the velocity potential ϕ are conjugate harmonic functions. If $z := x + iy$ denotes a point in the complex plane and $f(z) := \phi(z) + i\psi(z)$ then f is an analytic function of z. Consequently, in general, z is an analytic function of f, so that x and y satisfy the equations

$$x_{\phi\phi} + x_{\psi\psi} = 0,$$
$$y_{\phi\phi} + y_{\psi\psi} = 0.$$

If either the stream-function or the velocity potential is given on each part of $\partial\Omega$ then the mapping

$$(x,y) \rightarrow (\phi,\psi)$$

maps Ω onto a known region of the $\phi\psi$-plane.

This method was frequently used in the 1960's. A variant, in which $\log(z)$ is used as the new dependent variable was also frequently used in numerical calculations, and indeed this transformation was used in 1907 in analytical work of Levi-Civita.

(b) *The Lagrangian equations in fluid flow.* For example Bach and Hassager (1985).

2. *Complicated Analytic Transformations*

An example of such a transformation is given by the *isotherm migration method* for Stefan problems which has been used a great deal by Crank and his collaborators. The method is extremely reasonable, in that the temperature u is taken to be an independent variable. The governing equation is thus transformed into a partial differential equation for one space variable, which is, unfortunately, nonlinear. Despite this disadvantage the isotherm migration method has been successfully applied to many Stefan problems.

3. *Brute-Force Transformations*

In many cases, the FB or MB has been mapped onto a known surface by means of a transformation which involves the unknown UB. As a result, the governing differential equation becomes an equation, the coefficients of which depend on the UB. At first sight this seems fraught with danger, but there are many successful applications.

A particular example of this approach is known as the Landau transformation and is widely used for one-dimensional Stefan problems.

6. ANALYTICAL TRANSFORMATIONS

There is an extensive literature on FBPs in the plane with the equation $Au=0$ being either the Laplace equation or, more rarely, the biharmonic equation. A variety of analytical techniques have been developed, the most important being ingenious applications of conformal mapping. Indeed the mathematical theory of FBPs began with the work of Helmholtz in 1868 who gave the first mathematical formulation of a fluid flow FBP and then showed how it could be solved using conformal mapping methods.

This work has several goals:

- To obtain analytical solutions in closed form.
- To derive qualitative information about the solution, such as that the UB is convex.
- To derive new formulations of the problem which are amenable to analytical and numerical techniques.

While all the objectives are of general interest in that they provide welcome additional information, it is the last which is of greatest interest to the numerical analyst. There are two points which we wish to emphasise and illustrate by examples:

- It is very worthwhile for special problems to consider using special methods.
- The steady development of software for symbolic manipulation means that many classical methods can be implemented on the computer.

EXAMPLE 6.1

In 1834 John Scott Russell observed a wave, consisting of a single hump which progressed unchanged in form. Russell called it the *Great Primary Wave* but we now call such waves *solitary waves*.

The mathematical problem requires the determination of a

harmonic function ϕ (the velocity potential) and a function ζ (which defines the surface of the wave).

We may postulate that

$$\begin{aligned} \zeta(x) &= h + \sum_{k=1}^{\infty} a_k \varepsilon^k \psi_k \\ \phi(x,y) &= x + \sum_{k=1}^{\infty} b_k \varepsilon^k \phi_k \end{aligned} \tag{6.1}$$

where ε is a perturbation parameter, h (a constant) is the undisturbed height, a_k and b_k are coefficients, and ψ_k and ϕ_k are prescribed functions, ϕ_k being analytic. The coefficients a_k, b_k should then be determined (if possible) so as to satisfy the boundary conditions imposed upon ϕ and ζ.

This approach is a standard one in applied mathematics. For water waves it has been used for almost a century to prove the existence of small amplitude waves and to obtain numerical results. In particular, the maximum amplitude $A_{\max}$ of solitary waves has been computed by many authors. In a recent paper, Pennell and Su (1984) computed the first 17 coefficients of a series expansion and then estimated $A_{\max}$. The computations involved: manipulation of the series expansions; solution of a linear algebraic system of 17 equations; and finally a series transformation to accelerate convergence.

Some of the estimates for $A_{\max}$ which have been obtained over the past century are (not in order of publication):

0.827 , 1.0 , 0.83322 , 0.78 , 0.827 , 0.83.

The interested reader can find the publication dates in the paper of Pennell and Su (1984).

EXAMPLE 6.2

In 1922 the Russian mathematician A.I. Nekrasov showed that the problem of periodic two-dimensional water waves of finite height could be reformulated as a nonlinear integral equation. It has since been found that several similar problems can also be reformulated as integral equations. For the problem of solitary waves which was described in Example 6.1 one obtains (Amick and Toland (1979)) the equation:

$$\theta(s) = (1/3) \int_{-\pi/2}^{+\pi/2} \left\{ \frac{k(s,t) \sec(t) \sin(\theta(t))}{(1/\mu) + \int_0^t \sec(w) \sin(\theta(w))\, dw} \right\} dt, \quad (6.2)$$

where $k(s,t)$ is a known function.

The equation (6.2) has been used to prove results about the existence and uniqueness of solitary waves (Amick and Toland (1979)). Similar equations have been solved numerically (Conway and Thomas (1974)).

EXAMPLE 6.3

The rigorous mathematical theory of FBPs began with the work of Helmholtz in 1868. Helmholtz formulated the problem of a two-dimensional fluid jet as a FBP and showed how an analytic solution could be found by the ingenious use of conformal mapping techniques. This work was extended by Kirchhoff in 1869. There followed an enormous number of papers in which a great variety of FBPs were solved in closed form. Even today there are occasional papers in which new FBPs are solved in closed form using these ideas.

Recently, Elcrat and Trefethen (1986) have shown how the ideas of Helmholtz and Kirchhoff can be made the basis of an effective numerical method. Their method makes use of efficient numerical methods for evaluating integrals similar to those occurring in Schwarz-Christoffel maps.

In the above examples we have tried to show that many old ideas can be rejuvenated and made the basis of numerical methods. While the methods are necessarily rather special, because they depend on the specific problems, they nevertheless address questions which are still of interest.

7. VARIATIONAL INEQUALITIES

The theory of variational inequalities has been applied to FBPs with immense success. There are several important advantages:

- There is often a very natural physical idea underlying the variational inequality formulation of a FBP.
- There is an extensive mathematical literature on variational inequalities and in many cases the first existence proofs for certain FBPs have been obtained using variational inequalities.
- There are efficient numerical methods for solving variational inequalities, and in many cases the convergence of the numerical approximations to the exact solution has been established.

Given the advantages of variational inequalities it might seem reasonable to devote considerable attention to them. However, there are excellent treatments of the various aspects of variational inequalities already available in the literature. We will therefore merely give a simple illustrative example and give references for the interested reader.

EXAMPLE 7.1

We consider the problem stated in Example 1.1. The function u is defined only on the interval $[0,\tau]$. We extend the domain of definition of u to the entire interval $[0,2]$.

For convenience we denote by u the function as thus extended, since no confusion can arise:

$$u(x) := \begin{cases} u(x), & \text{if } x \in [0,\tau] \\ 0, & \text{if } x \in (\tau,2]. \end{cases}$$

It can be proved that:

$$(Lu)(x) := -u_{xx} + (x-1) \geqslant 0, \quad 0 \leqslant x \leqslant 2, \tag{7.1}$$

$$(Lu)(x) \cdot u(x) = 0, \quad 0 \leqslant x \leqslant 2, \tag{7.2}$$

$$u(x) \geqslant 0, \quad 0 \leqslant x \leqslant 2. \tag{7.3}$$

The equations (7.1)–(7.3) are simple examples of a *linear complementarity problem*, and can be shown to be equivalent to a variational inequality.

The first applications of variational inequalities were to the problem of cavities in journal bearings, and the problem of elastic-plastic torsion, and were made by the engineers Christopherson and Southwell using intuitive concepts. The rigorous mathematical theory was initially developed by Fichera, Stampacchia and Lewy in the 1960s. A further impetus was provided by the work of Baiocchi who showed that many porous flow FBPs could be transformed into variational inequalities using an ingenious transformation which we now call the Baiocchi transformation. There now exists a well-developed theory with many applications.

A particularly attractive feature of variational inequalities is that efficient numerical methods are available for their solution. In conjunction with various collaborators, the author has for example been able to apply multigrid methods and alternating direction methods, and has also found it possible to make efficient use of both the Cyber 205 and the DAP.

For further information about variational inequalities see:

- *Mathematical theory*. Kinderlehrer and Stampacchia (1980), Elliott and Ockendon (1982), Friedman (1982).
- *Numerical analysis*. Glowinski, Lions and Tremolieres (1981).

8. ENTHALPY FORMULATION

The *enthalpy method* is a method which has been used with great success for solving phase-change MBPs such as the Stefan problem considered in Example 1.3. The enthalpy method is based upon two essential mathematical ideas:

1. A new independent variable, the enthalpy H, is introduced:

$$H(u)\begin{cases} = u & , \text{ if } u < 0, \\ \in [0, L] & , \text{ if } u = 0, \\ = u + L & , \text{ if } u > 0. \end{cases}$$

H is thus a multi-valued function of u.

2. A differential equation for the enthalpy is derived. Physically, the enthalpy represents the total heat content. Since the heat equation represents the balance between the change in heat content at a point and the flux of heat towards that point, it is plausible that H satisfies the equation:

$$\frac{\partial H}{\partial t} = \frac{\partial^2 H}{\partial x^2} \tag{8.1}$$

However, equation (8.1) must be treated with some caution because of the jump in the value of H at $u=0$.

As with variational inequalities, there exists a very satisfactory theory for the enthalpy method. Solomon, Alexiades and Wilson (1985) unreservedly recommend the enthalpy method, although other workers have expressed concern at the low accuracy with which the UB is approximated. A particularly important aspect of the enthalpy method is that it is able to handle so-called *mushy regions*. These are regions in which the material being considered consists of a mixture of two phases. For water and ice, such a region may be thought of as *slush*.

9. SUMMARY

The past decade has seen steady progress in the numerical solution of FBPs:

- Variational inequalities and the enthalpy method are now well established and understood, and they provide generally satisfactory theoretical and numerical results.
- One-dimensional MBPs can be readily solved by front-tracking methods.
- Trial-free-boundary methods are proving very successful in coping with two-dimensional viscous flow problems.
- The difficulties associated with solving problems on general regions have been successfully overcome by using body-fitted curvilinear coordinates, integral equations, or variable finite elements.

Nevertheless, there are still serious shortcomings in our capabilities and knowledge:

1. Only a very small number of three-dimensional FBPs have been solved numerically.

2. Similarly, there are numerous results for MBPs in one space dimension, but only a small number in two or more space dimensions.

3. Apart from the excellent results for variational inequalities and the enthalpy method, there are hardly any convergence results for numerical methods for FBPs.

4. While the method of variational inequalities and the enthalpy method are in general excellent methods, they do not give very accurate approximations to the UB.

5. There are many physical FBPs where present computing resources are stretched to the limit:

(a) Problems with complicated structures, such as clusters of bubbles or dendrite growth.

(b) Problems where the UB changes abruptly. For example:

(1) Breaking waves.

(2) Bursting bubbles.

(c) Problems where either the governing equation is very complicated, or the UBP merely forms one component of larger problem.

6. In the introduction it was stated that most UBPs are well-behaved. This is true, but it may be expected that in the future more difficult problems will be considered. An example is the blow-up of a Hele-Shaw cell subject to suction (Howison, Ockendon and Lacey (1985)), which is attracting increasing interest.

REFERENCES

Amick, C.J. and Toland, J.F. (1979), "Finite-amplitude water waves", Technical Summary Report No. 2012, Mathematics Research Center, University of Wisconsin, Madison, Wisconsin, USA.

Bach, P. and Hassager, O. (1985), "An algorithm for the use of the Lagrangian specification in Newtonian fluid mechanics and applications to free-surface flow", *J. Fluid. Mech.*, *152*, pp.173-190.

Barbu, V. (1984), *Optimal Control of Variational Inequalities*, Pitman (Boston).

Baiocchi, C. and Capelo, A. (1984), *Variational and Quasivariational Inequalities*, John Wiley and Sons (Chichester).

Bensoussan, A. and Lions, J.-L. (1984), *Impulse Control and Quasi-Variational Inequalities*, Gauthier-Villars (Paris).

Birkhoff, G. and Zarantonello, E.H. (1957), *Jets, Wakes and Cavities*, Academic Press (New York).

Bonnerot R. and Jamet P. (1979), "A third order accurate discontinuous finite element method for the one-dimensional Stefan problem", *J. Comput. Phys.*, *32*, pp.145-167.

Bossavit, A., Damlamian, A. and Fremond M. (eds.) (1985), *Free Boundary Problems: Applications and Theory*, Vols. 3 and 4, Pitman (Marshfield, Massachusetts).

Chipot, M. (1984), *Variational Inequalities and Flow in Porous Media*, Springer (New York).

Christov, C.I. and Volkov, P.K. (1985), "Numerical investigation of the steady viscous flow past a stationary deformable bubble", *J. Fluid Mech.*, *158*, pp.341-364.

Conway, W.E. and Thomas, J.W. (1974), "Free streamline problems and the Milne-Thomson integral equation", *J. Math. and Phys. Sciences*, *8*, pp.67-92.

Crank, J. (1984), *Free and Moving Boundary Problems*, Clarendon Press (Oxford).

Crowley, A.B. and Elliott, C.M. (eds.) (1985), *Proceedings of Conference on Crystal Growth*; *IMA J. Appl. Math.*, *35*, pp.115-264.

Diaz, J.I. (1985), *Nonlinear Partial Differential Equations and Free Boundaries. Volume 1 - Elliptic Equations*, Pitman (Boston).

Elcrat, A.R. and Trefethen, L.N. (1986), "Classical free-streamline flow over a polygonal obstacle", *J. Comput. Appl. Math.*, *14*, pp.251-265.

Elliott, C.M. and Ockendon, J.R. (1982), *Weak and Variational Methods for Moving Boundary Problems*, Pitman (Boston).

Friedman, A. (1982), *Variational Principles and Free-Boundary Problems*, John Wiley and Sons (New York).

Glowinski, R., Lions, J.-L. and Tremolieres, R. (1981), *Numerical Analysis of Variational Inequalities*, North-Holland (Amsterdam).

Howison, S.D., Lacey, A.A. and Ockendon, J.R. (1985), "Singularity development in moving-boundary problems", *Quart. J. Mech. Appl. Math.*, *38*, pp.343-360.

Kelmanson, M.A. (1983), "Boundary integral equation solution of viscous flows with free surfaces", *J. Engrg. Math.*, *17*, pp.329-343.

Kinderlehrer, D. and Stampacchia, G. (1980), *An Introduction to Variational Inequalities and their Applications*, Academic Press (New York).

Lynch, D.R. (1982), "Unified approach to simulation on deforming elements with application to phase change problems", *J. Comput. Physics*, *47*, pp.387-411.

Meyer, G.H. (1981), "The method of lines and invariant embedding for elliptic and parabolic free boundary problems", *SIAM J. Numer. Anal.*, *18*, pp.150-164.

Monakov, V.N. (1983), *Boundary-Value Problems with Free Boundaries for Elliptic Systems of Equations*, Amer. Math. Soc. (Providence, RI).

Panagiotopoulos, P.D. (1985), *Inequality Problems in Mechanics and Applications*, Birkhauser (Boston).

Pennell, S.A. and Su, C.H. (1984), "A seventeenth-order series expansion for the solitary wave", *J. Fluid Mech.*, *149*, pp.431-443.

Pironneau, O. (1984), *Optimal Shape Design for Elliptic Systems*, Springer (New York).

Reinelt, D.A. and Saffman, P.G. (1985), "The penetration of a finger into a viscous fluid in a channel and tube", *SIAM J. Sci. Statist. Comput.*, *6*, pp.542-561.

Rubinstein, L. (1971), *The Stefan Problem*, American Math. Soc. (Providence, RI).

Ryskin, G. and Leal, L.G. (1984a), "Numerical solution of free-boundary problems in fluid mechanics. Part 1: The finite-difference technique", *J. Fluid Mech.*, *148*, pp.1-17.

Ryskin, G. and Leal, L.G. (1984b), "Numerical solution of free-boundary problems in fluid mechanics. Part 2: Buoyancy-driven motion of a gas bubble through a quiescent liquid", *J. Fluid Mech.*, *148*, pp.19-35.

Ryskin, G. and Leal, L.G. (1984c), "Numerical solution of free-boundary problems in fluid mechanics. Part 3: Bubble deformation in an axisymmetric straining flow", *J. Fluid Mech.*, *148*, pp.37-43.

Solomon, A.D., Alexiades, V. and Wilson, D.G. (1985), "Moving boundary problems in phase change models. Current research questions", *SIGNUM Newsletter (ACM Special Interest Group Num. Anal.)*, *20*, pp.8-12.

Tanner, R.I. (1984), "Problems and progress in polymer extrusion studies", in *Numerical Analysis of Forming Processes*, Pittman, J.F.T., Zienkiewicz, O.C., Wood, R.D. and Alexander, J.M., eds., John Wiley and Sons (Chichester), pp.285-305.

Colin W. Cryer
Institut für Numerische und Instrumentelle Mathematik
Westfaelische Wilhelms-Universität zu Muenster
Einsteinstrasse 62
D-4400 Münster
Germany FDR

21 Multigrid Methods for Elliptic Equations

J. WALSH

ABSTRACT

The systematic use of a hierarchy of grids in finite-difference and finite-element approximation has led to very efficient methods for solving elliptic problems. A particular advantage of this multigrid approach is that nonlinearities can often be incorporated directly into the procedure. The constituents of the method are smoothing operators, residual calculations, and transfers between grids, and these can be analysed in detail for model problems using Fourier or matrix methods. Algorithms are available for linear problems, but the treatment of boundary conditions and nonlinear terms is not always straightforward.

1. BASIC IDEAS OF MULTIGRID

The last "State of the Art" conference in 1976 included a survey of numerical methods for elliptic equations by Leslie Fox, who noted an early paper by Brandt (1973) on the use of multi-level grid techniques. As he remarked, this idea had been foreshadowed in the days of hand relaxation for elliptic equations, when practitioners of the art would often use coarser grids as an auxiliary device to speed up the reduction of the residuals on the fine grid. It was already clear then that the multigrid technique was very general, and could be used for both linear and nonlinear problems, but that the programming required was relatively complex (Fox, 1977). In the past ten years there has been extensive development of the theory and applications of multigrid methods, and there is now a considerable literature on the subject. I shall not attempt to make a complete survey here, but will concentrate on some of the main results.

There are many ways of using a number of levels of grid refinement in a calculation, and the problem for the user is to understand the options available and to select the appropriate technique for his own application. Some important convergence results have been proved, but as usual the practical method has developed well beyond the limits of the theory. The basic components of the multigrid method are fairly simple, and we will outline them here to introduce the later discussion.

Taking as an example the Poisson equation

$$\nabla^2 u = f, \tag{1.1}$$

the simple five-point discretization at the point P may be written as

$$u_N + u_W + u_S + u_E - 4u_P = h^2 f_P \tag{1.2}$$

on a square grid of side h (see Mitchell and Griffiths (1980) for the background). A classical iterative method of solution is successive over-relaxation (SOR), where a typical step is

$$u_P^{(l+1)} = u_P^{(l)} + \omega \times \text{Residual}\,, \tag{1.3}$$

$$\text{Residual} = \tfrac{1}{4}\left(u_N^{(l+1)} + u_W^{(l+1)} + u_S^{(l)} + u_E^{(l)} - 4u_P^{(l)} - h^2 f_P\right).$$

The aim of the iteration is to reduce the residual at every point to zero. In hand relaxation, the calculation was transferred to coarser grids in order to sweep the residuals to the boundary in fewer steps and hence to reduce them faster. In multigrid terms the residuals are expressed as a Fourier expansion (or eigenfunction expansion), with high-frequency modes which can be removed on the fine grid, and low-frequency modes which can be removed more effectively on coarser grids.

To make this explicit, suppose the equation (1.1) is defined in the unit square with Dirichlet boundary conditions.

Let G_k be the square grid of side h covering the region, where $Nh=1$. The eigenfunctions of the discrete operator in (1.2) may be written as

$$v_{p,q} = \sin(p\pi x)\sin(q\pi y), \quad 1 \leqslant p,q \leqslant N-1, \qquad (1.4)$$

and the residual (or the error) may be expanded in terms of these functions. Now the $v_{p,q}$ can be divided into two sets, high-frequency modes with $p,q \geqslant \frac{1}{2}N$ (assuming N even) which can only be represented on the grid G_k and low frequency modes with $p,q \leqslant \frac{1}{2}N$ which can be also represented on a coarser grid G_{k-1} with mesh-size $2h$. For many simple iterative schemes it is more efficient to reduce the low-frequency modes by computation on G_{k-1} rather than on G_k. Thus if we remove the high modes rapidly on G_k by some smoothing procedure, the residuals can be transferred to G_{k-1} for faster iteration. The situation is then repeated, and the calculation may be transferred to successively coarser grids $G_{k-1},\cdots,G_0$, where G_0 has very few points. The lowest error modes are removed very rapidly on G_0, and the fine-grid solution is reconstructed by transferring the corrections back to $G_1,G_2,\cdots,G_k$, with further smoothing at each stage.

To carry out this strategy we need a fast smoother on each grid. An extrapolated method such as SOR is not suitable because it is designed for fast overall convergence and not for rapid reduction of the higher error modes; if we take the optimum value of ω in (1.3) the dominant eigenvalues of the iteration are complex and of equal modulus, so the residuals behave erratically. For multigrid iteration it is more effective to use simpler methods such as Gauss-Seidel or variants of Jacobi. We also need operators to transfer the residuals between grids, without introducing or magnifying the low-frequency errors in going from coarse to fine grids. When all these operators have been chosen appropriately the resulting procedure is very fast, and the efficiency improves as the number of grids increases.

For a model problem such as the Dirichlet example above,

the computational cost of a multigrid calculation can be shown to be of the same order as the number of grid-points, i.e. $O(N^2)$ for the $N \times N$ grid. For this reason it is often said that multigrid methods are optimal for elliptic grid equations, because it is assumed that no method can have a lower asymptotic cost. For SOR and similar methods the cost is $O(N^3)$. These estimates are for calculating the solution to fixed arithmetic accuracy; if we solve the algebraic equations to the accuracy of the finite-difference approximation, which has error of $O(h^2)$, there is an additional factor $\log N$ in the above results.

The asymptotic estimate for the cost of multigrid may be compared with that for direct solution. Sparse matrix techniques are very expensive, but for simple problems such as Poisson or Helmholtz equations in a rectangle, direct solution can be carried out rapidly by fast Fourier transform and cyclic reduction methods. The cost of these procedures (implemented for example in FISHPAK) is $O(N^2 \log N)$, for a solution with accuracy close to the full word-length. Although this is asymptotically slower than multigrid, the actual cost of cyclic reduction may be less than that of multigrid for problems of moderate size, because the procedure is simpler.

Problems for which multigrid techniques are particularly useful include those with complex regions and large numbers of points, and nonlinear problems where the nonlinear terms can be handled directly in the iteration. In these cases we often get convergence at a cost roughly proportional to the number of points, if the procedure is carefully designed. Such results are very encouraging, and provide a motivation for trying to develop robust general algorithms in this area.

2. THEORETICAL DEVELOPMENTS

Let the system of algebraic equations which represents an elliptic problem on the grid G_k be

$$A_k \underline{u}_k = \underline{f}_k \,. \tag{2.1}$$

Suppose we carry out a few steps of the smoothing process on this grid and obtain an approximate solution $\bar{u}_k$ of (2.1). The residual defined by

$$\underline{r}_k = \underline{f}_k - A_k \underline{\bar{u}}_k \tag{2.2}$$

is transferred to the coarser grid G_{k-1} by a restriction operator R_k, which gives a mesh-function defined at the points of G_{k-1}. The simplest way of transferring a function to a coarser grid is to take the values at the common points of the two grids (injection), but we may also apply a smoothing operator at this stage. The equation on G_{k-1} defines a correction $\underline{e}_{k-1}$ to the solution, which satisfies

$$A_{k-1} \underline{e}_{k-1} = R_k \underline{r}_k \,. \tag{2.3}$$

The reverse transfer from the coarse grid to the finer one requires a prolongation (interpolation) operator P_k, and the solution of (2.3) is transferred to G_k by

$$\underline{u}_k = \underline{\bar{u}}_k + P_k \underline{e}_{k-1} \,. \tag{2.4}$$

In practice more than two grids are involved in the procedure; residuals are transferred after smoothing to $G_{k-2}, \cdots, G_0$, and corrections are transferred in the reverse direction back to G_k.

The analysis of convergence given by Brandt (1977) is based on the expansion of the residual in terms of local Fourier modes. For the example of Section 1 (Dirichlet problem in a rectangle) the Fourier modes are the same as the exact eigenfunctions, but this is not true in general. Brandt argues that smoothing is a local operation which has no effect a few steps away from the mesh-point concerned, and so a local expansion of the error is an adequate basis for analysis. Equations with variable coefficients can also be treated by freezing coefficients locally. However, the Fourier method does not deal with general boundary conditions, or with cases where the Fourier modes are coupled by nonlinear effects.

The effectiveness of the smoothing operator is measured by the damping factor which it applies to the higher modes, i.e. to those modes which cannot be represented on the next coarser grid. Suppose a particular mode in (1.4) is reduced by the factor $\mu_{p,q}$; then the overall smoothing factor on the grid G_k may be defined as

$$\bar{\mu} = \max\{|\mu_{p,q}| : \tfrac{1}{2}N \leqslant p,q \leqslant N-1\}, \tag{2.5}$$

which is the slowest rate of convergence for any of the higher modes. This definition assumes that the modes are uncoupled in the smoothing process. Some smoothing operators have the effect of coupling the modes in blocks, and in place of $\mu_{p,q}$ in (2.5) we have to consider the spectral norm of the amplification matrix for each block. If the reduction factors of successive smoothing steps are not independent, we take an average smoothing factor over the number of steps actually carried out.

For the point Gauss-Seidel method taken along rows, it may be shown that the smoothing factor for the Poisson operator is 0.5. This is not very satisfactory, and smaller factors are generally preferable so that only a few smoothing steps are needed on each grid (usually $\leqslant 3$). If $\bar{\mu}$ can be estimated accurately the number of steps can be pre-determined; alternatively the rate of convergence may be monitored during the calculation (at some additional cost).

The transfers between grids may be analysed similarly by considering their effect on local Fourier modes, either for individual components or for groups of components. Thus a complete cycle of multigrid operations is represented as a sequence of operations on typical components of the error, and this gives an indication of the overall error reduction achieved by one cycle.

The strategy for grid transfers may be either fixed or adaptive. If it is fixed, the transfers take place after a preset number of smoothing sweeps on each grid (typically two), and the equations are solved essentially to full accuracy on the

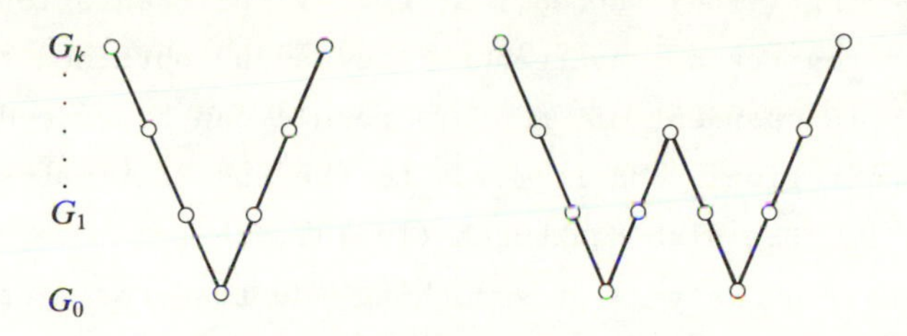

FIG. 1 Patterns of grid transfers

coarsest grid G_0. Figure 1 illustrates two patterns of grid transfers in a fixed strategy. If the strategy is adaptive, the transfer takes place when a suitable convergence criterion is satisfied, so that some norm of the residuals has to be calculated and tested on each sweep. The fixed strategy is usually preferred in general algorithms, and the evidence suggests that this is reasonably efficient for linear problems.

For a rigorous proof of convergence we have to consider the full linear operator which represents the computations of one cycle, rather than the local behaviour of the Fourier modes. A general treatment for rectangular regions is given by Wesseling (1980), the main assumptions of his proof being as follows:

(i) the smoothing process reduces the higher eigenfunctions by a factor independent of the grid;

(ii) any increase produced in the lower eigenfunctions by the smoothing process is limited;

(iii) the solution of the coarse-grid equation on G_{k-1} gives a good approximation to the solution on G_k.

To define the grid equation on G_{k-1} Wesseling uses the Galerkin approximation, which is given by

$$A_{k-1} = R_k A_k P_k \tag{2.6}$$

and similarly for $G_{k-2}, \cdots, G_0$. Thus the equation on the finest grid is obtained by a standard discretization, but the equations on other grids are derived from this by using the restriction and prolongation operators, without reference to the differential equation.

Wesseling (1980) showed that for elliptic equations on rectangular regions a convergence rate may be obtained that is essentially independent of h. This result can be extended to more general regions, and it leads to the cost estimates given in Section 1. See also Hackbusch (1981).

A detailed analysis of smoothing and transfer operators is given by Stüben and Trottenberg (1982). They consider the classical iterative methods for 5-point and 9-point difference equations on a rectangular grid, together with variants which are particularly effective for smoothing. These include the following

Jacobi with red-black ordering;

Line iteration with zebra ordering;

Alternating line zebra iteration.

The model problems discussed include Poisson's equation (1.1), the Helmholtz equation

$$\nabla^2 u - cu = f \tag{2.7}$$

and equations with strong anisotropy, such as

$$a\,\frac{\partial^2 u}{\partial x^2} + b\,\frac{\partial^2 u}{\partial y^2} = f\,, \quad 0 < a \ll b\,. \tag{2.8}$$

For the Helmholtz and Poisson equations a simple iteration such as red-black Jacobi is effective, but when the coordinate directions are distinguishable as in (2.8) the pointwise methods are unsatisfactory, and a line or zebra method in the direction of y is preferable. The overall convergence rate also depends on the choice of transfer operators.

The simple analysis of the smoothing and transfer procedures by Fourier methods does not give a proper treatment of the boundary conditions, except in the Dirichlet case. For general boundary conditions it is too complicated to analyse the eigenfunctions of the operator, and we have to use local behaviour as a guide to the choice of procedures. Brandt (1984) points out that the smoothness of the interior residuals may be lost at boundary points without special precautions, and suggests that additional sweeps around the boundary may be useful. These are not expensive, because the number of boundary points is small.

If G_k is a uniform rectangular grid, and the coarser grids $G_{k-1}, \cdots$ are formed by doubling the mesh-size, the points adjacent to the boundary move away from it as the mesh is coarsened. Thus the boundary condition is not well represented on the coarser grids, which is not satisfactory when the solution is strongly dependent on the boundary. A better strategy is to keep a fixed mesh-line close to the boundary on all grids. This gives non-uniform grids, of course, but it avoids the need for special smoothing procedures on the boundary.

3. PRACTICAL ALGORITHMS

It was quickly recognised that the multigrid method was potentially very efficient. However, it is not easy to obtain good convergence rates in practice without a lot of experimenting with different smoothers and transfer strategies. The programming is not simple, and a non-expert with a problem to solve may find that the time spent in developing the program outweighs any eventual gain in speed. The production of standard algorithms has therefore been an important factor in making the method more accessible. A general multigrid algorithm will usually not have optimal efficiency for a particular problem, but it will save a great deal of development effort, and it may also provide the starting point for designing an improved strategy to solve a large problem. We will outline here two successful algorithms of different types. Foerster and Witsch (1982) describe the program MGOO, which is one of the general algorithms developed at GMD Bonn. This treats an elliptic equation of the form

$$a \frac{\partial^2 u}{\partial x^2} + b \frac{\partial^2 u}{\partial y^2} - cu = f , \qquad (3.1)$$

with general boundary conditions

$$\alpha u + \beta \frac{\partial u}{\partial n} = \gamma \quad \text{(outward normal)}, \tag{3.2}$$

in a rectangular region. The coefficients may be constant or functions of x and y, so the program can be used for solving the linear equations in the inner loop of a nonlinear solver. To ensure that the problem is well-posed the following conditions are sufficient

$$a, b > 0, \quad c \geqslant 0, \quad \alpha\beta \geqslant 0. \tag{3.3}$$

These are not checked automatically, but there is a check for consistency in the case of a pure Neumann problem, where the right-hand sides of (3.1) and (3.2) must satisfy an additional condition.

The input data is the elliptic equation and the boundary conditions, and the program carries out the discretization as well as solving the resulting linear equations. This makes it very convenient for the user. The elliptic problem is replaced by a system of five-point difference equations on a rectangular grid, and the user has the option of applying various point or line smoothers. Transfers from coarse to finer grids are carried out with bilinear interpolation. The speed of computation is considerably better than that of classical iterative methods, and roughly similar to the speed of direct cyclic reduction methods.

Another multigrid algorithm, MGD1 (Wesseling 1982), starts from the algebraic equations rather than the differential problem. The input is a set of linear equations based on seven-point grid approximations on a rectangular grid. With the grid-points labelled as in (1.2), the additional points included are NW and SE. The seven-point operator is chosen in preference to a five-point form in order to allow cross-derivatives to be represented, and to maintain a uniform structure on the coarser grids.

The smoother in the MGD1 program is based on the incomplete LU factorization (ILU), which is very effective over a wide range of problems. It was developed as the basis of the strongly implicit iterative method of Stone (1968), but it is used here only for its smoothing properties. Suppose the discrete equations on the grid G_k are

$$A_k \underline{u}_k = \underline{f}_k \tag{3.4}$$

as before, where A_k is now a matrix with seven non-zero diagonals. If A_k is factorized into lower and upper triangular matrices which have the same sparsity pattern as A_k we cannot get exact LU factors; however, we can write

$$A_k + C_k = L_k U_k, \tag{3.5}$$

where C_k represents the correction terms which are needed to make L_k, U_k of the required form. By writing out the full equations in (3.5) it is easily seen that the non-zero elements of C_k can be restricted to two non-principal diagonal lines. The iteration for smoothing the residuals in equation (3.4) has the form

$$L_k U_k \underline{u}_k^{(l+1)} = \underline{f}_k + C_k \underline{u}_k^{(l)} \tag{3.6}$$

so that the calculation of $\underline{u}_k^{(l+1)}$ involves simple forward and backward sweeps over the mesh, with only three non-zero terms in the matrix for each sweep. This method has good smoothing properties, and the smoothing factors are such that usually only one sweep is needed on each grid.

Compared with simple point and line methods such as red-black Jacobi or alternating line iteration, there is rather more computation to be done in the ILU method. We have to set up the factors initially, and this makes it unsuitable for nonlinear problems in which the matrix A_k changes from cycle to cycle. However, it is fast and robust for linear problems.

In the MGD1 program the coarse-grid operators are obtained algebraically from the original matrix A_k so that the differential equation is not used at all after setting up the system

(3.4). The coarse-grid equations are based on the Galerkin approximation, as in (2.6), which defines the next coarser grid operator in terms of A_k and the transfer operators. The restriction operator uses local smoothing over six neighbouring points,

$$R_k(u_P)_k = \left\{u_P + \tfrac{1}{2}(u_N + u_S + u_W + u_E + u_{NW} + u_{SE})\right\}_k \qquad (3.7)$$

and the prolongation operator uses linear interpolation. With these forms, P_k and R_k are adjoint, so that

$$\sum_{G_k} (P_k \underline{u}_{k-1})^T \underline{v}_k = \sum_{G_{k-1}} u_{k-1}^T (R_k \underline{v}_k) . \qquad (3.8)$$

The definitions of P_k and R_k and the Galerkin approximation give successive grid operators $A_{k-1}, A_{k-2}, \cdots$ which retain the seven-point form when A_k has a seven-point or five-point structure. The initial discretization is based on a rectangular grid and the coarser grid operators may be interpreted as referring to coarser grids, but they are not the same as the operators derived by discretizing on these grids. The Galerkin method treats the boundary condition automatically, so that any refinement of the approximation needed near the boundary must be built in to the original equations.

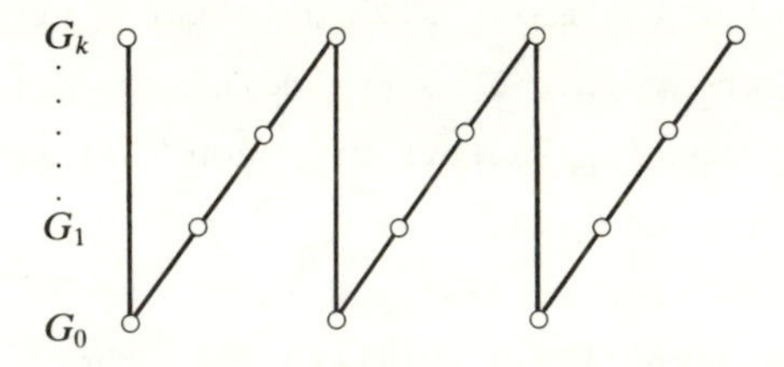

FIG. 2 Grid transfers in MGD1

The pattern of grid transfers used in Wesseling's program is exceptionally simple and is illustrated in Figure 2 (sawtooth cycle). The algorithm has been tested over a wide range of problems, and it is generally efficient and easy to use. It is faster than standard iterative methods, but slower than the cyclic reduction method for problems where this is applicable.

Tests by McCarthy (1983) showed that it is slower than the solution phase of a direct sparse matrix solver for Poisson's equation. However, the factorization phase of the sparse matrix solver is very much more expensive, and requires considerably more storage than multigrid.

4. EXTENSIONS OF MULTIGRID

We have discussed the multigrid method so far in terms of finite-difference equations, but it is clear that the same ideas can be applied to any type of discrete equations on a regular mesh, including those obtained from finite-element approximation. Now, one of the advantages of finite elements over finite differences is that the grid need not be regular; however, it is difficult to define grid coarsening on a completely irregular grid if we want to use the multigrid method. The problem becomes easier if we take a basic grid which is regular over the whole region and use local refinement in areas where the solution is expected to vary rapidly, for example near a corner or in a boundary layer.

For nonuniform grids of this type we need the approximate solution to the equations on intermediate grids as well as on the finest grid, so we work in terms of the solution rather than of corrections to it (this is called Full Approximation Storage). The equation for the correction on grid G_{k-1} (cf. eqn. (2.3)) is

$$A_{k-1}\underline{e}_{k-1} = R_k\underline{r}_k = R_k(\underline{f}_k - A_k\underline{\bar{u}}_k), \qquad (4.1)$$

where $\underline{\bar{u}}_k$ is the approximate solution on G_k. This is replaced by an equation for the actual solution on G_{k-1},

$$A_{k-1}\underline{u}_{k-1} = A_{k-1}(R_k\underline{\bar{u}}_k) + R_k(\underline{f}_k - A_k\underline{\bar{u}}_k), \qquad (4.2)$$

which has the same operator as (4.1) but a different right-hand side. The smoothing procedure is then applied to (4.2) as usual. The quantity which has to be smoothed and ultimately reduced to zero is the residual, not the solution. Thus it is the correction which is eventually transferred back to G_k by the relation

$$\underline{u}_k = \underline{\bar{u}}_k + P_k (\underline{u}_{k-1} - R_k \underline{\bar{u}}_k). \tag{4.3}$$

(Of course a full multigrid cycle normally involves more than two grids.)

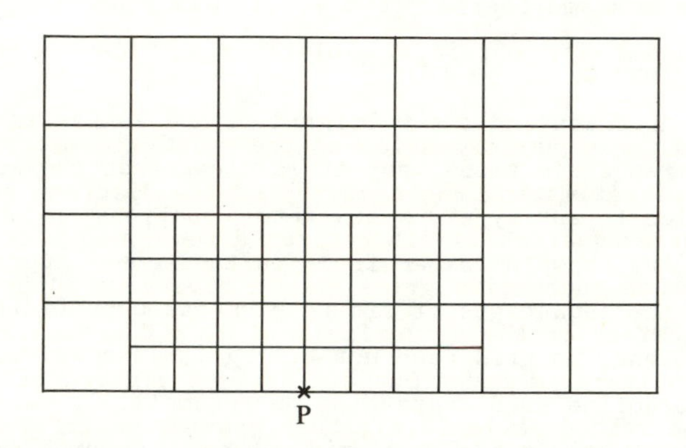

FIG. 3 Region with local refinements

Figure 3 illustrates a region covered by a regular grid with local refinement. Suppose the solution is known to have a singularity at P, and we want to improve the accuracy by reducing the mesh-size near P. The uniform grid G_k is taken over the whole region, and the finer grid G_{k+1} only in the neighbourhood of P. First we solve the equations on G_k, using a sequence of coarser grids in the usual way. The local solution is transferred to G_{k+1} by prolongation, and smoothing is then carried out on this grid, using finer grids if necessary. To correct the remainder of the solution the local solution is transferred back to G_k, and the equations on G_k are given by

$$A_k \underline{u}_k = \underline{f}_k \tag{4.4}$$

at points where the grid is not refined, and

$$A_k \underline{u}_k = A_k (R_{k+1} \underline{\bar{u}}_{k+1}) + R_{k+1} (\underline{f}_{k+1} - A_{k+1} \underline{\bar{u}}_{k+1}) \tag{4.5}$$

at points which are common to G_k and G_{k+1}. The complete cycle can then be repeated.

This procedure is simple in principle but not entirely straightforward to program, because of the need to keep track of the different types of mesh-points. A multigrid algorithm based on finite-element approximation with local grid refinement was described by Bank and Sherman (1981). It was found to give an efficient method for obtaining solutions of low to moderate accuracy.

Another important extension of the multigrid method is to the problem of estimating the accuracy of the discrete solution. If the elliptic equation is discretized directly on each grid, the solutions on successive grids should give some information about the accuracy of the approximations as a function of h. But if the coarse-grid equations are derived algebraically by a Galerkin-type method as in (2.6), they provide no information about the dependence of the error on the step-size.

Assuming the operators A_k are set up directly from the differential equation, the discrete equations on two grids can be used to give an estimate of the local truncation error (Brandt, 1984). The true error on grid G_{k-1} would be given by substituting the exact solution u into the equation

$$A_{k-1}\,\underline{u}_{k-1} = R_k\,\underline{f}_k\,, \tag{4.6}$$

but u is not known. The equation which is solved on G_{k-1} is (4.2); thus we can estimate the truncation error $\underline{\tau}_{k-1}$ on G_{k-1} by substituting $\underline{u}_k$ into

$$\underline{\tau}_{k-1} = A_{k-1}\,(R_k\,\underline{u}_k) - R_k\,(A_k\,\underline{u}_k)\,, \tag{4.7}$$

which involves essentially no additional calculation. If the truncation error of the approximation is known to be $O(h^p)$ say, we can extrapolate the quantity (4.7) by multiplying it by the factor $2^p/(2^p-1)$, as in Richardson's method. Using the extrapolated quantity in (4.2) will give an improved result on the coarser grid G_{k-1}, provided that the asymptotic assumptions are valid.

To obtain a better approximation on the fine grid G_k we can use the difference correction procedure (Fox, 1976). Suppose that a high-order representation of the differential equation is given by

$$A_k \underline{u}_k + C_k \underline{u}_k = \underline{f}_k \, , \tag{4.8}$$

where $C_k \underline{u}_k$ are high-order terms which were neglected in (2.1). For example, if A_k is based on the simple five-point formula (1.2) for Poisson's equation, $C_k \underline{u}_k$ might represent the fourth-order terms. Then (4.8) may be solved by the iteration

$$A_k \underline{u}_k^{(l+1)} = \underline{f}_k - C_k \underline{u}_k^{(l)} \, , \tag{4.9}$$

if it converges, and the result will be of higher order than the solution of (2.1). The advantage of using the form (4.9) is that only the simple operator A_k is involved in the algebraic equations.

This method (also called double discretization) is discussed in the multigrid context by Auzinger and Stetter (1982). Starting with $\underline{u}_k^{(0)} = \underline{0}$, we solve (4.9) using multigrid techniques to obtain $\underline{\bar{u}}_k^{(1)}$ say, which need not have a high algebraic accuracy. The next multigrid cycle starts from

$$A_k \underline{u}_k^{(2)} = \underline{f}_k - C_k \underline{\bar{u}}_k^{(1)} \, , \tag{4.10}$$

and the correction terms can be re-computed each time the procedure returns to the finest grid. Full approximation storage is not necessary if the difference correction is calculated only on G_k. Brandt (1984) suggests that the high-order terms should be used in calculating the residuals on the coarser grids, but this cannot be done on very coarse grids, because the correction terms (corresponding to high-order differences) will not be well-defined. The properties of the operator $(A_k + C_k)$ do not affect the smoothing and transfer procedures. It may be less sparse than A_k and may lack diagonal dominance without disturbing the algebraic convergence of the multigrid procedure for solving the linear equations (4.9). However the iteration in (4.9) is not guaranteed to converge, although it usually does so when the solution is smooth and the grid is fine enough.

The smoothing operators commonly used in multigrid algorithms generally work well for equations where the matrix has diagonal dominance. However this condition is by no means necessary for the success of the method. Even in cases where the operator is singular or nearly singular, we can adapt the procedure to obtain a solution.

Brandt and Taasan (1985) have proposed a method for dealing with nearly singular problems, for example the Helmholtz equation

$$\nabla^2 u + c^2 u = f \tag{4.11}$$

for certain ranges of the coefficient c^2. If the discrete operator has an eigenvalue close to zero with corresponding eigenfunction v_0, then the error on the fine grid will normally contain a component in the direction v_0 which will not be removed by the multigrid process. They suggest that v_0 should be estimated separately by solving the homogeneous problem and isolating the slowly decaying component. It can then be removed in the course of solving the discrete form of (4.11) by introducing a constraint which excludes v_0 from the coarse-grid equation.

5. NONLINEAR EQUATIONS

In the difference correction method described above, we use an extended operator in calculating the residual on G_k, while retaining a simple operator for the smoothing process. Similarly, in the case of nonlinear problems we can use the nonlinear operator to calculate the residual, with a simpler form for smoothing. Consider the boundary-value problem

$$F(u) = f \tag{5.1}$$

in a region R, where $F(u)$ is a nonlinear differential expression, with suitable boundary conditions (for example, u specified on the boundary of R). The standard Newton iteration uses a linearized form of (5.1),

$$F'(u^{(l)})(u^{(l)} - u^{(l-1)}) = f - F(u^{(l)}) , \qquad (5.2)$$

where $F'(u)$ is the Fréchet derivative, and the initial approximation $u^{(0)}$ is obtained by a separate calculation. If the linear equation (5.2) is discretized and solved by a multigrid method, we have a standard "inner-outer" iterative procedure. As usual, the calculation may be simplified by keeping the derivative $F'(u^{(l)})$ fixed for several steps of the outer iteration, although this reduces the order of convergence near the solution. Whether the derivative is re-calculated or not, a lot of the computational effort goes into solving (5.2) in the early stages of the iteration, when the linearized equation is not a good representation of the actual problem.

A more direct approach is to use the multigrid principle for the nonlinear residuals. Suppose the equation (5.1) is represented on the grid G_k by the discrete algebraic system

$$\underline{F}_k(\underline{u}_k) = \underline{f}_k , \qquad (5.3)$$

which has an approximate solution $\underline{\bar{u}}_k$. Then the residual (or defect) which is transferred to the next grid G_{k-1} may be calculated from the nonlinear equation (5.3) instead of the discrete form of (5.2), so that the defect is

$$\underline{d}_k = \underline{f}_k - F_k(\underline{\bar{u}}_k) . \qquad (5.4)$$

The equation to be solved on G_{k-1} is then

$$\underline{F}_{k-1}(\underline{u}_{k-1}) = \underline{F}_{k-1}(R_k \underline{\bar{u}}_k) + R_k \underline{d}_k , \qquad (5.5)$$

analogous to (4.2). This requires the full solution to be stored on all grids; we obtain an approximate solution on G_{k-1} from (5.5), but transfer a correction to $\underline{\bar{u}}_k$ back to G_k as in (4.3).

To carry out the smoothing process on each grid we have to find a smoothing operator for nonlinear equations of the form

$$\underline{F}_q(\underline{u}_q) = \underline{c}_q , \quad q = 0, 1, \cdots, k . \qquad (5.6)$$

This can be done by linearizing the algebraic system (5.6) and using the appropriate methods for a linear problem. But it may be more effective to treat the nonlinear equation directly at each grid-point.

For example, consider the simple nonlinear elliptic problem

$$F(u) \equiv \nabla^2 u - g(u) = 0 \tag{5.7}$$

in a region R, with u specified on the boundary. The differential part of $F(u)$ is linear, and if $g(u)$ is only mildly nonlinear we can carry out smoothing operations as for a Poisson equation (Auzinger and Stetter, 1982). Alternatively, the equation (5.7) may be represented at a grid-point by the linearized form

$$\nabla^2 u - \left(\frac{dg}{du}\right)_{\bar{u}} u = g(\bar{u}) - \left(\frac{dg}{du}\right)_{\bar{u}} \bar{u}, \tag{5.8}$$

where $\bar{u}$ is the current approximation to u. The discrete form of (5.8) gives a linear equation which can be solved for an improved value of u at the grid point. In general this is not the same as the inner-outer iteration mentioned above, because we are not trying to solve (5.8) globally, but to use it for smoothing the residual at a particular point.

It is clear that model problem analysis is not very helpful for general nonlinear equations. If we need estimates of the smoothing factors they have to be calculated from the linearized equations, but the representation of the residual in terms of Fourier modes is not so useful because there is interaction between the modes through the nonlinear terms. In practice it is usually necessary to experiment with various smoothing techniques in order to get an efficient program. However, numerical tests indicate that with sufficient care we can still get fast convergence in many cases (Brandt, 1984). In some respects the fully nonlinear multigrid iteration is easier to implement than the linearized iteration based on (5.2), and if point smoothing is used it does not require more than one solution vector on each grid. (For an ILU smoother, it would be necessary to linearize the whole system (5.6), and the process would probably be uneconomic.)

An important class of nonlinear problems for which multigrid methods have been successful is in fluid dynamics, for example the Navier-Stokes equation. For incompressible steady-state flow, this takes the form

$$\begin{aligned} \frac{1}{R}\nabla^2 u - u\cdot \operatorname{grad} u &= \operatorname{grad} p - F\,, \\ \operatorname{div} u &= 0\,, \end{aligned} \tag{5.9}$$

where u is the velocity, p the pressure, and R the Reynolds number. For large values of R the equations are relatively difficult to solve because of the nonlinear term $u\cdot \operatorname{grad} u$. The method often used is an iteration between the velocities and the pressure; the momentum equations in (5.9) are solved for a certain pressure distribution, and the pressure is then corrected by using the continuity equation (Patankar and Spalding, 1972).

A nonlinear multigrid approach enables us to take full account locally of the nonlinear terms, which gives a locally consistent solution. This seems to improve the rate of convergence, though the computation is still lengthy for large values of R. If boundary layers are present it is necessary to use locally-refined grids (or variable mesh spacing), and the multigrid process does not converge well when very coarse grids are included. This is because the solution cannot be represented properly on a small number of points.

Although it is unlikely that general algorithms can be constructed for nonlinear problems, it would be possible to provide an algorithmic framework within which various smoothing and transfer operators could be tried out. At present we have efficient linear algorithms which can be used in the inner loop of an iteration, and these may give some help in the development of a fully nonlinear method. The multigrid technique is particularly useful for very large problems, and we would expect to need special-purpose software for such cases.

REFERENCES

Auzinger, W. and Stetter, H.J. (1982), "Defect corrections and multigrid iterations", in *Multigrid Methods*, eds. Hackbusch, W. and Trottenberg, U., Lecture Notes in Mathematics, *960*, Springer-Verlag (Berlin), pp.327-351.

Bank, R.E. and Sherman, A.H. (1981), "An adaptive multi-level method for elliptic boundary-value problems", *Computing*, *26*, pp.91-105.

Brandt, A. (1973), "Multi-level adaptive techniques (MLAT) for fast numerical solution to boundary value problems", *Lecture Notes in Physics*, *18*, Springer-Verlag (Berlin), pp.82-89.

Brandt, A. (1977), "Multi-level adaptive solutions to boundary-value problems", *Math. Comp.*, *31*, pp.333-390.

Brandt, A. (1984), "Multigrid techniques: 1984 guide with applications to fluid dynamics", GMD-Studien, *85*, GMD, Bonn, W. Germany.

Brandt, A. and Taasan, S. (1985), "Multigrid method for nearly singular and slightly indefinite problems", ICASE Report 85-57, NASA Langley Research Center, Hampton, Virginia.

Foerster, H. and Witsch, K. (1982), "Multigrid software for the solution of elliptic problems on rectangular domains", in *Multigrid Methods*, eds. Hackbusch, W. and Trottenberg, U., Lecture Notes in Mathematics, *960*, Springer-Verlag (Berlin), pp.427-460.

Fox, L. (1977), "Finite-difference methods for elliptic boundary-value problems", in *The State of the Art in Numerical Analysis*, ed. Jacobs, D.A.H., Academic Press (London), pp.799-881.

Hackbusch, W. (1981), "On the convergence of multi-grid iterations", *Beiträge Numer. Math.*, *9*, pp.213-239.

McCarthy, G.J. (1983), "Investigations into the multigrid code MGD1", Report R 10889, AERE Harwell.

Mitchell, A.R. and Griffiths, D.F. (1980), *The Finite Difference Method in Partial Differential Equations*, John Wiley (London).

Patankar, S.V. and Spalding, D.B. (1972), "A calculation procedure for heat, mass and momentum transfer in three-dimensional parabolic flows", *Int. J. Heat and Mass Transfer*, *15*, pp.1787-1806.

Stone, H.L. (1968), "Iterative solution of implicit approximations of multi-dimensional partial differential equations", *SIAM J. Numer. Anal.*, *5*, pp.530-558.

Stuben, K. amd Trottenberg, U. (1982), "Multigrid methods: Fundamental algorithms, model problem analysis and applications", in *Multigrid Methods*, eds. Hackbusch, W. and Trottenberg, U., Lecture Notes in Mathematics, *960*, Springer-Verlag (Berlin), pp.1-176.

Wesseling, P. (1980), "The rate of convergence of a multiple grid method", in *Numerical Analysis*, ed. Watson, G.A., Lecture Notes in Mathematics, *773*, Springer-Verlag (Berlin), pp.164-184.

Wesseling, P. (1982), "A robust and efficient multigrid method", in *Multigrid Methods*, eds. Hackbusch, W. and Trottenberg, U., Lecture Notes in Mathematics, *960*, Springer-Verlag (Berlin), pp.614-630.

J. Walsh
Department of Mathematics
University of Manchester
Manchester M13 9PL
England

22 Galerkin Finite Element Methods and Their Generalisations

K. W. MORTON

ABSTRACT

The goal of optimum or near-optimum approximation in an integral norm is used as a theme to unify various generalisations of the Galerkin idea. Petrov-Galerkin and finite volume methods are considered for equilibrium problems and the relationship of Petrov-Galerkin and Taylor-Galerkin methods for evolutionary hyperbolic problems is shown to characteristic Galerkin and Lagrange Galerkin methods.

1. INTRODUCTION

Finite difference methods are essentially local in character, but very flexible. Their greatest triumph in the last ten years has been in shock modelling for nonlinear hyperbolic partial differential equations, of which much more will be said by Professor Osher (see Osher and Sweby (1986)). Shocks are very local phenomena and by exploiting the detailed structure of the differential equations, recently developed methods give sharp well-defined shocks without spurious oscillations on either side. (However it should be noted that these local properties are ensured by imposing a global inequality on the total variation.)

Finite element methods on the other hand are much more global in character and more disciplined in their approach. This has naturally delayed their development for problems exhibiting very localised phenomena, such as are described by hyperbolic equations. Progress has been very rapid however in the last few years and we will describe some of this in the present paper, with particular reference to general principles of constructing and analysing methods. To do so it is necessary to go back to the origins of the finite element approach in the Galerkin formulation.

A strictly finite different approach to solving a general differential equation of the form

$$Lu = f\,, \tag{1.1}$$

for solution u in terms of data f, is to define a mesh-point approximation u_h by an equation

$$L_h u_h = f_h\,, \tag{1.2}$$

where L_h is a difference approximation to L and f_h a mesh-point approximation to f. A finite element approach to the same problem is to define a function U from an approximation space S^h by equations of the form

$$\langle LU - f\,, MW\rangle = 0 \quad \forall W \in T^h\,, \tag{1.3}$$

where T^h is a space of test functions, M some operator and $\langle\cdot,\cdot\rangle$ an inner product over the space of independent variables, a common notation that we shall use for both scalar and vector arguments. It is fairly clear that, by a judicious choice of L_h and f_h in (1.2) or of M and T^h in (1.3), one could cover almost all methods in one formalism. But this would not be very illuminating in itself and not help in understanding the relative merits and key distinguishing features of the possible choices.

For example, the standard approach to an error analysis for (1.2) is to introduce a truncation error T_E defined by

$$T_E := L_h r_h u - f_h\,, \tag{1.4}$$

where r_h is the restriction operator mapping u onto its mesh-point values. Then the global error has to be estimated from

$$L_h r_h u - L_h u_h = T_E . \tag{1.5}$$

This requires establishing the stability of the procedure or, more precisely, obtaining a bound for the inverse of L_h. One could proceed in exactly the same way with (1.3): by introducing P^h as the projection operator from the solution space containing u into the approximation space S^h, we can define a truncation error by

$$T_E(W) := \langle LP^h u - f, MW\rangle \quad \forall W \in T^h \tag{1.6}$$

and then have

$$\langle LP^h u - LU, MW\rangle = T_E(W) \quad \forall W \in T^h . \tag{1.7}$$

This has the advantage over (1.5) that the original operator L rather than its approximation L_h is involved: moreover, if the projection operator is defined by $\langle P^h v - v, MW\rangle = 0, \quad \forall W \in T^h$, we can write from (1.3) and (1.7)

$$P^h LP^h u - P^h Lu = T_E \tag{1.8}$$

and use this as a guide to the choice of P^h (and thus of M and T^h) as was done by Cullen and Morton (1980).

However, there are severe disadvantages to designing a method on the basis of its truncation error when, through (1.5), (1.7) or (1.8), one has only a very approximate idea of the effect of this error on the closeness with which u_h or U approximates u. A much more direct means of comparison is highly desirable and such an approach is characteristic of the finite element formulation. Thus we have directly from (1.3), the error projection property

$$\langle Lu - LU, MW\rangle = 0 \qquad \forall W \in T^h . \tag{1.9}$$

Now suppose L is a linear, self-adjoint, elliptic operator with the associated symmetric, coercive bilinear form $a(\cdot,\cdot)$: that is, $a(u,v)$ is obtained from $\langle Lu, v\rangle$ by integrating by parts to obtain the same order of derivatives on u and v — see Strang and Fix (1973). Then with M chosen as the identity and T^h the same as S^h, (1.9) becomes

$$a(u-U, W) = 0 \qquad \forall W \in S^h, \tag{1.10}$$

which are equivalent to the Galerkin equations. There follows immediately the crucial approximation property

$$a(u-U\,, u-U) = \inf_{V \in S^h} a(u-V\,, u-V)\,. \tag{1.11}$$

That is, in the sense defined by the norm induced by $a(\cdot,\cdot)$, U is the best approximation to u from the approximation space S^h. By using appropriate comparison functions V, one can obtain sharp error estimates for $u-U$ not only in the $a(\cdot,\cdot)$ norm but also in other norms. Moreover, one can use (1.11) as the starting point for post-processing or recovery algorithms which yield approximations to u having other desired properties: one may require approximations having greater smoothness than members of S^h or satisfying a maximum principle or a convexity constraint.

This is such a richly rewarding approach that one should resort to looking at truncation errors only as a final crude alternative. Indeed, finite difference methods for elliptic and parabolic problems which are based on an integral form have long been preferred by many practical problem-solvers because they mirror global properties of the differential system, irrespective of their truncation errors; and they have led very naturally to present-day finite volume methods. Similarly, difference schemes for hyperbolic equations are often designed to reflect the desired treatment of Fourier modes, rather than minimising the truncation error; and the new developments referred to earlier were based on going back to first order methods, in preference to second order, because of their property of preserving monotonicity.

Thus in considering how finite element methods have broken

out from their early association and success with linear, self-adjoint, elliptic problems, we shall regard the attainment of an approximation property like (1.11) as the chief goal. We will pursue this through a number of different problem types. In the next section we will consider non self-adjoint elliptic equations, typified by diffusion-convection problems. We shall see that the finite element approach gives a much clearer answer to how the first and second derivative terms should be approximated in order to maintain second order accuracy uniformly; and it is precisely because the difference equations become ill-conditioned that truncation error arguments encouraging the use of central differencing are misleading.

We shall omit any consideration of parabolic equations and use a short description of finite volume methods, as they are applied to systems of first order equations which may be of elliptic or hyperbolic type, as our link to evolutionary hyperbolic equations in Section 3. There we shall trace the generalisation of Galerkin methods to their ultimate exploitation of hyperbolicity in characteristic Galerkin and Lagrange Galerkin methods.

No attempt has been made to make the list of references comprehensive: readers wishing to study the literature in some detail will find many further references in those which are cited, particularly Barrett and Morton (1984) and Morton (1985).

2. EQUILIBRIUM PROBLEMS

2.1 *Linear Elliptic Equations and Petrov-Galerkin Methods*

The first attribute to relax in going from (1.9) to (1.10) is the self-adjointness, and this has been studied the most thoroughly. In weak form, the differential equation for u in a real Hilbert space H and its Petrov-Galerkin approximation for $U \in S^h \subset H$ may be written

$$B(u,w) = \langle f,w \rangle \qquad \forall\, w \in H \tag{2.1a}$$

$$B(U,W) = \langle f,W \rangle \qquad \forall\, W \in T^h \subset H. \tag{2.1b}$$

Here $B(\cdot,\cdot)$ is a coercive, but not necessarily symmetric, bilinear form on $H \times H$ and the term Petrov-Galerkin is used to denote the fact that the *test space* T^h is generally different from the *trial space* S^h. When $B(\cdot,\cdot)$ is very unsymmetric, as in convection-dominated flow problems, the Galerkin case in which $T^h \equiv S^h$ can easily be shown to give the optimal order of accuracy as $h \to 0$. However, for practical values of h it can give singular equations and generally leads to extremely poor solutions of spuriously oscillating form. The key question then is how to choose T^h for a given $B(\cdot,\cdot)$ and S^h, in order to give reliable, accurate solutions.

Credit must be given to Zienkiewicz and Mitchell here (see Christie *et al* (1976) and Heinrich *et al* (1977)) for showing the way to the use of upwind test elements to produce equations of the same form as the exponentially-fitted difference equations of Allen and Southwell (1955). In abstract form we can interpret these as follows. Suppose we choose a symmetric form $B_S(\cdot,\cdot)$ equivalent to $B(\cdot,\cdot)$, in respect of which we wish to find the optimal approximation to u. Then by the Riesz Representation Theorem there is an operator R_S which converts the one bilinear form into the other through the identity

$$B(v,w) \equiv B_S(v,R_S w) \qquad \forall\, v,w \in H. \tag{2.2}$$

Now suppose that the test space T^h is chosen to have an approximation property

$$\inf_{W \in T^h} \|V - R_S W\|_{B_S} \leqslant \Delta_S \|V\|_{B_S} \qquad \forall\, V \in S^h \tag{2.3}$$

for an appropriately small constant $\Delta_S < 1$, where the norm is that induced by $B_S(\cdot,\cdot)$. Then we have the following result for U of (2.1b) — see Morton (1982) and Barrett and Morton (1984) for more details and a proof.

THEOREM. Suppose the test space T^h has the same dimension as the Galerkin test space and satisfies (2.3). Then there exists a unique solution U to the Petrov-Galerkin equations (2.1b) with an error bound

$$\|u-U\|_{B_S} \leq (1-\Delta_S^2)^{-\frac{1}{2}} \inf_{V\in S^h} \|u-V\|_{B_S}. \tag{2.4}$$

This clearly includes as a special case the self-adjoint Galerkin problem, in which R_S is the identity, $T^h = S^h$ and hence $\Delta_S = 0$. In the general case, to achieve the optimal approximation with $\Delta_S = 0$ for all data functions f, it is necessary to choose the test space such that

$$R_S T^h = S^h. \tag{2.5}$$

That is, for a given choice of symmetric form $B_S(\cdot,\cdot)$ there is a unique test space for each trial space which will always yield the optimal approximation in the sense of the norm induced by $B_S(\cdot,\cdot)$. Construction of such a test space is possible only in simple model problems: the importance of the theorem is that Δ_S can be estimated for practical test spaces used for real problems and hence their efficiency established – see Morton and Scotney (1985).

It is illuminating to present these results in more familiar operator notation and to illustrate them by reference to the diffusion-convection problems for which they were first developed. Let us suppose then that the operator L in (1.1) has the form $L_1^* L_2$ where L_1^* is the adjoint of L_1 and $L_1 \neq L_2$: suppose also that $B_S(\cdot,\cdot)$ is defined by means of an operator L_S so that

$$B(v,w) \equiv \langle L_2 v, L_1 w\rangle, \qquad B_S(v,w) \equiv \langle L_S v, L_S w\rangle. \tag{2.6}$$

If the trial space S^h is spanned by basis functions $\{\phi_i\}$ and the test space T^h by $\{\psi_i\}$, the Petrov-Galerkin equations (2.1b) for our problem

$$L_1^* L_2 u = f \tag{2.7a}$$

become

$$\langle L_2 U, L_1 \psi_i \rangle = \langle f, \psi_i \rangle \qquad \forall\, i, \tag{2.7b}$$

involving a stiffness matrix $K = \{K_{ij} := \langle L_1 \psi_i, L_2 \phi_j \rangle\}$. Furthermore, the identity (2.2) can be regarded as an equation for w, given $R_S w$, in a weak form and generated from the adjoint operator $L_2^* L_1$ of L. Thus to obtain the optimal test space from the trial space S^h through the relation (2.5), we would need to solve the adjoint equation for each $\phi_i \in S^h$

$$L_2^* L_1 \psi_i^* = L_S^* L_S \phi_i \tag{2.8}$$

to obtain basis function $\{\psi_i^*\}$ for the optimal test space.

The diffusion-convection problem (without depletion and, as implicitly assumed everywhere for simplicity, with homogeneous boundary conditions) is given by

$$-\underline{\nabla} \cdot (a \underline{\nabla} u - \underline{b} u) = f \qquad a > 0, \tag{2.9a}$$

and the two first order operators L_1 and L_2 can then be defined formally (that is, without defining their domains by specifying boundary conditions) by

$$L_1 v := a^{\frac{1}{2}} \underline{\nabla} v, \quad L_2 v := a^{\frac{1}{2}} \underline{\nabla} v - (\underline{b}/a^{\frac{1}{2}}) v. \tag{2.9b}$$

Here $\underline{b}$ is a velocity field and in (2.9a) u must be specified on that part of the boundary over which $\underline{b}$ flows into the region of solution. One consequence of (2.8) then is that the test function ψ_i^* must be specified where $\underline{b}$ leaves the region.

To go further requires a judicious choice of the operator L_S. There are two obvious choices, L_1 or L_2. The advantage of the former is that it is relatively easy to compute R_S, for from (2.2) we have

$$L_1^* L_1 R_S W = L_2^* L_1 W \tag{2.10}$$

and $L_1^* L_1$ is just a diffusion operator. Thus for any choice of T^h, error bounds of the form (2.4) can be obtained from values of Δ_S computed using (2.3). The effectiveness of any proposed test space can be checked in this way: in one dimension the bounds can sometimes be computed explicitly and optimised over free parameters; while even in two dimensions they can be well approximated — see Morton and Scotney (1985) for details of both cases. Furthermore, one can deduce directly some of the key features that the test space should have from considering (2.8). From (2.9b) we find that ψ_i^* should satisfy

$$\underline{\nabla}\cdot(a\underline{\nabla}\psi_i^*) + \underline{b}\cdot\nabla\psi_i^* = \underline{\nabla}\cdot(a\underline{\nabla}\phi_i), \qquad (2.11)$$

of course in a weak sense. Thus suppose $\{\phi_i\}$ are the familiar piecewise linear elements on triangles: then the right-hand side of (2.11) consists of a negative δ-function at the ith node. The corresponding ψ_i^* will therefore be sharply decreasing there in the direction of the velocity field $\underline{b}$. This implies immediately that if test functions $\{\psi_i\}$ are to be similarly localised positive functions, they should each be centred upwind of the corresponding node i.

The disadvantage of using L_1 is that in the limit of large $|b|/a$, when sharp boundary layers form, the corresponding norm (which is essentially the L^2 norm of the gradient) yields an inappropriate approximation goal; U may be almost as good an approximation as the optimal U_1^* but this may itself be very poor. For example, with continuous piecewise linears on a mesh of size h approximating a boundary layer of thickness $\varepsilon < h$, the error for U_1^* will be of the order $(\varepsilon^{-1} - h^{-1})$: and knowing that U has an error of the same order tells one virtually nothing about U if $\varepsilon \ll h$.

On the other hand, use of L_2 gives a norm

$$\langle L_2 v, L_2 v\rangle = \int [a|\underline{\nabla} v|^2 + (|\underline{b}|^2/a)\, v^2]\, d\Omega \qquad (2.12)$$

which becomes equivalent to the L^2 norm of the function itself as $|\underline{b}|/a \to 0$. This limit corresponds to a transition to hyperbolic equations, so it is natural and advantageous to use such a norm then. Unfortunately it is more difficult to compute R_S with this choice and therefore to assess the effectiveness of any particular test space in this norm by means of (2.3) and (2.4).

What one can do is to aim directly at using the optimal test functions ψ_i^* in (2.1b) to obtain the optimal approximation U_2^* given by

$$B(U_2^*, \psi_i^*) = \langle f, \psi_i^* \rangle, \tag{2.13a}$$

that is

$$\langle L_2 U_2^*, L_2 \phi_i \rangle = \langle f, \psi_i^* \rangle \quad \forall i. \tag{2.13b}$$

This is now a symmetric system of equations, in contrast to the general situation with (2.1b), but we still have to find an approximation to the right-hand side and in particular to $\psi_i^* = R_S^{-1} \phi_i$. From (2.8) we have directly in this case that

$$L_1 \psi_i^* = L_2 \phi_i + \underline{e}_i \tag{2.14a}$$

where $\underline{e}_i$ is in the null space of L_2^*. Thus, because of the simplicity of L_1, R_S^{-1} is this time easier to calculate than R_S and we have

$$\underline{\nabla} \psi_i^* = \underline{\nabla} \phi_i - (\underline{b}/a)\, \phi_i + \underline{e}_i \,. \tag{2.14b}$$

The upwind phenomenon is now even easier to deduce: clearly ψ_i^* must be decreasing in the direction of $\underline{b}$ at the point where ϕ_i has its peak. Moreover, suppose we compute a vector function $\underline{v}$ such that $\underline{\nabla} \cdot \underline{v} = -f$ or, rather more precisely,

$$\langle f, \psi_i^* \rangle = \langle \underline{v}, \underline{\nabla} \psi_i^* \rangle \quad \forall i. \tag{2.15}$$

Then we can rewrite (2.13) as

$$\langle L_2 U_2^*, L_2 \phi_i \rangle = \langle \underline{v}, \underline{\nabla} \phi_i - (\underline{b}/a)\, \phi_i + \underline{e}_i \rangle \quad \forall i. \tag{2.16}$$

Explicit expressions for ψ_i^* and $\underline{v}$ are given by Barrett and Morton (1984) in the one dimensional case and calculations based on approximating the pair of equations (2.15),(2.16) are given by Morton and Scotney (1985). In general, from such an approach one obtains an error bound of the form

$$\|u - U\|^2_{B_2} \leqslant \|u - U_2^*\|_{B_2} + |l(u - U_2^*)|^2 \tag{2.17}$$

where the norm is the B_S norm of (2.6) based on L_2 and $l(\cdot)$ is a linear functional: that is the error is augmented by an additive term rather than a multiplicative factor.

From the practical point of view, the direct Petrov-Galerkin approach (2.1b) using upwinded test functions has proved very useful and popular as it is readily extended to more general problems. Moreover, from a theoretical viewpoint the effectiveness of the approach is dramatically elucidated by applying the error bound (2.4) to the simple one-dimensional model problem (2.9) in which a, $\underline{b}$ are constants and the $\{\phi_i\}$ piecewise linear: for then the Galerkin method has an error factor which is proportional to the mesh Peclet number $\underline{b}h/a$ as this tends to infinity; while for the quadratic modification used by Heinrich *et al* (1977) to obtain $\{\psi_i\}$ the factor rapidly converges to $1.2383\cdots$ when analysed in the norm based on L_1.

2.2 *Test Functions, Difference Schemes and Finite Volume Methods*

In the next section we will consider nonlinear first order systems of equations, which may be elliptic or hyperbolic, and the finite volume methods which have been very successful in their solution. From many points of view, but particularly their analysis, these methods are closer to finite difference than to finite element methods. However, in this section we will try to open up some of the possibilities raised by the Petrov-Galerkin analysis of the last section in order to show some of their finite element aspects.

Let us consider the implications of that analysis for the diffusion-convection problem (2.9) in one dimension. The term upwinding is often used rather loosely and the idea introduced for various purposes, some of them misleading. For this problem a typical upwinded difference scheme would have the form

$$-\frac{\delta(a\delta U)_i}{h^2}+\frac{[(1-\alpha)\Delta_0+\alpha\Delta_-](bU)_i}{h}=f_i, \qquad (2.18)$$

where δ, Δ_0 are the usual half-step and full-step central difference operators and Δ_- the backward difference, or upwind difference because we will assume $b \geqslant 0$. Because the scheme is considered as centred at mesh point x_i, the introduction of the upwind difference is often made reluctantly, with comments about the order of accuracy being reduced from second order to first order: α is chosen just large enough to eliminate the oscillation which will otherwise occur in the solution (U_i) or, in the two-dimensional case, to give the diagonally dominant system of equations needed for the convergence of some of the commonly used iterative methods. This is despite the fact that thirty years ago Allen and Southwell (1955) correctly argued by exponential-fitting that taking $\alpha = \coth \frac{1}{2}\beta - (\frac{1}{2}\beta)^{-1}$ where $\beta = bh/a$ gives enhanced accuracy in the homogeneous case. Furthermore, in difference schemes such as (2.18) there is uncertainty as to where a and b should be evaluated when they depend on x, as the finite difference arguments are unclear. If one introduces the idea of fluxes into and out of a cell centred at x_i, the flux at $x_{i-\frac{1}{2}}$ for the centred scheme would be

$$-(a_{i-\frac{1}{2}}\,\delta U_{i-\frac{1}{2}})/h + \tfrac{1}{2}b_{i-\frac{1}{2}}(U_i + U_{i-1}), \qquad (2.19)$$

while for the fully-upwinded scheme the convective flux would presuably be replaced by $b_i U_i$. Use of $\alpha \neq 0,1$ requires some combination of these but the motivation is somewhat uncertain.

The Petrov-Galerkin arguments of Section 2.1 clarify all of these issues, with some surprising implications. The shapes of the various test functions are crucial and we start with those based on optimisation in the $\langle L_1 \cdot, L_1 \cdot \rangle$ norm: for this emphasises accuracy at the mesh points which is the objective of the difference schemes. The corresponding optimal test functions for two constant values of $\beta = bh/a$ are shown in Fig. 1(a). For this constant coefficient case they yield difference operators as on the left of (2.18) with the parameter α as for exponential fitting (indeed, the test functions are of exponential form). But the important point is that as $\beta \to \infty$ the centre of the scheme shifts from x_i to $x_{i-\frac{1}{2}}$, affecting the values of $f(x)$ that should be used and, with variable coefficients, the values of $a(x)$ and $b(x)$: in the limit we would obtain a scheme closer to

$$(bU)_i - (bU)_{i-1} = \tfrac{1}{2}h(f_i + f_{i-1})\ ; \qquad (2.20a)$$

and the difference of the test function from the square pulse over (x_{i-1}, x_i), which yields (2.20a), gives a diffusion operator of the form

$$h^{-2}[a_i(U_{i+1} - U_i) - a_{i-1}(U_i - U_{i-1})]. \qquad (2.20b)$$

Thus the convective terms are properly centred and the diffusion terms are "downwinded"!

These then are the optimal difference schemes for general source functions $f(x)$: they always give second order accuracy for any value of b and, in the particular case of constant a and b, give exact nodal values $\{U_i\}$ for any $f(x)$. Moreover, even when approximate test functions such as those in Heinrich *et al* (1977) are used instead of the optimal functions, one obtains very similar schemes and solutions whose error is within some 24% of that for the optimal choice.

It is easy to show when a, b are constants in (2.18), Dirichlet boundary conditions are imposed and there is an even number of intervals, that the matrix representing the operator on the left becomes progressively more and more ill-conditioned as β is increased if we take $\alpha = 0$. This is why the second-order truncation error obtained with central differences is not reflected in the quality of the approximation. On the other hand, the upwinded scheme always gives a well-conditioned matrix.

Localised test functions corresponding to optimisation on $\langle L_2\cdot, L_2\cdot\rangle$ are shown in Fig. 1(b): they are seen to spread over three neighbouring elements and are quadratic in form for constant a and b, being given by

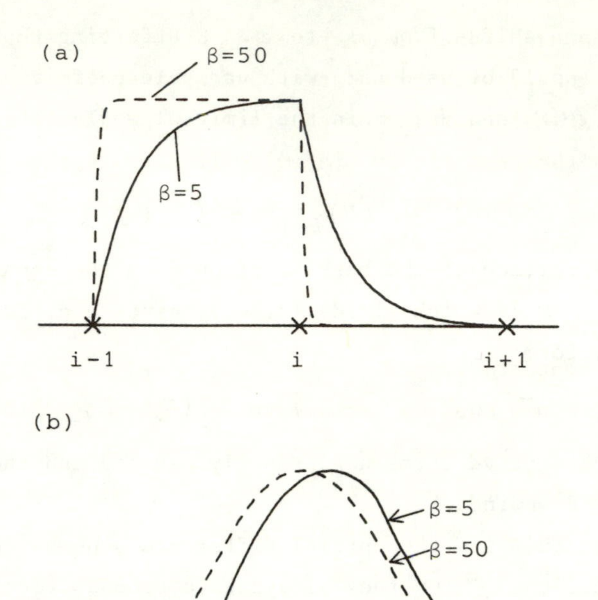

FIG. 1 Optimal test functions for diffusion-convection: (a) for the $B_1(\cdot,\cdot)$ norm; (b) for the $B_2(\cdot,\cdot)$ norm

$$\psi^*_{i-\frac{1}{2}}(x) := \phi_i(x) - \phi_{i-1}(x) + \int_x^1 (b/a)(t)\,[\phi_i(t) - \phi_{i-1}(t)]\,dt\,. \tag{2.21}$$

If we denote the average of $f(x)$ with this test function as weighting function by $\bar{f}_{i-\frac{1}{2}}$, the difference scheme that we obtain, in the limit of large β and in terms of a total flux F, will be of the form

$$F_i - F_{i-1} = h\bar{f}_{i-\frac{1}{2}}\,, \tag{2.22a}$$

where

$$F_i := b_i\left[\tfrac{1}{6}\,U_{i-1} + \tfrac{2}{3}U_i + \tfrac{1}{6}\,U_{i+1}\right] - a_i(2h)^{-1}\left[U_{i+1} - U_{i-1}\right]. \tag{2.22b}$$

Here we see that both convective and diffusive terms are fully centred at $x_{i-\frac{1}{2}}$, with the latter being averaged in order to match the natural centring of the former. The result is a scheme which gives the best fit by piecewise linears in the L^2 norm: it is again always second order accurate but has the usual L^2 oscillatory behaviour, with the poorest pointwise accuracy at the nodes. [This best-fit behaviour has been exploited by Barrett and Morton (1980, 1984) to estimate the thickness of sub-grid scale boundary layers.]

Suppose now that we go back to self-adjoint problems and, in particular, Laplace's equation. We know that the Galerkin formulation gives the best result in the energy norm but it is useful to consider what results can be obtained with other approaches. The direct use of alternative test functions can be handled simply by means of (2.3) and (2.4); what is of more interest here is a mixed finite element approach treating the equation as a first-order system. In one dimension $u_{xx} + f = 0$ can be written in the equivalent form

$$u_x - v = 0\,, \quad v_x + f = 0: \tag{2.23a}$$

and, if $U(x)$ is expanded in basis functions $\{\phi_i(x)\}$, the Galerkin equations for defining it can be written as

$$\langle U_x - V, \partial_x \phi_i \rangle = 0, \quad \langle V_x + f, \phi_i \rangle = 0 \quad \forall\, i\,, \tag{2.23b}$$

where we assume boundary conditions of the form $U = 0$ or $V = 0$ at each end. In this somewhat artificial construction the form of V is unimportant: indeed, in the Dirichlet problem it is undefined to within a linear function. However, if we wish to determine a function V from the equations, the most natural way is to introduce it as piecewise constant formed from the characteristic functions $\chi_{i-\frac{1}{2}}(x)$ of each interval: then it is easily checked that it is uniquely defined by decomposing the first set of equations of (2.23b) and writing instead

$$\langle U_x - V, \chi_{i+\frac{1}{2}} \rangle = 0\,, \quad \langle V, \partial_x \phi_i \rangle = \langle f, \phi_i \rangle \quad \forall\, i\,. \tag{2.23c}$$

This introduces the piecewise constants as test functions more explicitly than (2.23b) and completely identifies the Galerkin approximation with the compact first order difference scheme

$$U_i - U_{i-1} = hV_{i-\frac{1}{2}}, \quad V_{i+\frac{1}{2}} - V_{i-\frac{1}{2}} = h\bar{f}_i \qquad \forall\, i. \quad (2.23d)$$

Now let us move to two dimensions where such an identification seems to be much more difficult. We use a Cauchy-Riemann form

$$u_x + v_y + p = 0, \quad v_x - u_y - q = 0 \qquad (2.24)$$

which, with $f = p_x + q_y$, is equivalent to $u_{xx} + u_{yy} + f = 0$. Suppose that we use a bilinear form for $U(x,y)$ on a square mesh so that basis functions are $\phi_{ij}(x,y) = \phi_i(x)\,\phi_j(y)$. Then to write the defining Galerkin equations for an optimal approximation U^* in a first order form like (2.24), we have to write (where we are not assuming optimal approximation properties for the approximation $V^\dagger$ to v)

$$\langle U_x^* + V_y^\dagger + p, \partial_x \phi_{ij} \rangle = 0, \quad \langle V_x^\dagger - U_y^* - q, \partial_y \phi_{ij} \rangle = 0 \qquad \forall\, i,j, \qquad (2.25a)$$

where again we assume that either $U^* = 0$ or $V^\dagger = 0$ at each point on the boundary. If $V^\dagger$ is eliminated to give the Galerkin equations for U^*, one obtains a nine point difference scheme in which the central difference in the x-direction δ_x^2 has the familiar $(\frac{1}{6}, \frac{2}{3}, \frac{1}{6})$ weighting in the y-direction, and vice-versa: it is convenient to use the half-step averaging operators μ_x and μ_y to express this and so obtain

$$\tfrac{1}{3}\left[(1 + 2\mu_y^2)\,\delta_x^2 + (1 + 2\mu_x^2)\,\delta_y^2\right] U_{ij}^* = h^2 \bar{f}_{ij}. \qquad (2.25b)$$

However, with U_{ij} as the values of U at the element vertices, the natural finite volume approximation to (2.24) is obtained by representing V in the same way and taking inner products with the characteristic function for each element. The integrals of $\operatorname{div}(u,v)$ and $\operatorname{div}(v,-u)$ give total fluxes through the sides of the elements and hence yield the very compact difference approximations

$$\mu_y \delta_x U + \mu_x \delta_y V + h\bar{p} = 0, \quad \mu_y \delta_x V - \mu_x \delta_y U - h\bar{q} = 0 \tag{2.26a}$$

where we have dropped the clumsy subscripts $i-\frac{1}{2}$, $j-\frac{1}{2}$ referring to the centre of the element. Eliminating V gives the nine point difference scheme

$$\left[\mu_y^2 \delta_x^2 + \mu_x^2 \delta_y^2\right] U_{ij} = h^2 \bar{f}_{ij} := h(\mu_y \delta_x \bar{p} + \mu_x \delta_y \bar{q})_{ij} \tag{2.26b}$$

which may be compared with (2.25). The question now is whether we can estimate the accuracy of (2.26) without resorting to truncation errors, but instead comparing with the best fit in the Dirichlet norm which is provided by the Galerkin approximation (2.25). We have the following result.

THEOREM. If U^* and V^* are the Galerkin bilinear approximations to $\nabla^2 u + p_x + q_y = 0$ and $\nabla^2 v + p_y - q_x = 0$ (on a square) respectively, then the finite volume approximations U and V given by (2.26a) satisfy

$$\begin{aligned}\|u-U\|_1^2 + \|v-V\|_1^2 \leq \|u-U^*\|_1^2 + \|v-V^*\|_1^2 \\ + Kh\left[\|(U^*-U)_{xy}\|^2 + \|(V^*-V)_{xy}\|^2\right]^{\frac{1}{2}} \left[\|u-U\|_1^2 + \|v-V\|_1^2\right]^{\frac{1}{2}},\end{aligned} \tag{2.27}$$

where $\|\cdot\|_1$ denotes the Dirichlet norm and K is constant of the order of unity.

This is not an ideal result, and for that reason a proof is not given here. However, it is quite an illuminating intermediate inequality: if $(U^*-U)_{xy}$ could be bounded sufficiently sharply in terms of $\|U^*-U\|_1$, a bound of the desired form as in (2.4) follows; otherwise one obtains only an additive form as in (2.17). Moreover, for a bilinear function W,

$$h^2 W_{xy} = \delta_x \delta_y W \tag{2.28a}$$

and

$$\frac{1}{12} h^2 \|W_{xy}\|^2 = \|W_x - \bar{W}_x\|^2 = \|W_y - \bar{W}_y\|^2. \tag{2.28b}$$

Hence whether or not U^*-U satisfies the required bound depends entirely on the extent to which it suffers from "red-black" oscillations.

2.3 *Finite Volume Methods for Steady Compressible Flow*

The term *finite volume method* was introduced by Jameson and Caughey (1977) to describe highly effective methods that they devised for solving the full potential equation $\underline{\nabla}\cdot(\rho\underline{\nabla}\phi) = 0$, where $\rho \equiv \rho(|\underline{\nabla}\phi|)$, for exterior flow around an aircraft. This equation is appropriate to describing flows where compressibility effects are relatively unimportant: that is, predominantly subsonic flows where the equation is elliptic in form. More recently a great deal of interest has centred around solving the more accurate Euler equations which encompass both subsonic and supersonic flows and hence may have either an elliptic or hyperbolic character. We will therefore briefly describe some of this work as a link to the methods for evolutionary hyperbolic equations in the next section.

The Euler equations comprise a first-order system of conservation laws, describing the conservation of mass, momentum and energy. In their simplest form in two dimensions for $\underline{w} := (\rho, \rho u, \rho v)^T$, where ρ is the density and (u, v) the velocity components, they take the form

$$\frac{\partial \underline{f}(\underline{w})}{\partial x} + \frac{\partial \underline{g}(\underline{w})}{\partial y} = 0\,, \tag{2.29a}$$

where

$$\underline{f}(\underline{w}) := (\rho u,\ \rho u^2 + p,\ \rho uv)^T, \quad \underline{g}(\underline{w}) := (\rho v, \rho uv, \rho v^2 + p) \tag{2.29b}$$

and the pressure p is given as a function of the other variables. The flow region is divided into quadrilaterals by a curvilinear mesh fitted to the body about which the flow is to be calculated: integrating (2.29a) over an element e (i.e. using piecewise constant test functions) and applying the divergence theorem gives a boundary integral equation for each element

$$\int_{\partial e} \underline{f}dy - \underline{g}dx = 0\,. \tag{2.30}$$

There are two obvious schemes for discretising (2.30) which are shown in Fig. 2, and then the resulting systems of nonlinear

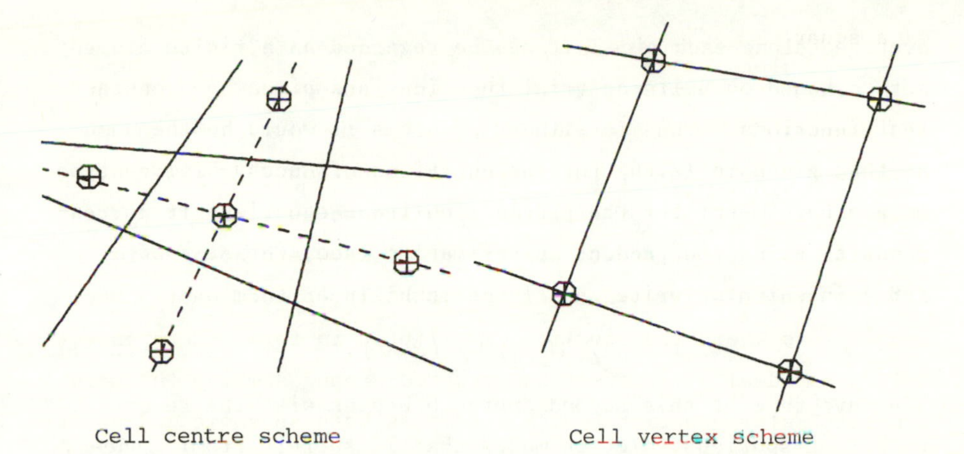

FIG. 2 Mesh layouts for the finite volume method

equations have to be solved. In the cell centre scheme the unknowns are referred to the centre of each element and the contribution to (2.30) from a side obtained from averaging across the side between the centres of the elements with the side in common. This was favoured in early methods because it gives a natural centring for an iteration scheme based on time-stepping: in the unsteady equations obtained from (2.29) by adding a term $\partial \underline{w}/\partial t$, the differencing for this term is just carried out at the centre of the element. The approach popularised by Jameson (1982) was to use iteration schemes based on Runge-Kutta methods and these have been studied in some detail – see, for instance, Kinmark and Gray (1984).

However, there are several disadvantages to the cell centre schemes which are discussed in Morton and Paisley (1986). At their heart lies the fact that the first order derivatives in (2.29) are mainly approximated by central differences extending over two element widths. The familiar result is a variety of spurious solution modes which have to be damped out to the detriment of accuracy.

The alternative cell vertex scheme shown on the right of Fig. 2 refers the unknowns to the vertices of the elements and gives a much more compact scheme with the contributions to (2.30) averaged along each side. It can be regarded as a finite element method based on bilinear trial functions and piecewise constant test functions. Thus for linear problems it would be the same as that given in (2.26) for the equations of Cauchy-Riemann type on a square mesh: for the present nonlinear equations it corresponds to using the product approximation (see, Christie *et al* 1981) in which we write the fluxes in bilinear form as

$$\sum \underline{f}(\underline{W}_{ij})\,\phi_{ij}(x,y). \tag{2.31}$$

The advantage of this second approach begins with the reduced number of spurious solution modes that it suffers from: indeed, the only one of note is the "red-black" oscillation whose control was required in order to develop a useful error bound from the inequality (2.27). Other advantages are demonstrated in Morton and Paisley (1986).

The apparent disadvantage of the cell vertex scheme lies in the lack of an obvious iteration to solve the resultant equations. However, again based on a time-stepping approach, Ni (1982) advocated the use of the Lax-Wendroff method: on a square mesh this has the form

$$\underline{W}^{n+1} = \underline{W}^{n} - (\Delta t/h)\,\mu_x\mu_y\underline{R} + \tfrac{1}{2}(\Delta t/h)^2\,[\mu_y\delta_x(A\underline{R}) + \mu_x\delta_y(B\underline{R})], \tag{2.32}$$

where $\underline{R}$ is the vector of residual flux differences obtained from (2.30), A and B are the Jacobian matrices for $\underline{F}$ and $\underline{G}$ respectively and μ_x, δ_x etc. are the averaging and difference operators introduced in the previous section. Thus the residuals from the four neighbouring elements are used to update the unknowns at a vertex. Despite the fact that these residual sums are used, one can show that if the boundary conditions are treated correctly then at convergence all the individual element residuals will be set to zero.

Multigrid methods are highly effective in achieving rapid convergence with either scheme for transonic flow problems, that is, those in which the bulk of the flow is subsonic but there are finite regions of supersonic flow near the body. However, rather carefully chosen damping terms are also necessary in both cases if problems with well developed shocks are tackled on a fixed mesh. The fact that the two-element difference in the cell centre schemes generates oscillations at a shock will come as no surprise — that the more compact cell vertex scheme does so (*albeit*, less strongly) needs closer analysis. For a one-dimensional problem of flow through a nozzle one can show that there is no problem if the nozzle has parallel sides near the shock but otherwise the variation in cross-section stimulates oscillations; in two dimensions with (2.29), so that there is no such forcing term in the differential equation, the oscillation is generated by the discretisation and is closely linked to the "red-black" oscillations already referred to.

In two dimensions this problem can often be eliminated by shock fitting. That is, the approximate locations of shocks are found from early stages of the iteration; then the mesh is adapted locally with one mesh line lying along each shock and the mesh made nearly uniform either side; finally, subsequent iterations modify the shock positions by making use of the Rankine-Hugoniot jump conditions which follow from the conservation laws (2.29). As with the boundary conditions, use needs to be made of the characteristic directions in determining the details of the algorithm but, as can be seen from Fig. 3 the result is greatly improved accuracy. Moreover, for the unfitted mesh result shown in the figure a good deal of parameter tuning is needed in the dissipation terms to obtain the shock in the "correct" position: very little dissipation is needed with shock fitting and the results are virtually independent of it. Details of these methods and their analysis can be found in Morton and Paisley (1986). The next steps are to see whether they can be extended to three dimensions and to the Navier-Stokes equations.

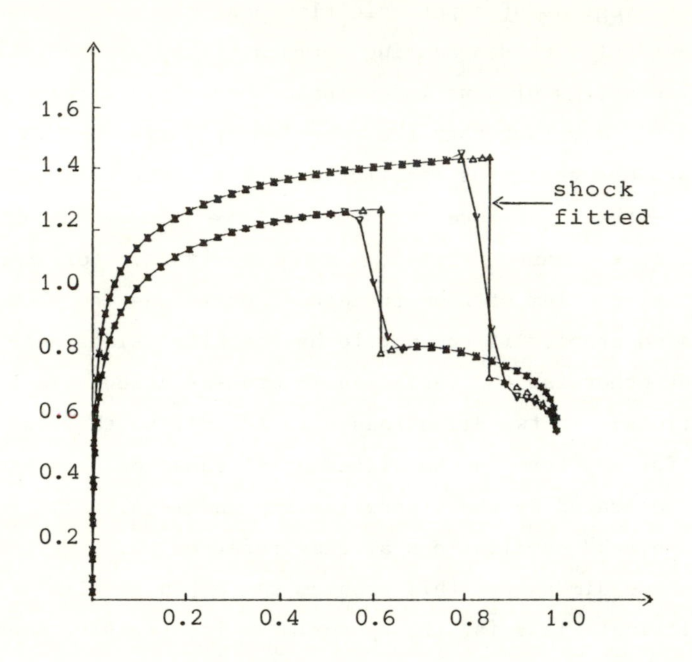

FIG. 3 Plot of Mach No. along surface Δ of NACA 0012 aerofoil at $M_\infty = 0.85$ and angle of attack $\alpha = 1°$

3. EVOLUTIONARY HYPERBOLIC PROBLEMS

3.1 *Galerkin, Petrov-Galerkin and Taylor-Galerkin Methods*

The starting point for most finite element approximations to time dependent problems has been the semi-discrete Galerkin formulation. That is, finite elements are used only in the space variables so that the approximation $U(x,t)$ to the true solution $u(x,t)$ of the equation

$$\partial_t u + L(u) = 0 \tag{3.1}$$

is expressed as

$$U(x,t) = \sum_{(j)} U_j(t)\,\phi_j(x) \tag{3.2}$$

where $\{\phi_j(x)\}$ are the basis functions spanning the trial space

S^h and $\{U_j\}$ are time-dependent coefficients. Then the Galerkin formulation leads to

$$\langle \partial_t U + L(U), \phi_i \rangle = 0 \qquad \forall \phi_i \in S^h, \tag{3.3}$$

which gives the system of ordinary differential equations

$$M\dot{\underline{U}} + \underline{K}(U) = \underline{0} \tag{3.4}$$

where M is the "mass" matrix $\{M_{ij} := \langle \phi_i, \phi_j \rangle\}$, $\dot{\underline{U}}$ is the vector of time-derivatives $\{\dot{U}_i\}$ and $\underline{K}$ is the "stiffness" vector $\{K_i := \langle L(U), \phi_i \rangle\}$.

Thus these problems are distinguished from the equilibrium hyperbolic problems of the previous section in that the time variable is treated differently from the space variable(s). One can argue that what is required here is a snapshot of the solution at successive instants of time: then (3.3) shows how $L(U)$ is projected onto S^h to give the best L^2 fit to the increments required to update U. If (3.4) is integrated accurately one might expect to maintain a close approximation to the best L^2 fit to u, and indeed this is so. For example, consider the model problem of linear advection on a uniform mesh, in which $L(u) \equiv a\partial_x u$ with constant a. Then if S^h is generated by B-splines of order s, the error introduced by integrating (3.4) is of order h^{2s} even though the approximation error at each time level (in particular that in the initial data) is of order h^s. For example, piecewise linears corresponding to $s=2$ give an approximation error of $O(h^2)$ but an evolutionary error of $O(h^4)$. This also holds for some nonlinear problems — see Cullen and Morton (1980) and references therein.

This remarkably high spatial accuracy is matched by other valuable properties. If $L(\cdot)$ is a conservative operator in the sense that $\langle L(v), v \rangle = 0$ for all v in the solution space, then the resulting conservation property for u also holds for the approximation U. In the same way various stability properties of u are carried over to U and the latter is much less plagued by the phenomenon of aliasing as a consequence of the Galerkin projection. The literature on these topics goes back twenty years and some of it was reviewed in my last State of the Art lecture — Morton (1977).

However, eventually one has to consider how the system (3.4) is to be integrated. It is then that matters deteriorate rapidly. Because $L(u)$ is normally very nonlinear there is considerable advantage in using explicit methods: also storage requirements for multi-dimensional problems point towards schemes with a small number of time-levels. Hence, suppose the linear advection problem is approximated by piecewise linears. The semi-discrete Galerkin formulation (3.3) generates the equations, in difference form,

$$(1+\tfrac{1}{6}\delta^2)\dot{U}_i + (a/h)\Delta_0 U_i = 0, \tag{3.5}$$

where $(1+\frac{1}{6}\delta^2)$ represents the familiar tridiagonal mass matrix for this case. As in the diffusion-convection problems of the previous sections, the central difference $\Delta_0 U_i$ is at the heart of the problems: if the one-step Euler method is used, stability limits the time-step to $\Delta t = O(h^2)$; if the leap-frog two-step method is used the CFL number $\nu := a\Delta t/h$ is limited to $1/\sqrt{3}$ rather than the expected $|\nu| \leqslant 1$ which could be obtained without the mass matrix in (3.5). However, even more seriously, all the attractive accuracy properties are lost for quite modest values of Δt and the two mesh-length oscillations inherent in (3.5) make the approximations unacceptable.

Initial attempts to overcome these difficulties used the Petrov-Galerkin approach which was so successful with the diffusion-convection problems. The idea followed in Morton and Parrott (1980) was to choose first a finite difference scheme for the time-differencing and then to choose a test space to give good accuracy in conjunction with it. Thus with $U_i(t)$ in (3.2) replaced by U_i^n for the time-level t_n, Euler time-stepping would give

$$\left\langle \frac{U^{n+1}-U^n}{\Delta t} + L(U^n)\,,\,\psi_i \right\rangle = 0 \quad \forall\,\psi_i \in T^h. \tag{3.6}$$

To choose ψ_i it was argued that, in the special case of linear advection with $\nu := a\Delta t/h = 1$, a best fit approximation at time-level n could be maintained with no further error at level t_{n+1} merely by translation through one mesh length. Special test functions $\chi_i(x)$ could be constructed with the same support as $\phi_i(x)$ to achieve this; and then in the general case one could use

$$\psi_i = (1-\rho)\,\phi_i + \rho\chi_i, \tag{3.7}$$

where ρ depends on ν such that $\rho(0)=0$ and $\rho(1)=1$. Satisfaction of the "unit CFL condition" led to Euler-Petrov-Galerkin methods EPGI and EPGII which were stable over the whole range $0 \leqslant \nu \leqslant 1$ for the linear advection problem and also highly accurate over that whole range, EPGI being second order and EPGII third order accurate. Similarly, one can construct LPG and CNPG schemes using leap frog and Crank-Nicolson time-stepping respectively: the former is fourth order accurate in both h and Δt and stable for $|\nu| \leqslant 1$.

The distinguishing features of these schemes is that their improved accuracy and stability mainly comes from modifications to the mass matrix. EPGI also results in a Lax-Wendroff spatial operator, becoming

$$\left[1+\tfrac{1}{6}(1-\nu)\,\delta^2\right]\left(U_i^{n+1}-U_i^n\right) + \nu\Delta_0 U_i^n + \tfrac{1}{2}\nu^2\,\delta^2 U_i^n = 0 \tag{3.8}$$

for the linear advection problem; while LPG has the form

$$\left[1+\tfrac{1}{6}(1-\nu^2)\,\delta^2\right]\left(U_i^{n+1}-U_i^{n-1}\right) + 2\nu\Delta_0 U_i^n = 0\,. \tag{3.9}$$

Thus, as $|\nu| \to 1$, the mass matrix becomes nearer to the identity so that the simple translation implied by the unit CFL condition can be achieved: in the limit they are identical to corresponding finite difference schemes. Such mass lumping, as it is called in the engineering literature, has been advocated by other authors to improve stability and also to speed up the solution of the equations for U^{n+1}, though this is no problem in practice if preconditioned conjugate gradient methods are used — see Wathen (1985).

However, this approach is not very systematic, it degrades the conservation properties of the Galerkin approach and it is more difficult to extend into two dimensions. The Taylor-Galerkin schemes on the other hand, though very similar, are much easier to generate for general problems. They are due to Donea (1984) and follow the same approach as that used to generate the Lax-Wendroff difference schemes. Thus in a Taylor series expansion

$$u(t+\Delta t) = u + \Delta t \partial_t u + \tfrac{1}{2}(\Delta t)^2 \partial_t^2 u + \tfrac{1}{6}(\Delta t)^3 \partial_t^3 u + \cdots \quad (3.10)$$

for the solution of the conservation law

$$\partial_t u + \partial_x f(u) = 0\,, \quad a(u) := \partial f/\partial u\,, \quad (3.11)$$

substitution of time derivatives by spatial derivatives is made as follows:-

$$\begin{aligned} \partial_t u &\to -\partial_x f \\ \partial_t^2 u &\to -\partial_x \partial_t f = -\partial_x(a\partial_t u) = \partial_x(a\partial_x f) \quad (3.12) \\ \partial_t^3 u &\to (\Delta t)^{-1}[(\partial_x(a\partial_x f))^{n+1} - (\partial_x(a\partial_x f))^n]\,. \end{aligned}$$

Applying the Galerkin projection to the result and rearranging gives the third order accurate scheme

$$\left\langle \frac{U^{n+1}-U^n}{\Delta t}, \phi_i \right\rangle + \tfrac{1}{6}\Delta t \left\langle (a\partial_x f)^{n+1} - (a\partial_x f)^n, \partial_x \phi_i \right\rangle + \left\langle \partial_x f^n, \phi_i \right\rangle + \tfrac{1}{2}\Delta t \left\langle (a\partial_x f)^n, \partial_x \phi_i \right\rangle = 0\,. \quad (3.13)$$

From this arrangement one sees that the last two terms give the finite element equivalent to the Lax-Wendroff spatial operator while the first, mass matrix, term is modified by the second (nonlinear) term which has the effect of giving some mass lumping. We have effectively used an Euler time-stepping approach here and, indeed, for linear advection (3.13) reduces to the EPGII scheme. Similarly, for leap-frog time-stepping one obtains LPG: but generally, and in particular for Crank-Nicolson, the approach does not reduce exactly to the Petrov-Galerkin schemes. The important point is the closeness in form: thus if the second term of (3.13) is linearised and a local value of a used in defining ν, it gives a term

$$\tfrac{1}{6}\nu^2\delta^2(U_i^{n+1} - U_i^n)/\Delta t \tag{3.14}$$

which is equivalent to the mass lumping term in (3.9); in practice such modifications to this term of (3.13) are frequently used – see Löhner *et al* (1985). On the other hand, it is much easier to establish conservation properties of the approximation for the form (3.13).

3.2 *Characteristic Galerkin Methods*

The unit CFL condition in the Petrov-Galerkin methods and the substitutions (3.12) in the Taylor-Galerkin methods force the resulting approximations to take account of the characteristics in the differential system. The test functions in the former case are shifted upwind relative to the trial functions just as in the diffusion-convection problems, the sign relation between diffusion and time derivative being fixed by considerations of well-posedness so that the sense of upwinding is consistent in the two types of problems. The characteristic Galerkin methods which we will now describe take this exploitation to its extreme.

Let us consider first the scalar conservation law (3.11). So long as the solution remains smooth, and the equation holds in the classical sense, it implies that the solution is constant along the characteristics $dx/dt = a(u)$, which must therefore be straight lines in the (x,t)-space. When the characteristics envelope a shock is formed, the equation holds only in a weak sense and the shock moves at a speed given by $[f]/[u]$, where $[\cdot]$ denotes the jump in value of the argument. However, even then one can still use the characteristics to define an approximate solution in the trial space S^h. Suppose $\tilde{u}^n(x)$ is an approximation to u at time level n, which may be a discontinuous function of x. Regard it as a continuous graph $[\tilde{u}^n, x]$ in the (x,u)-plane: and define a new graph $[\hat{E}\tilde{u}^n, x]$ through the "overturned manifold" evolutionary operator $\hat{E} = \hat{E}(\Delta t)$ defined through the mapping of the arbitrary graph $[v, x] \rightarrow [\hat{E}v, y]$ by

$$y = x + a(v(x))\,\Delta t \tag{3.15a}$$

$$(\hat{E}v)(y) = v(x)\,. \tag{3.15b}$$

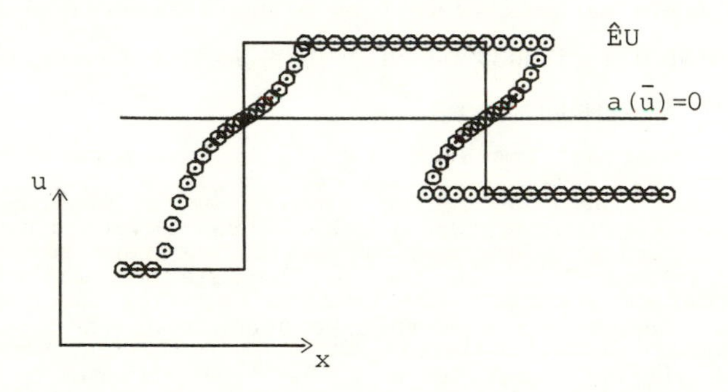

FIG. 4 Overturned manifold solution $[U, x] \rightarrow [\hat{E}U, y]$

The new graph may well correspond to a multivalued function of y (see Fig. 4) but projection into S^h gives U^{n+1} through the formula

$$\langle U^{n+1}, \phi_i \rangle = \langle \hat{E}\tilde{u}^n, \phi_i \rangle \qquad \forall \phi_i \in S^h \tag{3.16a}$$

$$= \int \tilde{u}^n(x)\, \phi_i(y)\, dy\,, \qquad y = x + a(\tilde{u}^n(x))\,\Delta t. \tag{3.16b}$$

Here integration is carried out along the graph: thus on a horizontal section $dy = dx$ and on a vertical section $dy = \Delta t\, da$. Important general properties of such a "transport collapse" approximation have been established by Brenier (1984). An alternative but equivalent formulation has been used by Morton (1982, 1983, 1985), which takes the form

$$\langle U^{n+1} - U^n, \phi_i \rangle + \Delta t \langle \partial_x f(\tilde{u}^n), \tilde{\Phi}_i^n \rangle = 0 \tag{3.17a}$$

where

$$\tilde{\Phi}_i^n(x) = \frac{1}{a(\tilde{u}^n(x))\Delta t} \int_x^y \phi_i(s)\,ds \qquad \forall \phi_i \in S^h, \tag{3.17b}$$

where we have assumed a projection property for $\tilde{u}^n$, namely

$$\langle U^n - \tilde{u}, \phi_i \rangle = 0 \qquad \forall \phi_i \in S^h. \tag{3.18}$$

Substitution of (3.17b) into (3.17a) and integration by parts (along the graph) readily leads back to (3.16). The advantage of (3.17) is that it shows the links to Petrov-Galerkin and Taylor-Galerkin methods as well as being a useful starting point for practical algorithms: we call this an *Euler characteristic Galerkin* or ECG method.

Some important features of (3.17) can be adduced by looking again at the linear advection problem on a uniform mesh. If B-splines of order s are used as basis functions for S^h, one obtains order of accuracy $2s-1$: that is, only one order lower than the semi-discrete Galerkin case. Moreover, time-steps of any magnitude can be used and the schemes are all unconditionally stable. This is because, while the mass matrix at the forward time spreads over $2s-1$ mesh-points arranged symmetrically either side of x_i, there are $2s$ mesh points involved at the old time level and they are spread symmetrically around the mesh-interval through which the characteristic drawn back from (x_i, t_{n+1}) passes at $t = t_n$. Thus the schemes are exact for any integer value of the CFL number. The two most useful are those for $s=1$ and 2: the former uses piecewise constants to obtain the first order scheme

$$U_i^{n+1} = U_{i-m}^n - \hat{\nu}\,\Delta_-\,U_{i-m}^n\,, \text{ where } \nu = m + \hat{\nu}, \quad \hat{\nu} \in (0,1]; \tag{3.19a}$$

the latter uses continuous piecewise linears for the third order scheme

$$(1 + \tfrac{1}{6}\delta^2)\,U_i^{n+1} = [(1+\tfrac{1}{6}\delta^2) - \hat{\nu}\,(\Delta_0 - \tfrac{1}{2}\hat{\nu}\,\delta^2 + \tfrac{1}{6}\,\hat{\nu}^2\,\delta^2\Delta_-)]\,U_{i-m}^n \tag{3.19b}$$

with the same definition of $\tilde{\nu}$ and m obtained from $\nu := a\Delta t/\Delta x$.

The scheme (3.19b) is comparable with the EPGII and Taylor-Galerkin scheme (3.13). In fact, they are members of a one-parameter family of schemes, that can be derived as ECG schemes using the mixed norm induced by the inner product

$$\langle u, v \rangle + \gamma^2 \langle \partial_x u , \partial_x v \rangle \tag{3.20}$$

for the projection stage. The leading term in the truncation error for these schemes when applied to the linear advection equation has the coefficient

$$\hat{\nu}^2(1-\hat{\nu})^2 + 12\,\hat{\nu}(1-\hat{\nu})(\gamma/\Delta x)^2 . \tag{3.21}$$

Thus it is clear that the L^2 projection gives the smallest error, with the Taylor-Galerkin schemes, for which $\gamma^2 = (a\Delta t)^2/6$, being four times less accurate.

Another important feature of the schemes that can be seen from this model problem is the value of the recovery relation (3.18). Suppose U^n is a B-spline of order s and $\tilde{u}^n$ is calculated as a spline of order $s+p$ from the equations (3.18). Then the resulting scheme for U^{n+1} has order of accuracy $2s+p-1$. For example, if quadratic splines are used to recover from piecewise constant approximations at each time step the resulting scheme is just the third order scheme (3.19b) obtained by using piecewise linears without the recovery stage. On the other hand, if recovery were carried out with piecewise linears the accuracy would be second order, rather than the first order scheme (3.19a) obtained without recovery. The value of using piecewise linears in this way is that a local parameter can be introduced so that the recovery is carried out adaptively according to the smoothness of the underlying data values and so as to preserve such properties as monotonicity.

Before describing this in more detail, let us return to the basic method (3.17). The fact that we can recover with higher order splines as necessary justifies taking piecewise constants as the basic trial space S^h. This can be based on a quite general mesh spacing: we define $\Delta x_i := x_{i+\frac{1}{2}} - x_{i-\frac{1}{2}}$ and $x_i := \frac{1}{2}(x_{i+\frac{1}{2}} + x_{i-\frac{1}{2}})$ in terms of the element boundaries $x_{i+\frac{1}{2}}$. Then with no recovery it is clear from (3.17) that the update $U^n \to U^{n+1}$ depends just on the jumps in $f(U^n)$ at the element boundaries. The resultant algorithm is quite straightforward even for general time steps; but if we assume that

$$-\min(\Delta x_{i-1}, \Delta x_i) \leqslant a(U_i^n)\,\Delta t \leqslant \min(\Delta x_i, \Delta x_{i+1}) \qquad \forall i \quad (3.22)$$

and that $f(\cdot)$ is convex, so that there is a unique "sonic point" $\bar{u}$ at which $a(\bar{u}) = 0$, we have the following very simple algorithm.

ALGORITHM. To obtain $\{U_i^{n+1}\Delta x_i\}$ from $\{U_i^n \Delta x_i\}$, for each k do the following:-

$$\textit{transfer} \ -\Delta t\,[f(U_k^n) - f(\bar{u})]$$

$$\textit{from} \quad \begin{bmatrix} U_{k+1}^n \,\Delta x_{k+1} \\ \\ U_k^n \,\Delta x_k \end{bmatrix} \quad \textit{to} \quad \begin{bmatrix} U_k^n \,\Delta x_k \\ \\ U_{k-1}^n \,\Delta x_{k-1} \end{bmatrix} \quad \textit{if} \quad a(U_k^n) \begin{bmatrix} > 0 \\ \\ < 0 \end{bmatrix}. \qquad (3.23)$$

On a uniform mesh this is exactly equivalent to the first order Engquist-Osher difference scheme (Engquist and Osher, 1980) and is illustrated in Fig. 5. Like other schemes based on the Brenier transport collapse operator, it has the valuable property that it satisfies an entropy inequality from which one can prove that it always converges to the correct weak solution of the differential equation as $\Delta t, \Delta x \to 0$, assuming only boundedness of the CFL number. Generalisation to systems of equations is achieved by decomposing the vectors U^n into characteristic fields (see Morton and Sweby, 1986). In two dimensions the transfers are based on the function discontinuities at the edges between elements.

Unfortunately, (3.23) is only first order accurate. What is required is a method which is at least second order accurate where the solution is smooth but, like (3.23) and the differential

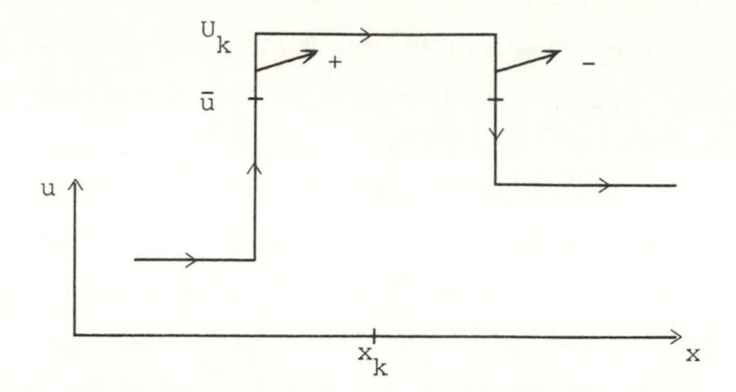

FIG. 5 Transfers of $-\Delta t[f(U_k)-f(\bar{u})]$ by integrating along the graph $[U,x]$

equation itself, maintains the monotonicity of a solution even when it is varying very rapidly. As Godunov (1957) showed, these are conflicting requirements for a difference scheme based on linear combinations of the function values $\{U_i^n\}$ and flux values $\{f(U_i^n)\}$. This is where adaptive recovery from piecewise constants (like flux-limited difference methods and flux-corrected transport methods) shows to advantage. If $\tilde{u}^n$ is monotone then this property is maintained by the transport collapse operator and the L^2 projection on to piecewise constants, so that U^{n+1} is monotone. Note, however, that this would not be maintained by L^2 projection on to continuous piecewise linears, which is why some form of damping or artificial viscosity is needed in schemes of that type. Hence the recovery by piecewise linears used by Morton and Sweby (1986), based on the use of the smoothness parameters

$$r_i := \left(\frac{\Delta_- U_i^n}{\Delta_- x_i}\right) \Big/ \left(\frac{\Delta_+ U_i^n}{\Delta_+ x_i}\right), \tag{3.24}$$

has the following key features:

(i) Sharp changes in the values of $\{r_i\}$ are used as indicators of the presence of shocks: these are then introduced as new element boundaries to replace those to the immediate left and right. In the update process these shocks are then moved with their correct shock speed.

(ii) Between the recovered shocks the jump at each element boundary $x_{i+\frac{1}{2}}$ is replaced by a linear section extending a fraction $\theta_{i+\frac{1}{2}}$ of the way to the mid points x_i, x_{i+1} of the elements either side. The parameters $\{\theta_{i+\frac{1}{2}}\}$ are chosen as large as possible conditional on the recovered solution $\tilde{u}^n$ given by (3.18) being such that each difference $\Delta_-\tilde{u}_i^n$ has the same sign as the corresponding $\Delta_- U_i^n$.

This procedure leads to all the desired properties since if $\theta_{i+\frac{1}{2}} = 1 \ \forall i$ one obtains a second order scheme. Moreover, the update algorithm is virtually unchanged from (3.23) if locally modified flux functions are introduced of the form

$$F_{i+\frac{1}{2}}(u) := f(u) + \tfrac{1}{2}\left(u - \tilde{u}^n_{i+\frac{1}{2}}\right)^2 \Big/ \left(\tilde{m}^n_{i+\frac{1}{2}} \Delta t\right), \qquad (3.25)$$

where $\tilde{m}^n_{i+\frac{1}{2}}$ is the slope of the linear section and $\tilde{u}^n_{i+\frac{1}{2}}$ the value at $x_{i+\frac{1}{2}}$. More details on this and other ECG algorithms can be found in Childs and Morton (1986).

4. CONCLUSIONS

In Sections 2 and 3 we have considered two very different problem types from a single viewpoint. In each case the Galerkin formulation has been generalised so as to yield an approximation as near optimal as possible, in a sense defined by an integral norm. It is this unification which is the main aim of the paper rather than any pretence to comprehensiveness: yet if the reader follows up the references given in the quoted papers he will obtain a good introduction to a very wide and active field. Several related developments should be mentioned as they could be considered as taking rather further the generalisations of the Galerkin method described here. One is the moving finite element method of Miller (1981) and Wathen and Baines (1985), in which not only the amplitudes $\{u_i^n\}$ but also the nodal positions $\{x_i^n\}$ are determined by Galerkin equations. Another is the wide variety of particle methods, which have recently been formulated by Raviart (1986) in such a way that they can be regarded as taking the methods of Section 3 to the extreme of using δ-functions as basis functions and having to use recovery techniques even more extensively. Finally, the Lagrange-Galerkin methods typified by Benqué *et al* (1982) and Douglas and Russell (1982) represent a compromise in which a flow speed is used in the same way as the characteristic speed in the characteristic Galerkin methods while terms other than the convective terms are treated by the simple Galerkin approach.

REFERENCES

Allen, D.N.deG. and Southwell, R. (1955), "Relaxation methods applied to determine the motion, in two dimensions, of a viscous fluid past a fixed cylinder", *Quart. J. Mech. and Appl. Math.*, *12*, pp.129-145.

Barrett, J.W. and Morton, K.W. (1980), "Optimal finite element solutions to diffusion-convection problems in one dimension", *Int. J. Numer. Meth. Engng.*, *15*, pp.1457-1474.

Barrett, J.W. and Morton, K.W. (1981), "Optimal Petrov-Galerkin methods through approximate symmetrization", *IMA J. Numer. Anal.*, *1*, pp.439-468.

Barrett, J.W. and Morton, K.W. (1984), "Approximate symmetrization and Petrov-Galerkin methods for diffusion-convection problems", *Comp. Meth. in Appl. Mech. Eng.*, *45*, pp.97-122.

Benqué, J.P., Labadie, G. and Ronat, J. (1982), "A new finite element method for the Navier-Stokes equations coupled with a temperature equation", in *Proc. 4th Intl. Conf. on Finite Element Methods in Flow Problems*, ed. Tadahiko Kavai, pp.295-301.

Brenier, Y. (1984), "Averaged multivalued solutions of scalar conservation laws", *SIAM J. Numer. Anal.*, *21*, pp.1013-1037.

Childs, P.N. and Morton, K.W. (1986), "On Characteristic Galerkin methods for scalar conservation laws", Oxford Univ. Comp. Lab. N.A. Rep. 86/5,

Christie, I., Griffiths, D.F., Mitchell, A.R. and Sanz-Serna, J.M. (1981), "Product approximation for nonlinear problems in the finite element method", *IMA J. Numer. Anal.*, *1*, pp.253-266.

Christie, I., Griffiths, D.F., Mitchell, A.R. and Zienkiewicz, O.C. (1976), "Finite element methods for second order differential equations with significant first derivatives", *Int. J. Numer. Meth. Engng.*, *10*, pp.1389-1396.

Cullen, M.J.P. and Morton, K.W. (1980), "Analysis of evolutionary error in finite element and other methods", *J. Comput. Phys.*, *34*, pp.245-268.

Donea, J. (1984), "A Taylor-Galerkin method for convective transport problems", *Int. J. Numer. Meth. Engng.*, *20*, pp.101-119.
Douglas, J. Jr. and Russell, T.F. (1982), "Numerical methods for convection dominated problems based on combining the method of characteristics with finite element or finite difference procedures", *SIAM J. Numer. Anal.*, *19*, pp.871-885.
Engquist, B. and Osher, S. (1981), "One sided difference equations for nonlinear conservation laws", *Math. Comp.*, *36*, pp.321-352.
Godunov, S.K. (1959), "A finite difference method for the numerical computation of discontinuous solutions of the equations of fluid dynamics", *Mat. Sb.*, *47*, pp.271-290.
Heinrich, J.C., Huyakorn, P.S., Mitchell, A.R. and Zienkiwicz, O.C. (1977), "An upwind finite element scheme for two-dimensional convective transport equations", *Int. J. Numer. Meth. Engng.*, *11*, pp.131-143.
Kinmark, I.P.E. and Gray, W.G. (1984), "One step integration methods with maximum stability regions", *Maths. and Comps. in Simul.*, *26*, pp.87-92.
Jameson, A. (1982), "Transonic aerofoil calculations using the Euler equations", *Proc. IMA Conf. in Num. Meth. in Aero Fluid Dynamics*, ed. Rose, P.L., Academic Press (London) pp.289-308.
Jameson, A. and Caughey, D.A. (1977), "A finite volume method for transonic potential flow calculations", AIAA Paper No.77-635.
Löhner, R., Morgan, K. and Zienkiewicz, O.C. (1984), "The solution of nonlinear hyperbolic equation systems by the finite element method", *Int. J. Num. Math. Fluids*, *4*, pp.1043-1063.
Miller, K. and Miller, R. (1981), "Moving finite elements", Parts I and II, *SIAM J. Numer. Anal.*, *18*, pp.1019-1057.
Morton, K.W. (1977), "Initial-value problems by finite difference and other methods", in *The State of the Art in Numerical Analysis*, ed. Jacobs, D.A.H., Academic Press (London), pp.699-756.
Morton, K.W. (1982), "Finite element methods for non-self-adjoint problems", in *Proc. SERC Summer School, 1981*, ed. Turner, P.R., Lect. Notes in Maths. 965, Springer-Verlag (Berlin), pp.113-148.
Morton, K.W. (1982), "Shock capturing, fitting and recovery", in *Eighth Int. Conf. on Numerical Methods in Fluid Dynamics*, ed. Krause, E., Lect. Notes in Physics 170, Springer-Verlag (Berlin), pp.77-93.
Morton, K.W. (1983), "Characteristic Galerkin methods for hyperbolic problems", in *Proc. Fifth GAMM Conf. on Numerical Methods in Fluid Dynamics*, eds. Pandolfi, M. and Piva, R., Vieweg and Sohn (Wiesbaden), pp.243-250.
Morton, K.W. (1985), "Generalised Galerkin methods for hyperbolic problems", *Comp. Meth. in Appl. Mech. and Engng.*, *52*, pp.847-871.

Morton, K.W. and Parrott, A.K. (1980), "Generalised Galerkin methods for first order hyperbolic equations", *J. Comp. Phys.*, *36*, pp.249-270.

Morton, K.W. and Paisley, M.F. (1986), "Finite volume methods and shock fitting for the Euler equations", to appear

Morton, K.W. and Scotney, B.W. (1985), "Petrov-Galerkin methods and diffusion-convection problems in 2D", in *Proc. Conf. MAFELAP 1984*, ed. Whiteman, J.R., Academic Press (London), pp.343-366.

Morton, K.W. and Sweby, P.K. (1984), "A comparison of flux-limited difference methods and characteristic Galerkin methods for shock modelling", to be published in *J. Comput. Phys.*

Ni, R.H. (1982), "A multi-grid method for solving the Euler equations", *AIAA. Jnl.*, *20*, pp.1565-1571.

Osher, S. and Sweby, P.K. (1986), "Recent developments in the numerical solution of nonlinear conservation laws", this volume, pp.681-701.

Raviart, P.A. (1986), "Particle numerical models in fluid dynamics", *Numerical Methods for Fluid Dynamics II*, eds. Morton, K.W. and Baines, M.J., O.U.P. (Oxford), pp.231-253.

Strang, G. and Fix, G.J. (1973), *An Analysis of the Finite Element Method*, Prentice-Hall (New York).

Wathen, A.J. (1985), "Attainable eigenvalue bounds for the Galerkin mass matrix", to appear

Wathen, A. and Baines, M.J. (1985), "On the structure of the moving finite element equations", *IMA J. Numer..Anal.*, *5*, pp.161-182.

K.W. Morton
Oxford University
Computing Laboratory
8-11 Keble Road
Oxford OX1 3QD
England

23 Recent Developments in the Numerical Solution of Nonlinear Conservation Laws

S. OSHER and P. K. SWEBY

ABSTRACT

The construction of finite difference shock capturing algorithms is described, including basic theory used for design. Particular emphasis is placed on the need for oscillation free solutions and the Total Variation Diminishing (TVD) property which ensures this. Such TVD schemes fall into two categories, post- and pre- processing. Details of both categories are given. The recent development of Essentially Non-Oscillatory (ENO) schemes is also described, with details of how they avoid the peak clipping, common with TVD schemes, by only approximately satisfying the TVD property.

1. INTRODUCTION

It is well known (see e.g. Lax (1972)) that solutions of the system of hyperbolic conservation laws

$$\underline{u}_t + \underline{f}(\underline{u})_x + \underline{g}(\underline{u})_y + \underline{h}(\underline{u})_z = 0 \tag{1.1}$$

develop discontinuities after finite time for arbitrary smooth data. These discontinuities represent shocks and contact discontinuities in, for example, supersonic flow around aircraft modelled by the Euler equations. It is therefore of great importance, when solving such systems numerically, to obtain high resolution representations of these features of the solution.

Much of the numerical work in this field is performed using finite difference techniques, which are the only methods considered here. In particular we are concerned with Shock Capturing techniques, which require no special 'fitting' of discontinuities.

First order accurate difference schemes are generally unsuitable for systems such as (1.1) due to the excessive smearing of discontinuities in the solution. Classical second order accurate schemes, however, are equally unsuited owing to the spurious oscillations which they generate in the neighbourhood of discontinuities. Such oscillations are not only esthetically undesirable but often lead to a breakdown in stability of the scheme and to non-physical effects such as negative densities.

The feature of these classical schemes which leads to unwanted oscillations is that they are based on constant coefficient stencils. In the last decade there has therefore been much work on overcoming such problems by the use of adaptive finite difference schemes where the coefficients and even the stencil depend on the current solution at each time step.

Such adaptive schemes fall into two main categories. Post-processing and Pre-processing. Included in the former category are the Flux Corrected Transport (FTC) schemes of Boris and Book (1973, 1975, 1979), Zalesak (1979) and the Flux Limiter class of schemes of Roe, van Leer and others, see e.g., Sweby (1984). The second category is comprised of such schems as van Leer's MUSCL (1979) and Colella and Woodward's Piecewise Parabolic Method (PPM) (1984).

In the next section we briefly summarise some of the basic theory used to design and evaluate adaptive difference schemes, while in Sections 3 and 4 we review some of the recent developments in the two categories of schemes mentioned above. The growth in this area is so great as to preclude an exhaustive survey in an article such as this and any ommisions are without prejudice. Finally in Section 5 we describe recent joint work by the first author with A. Harten, B. Engquist (UCLA) and S. Chakravarthy (Rockwell Science Centre) on a new scheme which extends the MUSCL concept.

2. BASIC THEORY

For ease of notation we restrict ourselves to considering the one-dimensional scalar version of (1.1), namely

$$u_t + f(u)_x = 0, \tag{2.1}$$

although most of the following theory carries over to the general case with scalar schemes readily extended to systems using Riemann solvers (Godunov (1959)) or approximate Riemann solvers such as those of Roe (1981a) or Osher and Solomon (1982). Extension to multi-dimensions is also (conceptually) easy.

A weak solution of (2.1) is one which satisfies the equation in the sense of distributions, that is $u \in L^\infty(\mathbb{R} \times [0,T])$ which satisfies

$$\iint \{u\phi_t(x,t) + f(u)\phi_x(x,t)\}\,dx\,dt + \int u(x,0)\,\phi(x,0)\,dx = 0 \tag{2.2}$$

for all test functions $\phi(x,t) \in C^\infty(\mathbb{R} \times [0,T])$ of compact support. Solutions in this sense give meaning to the equation in the presence of discontinuities in the solution.

We consider finite difference schemes which may be written in conservation form, i.e.

$$u_k^{n+1} = u_k^n - \lambda(h_{k+\frac{1}{2}}^n - h_{k-\frac{1}{2}}^n) \tag{2.3}$$

where $h_{k+\frac{1}{2}}^n = h(u_{k-1}^n, \cdots, u_{k+r}^n)$ is a consistent numerical flux function, i.e. $h(u,\cdots,u) = f(u)$, and $\lambda = \Delta t/\Delta x$ is the usual mesh ratio. Lax and Wendroff (1960) showed that if a scheme written in this form converges as $\Delta x \to 0$ (λ fixed) then it converges to a weak solution of (2.1).

Weak solutions of (2.1) are not unique, the classic example of non-uniqueness being for the inviscid Burgers' equation, for which $f(u) = \frac{1}{2}u^2$, where both

$$u(x,t) = \begin{cases} -1 & x < 0 \\ 1 & x > 0 \end{cases} \quad \text{and} \quad u(x,t) = \begin{cases} -1 & x < -t \\ x/t & -t < x < t \\ 1 & t < x \end{cases} \tag{2.4}$$

are weak solutions when the first solution is presented as initial data (Lax, 1973). We therefore need an extra criterion which will distinguish the correct 'physical' solution. The criterion used is that the correct solution should be the one obtained in the limit as $\varepsilon \downarrow 0$ in the viscous equation

$$u_t + f(u)_x = \varepsilon u_{xx}. \tag{2.5}$$

This criterion is known as entropy satisfaction (Lax (1973), Oleinik (1957)), so called since it often coincides with physical entropy. The entropy condition may be investigated by means of entropy functions and fluxes. Associated with any convex entropy function $U(u)$ is a corresponding entropy flux $F(u)$ defined by $F_u(u) = f_u(u)\,U_u(u)$. Multiplying (2.5) by $U_u(u)$, using the convexity of $U(u)$ $(U_{uu} > 0)$ and taking the limit as $\varepsilon \downarrow 0$ shows that the correct solution is the one satisfying

$$U_t(u) + F_x(u) \leqslant 0 \tag{2.6}$$

in the sense of distributions. Equality in (2.6) holds for smooth solutions $u(x,t)$ with inequality at discontinuities. If $f(u)$ is a convex function then, if (2.6) holds for a single convex entropy function it holds for all such entropy functions. However, for nonconvex $f(u)$ (2.6) must be verified for all $U(u)$ of the form $|u-c|$ where c is any constant (Kruzkov (1970)). (It is easily verified for the example above that the second solution of (2.4) is the correct physical solution for the problem.)

Conservative schemes of the form (2.3) which converge to a weak solution of (2.1) may likewise not converge to the entropy satisfying solution, and so it is necessary to prove an entropy inequality for the discretisation. One (somewhat limited) approach is to obtain an 'entropy in cell' inequality

$$\frac{U(u_k^{n+1}) - U(u_k^n)}{\Delta t} + \frac{H^n_{k+\frac{1}{2}} - H^n_{k-\frac{1}{2}}}{\Delta x} \leqslant 0 \tag{2.7}$$

where $H^n_{k+\frac{1}{2}} = H(u^n_{k-l},\cdots,u^n_{k+r})$ with $H(u,\cdots,u) = F(u)$ is the numerical entropy flux associated with the convex entropy function $U(u)$. If a semi-discrete approximation is being used, i.e.

$$\frac{\partial u}{\partial t} + \frac{1}{\Delta x}\left(h^n_{k+\frac{1}{2}} - h^n_{k-\frac{1}{2}}\right) = 0 \qquad (2.8)$$

then there exists a more useful entropy in cell equality (Osher (1984))

$$\frac{\partial U(u)}{\partial t} + \frac{1}{\Delta x}\left(H^n_{k+\frac{1}{2}} - H^n_{k-\frac{1}{2}}\right) = \frac{1}{\Delta x}\int_{u^n_k}^{u^n_{k+1}} U''(u)\left(h^n_{k+\frac{1}{2}} - f(u)\right) du. \qquad (2.9)$$

Then a sufficient condition for solutions to satisfy an entropy inequality is that the path-independent integral on the right hand side of (2.9) should be nonpositive.

If we insist that the integral of (2.9) is nonpositive for $U(u) = |u - c|$ for any c, then we must have

$$\operatorname{sgn}\left(u^n_{k+1} - u^n_k\right)\left(h^n_{k+\frac{1}{2}} - f(u)\right) \leqslant 0 \qquad (2.10)$$

for all u between u^n_k and u^n_{k+1}. This defines a class of entropy satisfying schemes, called E-schemes (Osher (1984) and see also Tadmor (1984) for fully discrete versions). The canonical case is Godunov's (1959) scheme,

$$\begin{aligned} h^G_{k+\frac{1}{2}} &= \min_{u_k \leqslant u \leqslant u_{k+1}} f(u), \quad u_k < u_{k+1} \\ h^G_{k+\frac{1}{2}} &= \max_{u_k \geqslant u \geqslant u_{k+1}} f(u), \quad u_k > u_{k+1}. \end{aligned} \qquad (2.11)$$

All other E-schemes, which include the important class of monotone schemes (see Hartman, Hyman and Lax, 1976), have numerical fluxes which lie below Godunov's – i.e. they have as much or more numerical viscosity as Godunov's scheme.

If a scheme is entropy satisfying and converges, then it does so to the correct physical solution (Le Roux, 1977). Not all schemes are entropy satisfying. For example, the Cole-Murman (Murman, 1974) scheme, whose numerical flux differs from that of Godunov's only at sonic points $\bar{u}$, where $f'(\bar{u})=0$, is not entropy satisfying: it will in fact produce the first solution in the Burgers' equation example (2.4) above. However, such non E-schemes are usually easily 'fixed' by correcting the flux to lie below that of Godunov's scheme.

E-schemes are only first order accurate but may be used as a basis for constructing higher order adaptive schemes — see Section 3 below. Even in their own right E-schemes can reap significant benefits over non E-schemes. For example, replacing the Cole-Murman scheme by the Engquist-Osher (1980a,b) scheme, an E-scheme, in solving the Transonic Small Disturbance equation gives greatly improved results in both unsteady flutter problems and convergence to steady state (Goorjian and Van Buskirk, 1981)

Finally we consider the conditions under which the discretisation possesses convergent subsequences as $\Delta x \to 0$ (λ fixed). Define the Total Variation (TV) of the discretised solution at time level n to be

$$TV(u^n) = \sum_k |u^n_{k+1} - u^n_k| . \tag{2.12}$$

If the total variation of the discrete solution is bounded as $\Delta x \to 0$, i.e.

$$TV(u_{\Delta x}(\cdot,t)) \leqslant cTV(u_0) \tag{2.13}$$

where $u_{\Delta x}(x,t)$ is the piecewise constant prolongation of the discrete values, u_0 is the initial data and c is a constant independent of Δx, then any refinement sequence $\Delta x \to 0$ (λ fixed) possesses a subsequence $\Delta x_j \to 0$ such that

$$u_{\Delta x_j} \xrightarrow{L_1} u(x,t) \tag{2.14}$$

where, due to the Lax-Wendroff Theorem, $u(x,t)$ is a weak solution of (2.1) (see Le Roux, 1977, 1981; Sanders, 1983; or Sweby and Bains, 1984; for full details). So we have that for a scheme satisfying (2.13) any sequence $u_{\Delta x}$, $\Delta x \to 0$ possesses a convergent subsequence in L_1 to a weak solution of the problem. If in addition all limit solutions satisfy an entropy inequality we have also uniqueness and the scheme is said to be convergent.

Harten (1983) crystallised the notion of Total Variation Diminishing (TVD) schemes, for which

$$TV(u^{n+1}) \leqslant TV(u^n). \tag{2.15}$$

This certainly leads to (2.13) and hence to convergence of subsequences, but it also has a very important practical effect. As the variation is non-increasing, this precludes the generation of spurious oscillations since creation of any new extrema, or growth of old extrema, must lead to an increase in the total variation of the solution which is not permitted by such schemes. That is, TVD schemes are guaranteed not to produce spurious oscillations. It is of interest to note that the TVD criterion in fact mimics a property of the analytic solution of (2.1), whose total variation is constant except at discontinuities where it decreases.

Harten (1983) also pointed out a simple algebraic criterion for a scheme to be TVD, namely, that if a scheme can be written in the form

$$u_k^{n+1} = u_k - C^n_{k-\frac{1}{2}}\Delta_- u_k^n + D^n_{k+\frac{1}{2}}\Delta_+ u_k^n \tag{2.16}$$

where the $C^n_{k+\frac{1}{2}}$ and $D^n_{k-\frac{1}{2}}$ may be data dependent, then sufficient conditions for the scheme to be TVD are the inequalities

$$\begin{gathered} 0 \leqslant C^n_{k+\frac{1}{2}} \\ 0 \leqslant D^n_{k+\frac{1}{2}} \\ C^n_{k+\frac{1}{2}} + D^n_{k+\frac{1}{2}} \leqslant 1, \end{gathered} \tag{2.17}$$

for necessary conditions see Jameson and Lax (1984).

For the case of a semi-discrete approximation (2.8), written in a similar form to (2.16), the TVD conditions are just the first two inequalities of (2.17), the third being a

CFL-like condition on the time-step. Note that addition of a physical viscosity, i.e. (2.5), enhances TVD properties by adding extra numerical viscosity to $C^n_{k+\frac{1}{2}}$ and $D^n_{k+\frac{1}{2}}$.

The TVD property is compatible with high order accuracy of the scheme except at extrema of the solution, $u_x = 0$, where the accuracy degenerates to first order and peaks are 'clipped'. It is this observation which provides the motivation for the work on 'nearly TVD' schemes described in Section 5.

In summary, we desire to design schemes which possess the following properties:

(0) Conservation form.

(1) TVD or TV bounded.

(2) High order accuracy in smooth regions of flow.

(3) Entropy satisfaction of limit solutions.

In the following sections we look at various schemes which have these properties.

3. POST-PROCESSING SCHEMES

We now look at post-processing schemes, that is those schemes which obtain high order accuracy and which are TVD by 'modifying' the solution obtained by a first order accurate scheme.

An early example of such schemes is the Flux Corrected Transport (FCT) method of Boris and Book (1973, 1975, 1979) and Zalesak (1979) which may be summarised as follows:

(1) Update the solution (via (2.3)) using a low order accurate scheme with numerical flux h^L.

(2) Compute an antidiffusive flux h^A as the difference between fluxes h^H of a higher order accurate scheme and the low order accurate scheme used in (1), i.e. $h^A = h^H - h^L$.

(3) Limit the antidiffusive flux in such a way that stage (4) below will not produce additional extrema not present in the data or the low order solution, i.e. form $h^C = Ch^A$ where $0 \leqslant C \leqslant 1$.

(4) Update the solution using (2.3) with h^C to obtain a corrected solution.

Around the same time that FCT appeared, van Leer (1973, 1974, 1977a) investigated the combining of two second order schemes to produce a second order monotonicity preserving (TVD) scheme. He used monotonicity preserving but non-conservative adaptations of the Lax-Wendroff and second order accurate upwind schemes combined to give a conservative monotonicity preserving version of Fromm's (1968) scheme, the weighting of the combination being a function of a parameter, θ, defined to be the ratio of consecutive nodal differences, i.e.

$$\theta_{k+\frac{1}{2}} = \frac{\Delta_+ u_k}{\Delta_- u_k} . \tag{3.1}$$

This parameter (or its reciprocal) plays a central role in most adaptive TVD schemes.

Other authors have also used the idea of 'switching' between two second order schemes to avoid oscillations. Warming and Beam (1976) switched between their second order upwind scheme and MacCormack's (1969) scheme obtaining almost oscillation-free solutions, while Roe (1981b) switched between second order upwind and Lax-Wendroff in such a way as to completely avoid oscillations. Le Roux (1981) meanwhile switched between Lax-Wendroff and a first order scheme to ensure no oscillations and entropy satisfaction.

This technique of 'switching' between various second order schemes may be expressed in the common framework of Flux Limiters (Sweby, 1984) akin to FCT. The procedure here is to add a limited antidiffusive flux to a first order accurate entropy satisfying scheme, thus obtaining a high resolution TVD scheme. The limiting is performed using a limiter $\phi(r)$ which is (in the linear case) a function of the ratio of consecutive cell gradients (in fact r is the reciprocal of van Leer's θ (3.1) for the linear equation and regular grids), and there is a well defined region in the (ϕ, r) plane (see Fig. 1) in which the limiter must lie in order to obtain a second order (except at extrema) TVD scheme. The main differences between Flux Limiters and FCT are that the former is a single step operation, the limiting being decided by the set $\{u_k\}$, and that Flux Limiters in fact allow enhancement of the antidiffusive flux, the upper bound on ϕ being 2.

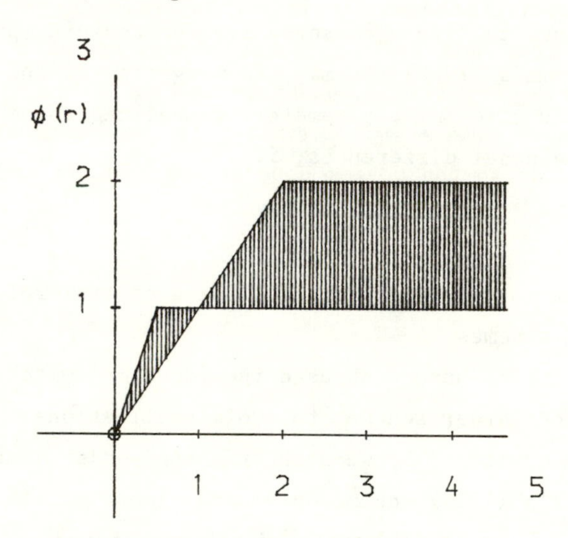

FIG. 1 Second order TVD limiter region

The general form of Flux Limiter schemes is

$$u^k = u_k - \lambda\Delta_- h^E_{k+\frac{1}{2}} \tag{3.2a}$$

$$- \lambda\Delta_-\{\phi(r^+_k)\,\alpha^+_{k+\frac{1}{2}}(\Delta f_{k+\frac{1}{2}})^+ - \phi(r^-_{k+1})\,\alpha^-_{k+\frac{1}{2}}(\Delta f_{k+\frac{1}{2}})^-\} \tag{3.2b}$$

where h^E is an E-scheme flux,

$$\begin{aligned} (\Delta f_{k+\frac{1}{2}})^+ &= -(h_{k+\frac{1}{2}} - f(u_{k+1})) \\ (\Delta f_{k+\frac{1}{2}})^- &= (h_{k+\frac{1}{2}} - f(u_k)), \\ \alpha^\pm_{k+\frac{1}{2}} &= \tfrac{1}{2}(1 \mp \nu^\pm_{k+\frac{1}{2}}), \\ \nu^\pm_{k+\frac{1}{2}} &= \lambda(\Delta f_{k+\frac{1}{2}})^\pm / \Delta u_{k+\frac{1}{2}} \end{aligned} \tag{3.3}$$

and

$$r_k^{\pm} = \left[\frac{\alpha_{k-\frac{1}{2}}^{\pm}(\Delta f_{k-\frac{1}{2}})^{\pm}}{\alpha_{k+\frac{1}{2}}^{\pm}(\Delta f_{k+\frac{1}{2}})^{\pm}} \right]^{\pm 1} .$$

(3.2a) is the first order part of the scheme while (3.2b) is the limited antidiffusive flux. (Note that the shorthand notation $v^k \equiv v_k^{n+1}$, $v_k \equiv v_k^n$ has been used here.)

Many adaptive schemes fit into this framework including those of van Leer and Roe mentioned above, as well as more recent work, e.g. Roe (1985), Chakravarthy and Osher (1985) and others. The most commonly used limiter is the 'minmod' limiter which corresponds to the bottom boundary of Fig. 1, and which may be written as

$$\phi_{\text{minmod}}(r) = \max(0, \min(1, r)) \tag{3.4a}$$

or alternatively as

$$\text{minmod}(x, y) = \begin{cases} x & : |x| < |y| \\ 0 & : xy < 0 \\ y & : |y| < |x| \end{cases} , \tag{3.4b}$$

where $r = x/y$.

The use of this limiter corresponds to Roe's original adaptive scheme which switches between the Lax-Wendroff and Warming and Beam second order upwind schemes while other limiters in the region of Fig. 1 can be considered as producing schemes which are weighted (data dependent) averages of these two constant coefficient schemes.

In the next section we look at pre-processing schemes in general and in Section 5 we examine in detail a recent development in these adaptive schemes.

4. PRE-PROCESSING SCHEMES

Godunov's scheme is derived by considering the nodal data u_k^n as cell averages over the cell $(x_k - \frac{1}{2}\Delta x, x_k + \frac{1}{2}\Delta x)$, i.e.

$$u_k^n = \frac{1}{\Delta x}\int_{x_{k-\frac{1}{2}}}^{x_{k+\frac{1}{2}}} u(x,t_n)\,dx\,. \tag{4.1}$$

In addition it is assumed that the underlying solution is piecewise constant giving the situation in Fig. 2. Godunov's scheme is then obtained by exactly solving the Riemann problem

$$\begin{aligned} &v_t + f(v)_x = 0 \\ &\text{with} \quad v_0 = \begin{cases} u_k^n & x < x_{k+\frac{1}{2}} \\ u_{k+1}^n & x > x_{k+\frac{1}{2}} \end{cases} \end{aligned} \tag{4.2}$$

at each cell interface $x_{k+\frac{1}{2}}$, the timestep Δt taken being restricted so that neighbouring solutions do not interact, and then averaging the solution $v(x,\Delta t)$ obtained over each cell to obtain the updated cell averages

$$u_k^{n+1} = \frac{1}{\Delta x}\int_{x_{k-\frac{1}{2}}}^{x_{k+\frac{1}{2}}} v(x,\Delta t)\,dx\,. \tag{4.3}$$

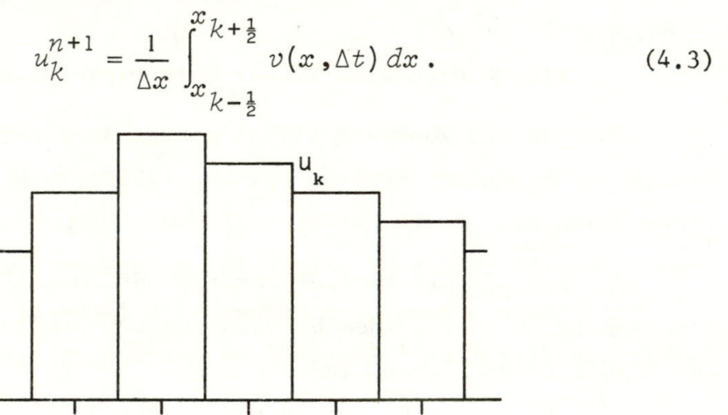

FIG. 2 Piecewise constant data representation

Pre-processing schemes are classed as those which obtain high resolution TVD schemes by the mechanism of taking different (other than piecewise constant) solution representations prior to solving a generalised Riemann problem at the cell interfaces and then averaging to obtain updated cell averages.

The canonical example here is van Leer's (1977b, 1979) MUSCL (Monotonic Upwind Scheme for Conservation Laws) which is an extension of the Godunov idea by using a piecewise linear discontinuous representation of the solution. The slopes of this representation are restricted in such a way as not to increase its variation (LeVeque and Goodman (1985) have proposed a similar scheme using the minmod limiter (3.4b) to achieve this restriction). The result is a representation of the form shown in Fig. 3. The generalised Riemann problem is solved (approximately) at the cell interfaces and the solution averaged.

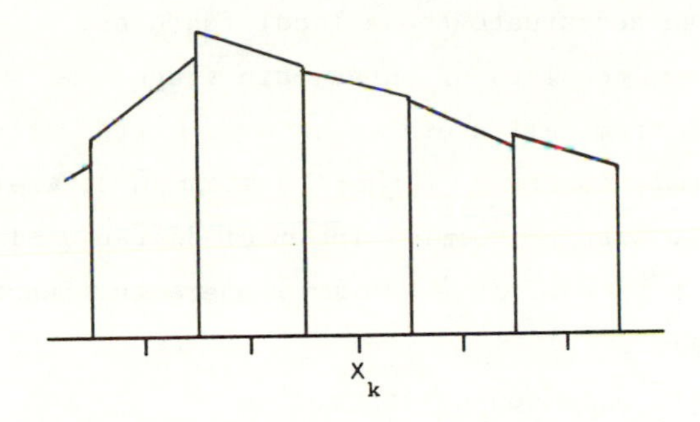

FIG. 3 Piecewise linear data representation

Colella and Woodward (1984) go one step further with their Piecewise Parabolic Method (PPM), taking a piecewise second order polynomial representation of the solution, again with a constraint on the choice of polynomial to ensure the TVD property.

All these schemes, in common with those of the previous section, suffer from 'clipping' of extrema, which is inevitable if the variation as defined by (2.12) is constrained to decrease. The work described in the next section relaxes this criterion to obtain Essentially Non-Oscillatory (ENO) schemes with arbitrary high interpolants which do not clip peaks.

5. ESSENTIALLY NON-OSCILLATORY SCHEMES

We now describe recent work by Harten, Osher, Engquist and Chakravarthy (1986) (see also Harten and Osher, 1985; and Harten, Engquist, Osher and Chakravarthy, 1986) on Essentially Non-Oscillatory Schemes. These schemes do not suffer from the excessive clipping of local extrema common with TVD schemes, while maintaining the TVD property to within $0(\Delta x^{p+1})$ where p is the order of the method which may be arbitrarily high.

The technique used is similar to that employed in the MUSCL and PPM schemes described in the preceding section except that the reconstruction of the function $u(x,t_n)$ from the cell averages $\{u_k^n\}$ is not required to damp the values of local extrema, as in the aforementioned schemes, and is even allowed occasionally to accentuate these local features.

The scheme consists of three main stages:

(i) Starting from cell averages $\bar{u}^n = \{u_k^n\}$ reconstruct the approximate function $u_{\Delta x}(x;t_n) = R(x;\bar{u}^n)$, where $R(x;\bar{u}^n)$ is a piecewise polynomial in x of degree $p-1$ satisfying

(1) $R(x;\bar{u}^n) = u(x,t_n) + 0(\Delta x^p)$ where the functions are smooth,

(2) $R(x;\bar{u}^n)$ is conservative, i.e.

$$\frac{1}{\Delta x}\int_{x_{k-\frac{1}{2}}}^{x_{k+\frac{1}{2}}} R(x;\bar{u}^n)\,dx = u_k^n ,$$

(3) $R(x;\bar{u}^n)$ is essentially non-oscillatory, i.e.

$$TV(R(\cdot;\bar{u}^n)) \leqslant TV(u(\cdot,t_n)) + 0(\Delta x^p).$$

(ii) Calculate the solution in the small, $u_{\Delta x}(\cdot,t+\tau)$, as the solution to the differential equation (2.1) with initial data $u_{\Delta x}(\cdot,t_n)$.

(iii) Update cell averages

$$u_k^{n+1} = \frac{1}{\Delta x}\int_{x_{k-\frac{1}{2}}}^{x_{k+\frac{1}{2}}} u_{\Delta x}(x,t+\Delta t)\,dx.$$

This produces a conservative scheme with arbitrary accuracy depending on the order p of the polynomial reconstruction.

Stage (i) is the key step in constructing Essentially Non-Oscillatory (ENO) schemes, and we now outline the procedure of obtaining such a non-oscillatory reconstruction $R(x;\bar{u}^n)$.

Let $I_m(x;w)$ be a piecewise polynomial of degree m interpolating a function $w(x)$ at the points x_k, i.e. $w(x_k) = I_m(x_k;w)$. We have

$$I_m(x;w) = q_{m,k+\frac{1}{2}}(x;w) \qquad x_k \leqslant x \leqslant x_{k+1} \tag{5.1}$$

where $q_{m,k+\frac{1}{2}}$ is a polynomial of degree m interpolating $w(x)$ at $m+1$ successive points $\{x_l\}$ including x_k and x_{k+1}. The important thing is to choose those points which use the 'smoothest' values of w. This is achieved as follows.

Define

$$q_{1,k+\frac{1}{2}} = w[x_k] + (x-x_k)\,w[x_k, x_{k+1}] \tag{5.2a}$$

where the $w[\cdot\cdot]$ are the divided differences

$$w[x_l] = w(x_l)$$

$$w[x_l, \cdots, x_{l+K}] = \frac{(w[x_{l+i}, \cdots, x_{l+K}] - w[x_l, \cdots, x_{l+K-1}])}{x_{l+K} - x_l}$$

and set

$$K^{(1)}_{\min} = k \;;\; K^{(1)}_{\max} = k+1\,. \tag{5.2b}$$

Then, given $q_{m-1,k+\frac{1}{2}}$, $K^{(m-1)}_{\min}$, $K^{(m-1)}_{\max}$ we construct $q_{m,k+\frac{1}{2}}$ as follows. Let

$$\begin{aligned} a_m &= w\left[x_{K^{(m-1)}_{\min}}, \cdots, x_{K^{(m-1)}_{\max}+1}\right] \\ b_m &= w\left[x_{K^{(m-1)}_{\min}-1}, \cdots, x_{K^{(m-1)}_{\max}}\right]; \end{aligned} \tag{5.3}$$

then if $|a_m| \geqslant |b_m|$ set

$$\begin{aligned} q_{m,k+\frac{1}{2}} &= q_{m-1,k+\frac{1}{2}} + b_m \prod_{K=K^{(m-1)}_{\min}}^{K^{(m-1)}_{\max}} (x-x_K) \\ K^m_{\min} &= K^{(m-1)}_{\min} - 1 \\ K^m_{\max} &= K^{(m-1)}_{\max}\,; \end{aligned} \tag{5.4a}$$

else if $|a_m| < |b_m|$ set

$$q_{m,k+\frac{1}{2}} = q_{m-1,k+\frac{1}{2}} + a_m \prod_{K=K_{\min}^{(m-1)}}^{K_{\max}^{(m-1)}} (x - x_K)$$

$$K_{\min}^{m} = K_{\min}^{(m-1)} \tag{5.4b}$$

$$K_{\max}^{m} = K_{\max}^{(m-1)} + 1 ;$$

i.e. look in the least oscillatory direction to choose interpolation points, e.g.

$$q_{1,k+\frac{1}{2}} = w(x_k) + \frac{1}{\Delta x}(x - x_k)(w(x_{k+1}) - w(x_k))$$

$$q_{2,k+\frac{1}{2}} = q_{1,k+\frac{1}{2}} + \frac{(x-x_k)(x-x_{k+1})}{2\Delta x^2} \begin{cases} w(x_{k+1}) - 2w(x_k) + w(x_{k-1}) \\ \text{or} \\ w(x_{k+2}) - 2w(x_{k+1}) + w(x_k) \end{cases}$$

$$q_{3,k+\frac{1}{2}} = \cdots\cdots .$$

It is easily shown that

LEMMA: For all piecewise smooth $w(x)$ *there exists a* Δx_0 *and* $u(x)$ *such that for all* $\Delta x < \Delta x_0$

$$I_m(x;w) = u(x) + O(\Delta x^{m+1})$$

$$TV(u) \leqslant TV(w) .$$

If we set, at time t_n, $w(x) = u(x,t_n)$ then

$$\bar{w}_k = u_k^n = \frac{1}{\Delta x} \int_{x_{k-\frac{1}{2}}}^{x_{k+\frac{1}{2}}} w(x)\, dx \tag{5.5}$$

and we may define

$$W(x) = \int_{x_0}^{x} w(y)\, dy \tag{5.6}$$

whence

$$W(x_{k+\frac{1}{2}}) = \sum_{i=i_0}^{k} \Delta x \bar{w}_i . \tag{5.7}$$

We now interpolate $W(x)$ at these points and approximate $w(x)$ (and hence $u(x,t_n)$) via

$$R(x;\bar{w}) = \frac{d}{dx} I_p(x;W) . \tag{5.8}$$

The desired properties ((1) to (3) above) of the reconstruction are easily verified.

Note that there is an alternative reconstruction from cell averages possible via Deconvolution: readers are referred to the references cited above for details.

It now remains to obtain the solution in the small of (2.1) with the reconstructed $u_{\Delta x}$ function as data, and to update the cell averages. The exact solution to the generalised Riemann problem, i.e. (2.1) with discontinuous polynomial data, can be difficult to compute. However a convergent power series may be obtained and stages (ii) and (iii) may be combined as follows.

Consider the sliding average

$$\bar{u}(x,t) = \frac{1}{\Delta x} \int_{-\Delta x/2}^{\Delta x/2} u(x+y,t)\,dy . \tag{5.9}$$

If $u(x,t)$ satisfies the conservation law (2.1) then $\bar{u}(x,t)$ satisfies

$$\frac{\partial}{\partial t}\bar{u}(x,t) = -\frac{1}{\Delta x}\left[f(u(x+\tfrac{1}{2}\Delta x,t)) - f(u(x-\tfrac{1}{2}\Delta x,t))\right] \tag{5.10}$$

or

$$\bar{u}(x,t+\Delta t) = \bar{u}(x,t) - \lambda\,[\hat{f}(x+\tfrac{1}{2}\Delta x,t;u) - \hat{f}(x-\tfrac{1}{2}\Delta x,t;u)] \tag{5.11}$$

where

$$\hat{f}(x,t;w) = \frac{1}{\Delta t}\int_0^{\Delta t} f(w(x,t+\tau))\,d\tau . \tag{5.12}$$

On a grid this becomes $(u_k^n \equiv \bar{u}(x_k,t_n))$

$$u_k^{n+1} = u_k^n - \lambda[\hat{f}(x_{k+\frac{1}{2}},t_n;u) - \hat{f}(x_{k-\frac{1}{2}},t_n;u)] \tag{5.13}$$

so to obtain the updated cell averages we approximate $\hat{f}$ of (5.12) to $O(\Delta x^R)$ and use this in (5.13).

The integral of (5.12) is replaced by a quadrature rule

$$\int_0^1 f(u(x_{k+\frac{1}{2}}, t+s))\, ds \approx A_0 f(u(x_{k+\frac{1}{2}}, t+s_0)) + \cdots + A_j f(u(x_{k+\frac{1}{2}}, t+s_j)) \quad (5.14)$$

$$\text{for} \quad 0 \leqslant s_0 < s_1 < \cdots < s_j \leqslant 1$$

wherein the f are replaced by the numerical flux of a 2-point E-scheme (e.g. Godunov), viz

$$f(u(x_{k+\frac{1}{2}}, t+s_r)) = h^E(u(x_{k+\frac{1}{2}}-, t+s_r), u(x_{k+\frac{1}{2}}+, t+s_r)) \quad (5.15)$$

where $u(x_{k+\frac{1}{2}}\pm)$ denotes limits taken from the left and right of the discontinuity in u at $x_{k+\frac{1}{2}}$.

Finally the values $u(x, t+\tau)$ are obtained by Taylor series expansion of $u_{\Delta x}$ $(=R(x\,;w))$ about (x,t) with the time derivatives of $u_{\Delta x}$ being replaced by spatial derivatives of f using the equation (2.1) in a Lax-Wendroff fashion, i.e. $u_t = -f_x$ etc.

The time accuracy of the scheme is determined by R and the quadrature rule whilst the spatial accuracy is determined by the order of the reconstruction p.

This approximation of $\hat{f}$ need not preserve the ENO property, however, due to the highly nonlinear nature of the interpolant, the method works very well numerically and readers are strongly urged to see the references cited for results. Preliminary results in two dimensions have also been outlined, see Harten (1986).

REFERENCES

Boris, J.P. and Book, D.L. (1973), "Flux corrected transport. I. SHASTA, a fluid transport algorithm that works". *J. Comput. Phys.*, *11*, pp.38-69.

Boris, J.P., Book, D.L. and Hain, K. (1975), "Flux corrected transport. II Generalisations of the method", *J. Comput. Phys.*, *18*, pp.248-283.

Boris, J.P. and Book, D.L. (1979), "Flux corrected transport. III Minimal-error FCT algorithms", *J. Comput. Phys.*, *20*, pp.397-431.

Chakravarthy, S.R. and Osher, S. (1985), "Computing with high resolution upwind schemes for hyperbolic equations", *Lecture Notes in Appl. Math.*, *22*, *Part 1*, eds. Engquist, B., Osher, S. and Somerville, R.C.J., AMS (Providence RI), pp.57-86.

Colella, P. and Woodward, P.R. (1984), "The Piecewise Parabolic Method (PPM) for gas-dynamical simulations". *J. Comput. Phys.* *54*, pp.174-201.

Engquist, B. and Osher, S. (1980a), "Stable and entropy satisfying approximations for transonic flow calculations", *Math. Comp.*, *24*, pp.45-75.

Engquist, B. and Osher, S, (1980b), "One-sided difference schemes and transonic flow", *Proc. National Academy of Sciences USA*, *77*, pp.3071-3074.

Fromm, J.E. (1968), "A method for reducing dispersion in convective difference schemes", *J. Comput. Phys.*, *3*, pp.176-189.

Godunov, S.K. (1959), "A finite difference method for the numerical computation of discontinuous solutions of the equations of fluid dynamics", *Mat. Sb.*, *47*, pp.271-295.

Goorjian, P.M. and van Buskirk, R. (1981), "Implicit calculations of transonic flow using monotone methods", AIAA paper 81-0331.

Harten, A. (1986), "Preliminary results on the extension of ENO schemes to two-dimensional problems", *Proc. International Conference on Hyperbolic Problems, Saint-Etienne,*

Harten, A., Hyman, J.M. and Lax, P.D. (1976), "On finite difference approximations and entropy conditions for shocks", *Comm. Pure Appl. Maths.*, *29*, pp.297-322.

Harten, A. and Osher, S. (1985). "Uniformly high-order accurate non-oscillatory schemes I", *SIAM J. Numer. Anal.*, to be published.

Harten, A., Osher, S., Engquist, B. and Chakravarthy, S.R. (1986), "Some results on uniformly high order accurate essentially non-oscillatory schemes", *J. App. Num. Math.*, to be published.

Harten, A., Engquist, B., Osher, S. and Chakravarthy, S.R. (1986), "Uniformly high order accurate non-oscillatory schemes III", UCLA Report, submitted to *J. Comput. Phys.*

Jameson, A. and Lax, P.D., (1984), "Conditions for the construction of multi-point total variation diminishing schemes", Princeton MAE Report 1650.

Kruzkov, S.M. (1970), "First order quasilinear equations in several independent variables", *Mat. Sb.*, *81* (*123*), pp.217-243.

Lax, P.D. (1973), *Hyperbolic Systems of Conservation Laws and the Mathematical Theory of Shock Waves*, SIAM Regional Conference Series, Lectures in Applied Mathematics 11.

Lax, P.D. and Wendroff, B. (1960), "Systems of conservation laws", *Comm. Pure Appl. Maths.*, *13*, pp.217-237.

Le Roux, A.Y. (1977), "A numerical conception of entropy for quasi-linear equations", *Math. Comput.*, *31*, pp.848-872.

Le Roux, A.Y. (1981), "Convergence of an accurate scheme for first order quasi-linear equations", *RAIRO Anal. Numer.*, *15*, pp.151-170.

Leveque, R.J. and Goodman, J.B. (1985), "TVD schemes in one and two space dimensions", *Lecture Notes in Appl. Math.*, *22*, *Part 2*, eds., Engquist, B., Osher, S. and Somerville, R.C.J, pp.57-62.

MacCormack, R.W. (1969), "The effect of viscosity in hypervelocity impact cratering", AIAA 12, paper 69-354.

Murman, E.M. (1974), "Analysis of embedded shockwaves calculated by relaxation methods", *AIAA J.*, *12*, pp.626-633.

Oleinik, O.A. (1957), "Discontinuous solutions of nonlinear differential equations", *Amer. Math. Soc. Transl. Ser. 2*, *26*, pp.95-171.

Osher, S. (1984), "Riemann solvers, the entropy condition, and difference approximations", *SIAM J. Numer. Anal.*, *21*, pp.217-235.

Osher, S. and Solomon, F. (1982), "Upwind difference schemes for hyperbolic systems of conservation laws", *Math. Comp.*, *38*, pp.339-374.

Roe, P.L. (1981a), "Approximate Riemann solvers, parameter vectors, and difference schemes", *J. Comput. Phys.*, *43*, pp.357-372.

Roe, P.L. (1981b), "Numerical algorithms for the linear wave equation", RAE Technical Report.

Roe, P.L. (1985), "Some contributions to the modeling of discontinuous flows". *Lecture Notes in Appl. Maths.*, *22*, *Part 2*, eds. Engquist, B., Osher, S. and Somerville, pp.163-193.

Sanders, R. (1983), "On convergence of monotone finite difference schemes with variable space differencing", *Math. Comp.*, *40*, pp.91-106.

Sweby, P.K. and Baines, M.J. (1984), "On convergence of Roe's scheme for the general nonlinear scalar wave equation", *J. Comput. Phys.*, *56*, pp.135-148.

Sweby, P.K. (1984), "High resolution schemes using flux limiters for hyperbolic conservation laws", *SIAM J. Numer. Anal.*, *21*, pp.995-1011.

Tadmor, E. (1984), "Numerical viscosity and the entropy condition for conservative difference schemes", *Math. Comp.*, *43*, pp.369-381.

van Leer, B. (1973), "Towards the ultimate finite difference scheme. I", *Lecture Notes in Physics*, *18*, pp.163-168.

van Leer, B. (1974), "Towards the ultimate finite difference scheme. II. Monotonicity and conservation combined in a second order scheme", *J. Comput. Phys.*, *14*, pp.361-370.

van Leer, B. (1977a), "Towards the ultimate finite difference scheme. III. Upstream-centred finite difference schemes for ideal compressible flow", *J. Comput. Phys.*, *23*, pp.263-275.

van Leer, B. (1977b), "Towards the ultimate finite difference scheme. IV. A new approach to numerical convection", *J. Comput. Phys.*, *23*, pp.276-299.

van Leer, B. (1979), "Towards the ultimate finite difference scheme. V. A second-order scheme to Godunov's method", *J. Comput. Phys.*, *32*, pp.101-136.

Warming, R.F. and Beam, R.M. (1976), "Upwind second order difference schemes and applications in aerodynamics", *AIAA J.*, *14*, pp.1241-1249.

Zalesak, S.T. (1979), "Fully multidimensional flux-corrected transport algorithms for fluids", *J. Comput. Phys.*, *31*, pp.335-362.

S. Osher
Department of Mathematics
University of California
Los Angeles
California 90024
U.S.A.

P.K. Sweby
Department of Mathematics
University of Reading
Whiteknights
Reading RG6 2AS
England

Index